Current Progress in Nucleic Acids Research

Current Progress in Nucleic Acids Research

Edited by Kaison Gibson

SYRAWOOD
PUBLISHING HOUSE
New York

Published by Syrawood Publishing House,
750 Third Avenue, 9th Floor,
New York, NY 10017, USA
www.syrawoodpublishinghouse.com

Current Progress in Nucleic Acids Research
Edited by Kaison Gibson

International Standard Book Number: 978-1-64740-387-4 (Hardback)

Cataloging-in-publication Data

Current progress in nucleic acids research / edited by Kaison Gibson.
p. cm.
Includes bibliographical references and index.
ISBN 978-1-64740-387-4
1. Nucleic acids. 2. Biomolecules. 3. Biochemistry. I. Gibson, Kaison.
QP620 .N83 2023
574.873 28--dc23

TABLE OF CONTENTS

PREFACE

Nucleic acids are large biomolecules that are necessary for all life forms including humans, animals and microorganisms. They are composed of a sugar molecule, nitrogen-containing base, and phosphate groups. Ribonucleic acid (RNA) and deoxyribonucleic acid (DNA) are the two primary types of nucleic acids. Each kind of nucleic acid has a unique structure and performs different functions in the cells. Nucleic acids play an important role in the storage and expression of genomic data. Cell-free nucleic acids (cfNAs) are a mixture of small segments of DNA and RNA present in biological fluids such as blood, saliva and urine. The cfNAs have the potential to act as valuable biomarkers for the diagnosis and prognosis of several diseases including autoimmune diseases, cardiovascular diseases, and cancer. They can also be used for prenatal diagnosis of genetic diseases. This book provides comprehensive insights into the current progress in nucleic acids research. A number of latest researches have been included to keep the readers updated with the latest concepts in this area of study.

After months of intensive research and writing, this book is the end result of all who devoted their time and efforts in the initiation and progress of this book. It will surely be a source of reference in enhancing the required knowledge of the new developments in the area. During the course of developing this book, certain measures such as accuracy, authenticity and research focused analytical studies were given preference in order to produce a comprehensive book in the area of study.

This book would not have been possible without the efforts of the authors and the publisher. I extend my sincere thanks to them. Secondly, I express my gratitude to my family and well-wishers. And most importantly, I thank my students for constantly expressing their willingness and curiosity in enhancing their knowledge in the field, which encourages me to take up further research projects for the advancement of the area.

Editor

1

Design of the Crosslinking Reactions for Nucleic Acids-Binding Protein and Evaluation of the Reactivity

Kenta Odaira, Ken Yamada, Shogo Ishiyama, Hidenori Okamura and Fumi Nagatsugi *

Institute of Multidisciplinary Research for Advanced Materials, Tohoku University, Sendai 980-8577, Japan; k.odaira@kobayashi.co.jp (K.O.); Ken.Yamada@umassmed.edu (K.Y.); rare.earth.144@gmail.com (S.I.); hidenori.okamura.b8@tohoku.ac.jp (H.O.)

* Correspondence: nagatugi@tohoku.ac.jp

Featured Application: Alkylation for nucleic acids-binding protein.

Abstract: Selective chemical reactions of biomolecules are some of the important tools for investigations by biological studies. We have developed the selective crosslinking reactions to form covalent bonds to DNA or RNA using crosslinking oligonucleotides (CFO) bearing reactive bases. In this study, we designed the cross-linkable 4-amino-6-oxo-2-vinyltriazine derivative with an acyclic linker (*acy*AOVT) to react with the nucleic acids-binding protein based on our previous results. We hypothesized that the *acy*AOVT base would form a stable base pair with guanine by three hydrogen bonds at the positions of the vinyl group in the duplex DNA major groove, and the vinyl group can react with the nucleophilic species in the proximity, for example, the cysteine or lysine residue in the nucleic acids-binding protein. The synthesized oligonucleotides bearing the *acy*AOVT derivative showed a higher reactivity than that of the corresponding pyrimidine derivative without one nitrogen. The duplex containing *acy*AOVT-guanine (G) formed complexes with Hha1 DNMT even in the presence of 2-mercaptoethanol. We expect that our system will provide a useful tool for the molecular study of nucleic acids-binding proteins.

Keywords: oligonucleotide; crosslink; nucleic acid binding protein; cytosine methyltransferase

---

1. Introduction

The binding of proteins to nucleic acids (DNA and RNA) is central to all aspects of gene expression regulation. DNA-binding proteins include the transcription factor to bind to specific DNA sequences and control the transcription rate of genetic information from the DNA to RNA [1,2]. Many DNA modifying enzymes are also included in the DNA binding protein, for example, repair enzymes [3,4] and epigenetic modification enzymes. DNA methyl transferase is one of the epigenetic modification enzymes and catalyses the transfer of the methyl group to DNA modifying the function of genes and affecting gene expression [5,6]. RNA-binding proteins have important functions in the post-transcriptional process, such as splicing regulation [7] and modulation of the mRNA translation. Recent studies have revealed that post-transcriptional gene regulation by non-coding RNAs is involved in most biological activities. The association of the RNA binding protein with ncRNA plays a crucial role in these biological functions [8–10]. The chemical tools for the control of the interaction between the nucleic acids-binding protein and nucleic acids have the potential for the development as the new strategy for artificial control of the gene expression.

Crosslinking reactions between nucleic acids and binding proteins are powerful tools for analyzing these interactions. Photoreactive functional groups, such as diazirines and benzophenones, were utilized for DNA or RNA-protein crosslinking [11–14], but lack selectivity for the target amino acid. Furthermore, due to their intrinsic reactivity, these groups react with water and other reactive chemical species resulting in low yields. Oligodeoxynucleotides (ODN) with reactive groups are exploited for their effective DNA-protein crosslinking reactions by the proximity effect of the specific DNA recognition with the binding protein. The reactive groups activated by oxidation, such as a diol [15] and furan [16], react with a proximal lysine or arginine in the protein or peptides. Hocek et al. reported that the ODN bearing a vinylsulfone amide [17] or chloroacetamide group [18] efficiently cross-linked with the p53 protein through alkylation of the cysteine in the proximal position with the reactive groups. Recently, they demonstrated that ODN with 2-vinylhypoxantine reacted with a thiol-containing minor groove binding peptide by proximity effect [19].

In our previous study, we reported the synthesis of the ODN bearing 4-amino-6-oxo-2-vinyltriazine with an ethyl linker (Et-AOVT: (**1**)) as a cross linkable derivative and an evaluation of its properties [20]. ODN showed a relatively high reactivity with the pyrimidine nucleobases of the complementary DNAs and the lowest reactivity to guanine. We hypothesized that the AOVT base would form a stable base pair with guanine (G) by three hydrogen bonds to be positioned at the vinyl group in the DNA major groove. In this study, we have designed an AOVT derivative with an acyclic linker (*acy*AOVT: (**2**)) as a cross-linkable probe of the nucleic acids-binding protein. The distance between the reactive base and sugar moiety of **2** is shorter than that of **1**, and consequently, the *acy*AOVT (**2**) should form a base pair with guanine (G) similar to the natural cytosine (C)-G pair. This assumption is supported by the molecular modeling shown in Figure 1. The molecular modeling revealed that the structure of the base pair of *acy*AOVT (**2**)-G and the natural C-G significantly overlap each other (Figure 1C). On the other hand, the structure of Et-AOVT (**1**)-G is subtly shifted from the natural base pair one (Figure 1B). The duplex DNA containing the *acy*AOVT (**2**)-G is expected to bind to the nucleic acid binding proteins similar to the natural duplex and form covalent linkages with the nucleophilic residues of the protein in the proximal position of the vinyl group (Figure 2).

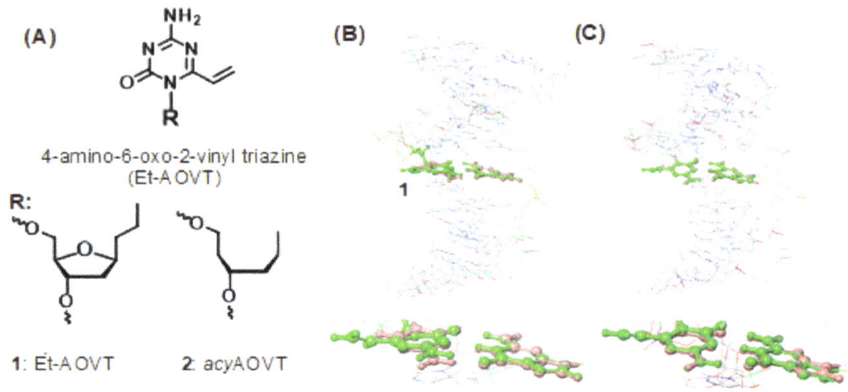

Figure 1. (**A**) Structures of cross-linkable derivatives; (**B**,**C**) Superimposed structures between the natural duplex (pink) and the duplex contained **1** (**B**) or **2** (**C**) (green).

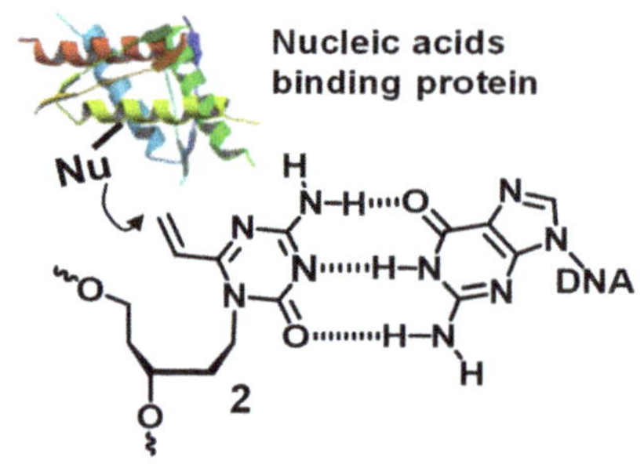

Figure 2. Design of the cross-linkable probe to nucleic acids-binding protein.

2. Materials and Methods

2.1. General

The ^{1}H-NMR spectra were obtained by 400 or 600 MHz spectrometers (Bruker, Billerica, MA, USA). The ^{1}H chemical shifts are described as δ values in ppm relative to acetone-d_6 (2.05 ppm), DMSO-d_6 (2.50 ppm), $CDCl_3$ (7.26 ppm), and tetramethylsilane (1H, 0.00 ppm). The ^{13}C NMR spectra were obtained by a Bruker 600 MHz spectrometer. The ^{13}C chemical shifts are described as the δ values in ppm relative to acetone-d_6 (29.84 ppm), DMSO-d_6 (39.52 ppm), and $CDCl_3$ (77.16 ppm). The ^{31}P-NMR spectra were recorded by a Bruker 500 MHz (202 MHz for ^{31}P). The multiplicity and qualifier abbreviations are as follows: s = singlet, d = doublet, t = triplet, q = quartet, quin. = quintet, sept. = septet, m = multiplet, br = broad. The electrospray ionization (ESI) mass spectra were recorded using a BioTOF II mass spectrometer or APEX III (Bruker Daltonics, Bruker, Billerica, MA, USA). The matrix-assisted laser desorption/ionization (MALDI-TOF) mass spectra were recorded Autoflex speed mass spectrometer and a laser at 337 nm in the negative mode using 3-hydroxypicolinic acid as the matrix or in the positive mode using 2,5-dihydroxybenzoic acid as the matrix. Thin-layer chromatography (TLC) was performed using silica gel 60 F_{254} pre-coated plates. Column chromatography was performed using silica gel 60 N (spherical, neutral, 100–210 μm, Kanto Chemical, Tokyo, Japan). Flash chromatography was performed using Kanto Chemical silica gel 60 N (spherical, neutral, 40–50 μm). The ultraviolet-visible (UV-vis) absorption spectra were recorded using a DU800 spectrometer (Beckman Coulter, Brea, CA, USA). The ODN synthesis was carried out using an automated DNA synthesizer (392 DNA/RNA synthesizer, ABI, Foster City, CA, USA) following the standard phosphoramidite chemistry. High performance liquid chromatography (HPLC) was performed using a Cosmosil 5C18MSII column (4.6 or 10 × 250 mm, Nacalai Tesque, Kyoto, Japan), a PU-986 pump (JASCO, Tokyo, Japan), JASCO 2075 UV detector and a JASCO 2067 column oven. pH measurements were measured by a Seven Easy pH meter (Mettler Toledo, Columbus, OH) using an 8220BNWP electrode. The denaturing polyacrylamide gel plates were visualized and quantified using a FLA-5100 Fluor Imager (Fujifilm, Tokyo, Japan). Anhydrous methanol, DMF, THF, CH_2Cl_2, dioxane, pyridine, DMSO, CH_3CN, toluene and THF were purchased from Wako Pure Chemical Industries, Ltd. (Tokyo, Japan). Unless otherwise noted, all the synthetic reactions were carried out at ambient temperature. The reactions requiring anhydrous conditions were achieved under an argon atmosphere in flasks dried under 1–3 mmHg. Commercially available reagents were obtained from Wako Pure Chemical Industries, Ltd., TCI, Inc., (Tokyo, Japan), Sigma-Aldrich Co. LLC (St. Louis, MO) and Kanto Chemical Co., Inc (Tokyo, Japan) and used without further purification. Human DNA (cytosine-5) methyltransferase (DNMT1) was obtained from New England Bio Labs Japan, Inc. (Tokyo, Japan)

2.2. Synthesis of the Nucleoside Derivatives

2.2.1. Synthesis of 4-amino-6-oxo-1-(3′,5′-*O*-t-butyldimethylsilyl-2′,4′-dideoxy-D-ribityl) triazine (**4**)

The linker **3** (1.3 g, 3.0 mmol), which was prepared by a modified literature procedure [21], was added to 5-aza cytosine Na salt (1.0 g, 7.5 mmol) in DMSO (30 mL). The mixture was stirred at 60 °C for 18 h. The resulting mixture was diluted with EtOAc and quenched with NH_4Cl. The product was extracted with EtOAc and the combined organic layer was washed with brine, dried over Na_2SO_4, filtered, and evaporated under reduced pressure. The resulting oil was purified by silica gel column chromatography ($CH_3OH/CHCl_3$, 1:100, then $CH_3OH/CHCl_3$, 1:50) to afford the corresponding N1-alkylation (772 mg, 1.74 mmol, 58%) and N3-alkylation (126 mg, 0.29 mmol, 10%) products.

N1-alkylation product: ^{1}H-NMR (400 MHz, $CDCl_3$) δ 7.90 (s, 1H), 6.98 (brs, 1H), 5.65 (brs, 1H), 3.97 (ddd, 2H, J = 12.0, 12.0, 6.0 Hz), 3.90–3.77 (m, 2H), 3.64 (dd, 2H, J = 6.4, 6.4 Hz), 1.99–1.60 (m, 6H), 1.37–1.23 (m, 1 H), 0.89 (s, 9H), 0.87 (s, 9H), 0.07 (s, 3H), 0.06 (s, 3H), 0.03 (s, 6H); ^{13}C-NMR (125 MHz, CDCl3) δ 166.7, 158.7, 154.5, 67.4, 59.6, 45.1, 40.2, 36.2, 26.12, 26.09, 18.4, 18.3, −4.2, −4.3, −5.14, −5.16; HRMS (ESI) calcd. for $C_{20}H_{43}N_4O_3Si_2^+$ $[M + H]^+$ *m/z* 443.2868, found *m/z* 443.2879.

N3-alkylation product: ^{1}H-NMR (600 MHz, CDC13) δ 8.34 (s, 1H), 5.90 (brs, 1H), 5.61 (brs, 1H), 4.42–4.35 (m, 1H), 4.05–4.01 (m, 1H), 3.70–3.64 (m, 2H), 1.99–1.95 (m, 1H), 1.89–1.84 (m, 2H), 1.7–1.67 (m, 2H), 0.871 (s, 9H), 0.867 (s, 9H), 0.046 (s, 3H), 0.027 (s, 3H), 0.025 (s, 3H), 0.018 (s, 3H); ^{13}C-NMR (125 MHz, CDC13) δ 170.4, 167.93, 167.88, 66.3, 64.5, 59.6, 40.3, 35.9, 25.9, 25.8, 18.2, 18.0, −4.6, −4.8, −5.36, −5.38; HRMS (ESI) calcd. for $C_{20}H_{43}N_4O_3Si_2^+$ [M + H]$^+$ *m/z* 443.2868, found *m/z* 443.2879

2.2.2. Synthesis of 4-amino-2-(2-octylthioethyl)-6-oxo-1-(3′,5′-*O*-t-butyldimethylsilyl-2′, 4′-dideoxy-D-ribityl) triazine (**6**)

A 28% aq NH_4OH (10 mL) solution was added to a solution of the N1-alkylation product **4** (172 mg, 390 μmol) in dioxane-methanol (10 mL, 1/1, *v*/*v*), then sealed with a glass stopper. The mixture was stirred at 55 °C for 24 h. The resulting solution was evaporated under vacuum at 43 °C. The resulting moist solid was co-evaporated with ethanol × 3 to afford guanylurea **5**. This product was directly used for the next step without further purification. The orthoester (597 mg, 1.95 mmol, purity = 91%) was added to a solution of guanylurea (169 mg, 390 μmol) in DMF (0.2 M, 1.95 mL). The mixture was stirred at 120 °C for 2 h. The resulting yellow solution was diluted with Et_2O. The product was extracted with Et_2O and the combined organic layer was washed with NH_4Cl aq., $NaHCO_3$ aq., and brine, dried over Na_2SO_4, filtered and evaporated. The crude product was purified by silica gel column chromatography ($CHCl_3$ then, $CHCl_3$ with 1% Et_3N, then $CH_3OH/CHCl_3$, 1:300 with 1% Et_3N) to afford **6** (198.3 μmol, 51%) as a yellow oil: ^{1}H-NMR (600 MHz, $CDCl_3$) δ 5.70 (brs, 1H), 5.17 (brs, 1H), 4.02–3.89 (m, 3H), 3.65 (dd, 2H, *J* = 6.6, 6.6 Hz), 2.95–2.88 (m, 3H), 2.54 (t, 2H, *J* = 7.8 Hz), 1.93–1.88 (m, 1 H), 1.79–1.64 (m, 4H), 1.59 (quin, 2H, *J* = 7.2 Hz), 1.39–1.25 (m, 10H), 0.91 (s, 9H), 0.89 (t, 3H, *J* = 7.2 Hz), 0.88 (s, 9H), 0.089 (s, 3H), 0.086 (s, 3H), 0.04 (s, 6H); ^{13}C-NMR (150 MHz, $CDCl_3$) δ 168.8, 165.3, 155.8, 67.6, 59.6, 41.6, 40.4, 35.8, 34.3, 33.0, 32.1, 29.9, 29.5, 29.2, 28.2, 26.19, 26.18, 22.9, 18.5, 18.3, 14.4, 0.26, −4.18, −4.26, −4.30, −5.1; HRMS (ESI) calcd. for $C_{30}H_{63}N_4O_3SSi_2^+$ [M + H]$^+$ *m/z* 615.4154, found *m/z* 615.4158.

2.2.3. Synthesis of 2-(2-octylthioethyl)-6-oxo-4-acetylamino-1-(3′,5′-*O*-t-butyldimethylsilyl-2′, 4′-dideoxy-D-ribityl) triazine (**7**)

AcCl (5.3 μL, 74.7 μmol) was added to a solution of **6** (15 mg, 24.9 μmol) in pyridine (250 μL). The mixture was stirred at rt for 19 h. The resulting solution was diluted with Et_2O and quenched with NH_4Cl. The product was extracted with Et_2O and the combined organic layer was washed with brine, dried over Na_2SO_4, filtered and evaporated. The crude product was purified by silica gel column chromatography ($CHCl_3$, then $CHCl_3$ with 1% Et_3N) to afford **7** (13.4μmol, 54%) as a colorless oil: ^{1}H-NMR (600 MHz, $CDCl_3$) δ 7.71 (br, 1H), 4.08–3.98 (m, 3H), 3.66 (t, 2H, *J* = 6.0 Hz), 3.65 (t, 2H, *J* = 6.0 Hz), 3.00 (t, 2H, *J* = 6.6 Hz), 2.91 (t, 2H, *J* = 6.6 Hz), 2.61 (s, 3H), 2.54 (t, 2H, *J* = 7.2 Hz), 1.95 (m, 1H), 1.81–1.64 (m, 3H), 1.58 (qt, 2H, *J* = 7.2, 7.2 Hz), 1.39–1.26 (m, 10 H), 0.92 (s, 9H), 0.91 (t, 3H, *J* = 7.2 Hz), 0.88 (s, 9H), 0.095 (s, 3H), 0.093 (s, 3H), 0.041 (s, 3H), 0.03 (s, 3H); ^{13}C-NMR (150 MHz, $CDCl_3$) δ 172.0, 170.2, 161.5, 154.8, 41.9, 40.0, 34.9, 34.1, 32.8, 29.7, 29.5, 29.2, 28.8, 27.7, 25.87, 25.86, 25.85, 22.6, 18.2, 18.0, 14.1, −4.52, −4.55, −5.38, −5.41; HRMS (ESI) calcd. for $C_{32}H_{64}N_4O_4SSi_2^+$ [M + H]$^+$ *m/z* 657.4246, found *m/z* 657.4246.

2.2.4. Synthesis of 2-(2-octylthioethyl)-6-oxo-4-acetylamino-1-(2′,4′-dideoxy-D-ribityl) triazine (**8**)

Boron trifluoride-diethylether complex (51 μL, 398 μmol) was slowly added to a solution of **7** (131 mg, 199 μmol) in CH_3CN (800 μL) at 0 °C. After the addition, the mixture was stirred at 0 °C for 2 h. The resulting solution was diluted with EtOAc, then quenched with phosphate buffer (pH = 6.0, 1 M). The product was extracted with EtOAc and the combined organic layer was washed with brine, dried over Na_2SO_4, filtered and evaporated. The crude product was purified by silica gel chromatography, eluting with $CH_3OH/CHCl_3$, 1:50, then 1:10 to give **8** (103 μmol, 52%) as a white solid: ^{1}H-NMR (600 MHz, $CDCl_3$) δ 7.80 (br, 1H), 4.38–4.33 (m, 1H), 4.00–3.96 (m, 1H), 3.91–3.88 (m, 1H), 3.85–3.81 (m, 2H), 3.15 (dt, *J* = 16.8, 1H, 7.2 Hz), 3.05 (dt, 1H, *J* = 16.8, 7.2 Hz), 2.93 (t, 2H, *J* = 7.2 Hz), 2.61 (s, 3H), 2.56 (t, 2H, *J* = 7.2 Hz), 1.95–1.90 (m, 1H), 1.82–1.71 (m, 3H), 1.59 (qt, 2H, *J* = 7.2, 7.2 Hz), 1.38–1.25 (m, 10H),

0.88 (t, 3H, J = 7.2 Hz); ^{13}C-NMR (150 MHz, $CDCl_3$) δ 171.9, 170.9, 161.7, 156.0, 68.1, 61.63, 61.59, 41.7, 38.0, 36.2, 34.1, 32.8, 31.8, 29.6, 29.2, 28.8, 27.9, 25.9, 22.6,14.1; HRMS (ESI) calcd. for $C_{20}H_{36}N_4O_4S^+$ $[M + H]^+$ m/z 429.2528, found m/z 429.2530.

2.2.5. Synthesis of 2-(2-octylthioethyl)-6-oxo-4-acetylamino-1-(5′-*O*-(4,4′-dimethoxytrityl)-2′, 4′-dideoxy-D-ribityl) triazine (**9**)

DMTrCl (58 mg, 170 μmol) was added to a solution of **8** (497 mg, 114 μmol) in pyridine (487 μL) at 0 °C. After the addition, the mixture was stirred at 0 °C for 1 h. The resulting mixture was diluted with CH_2Cl_2, then quenched with $NaHCO_3$. The product was extracted with CH_2Cl_2 and the combined organic layer was washed with brine, dried over Na_2SO_4, filtered and evaporated. The crude product was purified by chromatography on silica gel, eluting with $CHCl_3$ and 1% Et_3N, then $CH_3OH/CHCl_3$, 1:200 with 1% Et_3N to give **9** (80 μmol, 70%) as a colorless oil: ^{1}H-NMR (600 MHz, $CDCl_3$) δ 7.72 (br, 1H), 7.39–7.38 (m, 2H), 7.30–7.27 (m, 6H), 7.21–7.19 (m, 1H), 6.83–6.80 (m, 2H), 4.20 (dt, 1H, J = 13.2, 7.2 Hz), 4.04 (dt, 1H, J = 13.2, 7.2 Hz), 3.15 (dt, 1H, J = 16.8, 7.2 Hz), 3.06 (dt, 1H, J = 16.8, 7.2 Hz), 2.96 (t, 2H, J = 7.2 Hz), 2.60 (s, 3H), 2.54 (t, 2H, J = 7.2 Hz), 1.96–1.91 (m, 1H), 1.83–1.77 (m, 1H), 1.69–1.66 (m, 2H), 1.57 (qt, 2H, J = 7.2, 7.2 Hz), 1.37–1.26 (m, 10H), 0.87 (t, 3H, J = 7.2 Hz); ^{13}C-NMR (150 MHz, $CDCl_3$) δ 172.0, 170.8, 161.6, 158.5, 155.3, 144.5, 135.7 135.6, 129.88, 129.86, 128.0, 127.9, 126.9, 113.2, 86.9, 68.4, 62.4, 55.2, 42.1, 36.5, 35.5, 34.1, 32.6, 31.8, 29.5, 29.19, 29.17, 28.8, 25.9, 22.6, 14.1; HRMS (ESI) calcd. for $C_{41}H_{54}N_4O_6S^+$ $[M + H]^+$ m/z 731.3837, found m/z 731.3834.

2.2.6. Synthesis of 2-(2-octylthioethyl)-6-oxo-4-acetylamino-1-(3′-*N*,*N*-diisopropyl cyanoethyl-phosphoramidyl-5′-*O*-(4,4′-dimethoxytrityl)-2′,4′-dideoxy-D-ribityl) triazine (**10**)

DIPEA (83.2 μL, 14.3 μmol) was added to a solution of **9** (15.2 mg, 20.8 μmol) in CH_2Cl_2 (416 μL) and the mixture was cooled to 0 °C. $NCCH_2CH_2OP[(N(i\text{-}Pr)_2]Cl$ (9.3 μL, 41.6 μmol) was then added to the mixture. After the addition, the mixture was stirred at 0 °C for 1 h, diluted with CH_2Cl_2, then quenched with $NaHCO_3$. The product was extracted with CH_2Cl_2 and the combined organic layer was washed with brine, dried over Na_2SO_4, filtered and evaporated. The crude product was co-evaporated with toluene, then purified by chromatography on silica gel, eluting with EtOAc/hexane, 1:1 with 1% Et_3N, then 2:3 with 1% Et_3N to give **10** (15.0 μmol, 72%) as a colorless oil mixture of two diastereomers: ^{1}H-NMR (600 MHz, $CDCl_3$) δ 7.65 (br, 1H), 7.41 (d, 2H, J = 7.2 Hz), 7.31–7.27 (m, 6H), 7.21–7.18 (m, 1H), 6.83–6.81 (m, 4H), 4.22–4.04 (m, 3H), 3.82–3.49 (m, 4H), 3.79 (s, 6H), 3.22–3.15 (m, 2H), 3.10–2.95 (m, 2H), 2.91–2.86 (m, 2H), 2.64–2.61 (m, 1H), 2.62–2.61 (m, 3H), 2.55–2.50 (m, 2H), 2.48–2.46 (m, 1H), 2.01–1.77 (m, 4H), 1.59–1.54 (m, 2H), 1.39–1.25 (m, 10H), 1.16–1.03 (m, 12H), 0.88 (t, 3H, J = 7.2 Hz); ^{13}C-NMR (150 MHz, $CDCl_3$) δ 172.04, 172.00, 170.4, 170.2, 161.44, 161.40, 158.3, 154.9, 154.7, 145.10, 145.05, 136.32, 136.30, 136.29, 136.27, 130.0, 128.1, 128.0, 127.7, 126.7, 126.6, 118.0, 117.6, 113.0, 86.0, 77.2, 77.0, 76.8, 70.4, 70.3, 70.0, 69.9, 60.4, 60.0, 58.0, 57.8, 57.4, 57.3, 55.18, 55.15, 43.08, 43.05, 43.00, 42.97, 42.0, 41.9, 36.82, 36.80, 36.25, 36.23, 34.4, 34.3, 34.2, 34.04, 33.95, 32.82, 32.79, 31.8, 29.53, 29.52, 29.18, 29.16, 28.84, 28.83, 27.7, 25.9, 24.80, 24.75, 24.60, 24.56, 24.52, 24.50, 22.6, 21.1, 20.50, 20.45; ^{31}P- NMR (202 MHz, $CDCl_3$) δ 147.70, 146.72; HRMS (ESI) calcd. for $C_{50}H_{71}N_6O_7PS^+$ $[M + H]^+$ m/z 931.4915, found m/z 931.4919.

2.3. Synthesis of the Oligodeoxynucleotides Containing acyAOVT Derivatives

ODN1 and **2** were synthesized on a 1 μmol scale by an ABI 392 DNA/RNA synthesizer with standard β-cyanoethyl chemistry. 5′-Terminal dimethoxytrityl-bearing ODN**1** and **2** was removed from the solid support by treatment with 45 mM K_2CO_3-MeOH containing 10 mM 1-octanethiol (0.5 mL) and the residue was evaporated under reduced pressure. The crude product was purified by reverse phase HPLC with a C-18 column (Nacalai Tesque: Cosmosil 5C18-MS-II, 10 × 250 mm) by a linear gradient of 5–40%/25 min of acetonitrile in 0.1% TEAA buffer at the flow rate of 4 mL/min. The dimethoxytrityl group of the purified **ODN** was removed with 10% AcOH for 30 min and the mixture was additionally purified by reverse phase HPLC to afford **ODN1**; MALDI-TOF MS (m/z):

calcd. for $[M-H]^-$ m/z 4045.8746; found 4045.780; **ODN2**; MALDI-TOF MS (m/z): calcd. for $[M-H]^-$ m/z 6960.8115; found 6958.589.

To a solution of **ODN1** or **2** (70 μM) in ddH_2O was added a solution of MMPP (1 mM) in carbonate buffer (pH = 10) at room temperature. After 1 h, 5% AcOH was added to the mixture and the mixture was left for an additional 2 h to give **ODN3**. MALDI-TOF MS ODN (SOOct): calcd. for $[M-H]^-$ m/z 4061.8736, found 4061.755, **ODN5** (vinyl): calcd. for $[M-H]^-$ m/z 3899.5826, found 3899.304; **ODN6** (vinyl): calcd. for $[M-H]^-$ m/z 6814.5195, found 6811.219

ODN**7** was prepared by using 1.0 μM ODN**5** and 100 mM sodium thiomethoxide in 50 mM MES buffer (pH = 7.0) at 37 °C for 19 h. The reaction mixture was purified by reverse phase HPLC to afford ODN**7**: calcd. for $[M-H]^-$ m/z 3947.6856, found 3947.309.

2.4. *General Procedure for the Crosslinking Reactions*

The reaction was performed using 5.0 μM **ODN5** and 2.0 μM of the target DNA or RNA labelled by fluorescein at the 5′-end in a buffer of 100 mM NaCl and 50 mM MES buffer (pH = 7.0). The reaction was incubated at 37 °C for 1–24 h. The reaction was quenched with the addition of loading dye (95% formamide, 20 mM EDTA, 0.05% xylene cyanol, and 0.05% bromophenol blue). The cross-linked products were analyzed by a denaturing 20% polyacrylamide gel electrophoresis containing urea (7 M) with TBE buffer at 200 V for 1 h. The labelled bands were visualized and quantified using a FLA-5100 Fluor Imager.

2.5. *T_m Measurement*

All samples for the T_m measurements consisted of 100 mM NaCl, 25 mM MES buffer (pH = 7.0) and 1 μM Duplex. The T_m measurements were performed using a temperature controller. Both the heating and cooling curves were measured three times over the temperature range of 25 °C to 80 °C at 0.5 °C/min. The absorbance at 260 nm was recorded every 0.5 °C.

2.6. *Alkylation with ODN Probe to HhaI DNMT-1*

The duplex oligo (final 1 μM) in pH 7.5 reaction buffer (50 mM Tris-HCl, 10 mM EDTA, 5 mM 2-mercaptoethanol) was pre-incubated at 37 °C for 5 min. HhaI Methyltransferase was then added to the reaction mixture and the mixture was pre-incubated at 37 °C (or 4 °C) for another 5 min. SAM was added to the mixture and the mixture was incubated at 37 °C (or 4 °C) for 0.5–24 h. A 4.5 μL aliquot of the reaction mixture (4.5 μL: for silver stain, 12 μL: for Flamingo gel stain) was used every time. Loading buffer (6×Tris.HCl, SDS, glycerol, 0.05% bromphenol blue, 200 mM DTT (final 20 mM)) was added and the mixture was heated at 65 °C for 15 min. The samples were separated by 10% SDS-PAGE (0.025 M Tris, 0.192 M glycine, 0.1% SDS) at room temperature (10 mA/1.5 h then 20 mA/1.0 h). Visualization of the 5′- FAM oligonucleotide was performed by fluorescence imaging using an FLA-5100 Fluor Imager. Visualization of the protein was performed by silver stain or Flamingo gel stain.

3. Results

3.1. *Synthesis of the acyAOVT Nucleoside Phosphoramidite*

We planned to synthesize the C6-substituted 5-azacytosine by ring opening of the 5-azacytosine moiety forming guanylurea and subsequent ring closing reaction that follows a previous report (Scheme 1) [20]. In our previous study, the acyclic linkers-protected MOM or Bn group produced the coupling products in low yields. The acyclic side chain **3**-protected TBS group was synthesized from 2′-deoxy-D-ribose by a previous modified procedure [21]. The coupling reaction between the sodium salt of 5-azacytosine and the acyclic side chain compound **3** was done to form the N1- and N3- alkylated products. The reaction conditions were screened in order to optimize the formation of the N1-alkylated product **4** and a considerable amount of the N1-alkylation (58%) in DMSO at 60 °C was achieved with a N1:N3 ratio = 6:1. The formation of the glycosidic bond at the N1 position of 5-azacytosine was confirmed by HMQC and HMBC analyses.

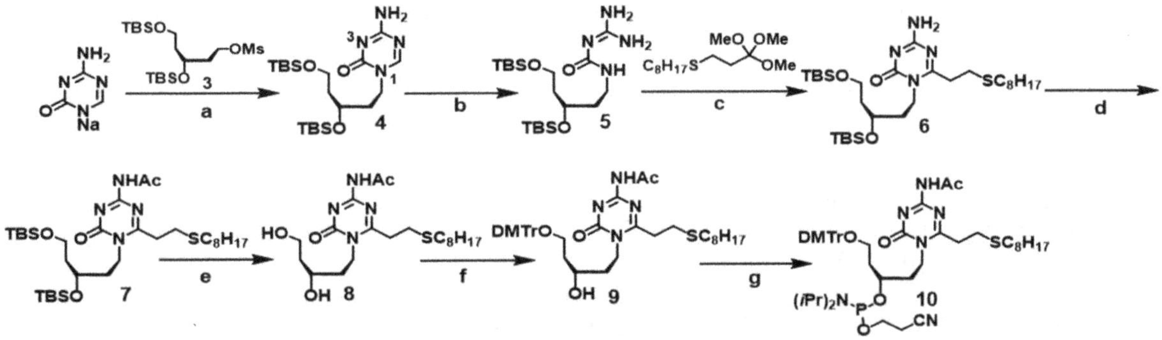

Scheme 1. *Reagents and conditions*: (**a**) 5-azacytosine sodium salt, DMSO, 60 °C, 18 h, 58%; (**b**) 7M NH_4OH, MeOH-1,4-dioxane (1:1), 55 °C, 24 h; (**c**) orthoester, DMF, 120 °C, 2 h, 51% (2 steps); (**d**) AcCl, pyridine, rt, 19 h, 54%; (**e**) $BF_3{\cdot}OEt$, CH_2Cl_2, 0 °C, 2 h, 52%; (**f**) DMTrCl, pyridine, 0 °C, 1 h, 70%; (**g**) 2-Cyanoethyl-*N*,*N*-diisopropylchloropho sphoramidite, DIPEA, CH_2Cl_2, 0 °C, 1 h, 72%.

The 5-azacytidine derivative **4** was treated with NH_3 to give the guanylurea intermediate **5**, which was condensed with the orthoester to afford the desired C6-octylthioethyl-5-azacytidine derivative **6** in 45% yield (2 steps). After the 4-*N*-acetylation and the deprotection of the TBS groups by BF_3OEt_2, the 5′-hydroxyl group was selectively protected with the DMTr group, then the 3′-hydroxyl group was phosphitylated to yield the *acy*AOVT nucleoside phosphoramidite **10**.

3.2. *Synthesis of the Oligonucleotides Containing acyAOVT and Evaluation of the Crosslinking Reactivity*

We synthesized two kinds of **ODNs1**, **2** by a DNA synthesizer using phosphoramidite **10** (Scheme 2). The synthesized ODNs were treated with 45 mM K_2CO_3/MeOH containing 10 mM 1-octanethiol to cleave them from the resin. After the DMTr-ON purification by reverse-phase (RP) HPLC, detritylation was carried out in an aqueous 10% AcOH solution at room temperature for 30 min. The obtained ODNs were further purified by RP-HPLC and were characterized by a MALDI-TOF MS analysis. The sulfide group of ODNs**1**, **2** was oxidized with magnesium monoperoxyphthalate (MMPP) and treated with aqueous 10% AcOH to give ODNs**5** and **6** according to the reported procedure [20]. The characterization of the ODNs was performed by MALDI-TOF MS.

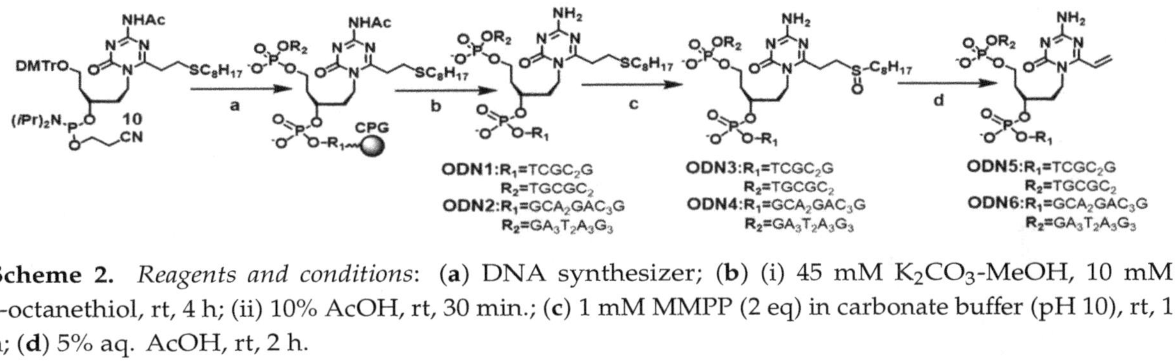

Scheme 2. *Reagents and conditions*: (**a**) DNA synthesizer; (**b**) (i) 45 mM K_2CO_3-MeOH, 10 mM 1-octanethiol, rt, 4 h; (ii) 10% AcOH, rt, 30 min.; (**c**) 1 mM MMPP (2 eq) in carbonate buffer (pH 10), rt, 1 h; (**d**) 5% aq. AcOH, rt, 2 h.

The crosslinking reactions using the reactive ODN**5** to initially the complementary target DNA**1** or RNA**1** labelled with fluorescein at the 5′ end were investigated under neutral conditions. The reaction mixture was analyzed by 20% polyacrylamide gel electrophoresis (PAGE) containing 7 M urea and the yields of the cross-linked product were calculated based on the fluorescent intensity of each band observed on the gels. Figure 3 shows a comparison of the reactivity of ODN**5** toward the different bases at the target site of the DNA**1** or RNA**1**. ODN**5** showed relatively high crosslinking reactivity to dC, dT, and dG, except for dA in the DNA. On the other hand, the crosslinked product was observed in high yields with rC and rU, and no significant products were observed with rG and rA in the RNA. The comparison of the reactivity to DNA or RNA with acyAOVT **2**, 4-amino-6-oxo-2-vinyl pyrimidine having an acyclic linker (**11:** acyAOVP) and Et-AOVT **1**, is shown in Figure 4. By comparison, acyAOVT **2** and acyAOVP **11** showed a significant difference in the reaction rate and the base selectivity to the target DNA and RNA (Figure 4A,B).

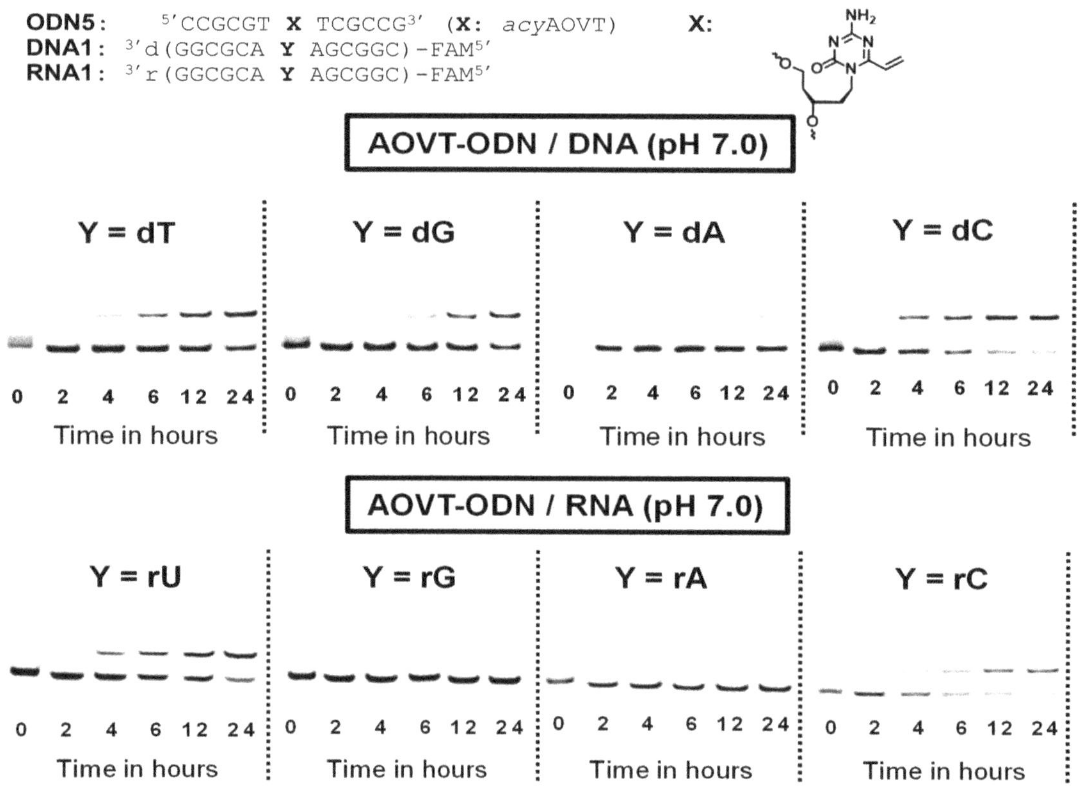

Figure 3. Denaturing gel electrophoresis of crosslink reaction to target DNA1 and RNA1 with **ODN5.** The reaction was performed in 50 mM MES-buffer (pH 7.0) containing 100 mM NaCl, **ODN5** (5 μM) and cDNAs or cRNAs (2 μM) at 37 °C.

The reaction rate with *acy*AOVT **2** was faster than that of *acy*AOVP **11**. The higher reaction rate with *acy*AOVT **2** might be attributed to the electron-withdrawing effect of the triazine increasing the reactivity of the 6-vinyl group of *acy*AOVT **2** than that of the pyrimidine-type *acy*AOVP **11**. Compared to the target base selectivity in **DNA1** with previously reported the *acy*AOVP **11** [21], AOVT **1** and **2** showed a drastically increased reactivity with the cytosine. On the other hand, the crosslinking yields to **RNA1** with AOVT **1** and **2** were significant higher to cytosine and uracil, and relatively lower to guanine than that of AOVP **11**. The reaction rate and selectivity to the pyrimidine bases (C, T and U) with *acy*AOVT **2** was comparable to that of Et-AOVT **1**. The reactivity to adenine (A) in DNA and RNA with **2** was significantly lower than that of Et-AOVT **1** (Figure 4A,C).

Next, we carried out a thermal denaturing study of the duplex containing the non-reactive *acy*AOVT. The **ODN7** bearing the non-reactive *acy*AOVT (SMe) was prepared by the addition of NaSMe to the vinyl derivative under weak acidic conditions for avoiding the decomposition of *acy*AOVT to guanylurea (Scheme 3).

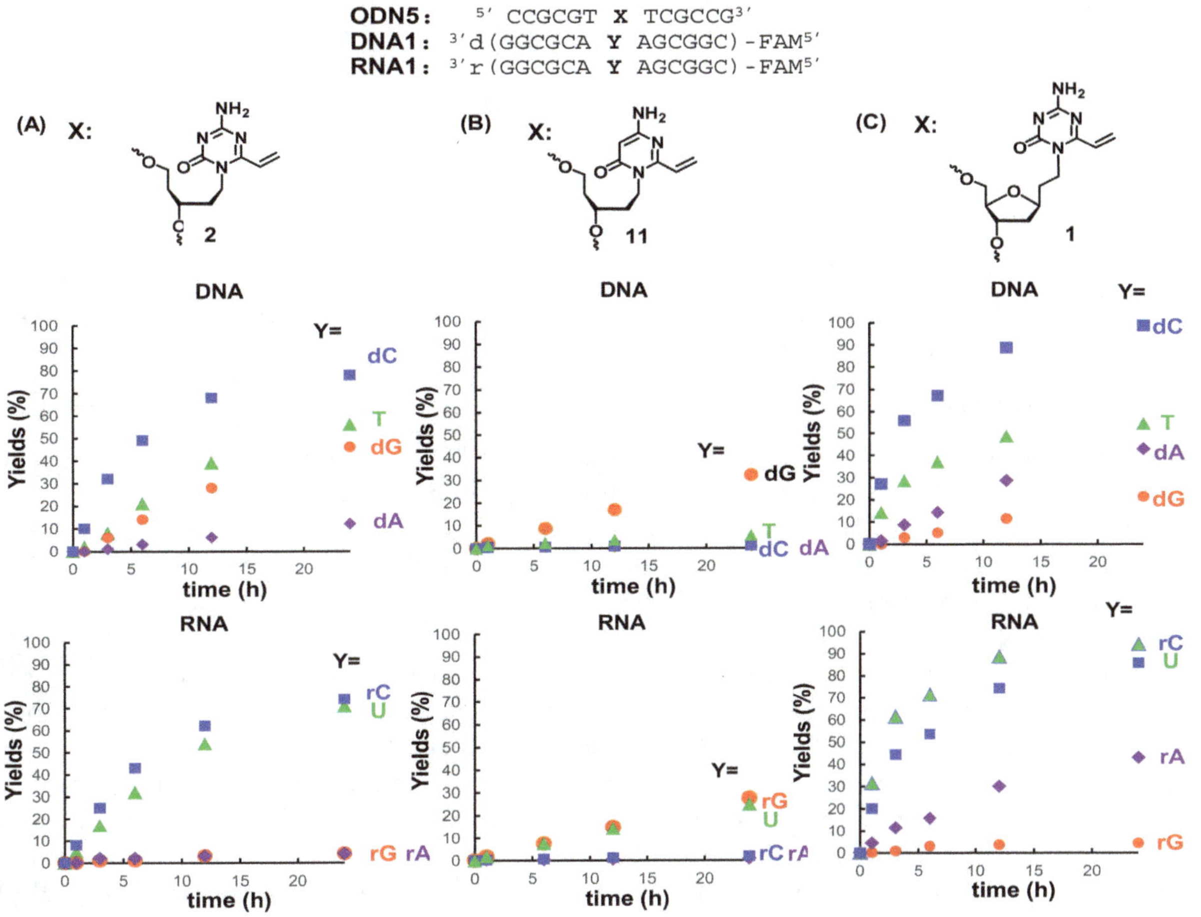

Figure 4. Comparison of the reaction yields calculated from the gel electrophoresis analysis of the cross-linking to target DNA**1** (**Y** = dT, dG, dC, dA) or RNA**1** (**N** = U, G, C, A) with *acy*AOVT (**2**) (**A**), *acy*AOVP (**11**) (**B**) and Et-AOVT (**1**) (**C**).

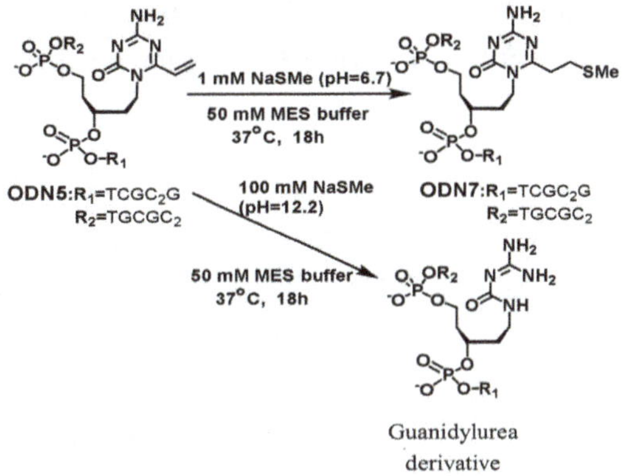

Scheme 3. Synthesis of the ODN bearing a stable precursor.

The melting temperature (*Tm*) of the **ODN7/DNA1** or **RNA1** is summarized in Figure 5. The *Tm* values of the **ODN7/DNA1** or **RNA1** were observed to be lower than that of the unmodified natural duplex (T_m = 67 °C). It should be noted that the *Tm* value for the duplex **ODN7/DNA1** (**Y** = dG) or **RNA1** (**Y** = rG) was 54 °C and 57 °C respectively, higher than the other duplexes. Taken together, our results of the T_m measurement suggest that AOVT might form stable triple hydrogen bonds with guanine to orient the C6-vinyl group of AOVT toward the opposite side of the complementary bases. Next, we attempted the crosslinking reactions to the nucleic acids-binding protein using the duplex DNA bearing AOVT derivative **2**.

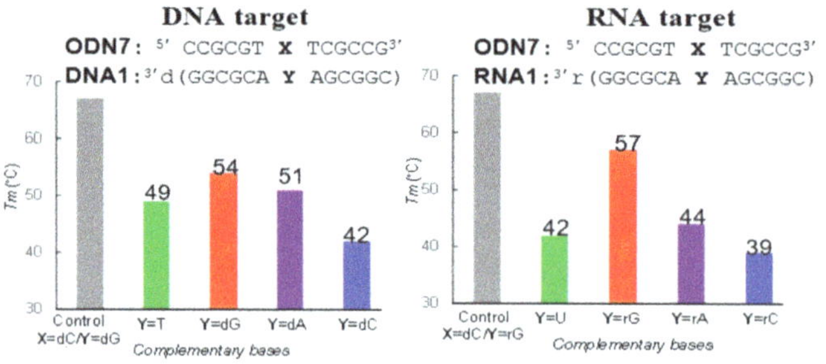

Figure 5. Comparative thermal denaturing analysis of duplexes **ODN7/DNA1** or **RNA1**. T_m values were measured in 50 mM MES buffer (pH 7.0) containing 100 mM NaCl.

3.3. Crosslinking Reactions to DNA Methyl Transferase

We chose cytosine-5-DNA methyl transferase (DNMT) as a model DNA binding protein. DNMT performs the methylation of cytosines for epigenetic modification of the genome that is involved in regulating many cellular processes [6]. In addition, aberrant methylation can cause many diseases, such as cancer [22–24]. The efficient inhibition of DNMT offers the possibility of interfering with the methylation process, which may allow control of the epigenetic regulation of cells and treat various cancers [25].

The mechanism of the cytosine-5 methylation with DNMT is illustrated in Figure 6A. A thiol of the Cys residue in the active site of DNMT acts as a nucleophile and attacks the C6 position of the cytosine to form an enzyme-linked intermediate. The resulting nucleophilic C5 position of the cytosine is then methylated by S-adenosyl-l-methionine (SAM). Subsequent abstraction of the proton at the 5 position and b-elimination provides the 5-substituted pyrimidine and active enzyme (Figure 6A) [26]. Based on this mechanism, the modified bases forming a covalent bond to DNMT were reported as irreversible inhibitors [27–31]. We expected that *acy*AOVT (**2**) replacement of the cytosine methylated by DNMT can react with the Cys residue in this enzyme (Figure 6B). In our investigation, we used the bacterial DNMT, i.e., the commercially-available Hha1 DNMT-1 from *Haemophilus haemolyticus.* Hha1 DNMT-1 methylated the internal 2′-deoxycytidine in the target sequence 5′-G<u>C</u>GC-3′, in which AOVT was substituted in place of the internal 2′-dC. We synthesized **ODN6**, which contained AOVT in the Hha1 DNMT-1 recognition sequence and evaluated the crosslink reactivity to DNA. The base selectivity was similar, but the reactivity to the target DNA was significantly lower compared to that of the previous sequence (Figure S12).

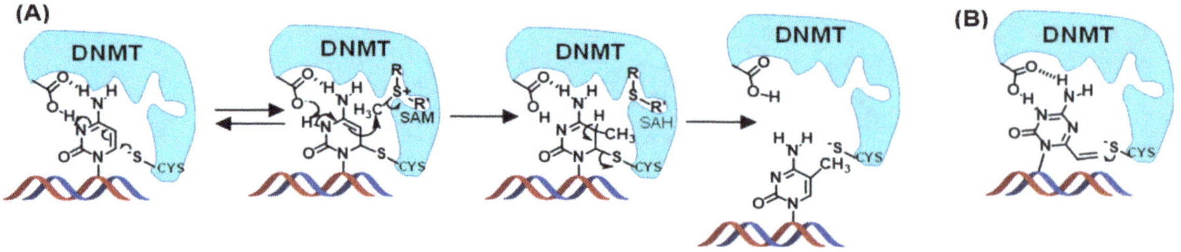

Figure 6. (**A**) Plausible mechanism for methylation by cytosine-5 methyltransferases (DNMTs) with SAM as a cofactor (**B**) Possible mechanism for covalent trapping with AOVT (**2**) derivative.

We selected DNA2 containing methyl cytosine as a target DNA because Dnmt1 methylates hemimethylated CpG sites on one strand of double-stranded DNA. The methylation conditions were determined using the duplex DNA between DNA**2** and DNA**3** contained cytosine instead of acyclic AOVT and the mono-methylated product in DNA**3** was confirmed by MALDI-TOF MS. The duplex DNA (**ODN6** and **DNA2**) was incubated with Hha1 DNMT-1 in the reaction buffer containing 5 mM 2-mercaptoethanol at 37 °C in the determined conditions. The reaction mixture was analyzed by sodium dodecyl sulfate −15% polyacrylamide gel electrophoresis (SDS−PAGE) and visualized with silver staining. The slower mobility band (*) was observed around 51kD and not observed with the non-reactive duplex DNA and Hha1 DNMT-1 (Figure 7A). We performed the reaction using the 5′ FAM-labelled **ODN6** and analyzed by SDS-PAGE. The slower mobility band stained silver matched the fluorescent band in the reaction using the 5′ FAM-labelled **ODN6** (Figure 7B) and this band was derived from the complex between Hha1 DNMT-1 and the duplex DNA. The quantification of the products was next performed by using the Flamingo^{TM} Gel staining in-gel fluorescence analysis. We confirmed the linear correlation between the protein amount and the fluorescence intensity. Thus, the complex between the protein and ODN can be quantified by measuring the fluorescence intensity on the gel (Figure 7C). The reactions were performed at 37 °C and 4 °C. The yields were obtained by quantification of the fluorescence intensity for each band and reached 77% at 37 °C after 24 h.

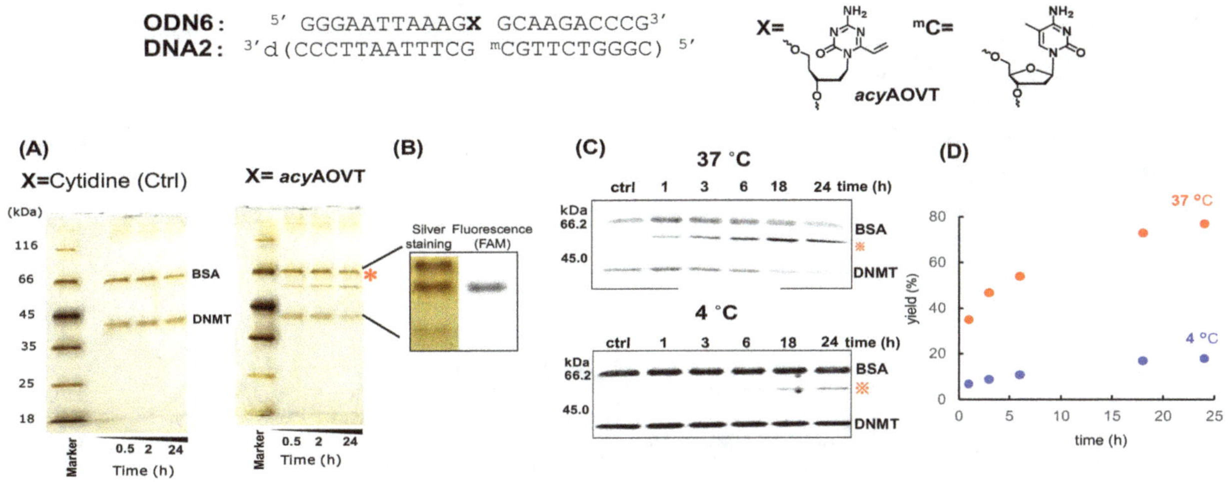

Figure 7. Evaluation of the crosslinking reaction between duplex DNA and Hha1 DNMT-1 (**A**) SDS-Page gel stained silver for analysis of reactions. Duplex DNA (**ODN6** (**X** = C or *acy*AOVT) and **DNA2**: 3 μM) and Hha1 DNMT-1 (1 μM) was incubated in the presence of SAM at 4 °C. (**B**) 5′-FAM labelled **ODN6** was used in the same reaction of (**A**) and the gel was visualized by silver staining and fluorescence. (**C**) SDS-Page gel stained Flamingo^{TM} for analysis of reactions. Duplex DNA (**ODN6** (**X** = C for control or *acy*AOVT) and **DNA2**: 1 μM) and Hha1 DNMT-1 (1 μM) was incubated in the presence of SAM at 4 °C or 37 °C. (**D**) Time course of the yields at 4 °C or 37 °C.

The yields significantly decreased at 4 °C (Figure 7D), suggesting that the reaction rate depends on the reaction temperature. The reactions were performed using ODN**6** and DNA**4** containing 4 kinds of bases (G, A, C and T) at the complementary site for *acy*AOVT. The crosslinking yields to these targets were similar to that observed with G (Figure S13). These results suggested that DNMT can bind to the mismatch sequences and make a transition to flip out of the target *acy*AOVT to induce the crosslinking reactions. Taken together, the results of the evaluation for the reaction to Hha1 DNMT-1 indicated that the *acy*AOVT in the duplex would react with Hha1 DNMT-1 even in the presence of 2-mercaptoethanol.

4. Discussion

We have synthesized ODN containing the *acy*AOVT base **2** that is expected to react with the nucleophilic residue of the nucleic acid-binding protein in the proximity of the vinyl group. The reactivity with *acy*AOVT to DNA or RNA was significantly higher than that of *acy*AOVP. The interesting result that only one nitrogen substitution on AOVP drastically increased the reactivity could be attributed to the electron withdrawing effect of the nitrogen atom. The *acy*AOVT (**2**) showed relatively high yields to dC, dT and dG in DNA and rC, rU in RNA. The highest reactivity to cytosine with the AOVT derivatives **1** and **2** might be due to the high nucleophilicity of the amino group in cytosine and to the flexible linker, which makes it possible to access the vinyl group in the amino group of the cytosine. The cross-linking yields to adenine in the DNA or RNA with *acy*AOVT **2** were significantly lower than that of Et-AOVT **1**. The vinyl group in *acy*AOVT **2** might be located away from the reactive site of adenine because of being positioned at similar natural base-pair by the short linker with **2**, resulting in the low crosslinking yields to adenine (Figure S14). The melting temperature (T_m) values of the duplex for **ODN7/DNA1(G)** or **RNA1(G)** were the highest in all the duplexes containing the other AOVT-base pairing. These results suggested that the duplex containing *acy*AOVT-G would be stabilized by forming three hydrogen bonding, resulting in orientation of the vinyl group in *acy*AOVT to the opposite side of the complementary bases (Figure 8).

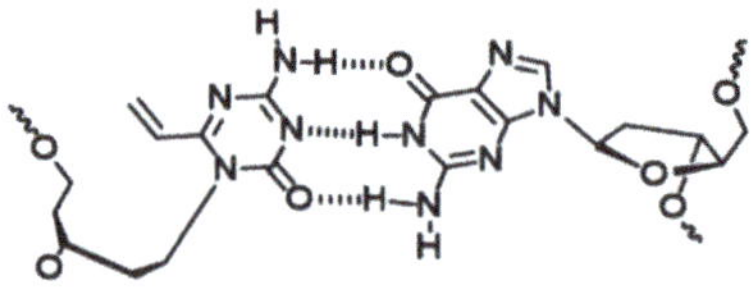

Figure 8. Hypothetical structure for *acy*AOVT and guanine in duplex DNA.

We attempted the reaction to the commercially-available DNMT using the duplex containing *acy*AOVT-G as a model reaction with the nucleic acids-binding protein. The results indicated that *acy*AOVT might react with the thiol group of the cysteine residue in DNMT in close proximity as shown in Figure 9.

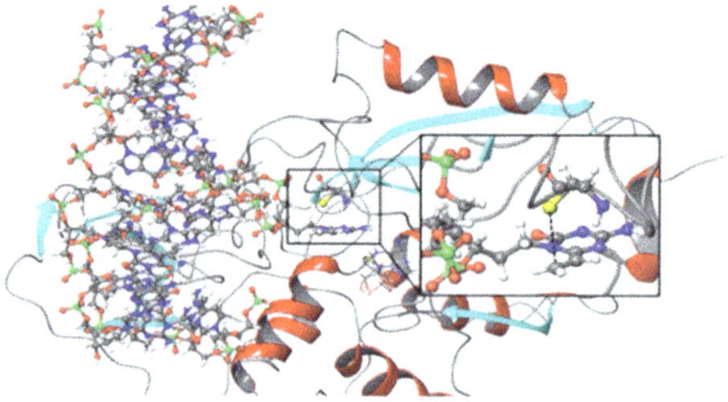

Figure 9. Predictive structure for reaction intermediate between *acy*AOVT and cysteine in *DNMT* (PDB-5MHT).

We expect that this *acy*AOVT-G base pairing system would provide a "cross-linkable duplex material" to react with the nucleic acid-binding proteins.

5. Conclusions

We have designed *acy*AOVT **2** as a reactive base to the nucleic acid-binding protein. The ODN containing **2** exhibited a higher reactivity to the pyridine bases (C, T and U) at the complementary site of **2** in duplexes DNA-DNA and DNA-RNA than that of *acy*AOVP. The results of the T_m measurements suggested that *acy*AOVT **2** can form a stable base pair with G, and consequently, the vinyl group of **2** might be located on the opposite side of the complementary bases. The preliminary results for the reaction to DNMT with the duplex DNA containing *acy*AOVT demonstrated the potential of *acy*AOVT-G base pairing system in the reaction to the nucleic acids-binding protein. This system can be used for the crosslinking reaction to the RNA binding protein. We expect that our system will provide a useful tool for the molecular study of the RNA binding protein in control of the RNA biological functions.

Author Contributions: K.O. and K.Y. conceived and designed the experiments. K.O. and S.I. performed the experiments. F.N. and H.O. wrote the study. F.N. supervised the project. All authors have read and agreed to the published version of the manuscript.

References

1. Bouhlel, M.A.; Lambert, M.; David-Cordonnier, M.H. Targeting transcription factor binding to DNA by competing with DNA binders as an approach for controlling gene expression. *Curr. Top. Med. Chem.* **2015**, *15*, 1323–1358. [CrossRef] [PubMed]
2. Todeschini, A.L.; Georges, A.; Veitia, R.A. Transcription factors: Specific DNA binding and specific gene regulation. *Trends Genet.* **2014**, *30*, 211–219. [CrossRef] [PubMed]
3. O'Brien, P.J. Catalytic promiscuity and the divergent evolution of DNA repair enzymes. *Chem. Rev.* **2006**, *106*, 720–752. [CrossRef] [PubMed]
4. Nevinsky, G.A. Structural, thermodynamic, and kinetic basis for the activities of some nucleic acid repair enzymes. *J. Mol. Recognit.* **2011**, *24*, 656–677. [CrossRef]
5. Gujar, H.; Weisenberger, D.J.; Liang, G.N. The roles of human DNA methyltransferases and their isoforms in shaping the epigenome. *Genes* **2019**, *10*, 171. [CrossRef]
6. Lyko, F. The DNA methyltransferase family: A versatile toolkit for epigenetic regulation. *Nat. Rev. Genet.* **2018**, *19*, 81–92. [CrossRef]
7. Taylor, K.; Sobczak, K. Intrinsic regulatory role of RNA structural arrangement in alternative splicing control. *Int. J. Mol. Sci.* **2020**, *21*, 5161. [CrossRef]
8. Janakiraman, H.; House, R.P.; Gangaraju, V.K.; Diehl, J.A.; Howe, P.H.; Palanisamy, V. The long (lncRNA) and short (miRNA) of it: TGF beta-mediated control of RNA-binding proteins and noncoding RNAs. *Mol. Cancer Res.* **2018**, *16*, 567–579. [CrossRef]
9. Ito, K.K.; Watanabe, K.; Kitagawa, D. The emerging role of ncRNAs and RNA-binding proteins in Mmitotic apparatus formation. *Non Coding RNA* **2020**, *6*, 13. [CrossRef]
10. Sternburg, E.L.; Karginov, F.V. Global approaches in studying RNA-binding protein interaction networks. *Trends Biochem. Sci.* **2020**, *45*, 593–603. [CrossRef]
11. Lercher, L.; McGouran, J.F.; Kessler, B.M.; Schofield, C.J.; Davis, B.G. DNA modification under mild conditions by Suzuki-Miyaura cross-coupling for the generation of functional probes. *Angew. Chem. Int. Ed.* **2013**, *52*, 10553–10558. [CrossRef] [PubMed]
12. Gerard-Hirne, T.; Thiebaut, F.; Sachon, E.; Desert, A.; Drujon, T.; Guerineau, V.; Michel, B.Y.; Benhida, R.; Coulon, S.; Saintome, C.; et al. Photoactivatable oligonucleotide probes to trap single-stranded DNA binding proteins: Updating the potential of 4-thiothymidine from a comparative study. *Biochimie* **2018**, *154*, 164–175. [CrossRef] [PubMed]
13. Smith, C.C.; Hollenstein, M.; Leumann, C.J. The synthesis and application of a diazirine-modified uridine analogue for investigating RNA-protein interactions. *RSC Adv.* **2014**, *4*, 48228–48235. [CrossRef]
14. Dziuba, D.; Hoffmann, J.E.; Hentze, M.W.; Schultz, C.A. Genetically Encoded Diazirine Analogue for RNA-Protein Photo-crosslinking. *Chembiochem* **2020**, *21*, 88–93. [CrossRef] [PubMed]
15. Wickramaratne, S.; Mukherjee, S.; Villalta, P.W.; Scharer, O.D.; Tretyakova, N.Y. Synthesis of sequence-specific DNA-protein conjugates via a reductive amination strategy. *Bioconj. Chem.* **2013**, *24*, 1496–1506. [CrossRef]
16. Carrette, L.L.G.; Morii, T.; Madder, A. Toxicity inspired cross-linking for probing DNA-peptide interactions. *Bioconj. Chem.* **2013**, *24*, 2008–2014. [CrossRef]
17. Dadova, J.; Orsag, P.; Pohl, R.; Brazdova, M.; Fojta, M.; Hocek, M. Vinylsulfonamide and acrylamide modification of DNA for cross-linking with proteins. *Angew. Chem. Int. Ed.* **2013**, *52*, 10515–10518. [CrossRef]
18. Olszewska, A.; Pohl, R.; Brazdova, M.; Fojta, M.; Hocek, M. Chloroacetamide-linked nucleotides and DNA for cross-linking with peptides and proteins. *Bioconj. Chem.* **2016**, *27*, 2089–2094. [CrossRef]
19. Matyasovsky, J.; Hocek, M. 2-Substituted 2′-deoxyinosine 5′-triphosphates as substrates for polymerase synthesis of minor-groove-modified DNA and effects on restriction endonuclease cleavage. *Org. Biomol. Chem.* **2020**, *18*, 255–262. [CrossRef]
20. Yamada, K.; Ishiyama, S.; Onizuka, K.; Nagatsugi, F. Synthesis and properties of cross-linkable DNA duplex using 4-amino-2-oxo-6-vinyl-1,3,5-triazine. *Tetrahedron* **2017**, *73*, 1424–1435. [CrossRef]
21. Kusano, S.; Ishiyama, S.; Lam, S.L.; Mashima, T.; Katahira, M.; Miyamoto, K.; Aida, M.; Nagatsugi, F. Crosslinking reactions of 4-amino-6-oxo-2-vinylpyrimidine with guanine derivatives and structural analysis of the adducts. *Nucl. Acids Res.* **2015**, *43*, 7717–7730. [CrossRef] [PubMed]

22. Robertson, K.D. DNA methylation and human disease. *Nat. Rev. Genet.* **2005**, *6*, 597–610. [CrossRef]
23. Jones, P.A.; Baylin, S.B. The epigenomics of cancer. *Cell* **2007**, *128*, 683–692. [CrossRef] [PubMed]
24. Pfeifer, G.P. Defining driver DNA methylation changes in human cancer. *Int. J. Mol. Sci.* **2018**, *19*, 1166. [CrossRef]
25. Zhou, Z.H.; Li, H.Q.; Liu, F. DNA Methyltransferase inhibitors and their therapeutic potential. *Curr. Top. Med. Chem.* **2018**, *18*, 2448–2457. [CrossRef] [PubMed]
26. Klimasauskas, S.; Kumar Roberts, R.J.; Cheng, X. Hhal methyltransferase flips its target base out of the DNA helix. *Cell* **1994**, *76*, 357–369. [CrossRef]
27. Shigdel, U.K.; He, C. A New 1′-methylenedisulfide deoxyribose that forms an efficient cross-link to DNA cytosine-5-methyltransferase (DNMT). *J. Am. Chem. Soc.* **2008**, *130*, 17634–17635. [CrossRef]
28. Sato, K.; Kawamoto, K.; Shimamura, S.; Ichikawa, S.; Matsuda, A. An oligodeoxyribonucleotide containing 5-formyl-2′-deoxycytidine (fC) at the CpG site forms a covalent complex with DNA cytosine-5 methyltransferases (DNMTs). *Bioorg. Med. Chem. Lett.* **2016**, *26*, 5395–5398. [CrossRef]
29. Sato, K.; Kunitomo, Y.; Kasai, Y.; Utsumi, S.; Suetake, I.; Tajima, S.; Ichikawa, S.; Matsuda, A. Mechanism-based inhibitor of DNA cytosine-5-methyltransferase by a SNAr reaction with an oligodeoxyribonucleotide containing a 2-amino-4-halopyridine-C-nucleoside. *Chembiochem* **2018**, *19*, 865–872. [CrossRef]
30. Utsumi, S.; Sato, K.; Ichikawa, S. Insight into the recognition mechanism of DNA cytosine-5 methyltransferases (DNMTs) by incorporation of acyclic 5-fluorocytosine (C-F) nucleosides into DNA. *Bioorg. Med. Chem. Lett.* **2018**, *28*, 2189–2194. [CrossRef]
31. Wildenhof, T.M.; Schiffers, S.; Traube, F.R.; Mayer, P.; Carell, T. Influencing epigenetic information with a hydrolytically stable carbocyclic 5-aza-2′-deoxycytidine. *Angew. Chem. Int. Ed.* **2019**, *58*, 12984–12987. [CrossRef]

2

De Novo Nucleic Acids: A Review of Synthetic Alternatives to DNA and RNA that could Act as Bio-Information Storage Molecules

Kevin G Devine [1] **and Sohan Jheeta** [2,*]

[1] School of Human Sciences, London Metropolitan University, 166-220 Holloway Rd, London N7 8BD, UK; K.Devine@londonmet.ac.uk

[2] Network of Researchers on the Chemical Evolution of Life (NoR CEL), Leeds LS7 3RB, UK

* Correspondence: sohan@sohanjheeta.com

† This paper is dedicated to Professor Colin B Reese, Daniell Professor of Chemistry, Kings College London, on the occasion of his 90th Birthday.

Abstract: Modern terran life uses several essential biopolymers like nucleic acids, proteins and polysaccharides. The nucleic acids, DNA and RNA are arguably life's most important, acting as the stores and translators of genetic information contained in their base sequences, which ultimately manifest themselves in the amino acid sequences of proteins. But just what is it about their structures; an aromatic heterocyclic base appended to a (five-atom ring) sugar-phosphate backbone that enables them to carry out these functions with such high fidelity? In the past three decades, leading chemists have created in their laboratories synthetic analogues of nucleic acids which differ from their natural counterparts in three key areas as follows: (a) replacement of the phosphate moiety with an uncharged analogue, (b) replacement of the pentose sugars ribose and deoxyribose with alternative acyclic, pentose and hexose derivatives and, finally, (c) replacement of the two heterocyclic base pairs adenine/thymine and guanine/cytosine with non-standard analogues that obey the Watson–Crick pairing rules. This manuscript will examine in detail the physical and chemical properties of these synthetic nucleic acid analogues, in particular on their abilities to serve as conveyors of genetic information. If life exists elsewhere in the universe, will it also use DNA and RNA?

Keywords: non-standard nucleic acids; sugar-phosphate backbone; pentose sugars; hexose derivatives; phosphate group replacement; alien life forms

---

1. Introduction

Life on Earth uses three key biopolymers, namely nucleic acids, proteins and polysaccharides; each of which possesses intrinsic structural features. Nucleic acids are polymers comprised of heterocyclic aromatic bases appended to a sugar-phosphate backbone, held together by phosphodiester bonds. Proteins are polymers of amino acids linked via amide bonds of dubbed as peptide bonds $[-C(=O)N(-H)-]$ and polysaccharides are polymers of carbohydrates linked via acetal ether bonds. These unique chemical features have long fascinated organic chemists and stimulated the most creative minds among them to question nature's choices, and, indeed, design and test alternatives using the power of laboratory-based synthetic organic chemistry. This paper will focus entirely upon re-designed nucleic acids, which feature three key structural modifications of their natural counterparts: (a) replacement of the phosphate moiety with an uncharged analogue, (b) replacement of the pentose sugars ribose and deoxyribose with alternative acyclic, pentose and hexose derivatives and, finally, (c) replacement of the two heterocyclic base pairs adenine/thymine and guanine/cytosine with non-standard analogues that obey (or disobey) the Watson–Crick pairing rules. As will be shown, the results are indeed intriguing and have profound consequences for the development of artificial Darwinian chemical systems, and the discovery of life, if it exists, elsewhere in the Universe. This is an example of how the *synthesis paradigm* can drive discovery and understanding in ways that *analysis* of the natural world alone cannot. Just as the most skilled mechanical, electrical and software engineers can build modern automobiles, aircraft and super-computers, creations whose intricate inner workings they fully understand, so, it is hoped, the new generation of synthetic biologists will be able to manufacture, and thus fully understand, artificial life forms, built from different biopolymers to those found in nature.

1.1. Nucleic Acid Structure

Nucleic acids are biopolymers that are built from nucleotides. The latter consist of three molecular components; a heterocyclic aromatic base (also known as a nucleobase), a five-atom ring sugar, and a phosphate unit that connects the sugars together and forms the alternating sugar-phosphate backbone. The structures of the nucleobases, the two sugars and the nucleobase-sugar conjugates, which are known as nucleosides, are shown in Figure 1. DNA differs from RNA in two distinct ways: the sugar is 2′-deoxyribose instead of ribose, and the pyrimidine base thymidine has a methyl group attached at the 5-position where uracil has hydrogen and so technically speaking the thymidine is a 5′-methyluracil base.

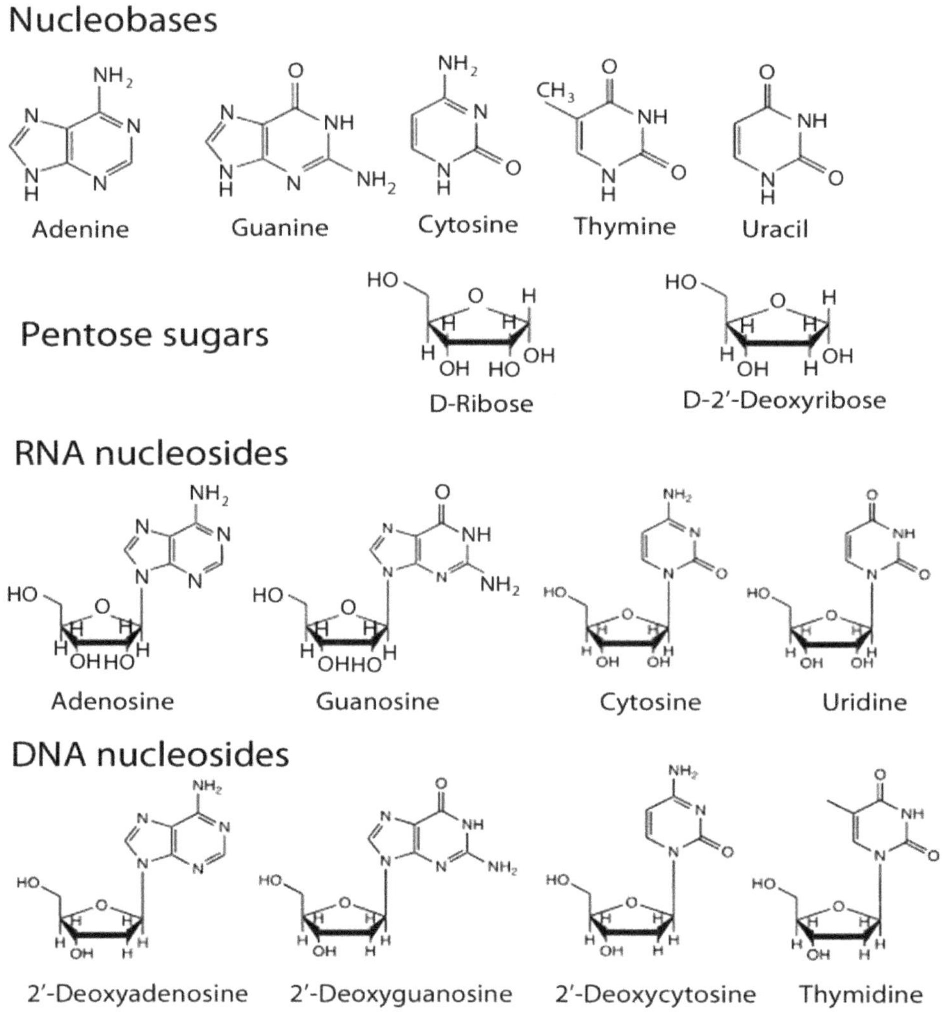

Figure 1. The molecular structures of the nucleobases, sugars and nucleosides found in DNA and RNA.

The numbering systems are distinct for the nucleobases and sugars, with the latter using affixed "primed" numbers 1′-5′ to distinguish them (Figure S1a, in the Supplementary Materials). Oligonucleotides, or polynucleotides, are polymers made from nucleosides that are linked via their 3′- and 5′-oxygen atoms by phosphate groups (Figure S1b, in the Supplementary Materials). The sequence of bases is read from the 5′-end to the 3′-end (i.e., 5′ → 3′).

The base sequence of an oligonucleotide is important because genetic information is stored in the sequence of these bases in a DNA (or RNA) molecule. The key storage unit for genetic information in most organisms is not a single-stranded DNA oligomer, instead, it is two complementary strands. These strands are held together by base-pairs on opposite strands which follow two complementary principles: size and hydrogen-bonding complementarity. In size complementarity, a large 9-atom bicyclic purine base (adenine and guanine) pairs with a small, 6-atom ring pyrimidine base (uracil/thymine and cytosine). In hydrogen-bonding complementarity, hydrogen bond donors (N-H bonds) on one base interact with the hydrogen bond acceptors generally with lone pairs of electrons on N or O atoms on its partner in the opposite strand. In this way, an adenine on one strand pairs with a thymine (or uracil in RNA) in another, and a guanine likewise pairs with a cytosine (Figure 2 panel a).

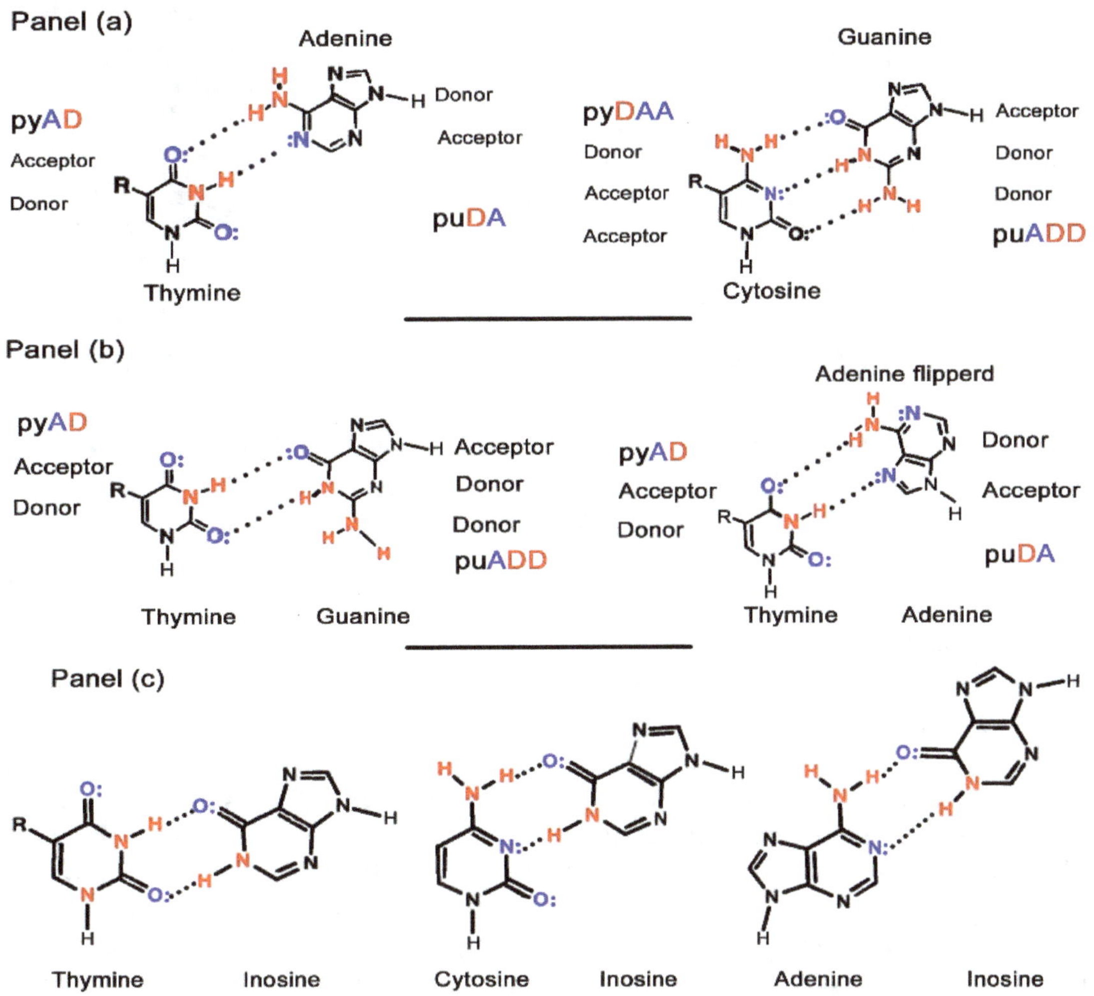

Figure 2. Panel (**a**): the base-pairs between T–A and C–G revealing both the size and hydrogen-bonding complementarities displayed by these pairs. Panel (**b**): The wobble base-pairing of T and G (left), and the Hoogsteen base-pairing of T and A (right). Panel (**c**): wobble base-pairings between thymine and inosine (left), cytosine and inosine (centre), and adenine and inosine (right).

Two complementary strands held together by their base-pairs is known as a duplex. The two complementary single strands are oriented "anti-parallel" with respect to each other; a strand whose direction is 5′→3′ pairs with its' 3′→5′ complement. The duplex nature of DNA was first revealed by the X-ray crystallographic data obtained by Rosalind Franklin and Maurice Wilkins in 1953. From this data, Francis Crick and James Watson were able to construct models, which showed that the two complementary strands were held together by hydrogen bonds between base partners, forming a double-helical spiral staircase in which the base-pairs are the steps and the sugar-phosphate backbone the handles [1]; Figure 3 panels a and b). The T–A and C–G hydrogen bonding interactions thus became known as Watson–Crick base-pairs, and their elucidation of the double helix nature of DNA earned them the 1964 Chemistry Nobel Prize, along with Maurice Wilkins (Rosalind Franklin had by then passed away in 1958).

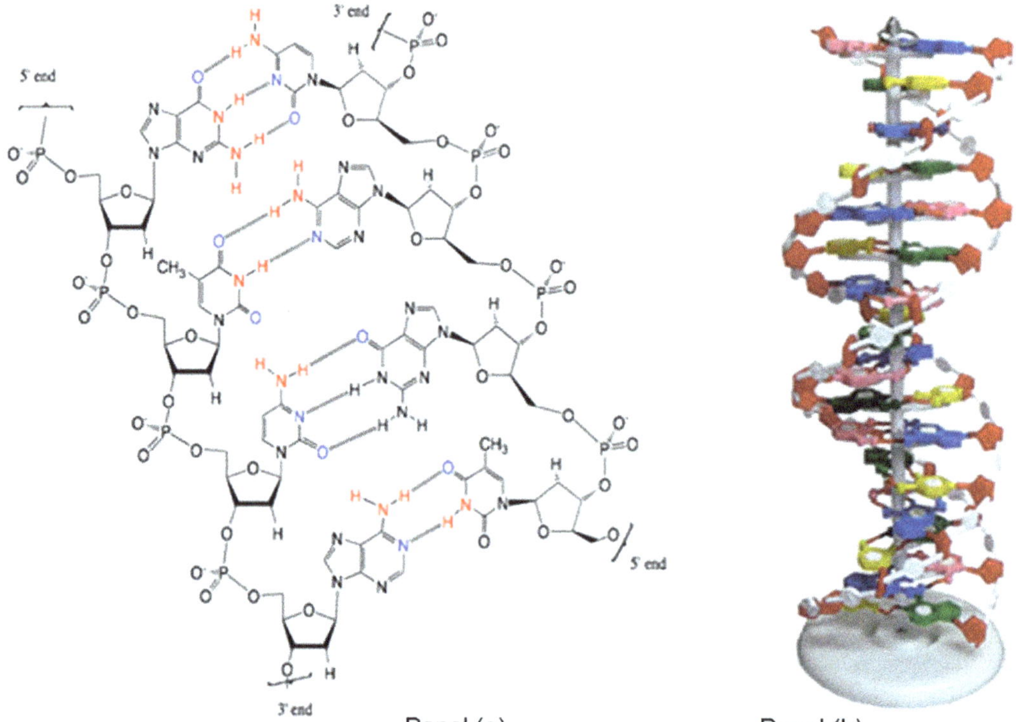

Panel (a) Panel (b)

Figure 3. The duplex between two Watson–Crick base-paired complementary DNA strands panel (**a**), is arranged as a double-helix in which the base-pairs are the steps and the sugar-phosphate backbone the handles panel (**b**).

Two non-Watson–Crick base pairings also occur and are known respectively as wobble and Hoogsteen pairings (Figure 2 panel b). The "wobble" pairs occur between thymine/uracil and guanine, thymine and inosine, cytosine and inosine and adenine and inosine. Inosine, which occurs only in transfer RNA (tRNA), is an analogue of guanosine that lacks the 2-NH_2 and its wobble pairings help maintain the 3-D structure of tRNAs and explain the non-specificity of the genetic code (Figure 2 panel c).

Hoogsteen base-pairing occurs between thymine and adenine in which the N–H of thymine interacts with the N-7 of adenine, with the latter having "flipped" through 180° [2]. The biological significance of Hoogsteen pairings has still not yet been fully elucidated, but they may be important for stimulating mutations and for the stabilization of tRNA conformations [3].

It takes much more than just hydrogen-bonded base-pairing to hold the double-helical DNA duplex together. The base pairs are aromatic heterocycles and as such, are relatively hydrophobic. They interact with each other through a complex combination of van der Waals and electrostatic intermolecular forces known as "base-stacking" and thus readily "stack" on top of each other (Figure 3 panel b). This "base-stacking" removes these hydrophobic components away from water and into a more hydrophobic environment; this process is entropy-driven and drives duplex formation.

There are two major regular DNA duplex structures known as A-DNA and B-DNA. B-DNA is the form that predominates in the cellular nucleus and has ten base pairs per turn of the helix with the base pairs perpendicular to the helix axis. Two distinct structural features can be seen, a wide major groove and a narrow minor groove. A-DNA is a more tightly packed structure with 11 base pairs per helical turn and these base pairs are tilted at a 20° angle to the helix axis. In A-DNA, the major groove is narrow but deep, and the minor groove broad and shallow. Another type of DNA and RNA duplex conformer, known as Z-DNA and Z-RNA, was initially found in crystallised oligomers that contained alternate G–C sequences, in solutions containing high concentrations of salts (NaCl, $MgCl_2$) or ethanol. Z-DNA is a left-handed double helical structure with ten base pairs per turn, in which the base pairs are roughly parallel to the helix axis. The phosphate backbone has a zig-zag appearance, and the minor groove is very deep and narrow, and the major groove extremely shallow. Recent studies have revealed the role of Z-DNA and Z-RNA in human diseases [4].

Moreover, both DNA and RNA possess a polyanionic backbone in which a negative charge is located on a phosphate oxygen, and there are seven atoms between each charge. This repeating polyanionic array renders all nucleic acids highly water-soluble, and gives them their predominant physical property, as Coulombic repulsions prevent intra-strand interactions and folding. The nucleic acids tend to stretch out in solution to minimize intra-strand repulsions, and this intrinsic property enables duplex formations to occur between two complementary strands; the relatively stretched single strands, upon interaction, can then arrange themselves into the minimum energy duplex conformation, driven by base-stacking, in which Watson–Crick base-pairings occur.

Furthermore, the polyanionic backbone dominates the physico-chemical properties of nucleic acids to such an extent that they are, essentially, independent of their base sequences. All nucleic acids containing the same number of bases (and hence negatively charged phosphates) will, regardless of their base sequences, move along a gel in an applied electric field at the same rate towards the anode. This behaviour is unique to nucleic acids and is what renders them fit to act as genetic polymers, which enable Darwinian evolution to take place.

1.2. Genetic Molecular Constraints

For any biopolymer to act as an efficient chemical storage system and conveyor of genetic information, it must have the following two important properties: first, it must be readily replicable, through the use of protein enzymes as in modern biology, and ribozymes or "self-replicating RNAs" in the precursor RNA world; second, it must be within a low tolerable rate of mutations; structural changes cannot compromise replication fidelity. In summary, a genetic biopolymer must be "Capable of Surviving Modifications in Constitution without Loss of Properties Essential for Replication" and this statement generates a very appropriate acronym: COSMIC-LOPER [5].

COSMIC-LOPER behaviour is unique to nucleic acids. In DNA and RNA, changing the base sequence does not change either the physical properties or chemical reactivity. This is certainly not the case for proteins, where changing even one amino acid in a protein sequence profoundly alters both the physical properties and the reactivity. Perhaps the best-known example of this is haemoglobin, a four-chain quaternary protein with four haem units responsible for the transport of molecular oxygen in the blood of vertebrates. A mutation which replaces the hydrophilic amino acid, glutamate, with the hydrophobic, valine, at position 6 in the B1 chain results in a much less soluble protein and is responsible for the debilitating disease of sickle cell anaemia.

In the enzyme ribonuclease, which cleaves RNA molecules, substitution of any of the four catalytic amino acids histidines at positions 12 and 119; lysine at position 41, and aspartic acid at position 121 renders this enzyme completely inactive.

1.3. Catalytic Nucleic Acids and the Origin of Life

An examination of contemporary biology encounters a classical "chicken and egg" paradox. Polymeric nucleic acids may contain all the information required to construct a protein, but proteins called polymerases are needed to catalyse nucleic acid synthesis. The big question is, how did it all begin? Did proteins and nucleic acids arise and encounter each other spontaneously (which is highly unlikely based upon statistical probability), or was there an initial biopolymer that could perform both catalysis and replication, in other words, was the first living entity a biopolymer that was self-replicating? Life can be defined as "a self-sustaining chemical system capable of undergoing Darwinian evolution". This clearly distinguishes living chemistry from other physico-chemical processes like crystal growth. In the latter, for example, the addition of small crystals of sodium chloride to a saturated solution of this salt can seed the formation of more crystals. This may be considered as reproduction, but this system cannot undergo Darwinian evolution as there is no possible way that the structure of sodium chloride can be changed (by replacing either the sodium or the chloride ions) without interfering with the seeding of more crystals of this salt from solution. What is essential to the process of Darwinian evolution is that a small change needs to occur in structures, which may enable the replicating process to improve by just a tiny amount in future progeny. Conversely, such changes may result in the demise of the same structures. These small changes lie at the heart of the concept of Darwinian evolution, without which life on Earth is not possible, as exemplified by the non-biological replicator, sodium chloride above. These changes are often referred to as selection pressures and are due to ever-fluctuating environmental challenges like changes in temperature, pressure and/or pH.

Supporters of the "one-biopolymer" as the earliest form of life were given a big boost in the early 1980s with the discovery that RNA, in addition to acting as a chemical code repository (*cf* mRNA and genomes of Retroviruses), exhibits catalytic behaviour; for this discovery, Sydney Altmann [6–8] and Tom Cech [9] received the 1990 Nobel Prize in Chemistry. The catalytic activity of RNA molecules, referred to as "ribozymes" (*cf* enzymes), included cleavage and splicing of RNA inter-nucleotide bonds, as well as, more importantly, peptide bond formation.

Elucidation of the structure of the ribosome has revealed that protein synthesis, via peptide bond formation, is entirely carried out by non-coding ribosomal RNA (rRNA) molecules (*cf* mRNA, which is referred to as coding). The proteins present in the ribosome appear to act as scaffolds. This has led to the idea that an "RNA world", whose organisms were effectively self-replicating RNA molecules followed later by organisms whose genomes consisted of RNA that was synthesized and processed entirely by proto-ribozymes, preceded the contemporary DNA-protein system.

Proteins are of course much better catalysts than ribozymes on account of the richer and more diverse functionalities to be found on their amino acid side chains, and the absence of a charged backbone, which tends to impede close-knit folding as a result of Coulombic repulsion. Ribozymes also require relatively high divalent metal ion concentrations (Mg^{2+}, Mn^{2+}, Pb^{2+}) to function effectively. In addition to bringing reacting groups close together through ligand-binding, it is also believed that the high ionic strength is a prerequisite to overcome the Coulombic repulsion between the phosphate anions in the backbone, and thus enable a degree of folding, similar to that observed in tRNAs though unseen in genomic DNA and RNA.

1.4. Catalytic DNA Molecules: DNAzymes (Deoxyribozymes)

In 1994, the first catalytic DNA molecules, DNAzymes or deoxyribozymes, were reported by Breaker and Joyce [10]. These catalysts are purely artificial and are produced using a technique known as "Selective Evolution of Ligands by Exponential Enrichment (SELEX)". Most deoxyribozymes are single-stranded 30–60 mer DNA molecules and have found widespread applications as catalysts for the cleavage and ligation of phosphodiester bonds in RNA, DNA and amino acid substrates (serine and tyrosine-phosphates); cleavage and ligation of ester bonds, and even a Diels–Alder cycloaddition reaction [11]. Like ribozymes, deoxyribozymes require relatively high divalent metal (e.g., Pb^{2+}) salt solutions for optimal efficiency (Figure 4 panel a).

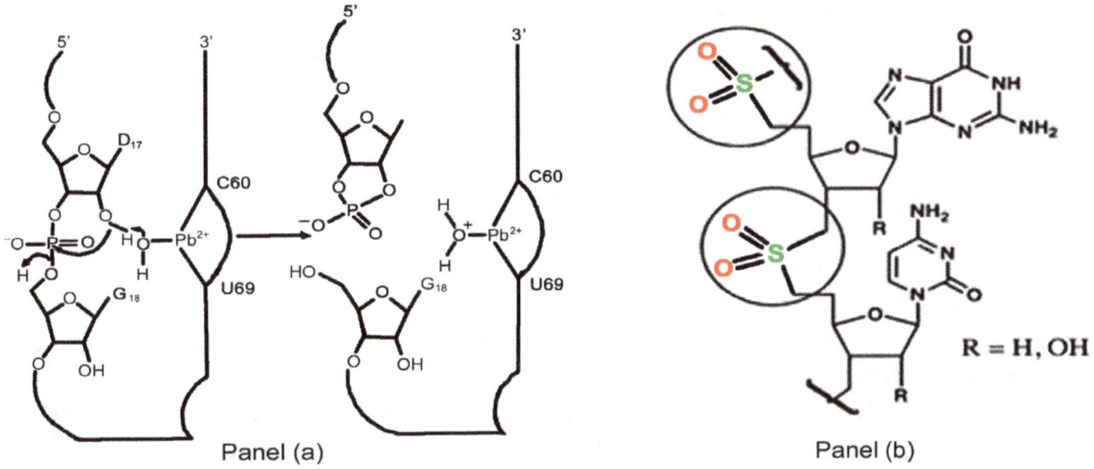

Figure 4. Panel (**a**): a proposed mechanism for the cleavage of a phosphodiester bond in yeast phenylalanine tRNA. The Pb^{2+} is bound to a specific site on the RNA around the bases U59 and C60. The active species is probably $(PbOH)^+$ which has a pKa of about 7.0. Panel (**b**): oligosulfones, which contain a dimethylene sulfone group in place of phosphate, no longer form duplexes with Watson–Crick base-pairings but bend and fold more like proteins.

2. Synthetic Organic Modifications of Nucleic Acid Structure

During the past three decades, leading organic chemists have synthesized novel nucleic acids and studied their properties. These novel nucleic acids differ from their natural counterparts in three key features: replacement of the polyanionic backbone with uncharged analogues; replacement of the ribose and 2′-deoxyribose sugars with alternative acyclic, pentose and hexose derivatives; and replacement of the standard base-pairs adenine-thymine and guanine-cytosine with non-standard analogues, including some that don't obey the two "complementarity" principles observed in DNA and RNA.

2.1. Nucleic Acids with Uncharged Backbones

The initial need for nucleic acid analogues with uncharged backbones was driven by a new therapeutic concept in the mid to late 1980s; the "anti-sense" approach, whose idea was straightforward and logical [12]. Many diseases, in particular cancer and viral infections such as AIDS and SARS, arise from the presence of unwanted DNA (oncogenes or viral DNA sequences) that are transcribed into unwanted messenger RNA, which gets translated into unwanted proteins. Most current drugs work by targeting and binding to undesired proteins. The anti-sense concept went back a stage further and targeted the undesired mRNA that preceded the undesired protein. Often the sequence of the undesired mRNA is known, so the best entity for binding would obviously be a single-stranded oligonucleotide complementary to a sequence of 15–20 bases on the mRNA, as the probability of such a sequence occurring anywhere else in the human genome is about one in ten trillion (and noting that there are only 3 billion base-pairs in the human genome). The resulting mRNA/oligonucleotide complex cannot interact with tRNA molecules on the ribosome, translation is therefore impeded, and RNase enzymes degrade the complex, so the undesired protein is not synthesised.

Administering a natural DNA molecule that is complementary to the 20 base sequence on the mRNA (referred to as an anti-sense molecule because it interacts with the coding sequence on mRNA which is a copy of the "sense" strand of the original gene sequence on DNA) has one severe drawback; human serum contains enzymes known as nucleases, which degrade DNA, so it is unlikely that enough of the DNA oligonucleotide would reach its target cell. It is the phosphodiester backbone, which is the target of these enzymes. Many researchers believed that the answer was to make analogues of DNA in which the sugar-phosphate backbone was replaced with neutral, often non-polar analogues, which would withstand degradation by nucleases. Much of this work began in the mid to late 1980s, before work performed by Steven Benner and others realized the importance of retaining a polyanionic backbone to preserve the Watson–Crick recognition properties of new oligonucleotide analogues. Many different backbones were made, and virtually all failed to show sequence-specific base pairing, the pre-requisite for use as antisense agents.

In the early to mid-1990s, work performed in Steven Benner's laboratories, (both in Florida and previously at the ETH in Zurich, Switzerland) has revealed the importance of the polyanionic backbone. In a remarkable feat of skilled organic syntheses rivalling that of any complex natural product, Benner's group constructed nucleic acid analogues in which the charged phosphate $[-O-P(=O)O^{-}-O-]$ is replaced by a polar but uncharged dimethylene sulfone $[-CH_2-SO_2-CH_2-]$ function [13]. These "oligosulfones" possess physico-chemical properties that are very different from their natural counterparts. In particular, they can bend and fold up in a manner more analogous to proteins, disrupting both base-stacking and Watson–Crick hydrogen bonding between two complementary strands, which, significantly, will not form duplexes (Figure 4 panel b). There are other examples in the literature where other groups have, likewise, replaced the charged phosphates with uncharged groups and observed similar results. This is a classic example of how a quintessential "blue skies" research enterprise could have guided and informed an important "bio-medicinal" one.

One notable exception to the observed properties of uncharged nucleic acid analogues like oligosulfones was displayed by the peptide nucleic acids (PNAs), developed by Peter Nielsen and his co-workers at the Panum Institute in Copenhagen, Denmark in the early 1990s [14]. These analogues differed significantly from their natural counterparts. The sugar-phosphate backbone was replaced by a linear polyamide N-ethyl-glycinyl entity (Equation 1) linked to the bases via a $[CH_2-CO]$ amide bond to the glycinyl nitrogen (N^2).

$$\left[-N^1H-CH_2-CH_2-N^2H-CH_2-C(=O)-\right] \qquad (1)$$

These novel analogues contain the same number of backbone atoms (6), and atoms linking the backbones to the bases (3), like their natural counterparts (Figure S2, in the Supplementary Materials); the N- and C-terminals corresponding to the 5′- and 3′-ends of DNA respectively. Poly-PNAs were readily synthesized using conventional solid-phase peptide chemistry [15]. Remarkably, the Danish researchers found that complementary PNAs formed stable duplexes, forming Watson–Crick base-paired double helices in a similar manner to DNA and RNA [16]. Moreover, PNA strands also formed stable duplexes, via Watson–Crick base-pairings, with their DNA and RNA counterparts [17,18]. However, PNA sequences containing more than 25 bases were found to be insoluble in water, so despite some remarkably similar attributes to natural nucleic acids, increasing insolubilities shown by all long PNAs (>25-mers) prevents them from becoming uncharged, alternative COSMIC-LOPERS.

2.2. *The Importance of Phosphates*

In a landmark 1987 paper in Science entitled "Why nature chose phosphates?", the late Frank Westheimer applied critical physical chemistry analysis to attempt to answer the question [19]. In addition to the diesters in DNA and RNA, other phosphate esters and phosphate anhydrides play essential roles in biochemistry: coenzymes; energy storage (adenosine triphosphate, ATP); creatine phosphate, CP; phosphoenolpyruvate and in metabolic products (e.g., glucose-6′-phosphate and fructose 1,6-diphosphate). It is clearly of great importance for all living entities, from bacteria to blue whales, to retain these essential molecules within their cell membranes, which are dominated by relatively hydrophobic substances like fatty acids. All of these essential compounds are esters or diesters of the trivalent phosphoric acid, which has three ionisable O-H bonds. The first pKa of phosphoric acid and that of mono and diesters is approximately 2, meaning that all the phosphate-containing molecules will be ionized at physiological pH (7.4), and will therefore be trapped within the lipophilic cellular membranes.

Nature's choice of phosphoric acid instead of alternatives like arsenic, citric, glutamic and silicic acids primarily arises from the fact that phosphate esters are much more stable than their alternative counterparts. Arsenic occurs just below phosphorus, in group 5 of the periodic table and forms many analogous compounds with the same valencies as phosphorus. The first pKa of arsenic acid, 2.19, is very close to that of phosphoric acid. However, arsenate esters are much less stable than their phosphorus counterparts; the di-isopropyl ester is completely hydrolysed in less than 2 min in water at room temperature, whereas the half-time of hydrolysis ($t_{1/2}$) of dimethyl phosphate, in 1M NaOH solution at 110 °C, is about 24 h. Even though arsenate diesters possess a negative charge at physiological pH, the larger arsenic atom is more easily accessible to attacking nucleophiles like water, and so arsenate esters are both kinetically and thermodynamically unstable with respect to aqueous hydrolysis. Moreover, phosphorus is geologically much more abundant, at 1.0 ppth (i.e., 1.0 mg/g) in the Earth's crust (0.1%) compared to arsenic, which occurs at 5 ppm.

Similarly, the diesters and triesters of silicic acid are much less stable than those of phosphoric acid, despite the much greater geological abundance of silicon. Furthermore, silicic acid is a much weaker acid, with a pKa of 9.50, so it would be mostly unionized at physiological pH, rendering the electrophilic silicon atom much more vulnerable to nucleophilic attack by water molecules than its phosphorus counterpart.

The only viable trivalent organic acid, citric acid, also forms esters that are much less stable than those of phosphate. The first pKa, of the central carboxylic acid, is 2.92 and so is quite similar to that of phosphoric acid at 2.00 and would thus be ionized at physiological pH. However, a citrate nucleotide would have a longer, 5-atom linker connecting the nucleosides, and the negative charge would provide very little effective shielding of the 5′- and 3′-ester linkers against nucleophilic attack by water. Nature's choice of anionic phosphate diesters as the backbone of her information storage systems is governed by the fundamental physico-chemical properties of phosphoric acid and its central phosphorus atom. While phosphate diesters and anhydrides are thermodynamically unstable with respect to hydrolysis, the negative charge and the relatively small size of the phosphorus atom (*cf* arsenic) afford them more protection from rapid nucleophilic attack by water, enabling them to persist in an aqueous environment for much longer periods of time than their arsenic, silicon and carboxyl ester counterparts (Figure 5).

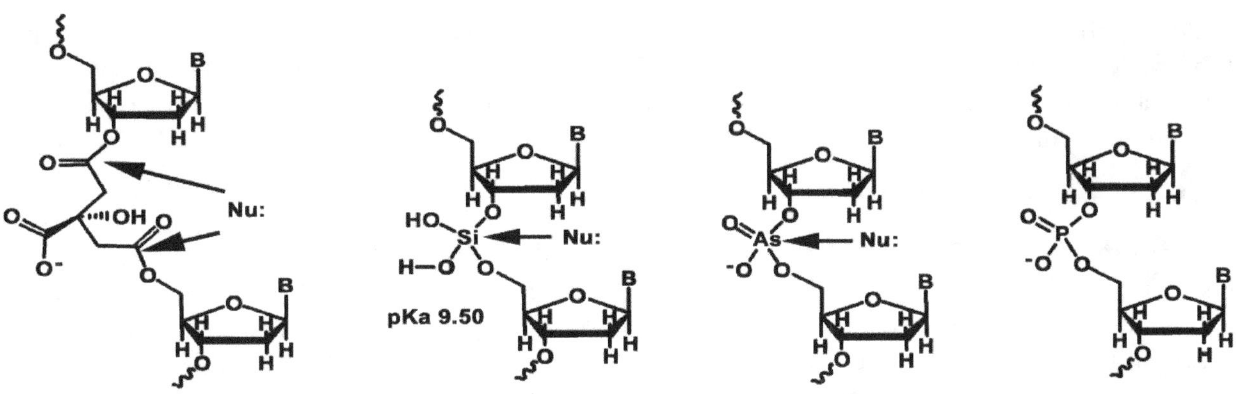

Figure 5. The phosphodiester linkage in nucleic acids is much more resistant to attack by nucleophiles (Nu:) than the alternative arsenate diesters, silicate diesters and citrate diesters.

2.3. Nucleic Acids Containing Acyclic and Non-Ribose Sugar-Phosphate Backbones

2.3.1. Acyclic 3′-1′-glycerol Phosphates

In the late 1980s, Steven Benner's group (then at the ETH in Zurich) pondered what effect the replacement of ribose sugars with a more flexible open-ended linker would have on the stability of nucleic acid helices. Nucleotides containing the more flexible 3′-1′-glycerol linker were synthesized and incorporated at various positions into synthetic oligodeoxynucleotides nine bases in length, containing the sequence 5′-CTTTTTTTG-3′. The complementary sequences 3′-GAAAAAAAC-5′ were also prepared as they form duplexes held together through Watson–Crick pairing (Figure 6 panel a). The UV-melting temperatures T_m [20] were measured for the standard duplex (which contains only deoxyribose-phosphate) and then for several other duplexes in which one, then two deoxyribose-phosphates were replaced by the more flexible 3′-1′ glycerol units, denoted by t (Figure 6 panel b, red arrows). The results are depicted in Figure 6 and Figure S3 (in the Supplementary Materials). Replacement of one deoxyribose sugar with a glycerol unit lowers the melting temperature by 15 °C. The substituting of two deoxyribose sugars depresses the melting point by 27 °C, and by 29 °C if the two glycerol units are adjacent.

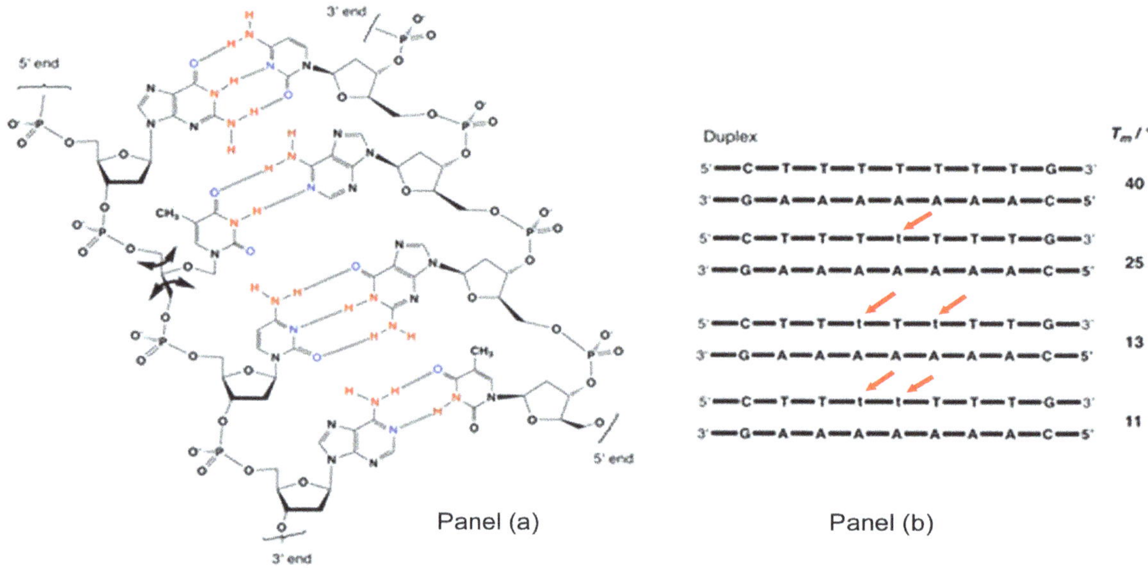

Figure 6. The more flexible 3′-1′-glycerol linker disrupts base-pairing and duplex formation in comparison to the more rigid 2′-deoxyribose panel (**a**), as revealed by the much lower melting temperatures of the duplexes containing one or two of these linkers panel (**b**).

Clearly, this more flexible backbone, with free rotation about the C3′–C2′ and C2′–C1′ bonds, destabilizes duplexes (Figure 6 panel a, thick black arrows). This is because it is thermodynamically unfavourable to constrain it in a relatively rigid structure like a DNA helix, and it indicates that oligonucleotides made from this linker, or Glycerol Nucleic Acids, GNAs, could not be COSMIC-LOPERS.

2.3.2. Hexose Sugar-Phosphates

Beginning in 1986, Albert Eschenmoser's group, also based at ETH in Zurich, carried out an intensive research programme; synthesizing and studying analogues of DNA and RNA that contain 6-membered ring sugars i.e., hexoses, instead of ribose and deoxyribose [21]. Nucleic acids containing six different hexoses were synthesized, five of which were the naturally occurring D-allose, D-altrose, D-mannose, D-glucose and D-pyranosylribose, and one dideoxy derivative, 2′,3′-dideoxyallose (a hexose equivalent of 2′-deoxyribose) (Figure 7).

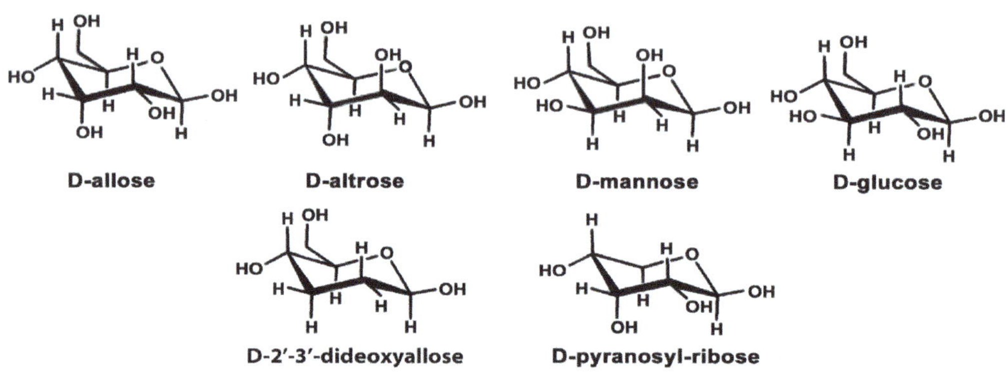

Figure 7. The six hexose sugars used by Eschenmoser's group to make hexose-nucleic acids.

They discovered that the hexopyranosyl-(6′-4′) systems, namely (6′-4′)-altro-pyranosyl-, (6′-4′)-allo-pyranosyl-, (6′-4′)-manno-pyranosyl- and (6′-4′)-gluco-pyranosyl showed Watson–Crick base pairing that was greatly inferior in terms of strength and selectivity compared to natural DNA and RNA. In fact, the (6′-4′)-gluco-pyranosyl system does not form duplexes in which complementary sequences exhibit Watson–Crick base-pairing [22]. The reason for this is intra-strand steric hindrances, which occur in the pairing conformations, caused by the 2′- and 3′-OH groups, clashing with the backbone and the edges of the bases respectively. This steric hindrance is especially severe in the (6′-4′)-glucopyranosyl system; graphically displaying the reason why nature's most abundant carbohydrate, "D-glucose, is not a component of its information storage system"; too much steric hindrance caused by "too many atoms" [23] (Figure S4, in the Supplementary Materials).

In marked contrast, however, the (4′-2′)-ribo-pyranosyl system, (which is the pentapyranosyl isomer of RNA) denoted p-RNA, was shown to be superior to both DNA and RNA, in terms of both duplex stability and the fact that exclusively Watson–Crick base-pairing is observed (Figure S5, in the Supplementary Materials). However, neither Hoogsteen nor reverse Hoogsteen motifs have yet been observed in p-RNA duplexes, and p-RNA sequences will not form stable Watson–Crick duplexes with their DNA or RNA complements. But of all the hexose systems synthesized and studied by Eschenmoser's group, p-RNA is perhaps the most COSMIC-LOPER like.

Most surprisingly, the 6′-4′ (2′,3′)-dideoxyallopyranosyl system, which is the hexose version of 2′-deoxyribose, (and hitherto referred to as homo-DNA, being a homologue of DNA) displayed distinctly different base-pairing rules to naturally occurring DNA and RNA. Hoogsteen purine-purine pairings, A–A and G–G, were favoured over Watson–Crick A–T pairings, with the base-pairing energies decreasing in the following order: G–C>A–A=G–G>A–T (Figure S6, in the Supplementary Materials). Expanding the 2′-deoxyribose ring by one [-CH_2-] group to the hexose changes the orientation of the purine bases on the scaffold to such an extent that purine-purine interactions not observed in DNA become manifest and take precedence over Watson–Crick A–T pairings. But, as any good organic chemistry graduate will remind us, 6-membered aliphatic saturated rings are more rigid systems that prefer a "chair conformation" in which all bulky substituents take the "equatorial" positions, in which intra-ring steric repulsions are minimal. An extensive X-ray crystallographic analysis of homo-DNA duplexes (complementary sequences containing 2, 4, 6, 8, 10 and 12 bases), was carried out by Martin Egli's group in the early 2000s [24]. Their findings revealed that homo-DNA duplexes showed much less helicality and cannot form the more tightly twisted double helices that are observed in DNA and RNA. This appears to be a direct result of the presence of the more rigid hexose sugar, which almost always occurs in the chair conformation. This study also revealed significant differences between the backbone-base inclination angles observed in homo-DNA when compared to both DNA and RNA [25]. In B-form DNA, this angle is 0°; in A-form RNA it is −30° but in homo-DNA, it is +45°. These differences (+45° from DNA, +75° from RNA) prevent homo-DNA from forming complementary Watson–Crick duplexes with either DNA or RNA (*cf*, unlike PNAs; S6). DNA and RNA can form complementary cross-pairs because DNA can adopt the A-form helical geometry.

2.3.3. Alternative Ribose and Hexose-Sugar Phosphates; XNAs

In 2012 a significant breakthrough in alternative nucleic acids research was announced. An international team comprising molecular biologists and synthetic organic chemists, led by Phil Holliger and Vitor Pinheiro respectively, based at the Medical Research Council (MRC) Laboratory of Molecular Biology, Cambridge, UK, reported the discovery of synthetic nucleic acids, known as xeno-nucleic acids (XNAs) that were capable of replication and evolution, with the same fidelity as DNA and RNA [26]. The group synthesized and tested XNAs containing six different sugar derivatives; four pentoses and two hexoses. The pentoses were arabinose (the 2′-epimer of ribose), 2′-fluoro-arabinose, threo-furanose and 2′-O, 4′-C-methylene-β-ribose ("locked"-ribose), and the two hexoses were 1, 5-anhydrohexitol and cyclohexenose. Nucleic acids containing arabinose were termed ANAs; 2′-fluoro-arabinose, FANAs; threo-furanose, TNAs; locked-ribose, LNAs; 1,5-anhydrohexitol, HNAs; and cyclohexenose, CeNAs (Figure 8).

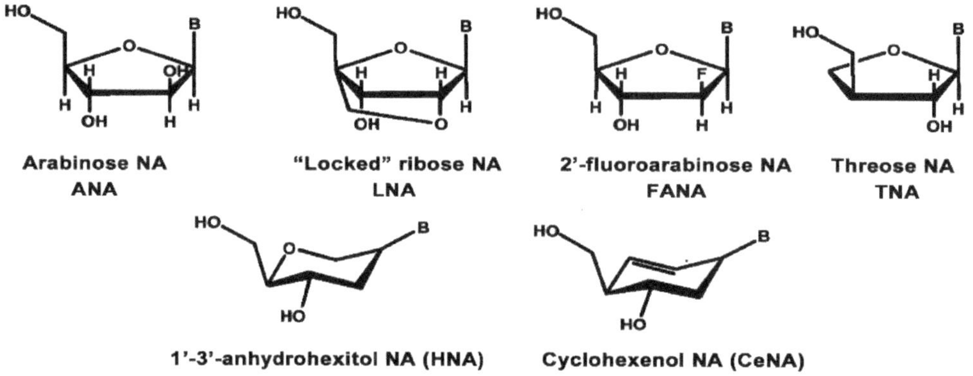

Figure 8. The six xeno-nucleosides, built from 4 pentoses and 2 hexoses used to construct xeno-nucleic acids, XNAs.

All six XNAs were shown to form complementary Watson–Crick paired helices to themselves, and their DNA and RNA complements, including the HNAs (Figure S7, in the Supplementary Materials), previously prepared by Piet Herdewijn as potential anti-sense agents [27]. This is quite remarkable as the 1, 5-anhydrohexitol is an isomer of 2′, 3′-dideoxyallose, used by Eschenmoser et al to construct homo-DNA, which, as has been discussed above, forms distinct duplexes with different pairing rules, and will not cross-pair with DNA or RNA.

Nucleic acid replication is facilitated by enzymes, known as polymerases. DNA amplification is routinely performed using thermostable enzymes in the polymerase chain reaction (PCR; [28]). This technique has revolutionized molecular biology, and found widespread applications in different fields, from forensics to palaeontology. A true test of an XNA, or indeed any nucleic acid variant's ability to replicate and evolve, requires that it can undergo PCR-amplification as with its natural counterpart. By employing a technique called in-vitro evolution [29], Holliger et al were able to synthesize specially evolved polymerases that accepted the XNA-triphosphates and incorporated them into new complementary XNAs and amplify XNA duplexes using PCR. Furthermore, some of these polymerases were able to synthesize XNAs from a complementary DNA template, and likewise synthesize DNA from an XNA template (Figure 9). All six XNAs showed high replication fidelities, ranging from 95% in LNAs to as high as 99.6% in HNA and CeNAs (Figure S7, in the Supplementary Materials).

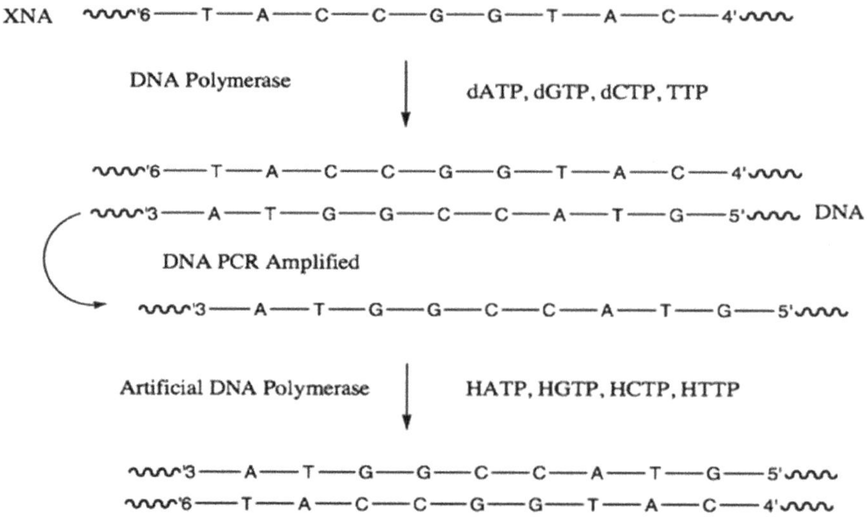

Figure 9. Specially evolved "Artificial" DNA polymerases enable XNA and DNA to act as templates for each other in PCR.

2.3.4. Catalytic XNAs: Xenoribozymes (XNAzymes)

In 2015, Holliger et al [30] reported the syntheses of a series of XNA oligomers, which they called XNAzymes, hereafter referred to as "xenoribozymes" that displayed catalytic behaviours similar to ribozymes. Their xenoribozymes used four of their non-standard sugar backbones, namely the previously mentioned ANA, FANA, HNA and CeNA.

Xenoribozymes were generated using an elaborate in-vitro selection method. Initially, xenoribozymes displaying RNA-cleavage ability (endonuclease) were prepared using all four backbones. The endonuclease activities included both intra-molecular (cis which refers to cleavage of an RNA segment in an RNA–XNA chimaera) and bimolecular (trans-cleavage of a phosphodiester bond in a bound RNA substrate). The FANAzymes displayed reaction rates comparable with analogous ribo- and deoxyribozymes, but the ANA-, CeNA- and HNAzymes all displayed rates that were 2–600 times slower.

2.4. Nucleic Acids Containing Non-Standard Nucleobases

The past three decades have seen the generation of synthetic nucleic acids having additional nucleobase "letters" that form additional nucleobase pairs that are distinct from the naturally occurring A–T/U and G–C. These new pairings can be divided into two different groups; conventional non-standard base pairs, developed by Steven Benner, use alternative hydrogen-bonding patterns to those observed in A–T/U and G–C; and unconventional non-standard base pairs, reported by the groups of Eric Kool, Ichiro Hirao and Floyd Romesberg, which are hydrophobic by nature and cannot interact by hydrogen-bonding.

The presence of additional nucleobase pairs expands the sequence and functional diversity of nucleic acids. A six-letter genetic alphabet has 6^n different sequences of length n and an eight-letter alphabet 8^n. These varied expanded genetic systems have led to the development of new molecular biology tools, clinical diagnostic kits, and, very recently, artificial Darwinian genetic systems and the potential for the development of synthetic life.

2.4.1. Nucleic Acids Containing Conventional Non-Standard Bases

Beginning in the late 1980s, the group of Steven Benner began an intensive research programme to develop an expanded genetic information system (AEGIS) that incorporated additional nucleobase letters into nucleic acids. By carefully examining the structures of the standard nucleobases A, C, G and T/U, they realized that subtle re-arrangement of the embedded ring nitrogens and the appended amino and keto oxygen functionalities could generate 8 new nucleobases; arranged as 4 new base-pairs which obeyed the Watson–Crick paradigm; hydrogen bonding between acceptors and donors, and large purines binding with small pyrimidines. As these new bases display different hydrogen-bonding arrangements they can only pair with their complementary partners in the same way that adenine only pairs with thymine or uracil: and guanine only pairs with cytosine. Benner's group were intrigued as to why nature uses only 4 nucleobases and pondered whether additional nucleobase letters were used in the precursor RNA world and whether alien biochemistries might utilize different nucleobases, and possibly even expanded genomes.

Iso-cytosine and iso-guanine are isomers of the naturally occurring cytosine and guanine bases respectively, in which the exocyclic amino and keto oxygens are inter-converted. Xanthine is the product of hydrolysis of guanine; it occurs naturally as the biosynthetic precursor to both theobromine and caffeine. Its partner, 2,4-diamino-pyrimidine, is linked to deoxyribose or ribose sugars via a carbon atom, and is a C-nucleoside, and has been given the trivial name *kappa* (K). The two other non-standard pyrimidines: 3-methyl-6-amino-pyrimidin-2-one, referred to as S, and 6-amino-5-nitro-1H-pyridin-2-one, known as Z, also form C-nucleosides (Figure 10).

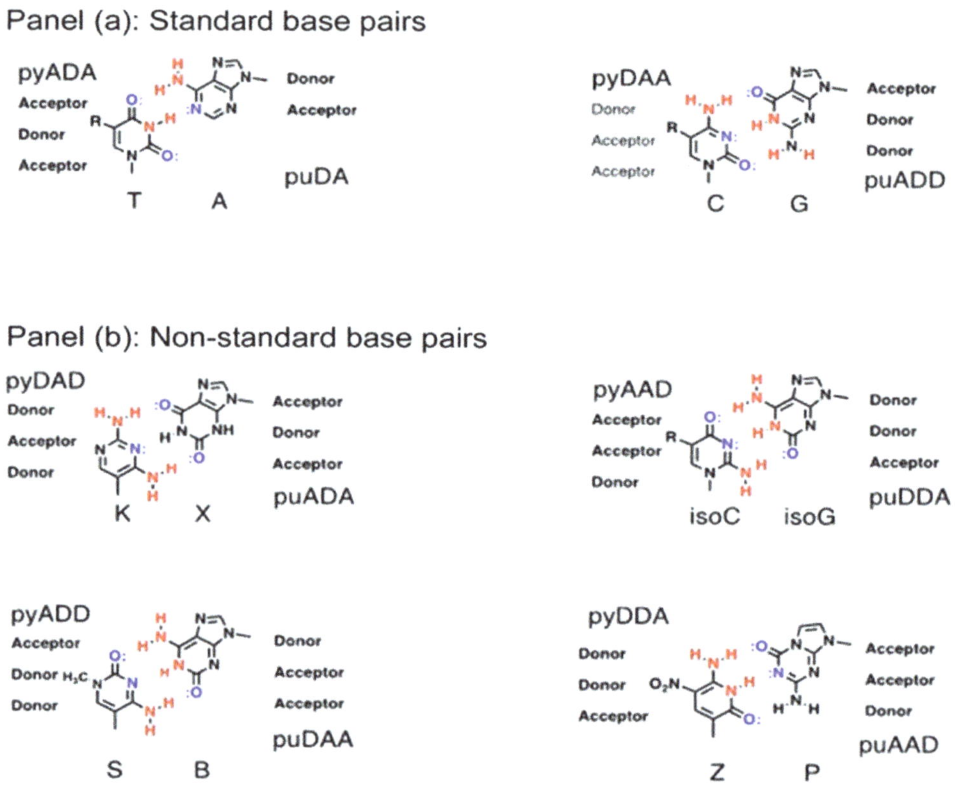

Figure 10. Panel (**a**) shows standard base pairs in contemporary biology. Panel (**b**) shows an Expanded Genetic Information System containing 8 non-standard bases arranged as 4 non-standard pairs; K–X, isoC–isoG, S–B and Z–P.

Synthetic DNA and RNA duplexes containing one or more non-standard base pairs were prepared and extensively studied and shown to be at least as stable as their natural counterparts. Indeed, in a 12-mer duplex (Figure S8, in the Supplementary Materials), when a G–C pair is replaced with an isoG–isoC, the melting temperature remains the same at 52 °C. When G–C is replaced by K-X, the melting temperature increases to 58 °C, and to 62 °C when replaced by the Z–P pair. Because the non-standard bases will not cross-pair with any of the standard bases, oligonucleotides containing them tend not to form duplexes with standard oligonucleotides that possess a similar standard base sequence. Cross-pairing of oligonucleotides that contain similar but mismatched sequences is a common problem with diagnostic kits that utilize nucleic acid hybridization assays and can generate many false-positive signals, thus greatly reducing their detection efficacies.

Clever use of oligonucleotides containing non-standard base pairs can greatly reduce these problems, and greatly improve the efficacies and detection limits of nucleic acid diagnostics [31]. In 1995 Bayer launched a branched DNA (bDNA) Quantiplex diagnostics probe, whose schematic setup is shown in (Figure S9, in the Supplementary Materials). The chip contains a series of oligonucleotide capture probes that are complementary to 15–20 bases on the target nucleic acid, e.g., mRNA from HIV or HBV. A set of chimeric oligonucleotides containing both standard and isoC/isoG rich sequences then bind to their complements on both the captured analyte (via the standard base sequence), and with non-standard bDNA sequences via the isoC/isoG containing sequences, anchoring them to the chip. Finally, another set of isoC/isoG oligonucleotides containing fluorescein-labelled isoG conjugates are then captured through complementary binding to sequences on the bDNA components. This incorporation of non-standard (isoC/isoG) bases into the parts of the hybridization complex that does not interact with the analyte greatly improves the detection efficacy as it prevents contaminant nucleic acids (from the sample) from binding and disrupting the signalling; thus, greatly improving the affinity of the system for the target analyte. This probe can detect as few as 50 viral mRNA molecules per ml of blood from HIV and HCV patients and is widely used to monitor the progress of these patients on both existing and new antiviral drug treatments [32,33]. More recent versions of this diagnostic system have been used to detect viral RNA from Zika and coronaviruses [34].

DNA polymerases, both wild-type and artificial, have been widely used to construct oligonucleotides containing all of the non-standard base pairs, placing them opposite their partners in the template strands with >95% efficiency. Following on from this, PCR systems capable of replicating AEGIS nucleic acids, including those with several repeating sequences of non-standard bases, were also developed. These systems worked best with the Z–P and S–B pairs. However, the isoC–isoG pair was shown to possess two significant disadvantages. Firstly, isoG has a minor tautomer that enables it to pair with thymine, and so after successive rounds of PCR, the isoC–isoG base pair is progressively edited out, replaced by the standard A–T pair (Figure S10, in the Supplementary Materials). Secondly, the iso-cytidine nucleoside is highly acid-sensitive, rendering the syntheses of isoC rich oligonucleotides difficult. Such inherent issues with the isoC–isoG pair, like those observed for the 6-membered homo-DNA, display clearly why nature's nucleic acids don't contain them (Figure S11, in the Supplementary Materials) [35].

In early 2019, the Benner group, in collaboration with four other leading research groups, announced the discovery of "hachimoji" DNA and RNA [36]; a genetic system with eight letters (hachimoji being the Japanese word for eight). The 8-letter nucleic acids contained the Z-P and S–B pairs in addition to the standard pairs. Complementary 16-mer duplexes containing these two pairs in many varied sequences were crystallized and analyzed and found to adopt the DNA B-form, with similar major and minor groove widths. Benner remarked that these "hachimoji DNAs met the Schrodinger requirement for a living system, forming the same 'aperiodic crystal' regardless of the sequences".

Hachimoji DNA sequences were then transcribed into hachimoji RNAs using a T7 RNA polymerase variant (Y639F H784A P266L, "FAL"). A hachimoji variant of the spinach fluorescent RNA aptamer was then designed, with one Z–P and one S–B pair incorporated, and transcribed from hachimoji DNA using the T7 RNA polymerase FAL variant. The standard aptamer folds and binds the ligand 3,5-difluoro-4-hydroxybenzylidene imidazolinone (DFHBI), which fluoresces green. The hachimoji variant was shown, by circular dichroism (CU) experiments to maintain the natural aptamer's folded structure. Further analyses confirmed the presence of the non-standard base pairs in the correct positions. In addition, the substitution of U at position 50 with Z yielded, as expected, an aptamer variant that did not fluoresce in the presence of the ligand, DFHBI, as the Z is close enough to the bound ligand to enable quenching of its fluorescence. This is therefore the first Darwinian genetic system built from eight rather than the standard four letters, and its' potential applications are limitless.

In 2018, Benner and co-workers reported the discovery of two new DNA-like systems that supported hydrogen-bond molecular recognition on a backbone that violated the size complementarity, small pyrimidine pairs with large purine, paradigm [37]. These two new nucleic acid systems are referred to as "skinny" where the base pairs are small 6-membered ring pyrimidines and "fat" where they are large 9-membered ring purines. Both skinny and fat pairs obey Watson–Crick hydrogen-bond complementarity (Figure S12, in the Supplementary Materials). Three pyrimidine pairs are possible, namely S–Z, T–K/K' and C–V—where S and Z are the non-standard pyrimidines that were previously used in conventional non-standard base pairings with their purine partners B and P respectively. Naturally occurring thymine pairs with both kappa (denoted here by K') and 3-nitro-2,6-diaminopyridine (K). Two 'fat' pairs are possible as in D–X and B–P—where D is diaminopurine, the 2'-amino derivative of adenine; and X is 7-deazaxanthosine and was preferred over xanthosine, as it lacks the 7-nitrogen, preventing D–X Hoogsteen interactions from occurring. B is isoguanosine as its N-1 tautomer.

Melting temperature studies of both fat and skinny 15-mer duplexes clearly revealed that they do form the expected Watson–Crick hydrogen-bonded systems. The skinny duplexes had sharp 'melting' temperatures (T_m) that averaged 58.3 °C, higher than those of standard Watson–Crick analogues (~36 °C), even when the A residues in the latter were replaced by D (~45 °C) which forms

3 hydrogen-bonds to T. The fat duplexes also displayed sharp melting but at higher temperatures (~77.7 °C). Both skinny and fat DNA systems disfavoured mismatches, which were seen to lower the melting temperatures by 3–5 °C, which is similar to what is observed in standard DNA duplexes (Figure S13, in the Supplementary Materials).

X-ray crystallography studies were performed on 16-mer duplexes containing four central skinny (5′-CTTATA**KKTT**TATAAG) and six central fat (5′-CTTAT**XXXDDD**ATAAG) complementary base pairs. The obtained data confirmed that the skinny and fat regions do indeed interact via Watson–Crick hydrogen bonding, and although the duplex geometry in these regions was neither A- nor B-form, the standard portions of the duplexes exhibited the B-form.

2.4.2. Unconventional Non-Standard Bases

"Unconventional" non-standard nucleobases refer to hydrophobic aromatic analogues of the standard 4 bases that lack hydrogen bond acceptor and donor functional groups, and so are unable to interact via Watson–Crick hydrogen-bonded base pairings. The "nonpolar nucleoside isosteres" concept, was introduced by Eric Kool in 1994 [38] with the introduction of 2,4-difluorotoluene, F, which is a near-perfect isostere (shape mimic) for naturally occurring thymine (Figure S14, in the Supplementary Materials). Kool's group also used other toluene-nucleoside derivatives as thymine replacements and indoles and benzimidazoles as purine substitutes. The inability of difluorotoluene to partake in Watson–Crick hydrogen bonding with adenine was revealed, unsurprisingly, by lower melting temperature values for complementary duplexes which contained A–F in place of A–T or G–C (Figure S15, in the Supplementary Materials). Despite possessing three lone pairs of electrons, fluorine hardly ever acts as a hydrogen bond acceptor in organic molecules [39]. The highly electronegative F atom holds these lone pairs very tightly, and in fact, organic molecules rich in fluorine, fluorocarbons, are even more hydrophobic than their hydrocarbon counterparts. Very surprisingly, Kool et al. then revealed that difluorotoluene could act as a replacement for thymine during the syntheses of complementary DNA duplexes by several A- and Y-family type DNA polymerases [40]. The DNA pol I, Klenow fragment (Kf) from *E. coli* inserted dATP opposite F and dFTP opposite A in DNA primer-templates with similar efficiency as the natural A–T base pair. Full-length products were obtained from templates containing up to eight F residues, although this did not work when two or more Fs were in sequence. Similar results were obtained with other toluene derivatives such as 2, 4-dichloro-, 2-fluoro 4-chloro- and 2-bromo 4-chloro-toluene. These results attracted much scrutiny from the biochemical community, and much criticism [41], to which Eric Kool responded [42] most eloquently. Prior to this work, the widely held view about enzyme-catalysed DNA replication was that hydrogen bonding was the dominant factor that governed how the enzyme placed the correct base in the synthesis strand opposite its partner in the template. Kool's work strongly suggests that this is not so and that steric effects, governed by base shape, is the paramount source of DNA replication fidelity, and indeed, since 2002, editions of the classic biochemistry text of Stryer have cited these findings [43]. In 1998, Kool and his coworkers reported the first hydrophobic base-pair (Figure S16, in the Supplementary Materials), difluorotoluene-4-methylbenzimidazole (F–Z′) [44]. 4-Methylbenzimidazole is referred to here as Z′ to avoid confusion with the Benner group base 6-amino-5-nitro-1H-pyridin-2-one, which is also denoted as Z. 4-Methylbenzimidazole is a non-polar isostere of adenine lacking the 3-nitrogen. This study revealed that DNA pol I Klenow fragment (Kf) efficiently inserted F and Z′ opposite each other, as well as F–A and Z′–T and was the first to show that a purely hydrophobic base-pair could be replicated in this way. This inspired a flurry of activity in unconventional hydrophobic non-standard base-pair research, led by the groups of Ichiro Hirao and Floyd Romesberg.

In 2006, Hirao's group reported the development of an expanded genetic system using a hydrophobic base pair between 7-(2-thienyl)-imidazo[4,5-b]pyridine (Ds) and pyrrole-2-carbaldehyde (Pa). This pair maintained the Watson–Crick shape complementarity, with Ds acting as an isostere of adenine, and Pa an isostere of thymine. Initial problematic misincorporations, of dDsTP opposite template Ds, and dATP opposite template Pa were overcome by using the respective γ-amidotriphosphates. This led to the first-ever PCR amplification of an expanded genetic system containing a hydrophobic base pair, using vent

DNA polymerase (exo) which proceeded with very high efficiency [45]. The same paper reported the successful transcription of DsTP and PaTP into RNA, with very high selectivity and efficiency, using T7 RNA polymerase. However, the presence of two or more of these unnatural bases in sequence in primer-template results in greatly diminished yields of full-length product, and PCR will not replicate such primers. Subsequently, another pyrrole derivative, 2-nitro-4-propynylpyrrole, known as Px, was shown to be as efficient in PCR systems as Pa but without the cross-pairing problems inherent to the latter, and replaced it as the smaller hydrophobic base [46]. The Hirao group's principal application of their hydrophobic base pair has been on the development of novel DNA aptamers with enhanced binding efficacies for proteins (Figure S17, in the Supplementary Materials), arising from hydrophobic interactions between Ds and Px with hydrophobic amino acid side chains [47].

Beginning in 1998 at the Scripps Institute in La Jolla, the group of Floyd Romesberg began an extensive program aimed at the development of an artificial genetic system containing hydrophobic non-standard bases [48]. Many different hydrophobic base pairs were synthesized, incorporated into DNA templates, and the synthetic efficacies of many naturally occurring, and artificial DNA polymerases were assessed. Two base pairs, dNaM-d5SICS and dNaM-dTPT3, proved to be superior and were incorporated into synthetic strands opposite their template partners with the same efficacies as the naturally occurring base pairs (Figure S18, in the Supplementary Materials). However, the presence of two or more in sequence reduced the synthetic efficiency.

X-ray crystallographic studies on complementary duplexes containing these pairs clearly revealed that they interact via hydrophobic stacking, and this distorts the helicity of the duplex when two or more are in sequence, and this has profound effects on the ability of polymerases to produce full-length complementary products.

PCR systems were developed that successfully amplified primers containing these non-standard base pairs, with the same efficiency as the natural pairs following 100 or more cycles [49]. However, they did not work when two or more of these non-standard bases are in sequence in the template. In a breakthrough in 2013, the d5SICS-dNaM pair were then inserted into a bacterial plasmid which was taken up by a strain of *E. coli* which was shown to replicate only in the presence of the unnatural nucleoside triphosphates d5SICS-TP and dNaM-TP in addition to the natural ones; dATP, dCTP, TTP and dGTP. This was the first-ever report of an organism that could grow and replicate with an unnatural base pair in its genome [50].

The Romesberg group continued work on the development of semi-synthetic organisms (SSO) that can use a 6-letter genetic alphabet to make proteins containing unnatural amino acids [51,52]. Using the dTTP3-dNaM pair, they cleverly engineered an *E. coli* containing the unnatural base pair in its' genome, mRNA and tRNA. Use of tRNA-loading enzymes that were tolerant of unnatural anticodons was vital to the success of this project. Beginning with the *E. coli* tRNA synthetase that loads serine onto tRNA, they inserted serine into a position on the green fluorescent protein (GFP) via a codon containing one of the non-standard bases paired with its partner on the anticodon. They then engineered a modified *E. coli* whose gene for tRNA synthetase was replaced with a gene for an enzyme from a different microbe, *Methanosarcina barkeri* which uses the rare amino acid pyrrolysine, and the expressed tRNA synthetase added pyrrolysine into the same position on GFP. They then went a step further and inserted a gene from the archaea *Methanococcus jannaschi* for a tRNA synthetase which adds the even rarer amino acid 4-azido phenylalanine, and the expressed enzyme added 4-azido phenylalanine at the same position in GFP, encoded by the non-standard base containing codon.

This work paves the way for the development of both semi and fully synthetic microbes, containing an expanded 6-letter genetic alphabet, that can be engineered to produce novel proteins containing non-standard amino acids, both rare and synthetic, with great potential for the development of new materials and therapeutics.

3. Conclusions

This article has summarized the discoveries and developments in de novo nucleic acid research going back just over 30 years. The results have greatly increased our understandings of naturally occurring nucleic acids DNA and RNA and have expanded the scope of molecular structures that can support life, both here on Earth in new, artificial Darwinian systems, and potentially elsewhere in the Universe. In this respect we can indeed state that life may not necessarily be limited to the nucleic acid building blocks seen in terran DNA and RNA; alternatives to ribose and deoxyribose, like arabinose, threo-furanose, locked-ribose, 1,5-anhydrohexitol and cyclohexenose, are feasible. Non-standard nucleobase pairs like Z–P and S–B which display Watson–Crick size and hydrogen bonding complementarities, could replace the naturally occurring A–T and G–C pairs or indeed occur together with them in alien organisms with expanded genetic alphabets. In addition, non-standard nucleobase pairs which follow Watson–Crick hydrogen bonding but not size complementarity are possible; three "skinny" pyrimidine-pyrimidine pairings; S–Z, T–K/K′ and C–V; and two "fat" purine-purine pairs; D–X and B–P. Furthermore, unconventional hydrophobic base pairs like Ds-Px, 5SICS-NaM and TPT3-NaM might be possible, as long as they are separated by standard base-pairs and not adjacent in a sequence. The possibility of exclusively hydrophobic base pairs occurring in an alien genetic system, which has arisen in a non-aqueous solvent, is also conceivable.

We may safely conclude then that extraterrestrial life if it has ever arisen elsewhere in the Universe, might not necessarily use DNA or RNA, but, in an aqueous environment like Earth, it will have a sugar-phosphate backbone in its genetic information storage biopolymers.

Supplementary Materials: Figure S1. (a) The numbering system used for nucleosides, in which the sugars use primed numbers 1′-5′ in order to distinguish them from the bases. (b) A section of an oligonucleotide strand; RNA when X=OH and R=H; DNA when X=H and R=CH_3. The sequences shown here are GUCA (RNA) or GTCA (DNA). Figure S2: the structure of peptide nucleic acid (PNA) compared to DNA. Figure S3: 3′-1′-glycerol nucleic acids (GNA) are flexible linker analogs of DNA. Figure S4: nucleic acids containing D-allose, D-altrose, D-mannose and D-glucose cannot form Watson–Crick base-paired duplexes because of severe intra-strand steric clashes, arising from interactions between the 2′-OHs and the edges of the bases (right arrows) and the 3′-OHs and the backbone (left arrows). Figure S5: 4′-2′-Pyranosyl RNA forms very stable duplexes with exclusively Watson–Crick base-pairings. Figure S6: D-2′-3-Dideoxyallose nucleic acid (homo-DNA) duplexes display distinct Hoogsteen type base pairing rules very different to those occurring in DNA and RNA, with A-A and G-G base pairs that are more stable than A-T; pairing rules in order of stability G–C > A–A~G–G > A–T; 6′-$(A)_3(G)_3$-4′ duplexes have a melting temperature of 34 °C. Figure S7: XNAs like HNA form complementary Watson–Crick base paired duplexes; with themselves and complementary DNA and RNA sequences. Figure S8: comparison of melting temperatures (T_m) of 12-mer duplexes containing a non-standard base pair X-Y. Figure S9: oligonucleotides containing the non-standard base pair isoC-isoG are used in a Bayer bDNA diagnostic system and greatly improve the detection limits for viral mRNA and other disease-related nucleic acid analytes. Figure S10: isoG has a minor tautomer that pairs with T, resulting in a replacement of the isoG-isoC pair with T-A following PCR amplification. Figure S11: PCR amplification of a 6-letter template containing two adjacent non-standard bases (P-P) [35]. Figure S12: pyrimidine-pyrimidine pairs obey Watson–Crick hydrogen-bonding complementarity, via S-Z, T-K and C-V pairs to generate 'skinny' DNA-like duplexes; purine-purine pairs do likewise, via D-X and B-P pairs to generate 'fat' DNA-like duplexes. Figure S13: melting temperatures of complementary 'fat' and 'skinny' and standard DNA duplexes. Figure S14: difluorotoluene is a very good shape mimic for thymine, even though it is much less polar and cannot engage in hydrogen bonding. Figure S15: replacement of thymine (T) with difluorotoluene (F) in an otherwise complementary 12-mer duplex lowers the melting temperature (T_m) significantly. Figure S16: The first reported hydrophobic base pair, difluorotoluene (F)-4-methylbenzimidazole (Z′). Figure S17: The hydrophobic base pairs Ds-Pa and Ds-Px, developed by the Hirao group. Figure S18: The hydrophobic base pairs d5SICS-dNaM and dTPT3-dNaM, developed by the Romesberg group.

Author Contributions: Original concept and manuscript draft, K.G.D. Proofreading, editing and image preparation, S.J. All authors have read and agreed to the published version of the manuscript.

References

1. Watson, J.D.; Crick, F.H.C. Molecular structure of Nucleic Acids. *Nature* **1953**, *171*, 737–738. [CrossRef]
2. Hoogsteen, K. The crystal and molecular structure of a hydrogen-bonded complex between 1-methylthymine and 9-methyladenine. *Acta Crystallogr.* **1963** *16*, 907–916. [CrossRef]

3. Zagryadskaya, E.I.; Doyon, F.R.; Steinberg, S.V. Importance of the reverse Hoogsteen base pair 54-58 for tRNA function. *Nucleic Acids Res.* **2003**, *31*, 3946–3953. [CrossRef]
4. Herbert, A. Z-DNA and Z-RNA in human disease. *Nat. Commun. Biol.* **2019**, *2*, 1–10. [CrossRef] [PubMed]
5. Benner, S.A. *Size Limits of Very Small Microorganisms: Proceedings of a Workshop*; Steering Group on Astrobiology of the Space Studies Board; National Research Council: Washington, DC, USA, 1999; pp. 126–135.
6. Altman, S. Aspects of biochemical catalysis. *Cell* **1984**, *36*, 237–239. [CrossRef]
7. Guerrier-Takada, C.; Gardiner, K.; Marsh, T.; Pace, N.; Altman, S. The RNA moiety of ribonuclease P is the catalytic subunit of the enzyme. *Cell* **1983**, *35*, 849–857. [CrossRef]
8. Guerrier-Takada, C.; Altman, S. Catalytic activity of an RNA molecule prepared by transcription in vitro. *Science* **1984**, *223*, 285–286. [CrossRef]
9. Zaug, A.J.; Cech, T.R. The intervening sequence RNA of Tetrahymena is an enzyme. *Science* **1986**, *231*, 470–475. [CrossRef]
10. Breaker, R.R.; Joyce, G.F. A DNA enzyme that cleaves RNA. *Chem. Biol.* **1994**, *1*, 223–229. [CrossRef]
11. Hollenstein, M. DNA Catalysis: The Chemical Repertoire of DNAzymes. *Molecules* **2015**, *20*, 20777–20804. [CrossRef]
12. Zamecnik, P.C.; Stephenson, M.L. Inhibition of Rous sarcoma virus replication and cell transformation by a specific oligodeoxynucleotide. *Proc. Natl. Acad. Sci. USA* **1978**, *75*, 280–284. [CrossRef] [PubMed]
13. Richert, C.; Roughton, A.L.; Benner, S.A. Nonionic Analogs of RNA with Dimethylene Sulfone Bridges. *J. Am. Chem. Soc.* **1996**, *118*, 4518–4531. [CrossRef]
14. Nielsen, P.; Egholm, M.; Berg, R.; Buchardt, O. Sequence-selective recognition of DNA by strand displacement with a thymine-substituted polyamide. *Science* **1991**, *254*, 1497–1500. [CrossRef] [PubMed]
15. Egholm, M.; Buchardt, O.; Nielsen, P.E.; Berg, R.H. Peptide nucleic acids (PNA). Oligonucleotide analogs with an achiral peptide backbone. *J. Am. Chem. Soc.* **1992**, *114*, 1895–1897. [CrossRef]
16. Rasmussen, H.; Kastrup, J.S.; Nielsen, J.N.; Nielsen, J.M.; Nielsen, P.E. Crystal Structure of a peptide nucleic acid (PNA) duplex at 1.7 Å resolution. *Nat. Struct. Biol.* **1997**, *4*, 98–101. [CrossRef]
17. Brown, S.C.; Thomson, S.A.; Veal, J.M.; Davis, D.G. NMR solution structure of a peptide nucleic acid complexed with RNA. *Science* **1994**, *265*, 777–780. [CrossRef]
18. Eriksson, M.; Nielsen, P.E. PNA-nucleic acid complexes. Structure, stability and dynamics. *Nat. Struct. Biol.* **1996**, *3*, 410–413. [CrossRef]
19. Westheimer, F.H. Why nature chose phosphates. *Science* **1987**, *235*, 1173–1178. [CrossRef]
20. Schneider, K.C.; Benner, S.A. Oligonucleotides containing flexible nucleoside analogs. *J. Am. Chem. Soc.* **1990**, *112*, 453–455. [CrossRef]
21. Eschenmoser, A.; Loewenthal, E. Chemistry of potentially prebiological natural products. *Chem. Soc. Rev.* **1992**, *21*, 1–16. [CrossRef]
22. Pitsch, S.; Wendeborn, S.; Jaun, B.; Eschenmoser, A. Why Pentose-and Not Hexose-Nucleic Acids? Part VII. Pyranosyl-RNA ('p-RNA'). Preliminary communication. *Helvitica Chim. Acta* **1993**, *76*, 2161–2183. [CrossRef]
23. Eschenmoser, A. Towards a chemical etiology of the structure of nucleic acids. *Chem. Biol.* **1994**, *1*. [CrossRef]
24. Egli, M.; Pallan, P.S.; Pattanayek, R.; Wilds, C.J.; Lubini, P.; Minasov, G.; Dobler, M.; Leumann, C.J.; Eschenmoser, A. Crystal structure of homo-DNA and nature's choice of pentose over hexose in the genetic system. *J. Am. Chem. Soc.* **2006**, *128*, 10847–10856. [CrossRef] [PubMed]
25. Egli, M.; Lubini, P.; Pallan, P.S. The long and winding road to the structure of homo-DNA. *Chem. Soc. Rev.* **2007**, *36*, 31–45. [CrossRef] [PubMed]
26. Pinheiro, V.B.; Taylor, A.I.; Cozens, C.; Abramov, M.; Renders, M.; Zhang, S.; Chaput, J.C.; Wengel, J.; Peak-Chew, S.-Y.; McLaughlin, S.H.; et al. Synthetic Genetic Polymers Capable of Heredity and Evolution. *Science* **2012**, *336*, 341–344. [CrossRef]
27. Van, A.A.; Verheggen, I.; Hendrix, C.; Herdewijn, P. 1,5-Anhydrohexitol Nucleic Acids, a New Promising Antisense Construct. *Angew. Chem. Int. Ed.* **1995**, *34*, 1338–1339. [CrossRef]
28. Mullis, K.; Faloona, F.; Scharf, S.; Saiki, R.; Horn, G.; Erlich, H. Specific Enzymatic Amplification of DNA In Vitro: The Polymerase Chain Reaction. *Cold Spring Harb. Symp. Quant. Biol.* **1986**, *51*, 263–273. [CrossRef]
29. Arnold, F.H. Directed evolution: Creating biocatalysts for the future. *Chem. Eng. Sci.* **1996**, *51*, 5091–5102. [CrossRef]
30. Taylor, A.I.; Pinheiro, V.B.; Smola, M.J.; Morgunov, A.S.; Peak-Chew, S.; Cozens, C.; Weeks, K.M.; Herdewijn, P.; Holliger, P. Catalysts from synthetic genetic polymers. *Nature* **2015**, *518*, 427–430. [CrossRef]

31. Collins, M.L.; Irvine, B.; Tyner, D.; Fine, E.; Zayati, C.; Chang, C.-A.; Horn, T.; Ahle, D.; Detmer, J.; Shen, L.-P.; et al. A branched DNA signal amplification assay for quantification of nucleic acid targets below 100 molecules/ml. *Nucleic Acids Res.* **1997**, *25*, 2979–2984. [CrossRef]
32. Gleaves, C.A.; Welle, J.; Campbell, M.; Elbeik, T.; Ng, V.; Taylor, P.E.; Kuramoto, K.; Aceituno, S.; Lewalski, E.; Joppa, B.; et al. Multicenter evaluation of the Bayer VERSANT™ HIV-1 RNA 3.0 assay: Analytical and clinical performance. *J. Clin. Virol.* **2002**, *25*, 205–216. [CrossRef]
33. Elbeik, T.; Surtihadi, J.; Destree, M.; Gorlin, J.; Holodniy, M.; Jortani, S.A.; Kuramoto, K.; Ng, V.; Valdes, R.; Valsamakis, A.; et al. Multicenter Evaluation of the Performance Characteristics of the Bayer VERSANT HCV RNA 3.0 Assay (bDNA). *J. Clin. Microbiol.* **2004**, *42*, 563–569. [CrossRef] [PubMed]
34. Glushakova, L.G.; Sharma, N.; Hoshiko, S.; Bradley, A.; Bradley, K.M.; Yang, Z.; Benner, S.A. Detecting respiratory viral RNA using expanded genetic alphabets and self-avoiding DNA. *Anal. Biochem.* **2015**, *489*, 62–72. [CrossRef] [PubMed]
35. Benner, S.A.; Kim, H.-J.; Yang, Z. *RNA Worlds: From Life's Origins to Diversity in Gene Regulation*; Cold Spring Harbor Laboratory Press: Cold Spring Harbor, NY, USA, 2011.
36. Benner, S.A.; Hoshika, S.; Leal, N.A.; Kim, M.-J.; Kim, M.-S.; Kim, H.-J.; Karalkar, N.B.; Bates, A.M.; Watkins, N.E., Jr.; SantaLucia, H.A.; et al. Hachimoji DNA and RNA: A genetic system with eight building blocks. *Science* **2019**, *363*, 884–887.
37. Benner, S.A.; Hoshika, S.; Singh, I.; Switzer, C.; Molt, R.W.; Leal, N.A.; Kim, M.-J.; Kim, M.-S.; Kim, H.-J.; Georgiadis, M.M. "Skinny" and "Fat" DNA: Two New Double Helices. *J. Am. Chem. Soc.* **2018**. [CrossRef]
38. Schweitzer, B.A.; Kool, E.T. Aromatic Nonpolar Nucleosides as Hydrophobic Isosteres of Pyrimidines and Purine Nucleosides. *J. Org. Chem.* **1994**, *59*, 7238–7242. [CrossRef]
39. Dunitz, J.D.; Taylor, R. Organic Fluorine Hardly Ever Accepts Hydrogen Bonds. *Chem. A Eur. J.* **1997**, *3*, 89–98. [CrossRef]
40. Liu, D.; Moran, S.; Kool, E.T. Bi-stranded, multisite replication of a base pair between difluorotoluene and adenine: Confirmation by 'inverse' sequencing. *Chem. Biol.* **1997**, *4*, 919–926. [CrossRef]
41. Evans, T.A.; Seddon, K.R. Hydrogen bonding in DNA—A return to the status quo. *Chem. Commun.* **1997**, *21*, 2023–2024. [CrossRef]
42. Kool, E.T.; Sintim, O.H. The difluorotoluene debate—A decade later. *Chem. Commun.* **2006**, *35*, 3665–3675. [CrossRef]
43. Berg, J.M.; Tymoczko, J.L.; Stryer, L. *Biochemistry*, 5th ed.; W.H. Freeman: New York, NY, USA, 2002; 751p.
44. Morales, J.C.; Kool, E.T. Efficient replication between non-hydrogen-bonded nucleoside shape analogs. *Nat. Genet.* **1998**, *5*, 950–954. [CrossRef] [PubMed]
45. Hirao, I. Unnatural base pair systems for DNA/RNA-based biotechnology. *Curr. Opin. Chem. Biol.* **2006**, *10*, 622–627. [CrossRef] [PubMed]
46. Yamashige, R.; Kimoto, M.; Takezawa, Y.; Sato, A.; Mitsui, T.; Yokoyama, S.; Hirao, I. Highly specific unnatural base pair systems as a third base pair for PCR amplification. *Nucleic Acids Res.* **2012**, *40*, 2793–2806. [CrossRef] [PubMed]
47. Hirao, I.; Kimoto, M.; Lee, K.-H. DNA aptamer generation by ExSELEX using genetic alphabet expansion with a mini-hairpin DNA stabilization method. *Biochimie* **2018**, *145*, 15–21. [CrossRef]
48. Ogawa, A.K.; Wu, Y.; McMinn, D.L.; Liu, J.; Schultz, P.G.; Romesberg, F.E. Efforts toward the expansion of the genetic alphabet: Information storage and replication with unnatural hydrophobic base pairs. *J. Am. Chem. Soc.* **2000**, *122*, 3274–3287. [CrossRef]
49. Malyshev, D.A.; Pfaff, D.A.; Ippoliti, S.I.; Hwang, G.T.; Dwyer, T.J.; Romesberg, F.E. Solution Structure, Mechanism of Replication, and Optimization of an Unnatural Base Pair. *Chem. A Eur. J.* **2010**, *16*, 12650–12659. [CrossRef]
50. Malyshev, D.A.; Dhami, K.; Lavergne, T.; Chen, T.; Dai, N.; Foster, J.M.; Correa, I.R.; Romesberg, F.E. A semi-synthetic organism with an expanded genetic alphabet. *Nature* **2014**, *509*, 385–388. [CrossRef]
51. Zhang, Y.; Lamb, B.M.; Feldman, A.W.; Zhou, A.X.; Lavergne, T.; Lingjun, L.; Romesberg, F.E. A semisynthetic organism engineered for the stable expansion of the genetic alphabet. *Proc. Natl. Acad. Sci. USA* **2017**, *114*, 1317–1322. [CrossRef]
52. Zhang, Y.; Ptacin, J.L.; Fischer, E.C.; Aerni, H.R.; Caffro, C.E.; San Jose, K.; Feldman, A.W.; Turner, C.R.; Romesberg, F.E. A semi-synthetic organism that stores and retrieves increased genetic information. *Nature* **2017**, *551*, 644–647. [CrossRef]

3

Current Trends in Applications of Circulatory Microchimerism Detection in Transplantation

Hajnalka Andrikovics [1,2], Zoltán Őrfi [1], Nóra Meggyesi [1], András Bors [1], Lívia Varga [3,4], Petra Kövy [1,3], Zsófia Vilimszky [1], Fanni Kolics [1], László Gopcsa [5], Péter Reményi [5] and Attila Tordai [2,6,*]

1 Laboratory of Molecular Genetics, Central Hospital of Southern Pest National Institute of Hematology and Infectious Diseases, 1097 Budapest, Hungary
2 Department of Pathophysiology, Semmelweis University, 1089 Budapest, Hungary
3 School of PhD Studies, Semmelweis University, 1085 Budapest, Hungary
4 Hungarian National Blood Transfusion Service, 1113 Budapest, Hungary
5 Department of Hematology and Stem Cell Transplantation, Central Hospital of Southern Pest National Institute of Hematology and Infectious Diseases, 1097 Budapest, Hungary
6 Department of Transfusion Medicine, Semmelweis University, 1089 Budapest, Hungary
* Correspondence: tordai.attila@med.semmelweis-univ.hu

Abstract: Primarily due to recent advances of detection techniques, microchimerism (the proportion of minor variant population is below 1%) has recently gained increasing attention in the field of transplantation. Availability of polymorphic markers, such as deletion insertion or single nucleotide polymorphisms along with a vast array of high sensitivity detection techniques, allow the accurate detection of small quantities of donor- or recipient-related materials. This diagnostic information can improve monitoring of allograft injuries in solid organ transplantations (SOT) as well as facilitate early detection of relapse in allogeneic hematopoietic stem cell transplantation (allo-HSCT). In the present review, genetic marker and detection platform options applicable for microchimerism detection are discussed. Furthermore, current results of relevant clinical studies in the context of microchimerism and SOT or allo-HSCT respectively are also summarized.

Keywords: microchimerism; solid organ transplantation; hematopoietic stem cell transplantation; genetic marker; single nucleotide polymorphism; deletion/insertion polymorphism

---

1. Introduction

Co-existence of genetically distinct/discordant tissues or cells is termed chimerism after the well-known Greek fusion creature. Based on the extent of the variant population, microchimerism is usually mentioned as a subgroup of chimerism in which the proportion of foreign cells or tissue does not exceed 1%. This proportion can readily be estimated in the circulation while quantitative comparison of major and minor populations is more challenging in cases of scattered cells or transplanted tissue.

Based on origin, naturally occurring and therapy related microchimerism can be distinguished. Naturally occurring microchimerism refers to the bidirectional interaction between the genetically distinct mother and fetus(es) resulting in the long-lasting presence of small numbers (well below 1%) of fetal cells in the mother and vice versa. The interesting immunological consequences of these coexistences have recently been elegantly reviewed [1] and is not discussed in the present review.

The term therapy related microchimerism describes situations primarily related to transplantations and occasionally transfusions. Subsequent to solid organ transplantations (SOT) a well-defined organ coexists with the recipient which rarely involves mixture of cells. In contrast, variable recipient–donor cell ratios are characteristic for patients treated by allogeneic hematopoietic stem cell transplantation (allo-HSCT).

The quantitative detection of chimerism in the circulation might serve as a biomarker with clinical relevance, which is based on polymorphic genetic structures (markers) similarly to those applied in forensic medicine for person identification. Large scale population studies such as the 1000 Genome Project characterized and classified millions of human variants in different populations allowing the identification of appropriate markers with high heterozygosity rate and discriminatory power. In the last decade, the development of highly sensitive, quantitative and accurate methods, e.g., digital PCR or next generation sequencing (NGS) enabled the precise quantitation of unprecedently low level molecular variants. As a result, the clinical application of these innovative technologies for microchimerism detection has started to enter the daily practice. In general, the paradigm of marker selection in transplantation dictates an allele choice as a marker that is not present in the dominant cell or molecule population: after SOT the donor should be marker positive and the recipient negative, while after allo-HSCT the opposite relationship is required.

Chimerism could be investigated on cellular or on cfDNA level. The latter consists mainly of small sized (around 150 base-pairs) double-stranded DNA fragments resulting from apoptosis, necrosis, immune-mediated cell damage, or release of nuclear DNA into the circulation predominantly from hematopoietic cells [2]. In the circulation, these fragments have a short half-life of approximately 1.5 h due to rapid hepatic and renal clearance [3].

In the present paper, we provide an extensive review of the available information about microchimerism in the context of various forms of transplantation and transplantation-related advanced therapies with a special emphasis on the available markers and the multitude of detection platforms in this high sensitivity range.

2. Polymorphic Genetic Markers Applicable for Chimerism Detection

2.1. Short Tandem Repeats (STR)

So far, more than 700,000 STRs (short tandem repeats) are characterized in human genome, for the most part they are considered as 'junk DNA', however some of them are located in protein coding region and associated with genetic disorders [4–6]. STRs—also referred to as microsatellites—consist of 2 to 6 nucleotide tandem repeats distributed throughout the genome. These loci are highly polymorphic and different alleles are determined by the varying number of nucleotide repeats (3 to 51 repeats). Short tandem repeat (STR) analysis is a standard and approved method applied in forensics and in hematopoietic chimerism quantification after allo-HSCT [7]. Nearly 100% informativity can be reached with a panel of 12–16 STR markers (Core STR Loci), including loci on sex chromosomes, that allow differentiation between individuals [8–10]. The major advantage of STRs is its multiallelic (up to 16 distinct variants) nature and consequential high informativity rate, which allows its application in all donor–recipient pairs or furthermore in transplants with multiple donors.

2.2. Single Nucleotide Polymorphism (SNP)

These represent the most frequently observed variations in the human genome, around 1/1000 basepairs [11]. In an average individual there are ten times more SNPs to be found compared to deletion insertion polymorphisms (DIP). Since the discrimination power of biallelic variants is lower compared to STRs, more loci have to be simultaneously tested to reach an acceptable informativity range. Reportedly, already seven SNP markers can be highly discriminative (97% informativity) in HLA-identical sibling pairs [12]. On the other hand, a study using a combined panel of markers described only 80% overall informativity with six SNPs, two null alleles (SRY, RhD) and two DIPs [13]. Almeida et al. calculated 81% informativity for eight biallelic SNP markers investigating 88 patient/donor pairs [14]. Fredriksson et al. used an array consisting of 51 SNPs to reach 100% informativity [15]. In the context of SOT, SNP assays have widely been used to detect donor-derived cfDNA (dd-cfDNA) amount in plasma with an early, method-focused application using digital PCR (see Section 3) in small sets of liver, kidney, and heart transplanted patients [16]. This approach was further developed and characterized in detail by Grskovic et al. [17] in the form of a commercial assay (CareDx, Brisbane, CA, USA) employing 266 SNPs and NGS (see Section 3) with a convincing validation in a large group of heart transplanted patients. The technical advances made possible a highly sophisticated approach of massive multiplexing allowing the utilization of more than 13,000 SNP markers [18], as well as an alternative commercial, NGS-based assay [19]. The enormous amount of information obtained from massively parallel SNP-typing by NGS also allows donor–recipient distinction in the absence of pre-testing potentially further simplifying the routine application [20,21].

2.3. Deletion Insertion Polymorphisms (DIP)

DIPs or INDELs are relatively short (up to about 50 bases in length), frequently biallelic variations located in non-coding regions of the genome displaying a lower mutation rate compared to STRs. As a consequence of the emergence of NGS technology and advanced software algorithms, the number of identified and characterized DIPs is rapidly increasing [22–24]. While searching the world's largest public human variation database (dbSNP Build 152) and applying a filter for intergenic human DIPs, not restrictive to the length of the polymorphism, nearly 9 million hits are found. Biallelic DIPs are suitable and reliable markers which are now commonly used for monitoring donor–recipient tissue proportions in transplantation as well in person-identification in forensic medicine [25,26]. To monitor chimerism after allo-HSCT, Alizadeh et al. utilized 19 specific markers located on 9 different chromosomes reaching an informativity level of approximately 90%. This approach has been later validated by others [27–29]. A further study extended this system with two additional DIP markers and optimized their assay for SYBR green with a possible discrimination of 94% [30]. Subsequently, Pereira et al. developed a DIP-based multiplex assay for human identification, which uses 38 biallelic markers characterized by high heterozygosity rates in distinct population groups, reaching an approximate informativity of 100% [26]. As an alternative to hybridization probes, Goh et al. analyzed dd-cfDNA after liver transplantation by targeting DIPs by the amplification primers themselves [31]. The successful application of as few as 10 DIPs along with digital PCR for the detection of dd-cfDNA in a small cohort of heart transplanted patients published as a less resource demanding approach [32]. Besides these publicly available marker collections, a large selection of commercial kits are also available, indicating the importance of the approach [33–37]. Combinations of DIPs and SNPs (a total of 26 markers) were also tested with a genotype identification rate of 97% [38] and with another set (a total of 29 markers) resulting an informativity of 97% [39]. The major advantage of DIPs is the potential to use these markers in high sensitivity assay such as quantitative PCR or digital PCR (see Section 3) to reliably detect microchimerism equally important in SOT and allo-HSCT.

2.4. Other Special Markers Used in Monitoring

Evidently, application of sex determining markers (e.g., SRY, AMELY, DFFRY, SMCY) are only useful and informative in sex-mismatched transplanted patients both in allo-HSCT and SOT. The amelogenin gene has an X and Y linked variant (AMELY and AMELX). The basis of variability in this case is a 6-basepair-long deletion segment found in X chromosome (AMELX). Utilization of Y specific markers, e.g., SRY, rely on the detection of a gene exclusively located on chromosome Y. Both sex-specific markers are informative in allo-HSCT when the recipient is a male and the donor is a female [2,40,41], while in SOT the expected situation is exactly the opposite. Similarly, RhD blood group (which is a 58 kilobase-long DIP, i.e., RhD-negativity is equal to a complete deletion) is also a straightforward choice in monitoring RhD positive allo-HSCT patients with RhD negative donor. The opposite is applicable in SOT (an RhD negative recipient with an RhD positive donor) [13,42]. Besides these, another informative marker set is the HLA which information for both donor and recipient is readily available prior to transplantation. Disparate HLA-markers were successfully used to detect elevations in dd-cfDNA plasma concentrations associated with lung transplant complications.

However, the major obstacle of the widespread use of this approach may be the need for the practically individual detection probe design depending on the actual HLA-disparities between the donor and recipient [43].

2.5. Guidelines for Marker Selection to Detect Clinical Chimerism

An ideal marker should be located in non-coding region ruling out the chance it is linked to genetic diseases, plays a role as genetic susceptibility factor or influences certain disease progression or response to therapy. It should be biallelic, differing by at least 2 basepairs, with a minimum allele frequency of at least 0.25 and a high level of heterozygosity (>0.4). Their chromosomal localization should be as distant as possible or on different chromosomes and they should also be located in sufficient distance from repetitive regions [25,26]. When designing SNPs as markers, loci with C/T and G/A polymorphisms are preferred to achieve more specific allele discrimination [38]. The recommended number of markers allowing sufficient informativity (>95%) for STR varies between 12 and 16, while for DIPs and SNPs a larger set is required (up to 52 markers). For follow-up diagnostics, more than one marker is recommended to avoid false negative results e.g., due to allele-dropout during the polymerase chain reaction (PCR) or due to the deletion of chromosomal region of the chosen marker in hematological malignancies [9,37]. Loss of chromosome Y was observed not only in hematological malignancies, but also in aging healthy males.

3. Comparison of Microchimerism Detection Techniques

As allo-HSCT became widespread, an imperative need arose to repeatedly discriminate recipient and donor cells in the circulation after allo-HSCT for which various techniques have been developed. Initially, these approaches had limited sensitivity in the range of 10% recipient cells allowing only to characterize chimerism and not microchimerism. The most commonly used chimerism-detection methods in this time period were XY fluorescence in-situ hybridization and polymerase chain reaction (PCR) based amplification of variable number of tandem repeats (VNTR) or short tandem repeats (STR). The PCR-based technologies more recently predominantly use fragment analyses (FA) making it possible reach a sensitivity range of 1% still unsuitable for the reliable detection of microchimerism (see the detailed review of Clark et al. [9]). Systematic comparison of three most widely used microchimerism detection technologies with reference to the gold-standard chimerism detection technique, fragment analysis is shown in Table 1.

Table 1. Comparison of chimerism/microchimerism laboratory detection techniques.

Technique	Fragment Analysis #	qPCR	dPCR	NGS
Targeted Genetic Variant	limited number of multiallelic markers (STR)	limited number of biallelic markers (SNP, DIP)	limited number of biallelic markers (SNP, DIP)	unlimited number of biallelic markers (SNP)
Limit of Detection	>1%	1–0.01%	1–0.01%	1–0.01%
TAT	short	shortest	short	longer
Equipment Cost	considerable	relatively lower	considerable	considerable
Allo-HSCT Marker number *	3	2	2	not relevant
Advantages	gold standard #; widespread application; large experience	high sensitivity; short TAT	high sensitivity; high precision	high number. of SNPs; simultaneous chimerism & MRD
Technical Limitations	stutter peak; preferential amplification; semi-quantitative	labor-intensive optimization; calibration curve; PCR inhibitors; duplicate low no. of variants	dependent on DNA concentration; low number of variants	infrastructure costs; longer TAT; bioinformatics; lack of standardization

Fragment analysis has been used for the longest time albeit due to its low sensitivity it is not suitable for microchimerism detection. * Minimum number of markers for allo-HSCT follow-up. Abbreviations: dPCR: digital PCR; DIP: deletion insertion polymorphism; HSCT: hematopoietic stem cell transplantation; MRD: measurable residual disease; NGS: next generation sequencing; qPCR: real time quantitative PCR; SNP: single nucleotide polymorphism; STR: short tandem repeat; TAT: turnaround time.

3.1. *Real-Time Quantitative PCR (qPCR)*

To increase the sensitivity of chimerism monitoring in transplanted patients, several studies proposed real-time quantitative PCR (qPCR) method to detect single nucleotide polymorphisms (SNPs) or short deletions/insertions (DIP). QPCR allows 0.1% sensitivity quantification of the minor genotype [12,25,28,37,38,44]. The vast majority of studies employ TaqMan technology requiring a hybridization probe labeled with two different fluorescent dyes: a reporter (FAM), and a quencher (TAMRA). When the probe is intact, fluorescent energy transfer occurs and the reporter fluorescence is readily absorbed by the quencher. In case of precise hybridization of the probe to its target, during the extension phase of the PCR cycle, the 5′-3′ exonuclease activity of the DNA-polymerase cleaves the TaqMan probe and releases the reporter dye, resulting in an increase of the reporter dye fluorescent signal [45]. Probes can be labeled with alternative, distinguishable reporter dyes allowing duplexing in a single reaction. The disadvantage of the qPCR technique is the requirement of labor-intensive optimization and the need for replicates (duplicates, triplicates) in each run for each target to provide the most accurate result [12,25].

3.2. *Digital PCR (dPCR)*

Digital PCR has initially been used to quantify low-copy number fetal DNA in maternal plasma [46]. This innovative approach is based on partitioning of the PCR in multiple nanoliter chambers or droplets, and after the amplification, each chamber/droplet is counted positive or negative for a specific polymorphism [47]. With increasing the number of partitions (e.g., performing replicates), sensitivity of the test can be improved. A further innovation of this technique, droplet digital PCR (ddPCR), is easily automated and does not require a dedicated PCR machine. In contrast to qPCR which allows a real time approach, dPCR is an end-point assay with the determination of the positive droplet fraction and Poisson statistics calculating the absolute number of starting copies making calibration curves unnecessary [47,48]. Due to the simultaneous presence of reference amplification in each reaction during dPCR, as well as a consequence of unprecedented precision, replicates are not needed in cases with target concentrations above the limit of detection. This method is less sensitive to inhibitors than fragment analysis or qPCR. The technique is highly suitable for the detection of rare events in the presence of high background. Performing the quality control test on artificial chimerism mixture samples Mika et al. found high correlation between the estimated and the observed percentage values for two discriminating markers. The standard deviation of the four-time repeated measurements of a dilution series with one discriminating marker ranged from 1.8% to 3.7%. Comparing dPCR to the 'gold-standard' STR by fragment analyzes indicated good agreement between these two techniques in the detection range of 83–100% of donor chimerism [49]. The limit of detection of dPCR was estimated to be as low as 0.008% allowing the reliable determination of microchimerism. Additionally, assay precision was acceptable also in ranges of microchimerism, with a variation coefficient of 16% at 1:999 dilution [36]. The dPCR technique proved to be more sensitive in detection of relapse after allo-HSCT compared to qPCR [35]. Similarly to allo-HSCT, dPCR has also been widely used in the context of SOT to detect dd-cfDNA in combination with DIPs [32,50], SNPs [3], copy number variations [51], or HLA-disparities [17,18,43,52].

3.3. *Next Generation Sequencing (NGS)*

Next generation sequencing (NGS) is a massively parallel or deep sequencing, where millions of small fragments of DNA are sequenced in parallel affording an unprecedentedly high sensitivity. Sequenced fragments are pieced together by mapping individual reads to the human reference genome by bioinformatic tools. Several different NGS platforms exist with different chemistries and techniques. NGS can be used to sequence individual genes, targeting the whole exome or even the entire genome [53]. This technique is more sensitive than fragment analysis (0.01% to 1% vs. >1%). Compared to STR by fragment analyzes, NGS showed an excellent correlation [54,55]. The unprecedented capacity of

NGS allows the simultaneous typing of large marker sets, predominantly SNPs, which has frequently been used to perform dd-cfDNA detection in various SOT settings. The large number of SNP markers applied clearly gives an unprecedented opportunity to distinguish two genomes, even in cases of unavailable donor pre-transplant DNA sample or in cases with multiple numerical chromosomal abnormalities. The main advantage of NGS is that it is applicable for simultaneous determination of chimerism or microchimerism as well as multiple disease specific markers to detect measurable residual disease (MRD). However, these techniques require substantial equipment background, reagent cost, trained bioinformatics expertise, standardization, and currently they are not always characterized with sufficiently short turn-around time to readily influence clinical decision making. The described approaches belonged either to non-commercially designed systems [17,18,52] or to commercially developed platform with an unequivocal intent of routine clinical application [56–58].

4. Sample Types, Preanalytics, and Isolation of Cell Free DNA

The volume of blood required for cfDNA analysis is usually between 6 and 10 mL. Concentrations in paired plasma and serum samples have revealed significantly higher cfDNA concentrations in serum than in plasma [59]. This is due to leukocyte damage during clotting in the serum tube with a single cell containing about 6 pg of DNA and adding up to 65 ng of DNA per mL of blood. As a consequence, serum samples are not advised, instead plasma from anticoagulated (predominantly EDTA) blood sample is used [60]. The use of tubes containing preservatives (PAXgene cfDNA tubes (Qiagen PreAnalytiX GmbH, Düsseldorf, Germany), Roche/Ariosa cfDNA collection tubes (Roche Diagnostics GmbH, Rotkreuz, Switzerland) and Streck BCT tubes (Streck Inc., Omaha, NE, USA, see [61]) to prevent hemolysis and to reduce the degradation of cfDNA is becoming increasingly common practice allowing an extended time to blood processing [62]. Interestingly, in the case of timely (up to 12 h at room temperature) processing, there was no significant difference between cfDNA concentrations in samples shipped in EDTA collection tubes versus samples shipped in cell-free DNATM blood collection tubes [63,64]. Plasma preparation typically involves two rounds of centrifugation: a classical, low speed centrifugation at 1000–2000 $\times g$ followed by a high speed centrifugation (16,000$\times g$) [65]. In cases of a single centrifugation step at 800 $\times g$, cfDNA concentrations were contaminated by genomic DNA [66].

Two main types of cfDNA extraction systems have frequently been used: column-based systems and magnetic beads systems. Studies comparing cfDNA extraction methods and kits revealed substantial differences in total DNA yield [67]. cfDNA samples may withstand three freeze-thaw cycles and storage at −20 °C. Storage time has no influence on specific sequences or mutations in cfDNA, and mutations can be detected several years after freezing plasma samples, but cfDNA content is decreasing with storage time, so quantification or characterization of cfDNA fragmentation, is preferred within 3 months after the extraction procedure [68].

Quantitation of extracted cfDNA may be performed with several techniques such as fluoro-spectroscopy, Qubit (Thermo Fisher Scientific Waltham, MA, US) fluorometry and qPCR. In addition, quality control of isolated cfDNA can be performed by fragment analysis by capillary electrophoresis. Based on fragment size, the relative amount of nucleosome-protected cfDNA fragments with 140–160 base pairs can be compared to degradation-fragments with higher molecular weight DNA [69,70]. Single-tube multiplex digital droplet PCR assays have also been described as alternative approaches to measure amplifiable DNA concentrations and fragment size of small amounts of cfDNA [62,71].

5. Role of Microchimerism in Solid Organ Transplantation

In the context of solid organ transplantations (SOT), the meaningful use of the term microchimerism is only possible for cfDNA in the circulation since intact graft cells are unlikely to survive in blood. Appearance of low amounts (usually less than 0.1%) of donor hematopoietic (passenger) cells in the recipient's circulation has been repeatedly observed after SOT and prompted speculations about a potential favorable role of these cells in graft acceptance. However, later clinical studies failed to find consistent associations and this cell based microchimerism also showed large inter-individual variations (see the excellent review by Eikmans et al. [1]).

After SOT, beside cells of donor origin, recently an increasing attention has turned towards dd-cfDNA since upon immune-mediated, apoptotic, or necrotic cell damage, its amount increases. This may then be translated to elevated levels of absolute amount or of relative proportions of dd-cfDNA. Such changes may serve as sensitive surrogate biomarkers to monitor the health of the transplanted organ (graft). This has outstanding significance as the overall frequency of graft related post-SOT complications—i.e., rejection—immunosuppressive treatment mediated toxicity or infection can reach 50% of all SOT cases [72]. For the timely alleviation of these complications (e.g., appropriate adjustments of immunosuppressive therapies), systematic utilization of appropriate biomarkers is essential. To this end, monitoring the health of the allograft by a diverse array of laboratory parameters as well as therapeutic drug monitoring of immunosuppressive drugs has traditionally been used to assist therapeutic decision making. In addition, monitoring and characterization of donor specific antibodies (DSA) have recently gained increasing attention as the major causative factor of antibody mediated rejection (ABMR). However, the direct translation of DSA-profile data into clinical decision making has been challenging. It has been reported that up to 80% of patients with DSA do not have ABMR and additional factors, such as type and strength of DSA, IgG subclass, and the ability to bind complement, are important further determinants [73]. Besides the above-mentioned, non-invasive monitoring options, graft biopsy procedures have traditionally represented the gold standard for graft-damage evaluation which are invasive and costly, limiting their use in clinical practice. A significant proportion of biopsies yield inadequate specimen and major complications also occur relatively frequently. Furthermore, biopsy results are often compromised by expert reader variance and can lead to delayed diagnosis of active rejection, after which irreversible organ damage may have occurred [74].

Monitoring of dd-cfDNA in plasma as a non-invasive indicator of graft damage after SOT was first suggested as early as in 1998 by the most outstanding expert of the field, Lo et al. [75] followed by several early studies focusing on SOT [76–79]. At this stage, the techniques applied did not readily allow the accurate detection of dd-cfDNA below 1% corresponding to true microchimerism. In addition, restricted availability of suitable markers for donor vs. recipient distinction also hampered cohort studies of larger consecutive series. With dramatic technical advancement (see Sections 2 and 3) both, the sensitivity and the informative marker availability problem has been solved allowing the completion of a number of cohort studies in various SOT settings. Studies on one hand attempted to characterize the precise behavior of dd-cfDNA in the early phase of SOT as well as to establish reference ranges documenting a sharp decline in the initial days and a proportion typically between 0.3% and 1.2% of total cfDNA in stable kidney transplant patients [16,19]. While for liver and transplants these proportion values are somewhat higher [16]. On the other hand, the majority of studies addressed the more exciting question namely how dd-cfDNA serves as a rejection biomarker. Several reviews have summarized these studies [72,80–83]. Out of these, in their recent systematic review, Knight et al. [84] published a large collection of relevant literature searches. However, no consistent distinction has been made between conference abstracts and peer-reviewed in extenso publications making it difficult to realize the number and extent of truly independent cohort studies. In addition, potential patient overlaps between cohorts are also impossible to identify in this manner.

Thus, we performed a focused literature overview of exclusively peer-reviewed papers addressing the potential role of plasma dd-cfDNA as a potential rejection marker in various SOT settings with an inclusion criteria of total cohort size equaling or exceeding $n = 50$ for kidney or $n = 20$ for other solid organs respectively. As shown in Table 2, a total of 15 studies met these criteria.

Table 2. Selected cohort studies focusing on the potential role of dd-cfDNA in SOT of various organs.

ID	Organ	Year	Author	Center/Country	Study Type	Study Design *	Total (n)	Rejection (n)
1	kidney	2017	Bloom [56]	Cedars-Sinai. LA, USA	prosp	biopsy	102	27
2	kidney	2018	Jordan [58]	Cedars-Sinai. LA, USA	retro	biopsy	87	16
3	kidney	2018	Sigdel [18]	UCSF, SF, USA	retro	consec	178	38
4	kidney	2019	Huang [57]	Cedars-Sinai LA, USA	prosp	biopsy	63	22
5	kidney	2019	Whitlam [51]	Melbourne, Australia	prosp	consec	55	13
6	heart	2011	Snyder [78]	Stanford, USA	retro	consec	112	12
7	heart	2014	DeVlaminck [85]	Stanford, USA	prosp	consec	65	n/a
8	heart	2016	Grskovic [17]	CareDX, Brisb., USA	retro	consec	101	27
9	heart	2018	Ragalie [86]	Wisconsin, USA	retro	biopsy	88	16
10	heart	2019	Macher [32]	Sevilla, Spain	prosp	consec	30	13
11	liver	2017	Schütz [87]	Gottingen, Germany	prosp	consec	107	17
12	liver	2019	Goh [50]	Melbourne, Australia	prosp	consec	20	3
13	lung	2015	DeVlaminck [88]	Stanford, USA	prosp	consec	51	n/a
14	lung	2018	Agbor-Enoh [89]	Stanford, NIH, USA	prosp	consec	157	34
15	lung	2019	Agbor-Enoh [90]	Stanford, NIH, USA	prosp	consec	106	n/a

* consecutive (consec): transplanted patients are consecutively included regardless of rejection; biopsy = biopsy-linked: sampling parallels biopsy indicated by organ-failure. Legend: Original studies were included if the cohort size equaled or exceeded $n = 50$ for kidney or $n = 20$ for other solid organs. Studies are listed according to organ transplanted then according to year published. Abbreviations: n/a: not available; prosp: prospective; retro: retrospective.

As an overall conclusion, significantly positive role was exclusively found across all organs for plasma dd-cfDNA as allograft injury biomarker, no study was found concluding the lack of association. In kidney transplant, the paramount prospective trial of Bloom et al. called "Circulating Donor-Derived Cell-Free DNA in Blood for Diagnosing Active Rejection in Kidney Transplant Recipients" (DART) had the strongest impact with subsequent positive decision about health insurance reimbursement in the US. By using a cut-off value of 1.0% of dd-cfDNA these authors found that this biomarker had stronger discriminating power between ABMR and absence of ABMR compared to cell mediated rejection versus lack that of [56]. The biopsy-linked patient selection design allowed the enrichment of patients with documented rejection episodes. Using essentially the same cohort additional aspects were also examined by the same center such as the joint diagnostic characteristics of dd-cfDNA and DSA [58] as well as biomarker performance details of dd-cfDNA with partial overlap with the original study [57]. The relevant UCSF and Australian studies indeed represent independent confirmations, albeit in smaller cohorts [18,51].

Heart transplant could be viewed as a pioneer of the field with an early study of a sizeable cohort from Stanford taking advantage of a large biorepository also containing plasma, allowing the selection more than a 100 sex-mismatched heart transplant cases [78] later confirmed in an independent prospective cohort using the novel NGS-based technique [85]. Among the further confirmatory studies a diagnostic method oriented study can be found by Grskovic et al. [17] as well as a pediatric heart transplant cohort examined with proprietary diagnostic [86] and a Spanish study using an alternative, technically less challenging diagnostic approach [32]. Regarding liver TX, besides an early case report [91] and small prospective cohorts [50,92], a larger ($n = 115$) multicentric study demonstrated significantly elevated dd-cfDNA levels in acute rejection compared to transplanted patients in stable phase or those with hepatitis C infection [87]. Similarly, in the field of lung TX, besides a few smaller studies, only the Stanford group was able to collect larger cohort after the initial study [85] with a multicentric prospective trial indicating markedly elevated dd-cfDNA prior to graft-rejection proven by biopsy [89] supplemented by a further study on essentially the same cohort [90].

Besides plasma, alternative test materials have also been used to monitor cfDNA after SOT. Sigdel et al. observed rather large inter-patient variations of cfDNA in urine samples of a sizeable ($n = 61$) kidney transplanted cohort with limited potential as a general diagnostic biomarker [93]. cfDNA was also measured in broncho-alveolar lavage (BAL) fluid after lung TX among 60 patients along with the chemokine CXCL10 and the combination of these markers significantly predicted allograft survival [52].

The most important limitation of plasma dd-cfDNA monitoring is its lacking specificity for rejection subtype, or specific histopathological alteration. However, due to its non-invasive character, the potentially rapid laboratory turn-around time, the increasing accessibility and the continuously decreasing cost, this approach may represent an early contributing diagnostic tool supplementing available classical indicators.

6. Role of Microchimerism in Allogeneic Hematopoietic Stem Cell Transplantation

6.1. *General Applications of Chimerism Monitoring in Allo-HSCT*

Chimerism analysis (detection of the relative ratio of donor and recipient hematopoiesis) from peripheral blood or bone marrow aspirates is a standard diagnostic procedure following allo-HSCT. The replacement of recipient cells by those of the donor called complete donor chimerism is an evidence for the successful engraftment and the reconstitution of hematopoiesis after allo-HSCT. After myeloablative conditioning, residual recipient hematopoiesis may indicate graft failure and/or persisting disease. Remaining or re-emerging recipient cells in the circulation, called mixed chimerism, after transplantation is mostly associated with relapse in cases of hematopoietic malignancies, although long term mixed chimerism without relapse has been described [94]. Persisting mixed chimerism is reported to reduce the graft versus leukemia (GvL) effect of alloreactive donor T cells [95–97]. In case of non-malignant disorders treated with non-myeloablative regimens, increasing mixed chimerism is an indicator of graft rejection.

The dynamics of serial chimerism monitoring facilitates therapeutic interventions such as modulation of immunosuppression, donor lymphocyte infusions, chemotherapy, or second transplantation both in myeloablative and in non-myeloablative settings [95]. Early, individualized interventions have the potential to balance between competing risks of graft failure, recurring disease, and the potentially life-threatening graft versus host disease (GvHD).

Analysis of STRs by fragment analysis (FA) is the most widely accepted choice of method for post-HSCT monitoring. Detailed recommendations are available for proper laboratory performance of STR chimerism monitoring [9]. The major disadvantage of STR chimerism monitoring is its low sensitivity (limit of detection: 1–5%). To circumvent this limitation, frequent monitoring (even weekly before day 200 after allo-HSCT in pediatric acute lymphoblastic leukemia), performing the analysis in subpopulations for the detection of lineage specific split chimerism are recommended, as acute leukemia relapse can occur within a relatively short time frame [98].

Chimerism monitoring provides a surrogate biomarker for tumor-specific measurable residual disease (MRD). Chimerism testing is suggested to be applied in combination with MRD monitoring in malignant hematopoietic diseases [99,100]. In several clinical settings, MRD assay is not available (due to the lack of identified disease specific somatic mutation) or clonal heterogeneity, in these cases chimerism testing remains the only follow-up marker after allo-HSCT.

6.2. *Relevance of Microchimerism Monitoring in Allo-HSCT*

In recent years, the sensitive detection of MRD (reaching at least 0.1% or deeper limit of detection) gained increasing attention, as patients with MRD-positive morphologic remission showed similar outcome to patients with active disease [99,101]. Introducing sensitive techniques like qPCR or dPCR enabled the reliable detection of microchimerism, which is defined as the presence of recipient cells below 1% in the circulation. Initially, this sensitivity was not a requirement for chimerism detection [102]. Earlier, the clinical utility of microchimerism detection after allo-HSCT was controversial and limited [39]. Low levels of undulating microchimerism might have been detected for years after allo-HSCT, which may not have been necessarily related to relapse. Studies focusing on the potential advantages and the clinical utility of sensitive chimerism detection are summarized in Table 3.

Several papers indicated that the dynamics of microchimerism was more closely correlated with relapse in acute leukemia. The reappearance of recipient DNA after complete donor chimerism is a warning sign, and increasing recipient chimerism in two consecutive samples reliably forecast relapse [28,103]. Stable or decreasing microchimerism levels might not indicate relapse. In acute leukemia, several lines of evidence regarding allo-HSCT chimerism monitoring suggest that sensitive methods (e.g., qPCR, dPCR) are superior compared to classic STR-fragment analysis in the prediction of hematological relapses: (i) the ratio of detected/all relapses for DIP/qPCR was 70–100% vs. 13–44% for STR/fragment analysis; (ii) the time difference in favor of DIP/qPCR vs. STR/fragment analysis was 26–90 days [13,27,34,44,96,104,105]. At the same time, the lack of recipient genotype in highly-sensitive chimerism analysis excludes the possibility of relapses with high certainty [28,103]. Systemic infections, especially virus infections, and acute GvHD were also variably associated with a slight increase in host microchimerism. Leukemia relapse is usually associated with a rapid expansion of host hematopoiesis [34,103,106].

Dynamics of chimerism also influence survival: complete donor and decreasing mixed chimerism present favorably, while increasing and stable mixed chimerism have adverse prognostic significance [105,107]. Recent studies indicated that not mixed chimerism, but also microchimerism after allo-HSCT show adverse overall survival compared to complete donor chimerism (mixed chimerism $p < 0.0002$; microchimerism $p = 0.0201$). The introduction of sensitive chimerism monitoring allows early detection of recipient hematopoiesis, early intervention, and the decrease of patients proportions with mixed chimerism above 1% (23% before and 15% after the introduction of sensitive chimerism test) [94].

Mixed chimerism, and also microchimerism, were found to strongly correlate with MRD-monitoring inasmuch, both biomarkers show similar time-course, when analyzed from the same sample type [29,96,100]. As a conclusion, chimerism including microchimerism monitoring is a reliable indicator of incipient acute myeloid leukemia (AML) relapse, especially in patients where no other specific molecular marker is available, which might affect up to 50% of allo-HSCT cases [34]. In case of chimerism monitoring, the lack of identifiable marker or subclonal heterogeneity do not alter the results however, cytogenetic alterations in the malignant clone resulting in the loss of the recipient allele carrying the respective marker may cause a false negative chimerism result.

Table 3. Potential advantages and clinical applicability of sensitive microchimerism detection in allo-HSCT based selected papers.

First Author	Year	Country	*n*	Marker *	Clinical Utility of Microchimerism Detection
Jiménez-Velasco [13]	2005	Spain	61	VNTR, DIP	qPCR superior to FA in relapse prediction: 88% vs. 44%, 20 days prior FA; rising host chimerism kinetics in relapse
Koldehoff [105]	2006	Germany	269	STR, SNP, FISH XYSRY	qPCR superior to PAGE in relapse prediction: 90 days earlier; adverse overall survival in rising kinetics; SRY qPCR superior to XY FISH in relapse prediction: 86% vs. 28%, 143 days prior relapse
Wiedermann [107]	2010	Germany	75	DIP, SRY	rising host chimerism kinetics in relapse
Chen [27]	2011	Taiwan	126	STR, DIP	qPCR superior to FA in relapse prediction
Horky [96]	2011	Czech	46	STR, DIP	qPCR superior to FA in relapse prediction: 87% vs. 39%, 26 days prior FA.
Bach [44]	2015	Germany	16	STR, DIP	qPCR superior to FA in relapse prediction: 95 days earlier
Jacque [28]	2015	France	85	DIP	stable qPCR complete chimerism = negative predictor for relapse; rising host chimerism kinetics in relapse

Table 3. *Cont.*

First Author	Year	Country	*n*	Marker *	Clinical Utility of Microchimerism Detection
Vicente [106]	2016	Brazil	41	DIP	early rising host chimerism kinetics in relapse
Ahci [34]	2017	Germany	30	STR, DIP	qPCR superior to FA in relapse prediction: 100% vs. 38%, 6 months prior relapse; rising or sustained host chimerism kinetics in relapse
Waterhouse [108]	2017	Germany	155	STR, SRY	dPCR superior to FA in relapse prediction: 90 days earlier
Cechova [94]	2018	Czech	474	STR, DIP	adverse overall survival in microchimerism compared to complete donor chimerism
Sellmann [103]	2018	Germany	71	DIP	stable qPCR complete chimerism = negative predictor for relapse; rising host chimerism positive predictor for relapse
Waterhouse [29]	2018	Germany	70	DIP, MRD	similar kinetics for chimerism and MRD
Tyler [37]	2019	USA	230	STR, DIP	combined STR-FA and DIP-qPCR in diagnostic algorithm; qPCR is more sensitive than FA
Valero-Garcia [35]	2019	Spain	28	DIP	dPCR superior to qPCR in relapse prediction

* STR, VNTR markers were tested with FA or PAGE; DIP, SNP, and SRY were tested with qPCR or dPCR. Legend: Original in extenso studies were included if clinical outcome of allo-HSCT (relapse, survival measures) was indicated in connection with chimerism levels below 1%. Studies are listed according to year published. Abbreviations: DIP: deletion-insertion polymorphism; FA: fragment analysis; FISH: fluorescence in situ hybridization; MRD: measurable residual disease; PAGE: poly-acrilamid gel-electrophoresis; qPCR: quantitative polymerase chain reaction; SNP: single nucleotide polymorphism; SRY: sex determining region Y; STR: short tandem repeat; VNTR: variable number tandem repeat.

6.3. Special Applications of Microchimerism Monitoring in Cellular Therapies

In hematology practice, special cellular therapies are becoming more and more frequent treatment options: examples are 'microtransplantation' in elderly AML patients, natural killer (NK) cell infusion or third-party donor antigen-specific T-cell infusions (virus specific T cells, VST) [109]. As these allogeneic cellular therapies are not intended to achieve durable engraftment, they frequently result in microchimerism not detectable with standard STR-based monitoring.

In a randomized, multicenter study of elderly AML patients, induction chemotherapy in combination with mobilized HLA-mismatched donor cells without GvHD-prophylaxis improved response rates with a low rate of mixed or complete donor chimerism (5/185) and severe acute GvHD (2/185). Microchimerism was detectable in 31% (8/26) sex-mismatched, informative transplants by SRY-qPCR (with up to 10 months follow-up) [110]. Microtransplantation was found to be inferior to HLA-matched sibling donor transplantation in allo-HSCT eligible patients for intermediate/high-risk AML in complete remission [111], but a case report described its utility as salvage therapy in refractory AML [112]. Haploidentical natural killer cell therapy is a safe and feasible therapeutic option in pediatric relapsed or refractory leukemia [113]. Adoptive transfer of VST cells aim to restore T-cell immunity and cure allo-HSCT patients with refractory, life-threatening viral infections like cytomegalovirus, Epstein–Barr virus, or adenovirus [114]. Kliman et al. described an ultrasensitive dPCR method capable to detect donor microchimerism after cellular therapies such as microtransplantation or VST [36].

6.4. Cell-Free DNA Chimerism Following Allo-HSCT

Plasma cfDNA markers as diagnostic applications have been intensively studied in the context of prenatal screening, solid tumors and SOT. The clinical usefulness of chimerism testing from plasma cfDNA in allo-HSCT has been sparsely reported. Initially, sex-mismatched allo-HSCT patients were screened to reveal the main cellular origin of plasma circulating cfDNA. Y-chromosome specific cfDNA was detected in plasma with a median of 6.9% in male patients with female donors, indicating the hematopoietic compartment as the main origin of plasma cfDNA [2].

Plasma cfDNA was found to be more sensitive than cellular chimerism in detecting relapse in leukemia patients with complete donor chimerism in polymorphonuclear cells. All patients with clinical relapse (16/84, 19%) had more than 10% of recipient cfDNA in the plasma. Altogether 60% of patients displayed various levels of mixed chimerism in plasma, including an additional 16 patients with >10% mixed chimerism level and without clinical relapse [115].

A further study indicated that mixed chimerism was detectable in cfDNA in a higher percentage of samples than in peripheral blood cells after allo-HSCT. Interestingly, plasma cfDNA-based microchimerism was capable of detecting isolated extramedullar relapses (central nervous system, $n = 3$), when in peripheral blood cells complete donor chimerism was observed [116]. Recipient derived cfDNA was elevated not only in relapse, but also in transplant related complications, especially in acute GvHD (aGvHD). The improvement of aGvHD symptoms during therapy coincided with the decrease in recipient cfDNA, in contrast, the lack of aGvHD amelioration was associated with stable or increasing levels of recipient cfDNA. This association suggested that recipient cell disruption in GvHD target organs can be a source of cfDNA. No correlation was found between mean recipient cfDNA percentage and aGvHD severity (grade I–II compared to grade III–IV), but organ specific involvements have not been described. Chronic GvHD (cGvHD) did not show an association with plasma cfDNA-based mixed chimerism in affected patients [116]. Plasma cell free mitochondrial DNA (mtDNA) was reported to be elevated in allo-HSCT patients. As an endogenous inflammatory signal, cell free mtDNA may activate a B-cell subset responsible for extensive cGvHD: the onset of cGvHD was accompanied with significantly elevated levels of plasma cell free mtDNA in patients with cGvHD [117].

7. Future Perspectives

In the context of SOT, additional independent, sufficiently large cohort studies are needed to further confirm the diagnostic value of microchimerism characterized by dd-cfDNA proportion in plasma. A further goal of these studies can be the targeting patients with subclinical graft damage for which currently the only diagnostic tool is the invasive organ biopsy. Applicability of dd-cfDNA based microchimerism detection as valuable information in the clinical decision-making process regarding actual immunosuppression should also be tested in a prospective setting. Routine application and comparisons of alternative detection techniques may also be attractive for future research since a considerable number of transplantation diagnostic facilities may not have access to the most advanced NGS-based determinations.

Although it is not widely accepted yet, monitoring of recipient cfDNA dynamics after allo-HSCT may be a valuable diagnostic tool to detect relapses at earlier time points, with special emphasis on extramedullary sites. Additionally, it can be utilized as a non-invasive biomarker for aGvHD activity. Further studies involving larger patient cohorts, as well as applying more sensitive methods than STR are necessary to address existing uncertainties regarding cfDNA chimerism testing in the allo-HSCT field.

Author Contributions: H.A. and A.T. initiated the project and worked out the main concept. Each co-author was responsible for the systematic literature search, evaluation of relevant publication regarding pre-defined subsections of the paper and writing an original draft. Overall wording of the manuscript was performed by H.A. and A.T. All authors evaluated, corrected, and consented to the final version.

References

1. Eikmans, M.; Van Halteren, A.G.S.; Van Besien, K.; Van Rood, J.J.; Drabbels, J.J.M.; Claas, F.H.J. Naturally acquired microchimerism. *Chimerism* **2014**, *5*, 24–39. [CrossRef] [PubMed]
2. Lui, Y.Y.; Chik, K.W.; Chiu, R.W.; Ho, C.Y.; Lam, C.W.; Lo, Y.M. Predominant hematopoietic origin of cell-free DNA in plasma and serum after sex-mismatched bone marrow transplantation. *Clin. Chem.* **2002**, *48*, 421–427. [PubMed]

3. Tanaka, S.; Sugimoto, S.; Kurosaki, T.; Miyoshi, K.; Otani, S.; Suzawa, K.; Hashida, S.; Yamane, M.; Oto, T.; Toyooka, S. Donor-derived cell-free DNA is associated with acute rejection and decreased oxygenation in primary graft dysfunction after living donor-lobar lung transplantation. *Sci. Rep.* **2018**, *8*, 15366. [CrossRef] [PubMed]
4. Lander, E.S.; Linton, L.M.; Birren, B.; Nusbaum, C.; Zody, M.C.; Baldwin, J.; Devon, K.; Dewar, K.; Doyle, M.; FitzHugh, W.; et al. Initial sequencing and analysis of the human genome. *Nature* **2001**, *409*, 860–921. [PubMed]
5. Mirkin, S.M. Expandable DNA repeats and human disease. *Nature* **2007**, *447*, 932–940. [CrossRef] [PubMed]
6. Willems, T.; Gymrek, M.; Highnam, G.; The Genomes Project, C.; Mittelman, D.; Erlich, Y. The landscape of human STR variation. *Genome Res.* **2014**, *24*, 1894–1904. [CrossRef] [PubMed]
7. Gettings, K.B.; Aponte, R.A.; Vallone, P.M.; Butler, J.M. STR allele sequence variation: Current knowledge and future issues. *Forensic Sci. Int. Genet.* **2015**, *18*, 118–130. [CrossRef]
8. Butler, J.M. Short tandem repeat typing technologies used in human identity testing. *Biotechniques* **2007**, *43*, ii–v. [CrossRef]
9. Clark, J.R.; Scott, S.D.; Jack, A.L.; Lee, H.; Mason, J.; Carter, G.I.; Pearce, L.; Jackson, T.; Clouston, H.; Sproul, A.; et al. Monitoring of chimerism following allogeneic haematopoietic stem cell transplantation (HSCT): Technical recommendations for the use of short tandem repeat (STR) based techniques, on behalf of the United Kingdom National External Quality Assessment Service for Leucocyte Immunophenotyping Chimerism Working Group. *Br. J. Haematol.* **2015**, *168*, 26–37.
10. Fan, H.; Chu, J.Y. A brief review of short tandem repeat mutation. *Genom. Proteom. Bioinform.* **2007**, *5*, 7–14. [CrossRef]
11. Wang, D.G.; Fan, J.B.; Siao, C.J.; Berno, A.; Young, P.; Sapolsky, R.; Ghandour, G.; Perkins, N.; Winchester, E.; Spencer, J.; et al. Large-scale identification, mapping, and genotyping of single-nucleotide polymorphisms in the human genome. *Science* **1998**, *280*, 1077–1082. [CrossRef] [PubMed]
12. Maas, F.; Schaap, N.; Kolen, S.; Zoetbrood, A.; Buno, I.; Dolstra, H.; De Witte, T.; Schattenberg, A.; Van de Wiel-van Kemenade, E. Quantification of donor and recipient hemopoietic cells by real-time PCR of single nucleotide polymorphisms. *Leukemia* **2003**, *17*, 621–629. [CrossRef] [PubMed]
13. Jiménez-Velasco, A.; Barrios, M.; Román-Gómez, J.; Navarro, G.; Buño, I.; Castillejo, J.A.; Rodríguez, A.I.; García-Gemar, G.; Torres, A.; Heiniger, A.I. Reliable quantification of hematopoietic chimerism after allogeneic transplantation for acute leukemia using amplification by real-time PCR of null alleles and insertion/deletion polymorphisms. *Leukemia* **2005**, *19*, 336. [CrossRef] [PubMed]
14. Almeida, C.A.; Dreyfuss, J.L.; Azevedo-Shimmoto, M.M.; Figueiredo, M.S.; De Oliveira, J.S. Evaluation of 16 SNPs allele-specific to quantify post hSCT chimerism by SYBR green-based qRT-PCR. *J. Clin. Pathol.* **2013**, *66*, 238–242. [CrossRef] [PubMed]
15. Fredriksson, M.; Barbany, G.; Liljedahl, U.; Hermanson, M.; Kataja, M.; Syvanen, A.C. Assessing hematopoietic chimerism after allogeneic stem cell transplantation by multiplexed SNP genotyping using microarrays and quantitative analysis of SNP alleles. *Leukemia* **2004**, *18*, 255–266. [CrossRef]
16. Beck, J.; Bierau, S.; Balzer, S.; Andag, R.; Kanzow, P.; Schmitz, J.; Gaedcke, J.; Moerer, O.; Slotta, J.E.; Walson, P.; et al. Digital Droplet PCR for Rapid Quantification of Donor DNA in the Circulation of Transplant Recipients as a Potential Universal Biomarker of Graft Injury. *Clin. Chem.* **2013**, *59*, 1732–1741. [CrossRef] [PubMed]
17. Grskovic, M.; Hiller, D.J.; Eubank, L.A.; Sninsky, J.J.; Christopherson, C.; Collins, J.P.; Thompson, K.; Song, M.; Wang, Y.S.; Ross, D.; et al. Validation of a Clinical-Grade Assay to Measure Donor-Derived Cell-Free DNA in Solid Organ Transplant Recipients. *J. Mol. Diagn.* **2016**, *18*, 890–902. [CrossRef]
18. Sigdel, T.; Archila, F.; Constantin, T.; Prins, S.; Liberto, J.; Damm, I.; Towfighi, P.; Navarro, S.; Kirkizlar, E.; Demko, Z.; et al. Optimizing Detection of Kidney Transplant Injury by Assessment of Donor-Derived Cell-Free DNA via Massively Multiplex PCR. *J. Clin. Med.* **2018**, *8*, 19. [CrossRef]
19. Gielis, E.M.; Beirnaert, C.; Dendooven, A.; Meysman, P.; Laukens, K.; De Schrijver, J.; Van Laecke, S.; Van Biesen, W.; Emonds, M.-P.; De Winter, B.Y.; et al. Plasma donor-derived cell-free DNA kinetics after kidney transplantation using a single tube multiplex PCR assay. *PLoS ONE* **2018**, *13*, e0208207. [CrossRef]
20. Gordon, P.M.K.; Khan, A.; Sajid, U.; Chang, N.; Suresh, V.; Dimnik, L.; Lamont, R.E.; Parboosingh, J.S.; Martin, S.R.; Pon, R.T.; et al. An Algorithm Measuring Donor Cell-Free DNA in Plasma of Cellular and Solid Organ Transplant Recipients That Does Not Require Donor or Recipient Genotyping. *Front. Cardiovasc. Med.* **2016**, *3*, 33. [CrossRef]

21. Sharon, E.; Shi, H.; Kharbanda, S.; Koh, W.; Martin, L.R.; Khush, K.K.; Valantine, H.; Pritchard, J.K.; De Vlaminck, I. Quantification of transplant-derived circulating cell-free DNA in absence of a donor genotype. *PLoS Comput. Biol.* **2017**, *13*, e1005629. [CrossRef] [PubMed]
22. Mullaney, J.M.; Mills, R.E.; Pittard, W.S.; Devine, S.E. Small insertions and deletions (INDELs) in human genomes. *Hum. Mol. Genet.* **2010**, *19*, R131–R136. [CrossRef] [PubMed]
23. The Genomes Project, C.; Durbin, R.M.; Altshuler, D.; Durbin, R.M.; Abecasis, G.R.; Bentley, D.R.; Chakravarti, A.; Clark, A.G.; Collins, F.S.; De La Vega, F.M.; et al. A map of human genome variation from population-scale sequencing. *Nature* **2010**, *467*, 1061. [CrossRef] [PubMed]
24. Deng, J.; Huang, H.; Yu, X.; Jin, J.; Lin, W.; Li, F.; Song, Z.; Li, M.; Gan, S. DiSNPindel: Improved intra-individual SNP and InDel detection in direct amplicon sequencing of a diploid. *BMC. Bioinform.* **2015**, *16*, 343. [CrossRef] [PubMed]
25. Alizadeh, M.; Bernard, M.; Danic, B.; Dauriac, C.; Birebent, B.; Lapart, C.; Lamy, T.; Le Prise, P.Y.; Beauplet, A.; Bories, D.; et al. Quantitative assessment of hematopoietic chimerism after bone marrow transplantation by real-time quantitative polymerase chain reaction. *Blood* **2002**, *99*, 4618–4625. [CrossRef] [PubMed]
26. Pereira, R.; Phillips, C.; Alves, C.; Amorim, A.; Carracedo, A.; Gusmao, L. A new multiplex for human identification using insertion/deletion polymorphisms. *Electrophoresis* **2009**, *30*, 3682–3690. [CrossRef] [PubMed]
27. Chen, D.-P.; Tseng, C.-P.; Wang, W.-T.; Wang, M.-C.; Tsai, S.-H.; Sun, C.-F. Real-time biallelic polymorphism–polymerase chain reaction for chimerism monitoring of hematopoietic stem cell transplantation relapsed patients. *Clin. Chim. Acta* **2011**, *412*, 625–630. [CrossRef] [PubMed]
28. Jacque, N.; Nguyen, S.; Golmard, J.L.; Uzunov, M.; Garnier, A.; Leblond, V.; Vernant, J.P.; Bories, D.; Dhédin, N. Chimerism analysis in peripheral blood using indel quantitative real-time PCR is a useful tool to predict post-transplant relapse in acute leukemia. *Bone Marrow Transplant.* **2014**, *50*, 259. [CrossRef] [PubMed]
29. Waterhouse, M.; Pfeifer, D.; Duque-Afonso, J.; Follo, M.; Duyster, J.; Depner, M.; Bertz, H.; Finke, J. Droplet digital PCR for the simultaneous analysis of minimal residual disease and hematopoietic chimerism after allogeneic cell transplantation. *Clin. Chem. Lab. Med.* **2018**, *57*, 641–647. [CrossRef]
30. Bai, L.; Deng, Y.M.; Dodds, A.J.; Milliken, S.; Moore, J.; Ma, D.D. A SYBR green-based real-time PCR method for detection of haemopoietic chimerism in allogeneic haemopoietic stem cell transplant recipients. *Eur. J. Haematol.* **2006**, *77*, 425–431. [CrossRef]
31. Goh, S.K.; Muralidharan, V.; Christophi, C.; Do, H.; Dobrovic, A. Probe-Free Digital PCR Quantitative Methodology to Measure Donor-Specific Cell-Free DNA after Solid-Organ Transplantation. *Clin. Chem.* **2017**, *63*, 742–750. [CrossRef] [PubMed]
32. Macher, H.C.; García-Fernández, N.; Adsuar-Gómez, A.; Porras-López, M.; González-Calle, A.; Noval-Padillo, J.; Guerrero, J.M.; Molinero, P.; Borrego-Domínguez, J.M.; Herruzo-Avilés, Á.; et al. Donor-specific circulating cell free DNA as a noninvasive biomarker of graft injury in heart transplantation. *Clin. Chim. Acta* **2019**, *495*, 590–597. [CrossRef] [PubMed]
33. Stahl, T.; Rothe, C.; Böhme, M.U.; Kohl, A.; Kröger, N.; Fehse, B. Digital PCR Panel for Sensitive Hematopoietic Chimerism Quantification after Allogeneic Stem Cell Transplantation. *Int. J. Mol. Sci.* **2016**, *17*, 1515. [CrossRef] [PubMed]
34. Ahci, M.; Stempelmann, K.; Buttkereit, U.; Crivello, P.; Trilling, M.; Heinold, A.; Steckel, N.K.; Koldehoff, M.; Horn, P.A.; Beelen, D.W.; et al. Clinical Utility of Quantitative PCR for Chimerism and Engraftment Monitoring after Allogeneic Stem Cell Transplantation for Hematologic Malignancies. *Biol. Blood Marrow Transplant.* **2017**, *23*, 1658–1668. [CrossRef] [PubMed]
35. Valero-Garcia, J.; González-Espinosa, M.d.C.; Barrios, M.; Carmona-Antoñanzas, G.; García-Planells, J.; Ruiz-Lafora, C.; Fuentes-Gálvez, A.; Jiménez-Velasco, A. Earlier relapse detection after allogeneic haematopoietic stem cell transplantation by chimerism assays: Digital PCR versus quantitative real-time PCR of insertion/deletion polymorphisms. *PLoS ONE* **2019**, *14*, e0212708.
36. Kliman, D.; Castellano-Gonzalez, G.; Withers, B.; Street, J.; Tegg, E.; Mirochnik, O.; Lai, J.; Clancy, L.; Gottlieb, D.; Blyth, E. Ultra-Sensitive Droplet Digital PCR for the Assessment of Microchimerism in Cellular Therapies. *Biol. Blood Marrow Transplant.* **2018**, *24*, 1069–1078. [CrossRef] [PubMed]
37. Tyler, J.; Kumer, L.; Fisher, C.; Casey, H.; Shike, H. Personalized Chimerism Test that Uses Selection of Short Tandem Repeat or Quantitative PCR Depending on Patient's Chimerism Status. *J. Mol. Diagn.* **2019**, *21*, 483–490. [CrossRef] [PubMed]

38. Gineikiene, E.; Stoskus, M.; Griskevicius, L. Single nucleotide polymorphism-based system improves the applicability of quantitative PCR for chimerism monitoring. *J. Mol. Diagn.* **2009**, *11*, 66–74. [CrossRef]
39. Qin, X.Y.; Li, G.X.; Qin, Y.Z.; Wang, Y.; Wang, F.R.; Liu, D.H.; Xu, L.P.; Chen, H.; Han, W.; Wang, J.Z.; et al. Quantitative assessment of hematopoietic chimerism by quantitative real-time polymerase chain reaction of sequence polymorphism systems after hematopoietic stem cell transplantation. *Chin. Med. J. (Engl.)* **2011**, *124*, 2301–2308.
40. Fehse, B.; Chukhlovin, A.; Kuhlcke, K.; Marinetz, O.; Vorwig, O.; Renges, H.; Kruger, W.; Zabelina, T.; Dudina, O.; Finckenstein, F.G.; et al. Real-time quantitative Y chromosome-specific PCR (QYCS-PCR) for monitoring hematopoietic chimerism after sex-mismatched allogeneic stem cell transplantation. *J. Hematotherapy Stem Cell Res.* **2001**, *10*, 419–425. [CrossRef]
41. Okano, T.; Tsujita, Y.; Kanegane, H.; Mitsui-Sekinaka, K.; Tanita, K.; Miyamoto, S.; Yeh, T.W.; Yamashita, M.; Terada, N.; Ogura, Y.; et al. Droplet Digital PCR-Based Chimerism Analysis for Primary Immunodeficiency Diseases. *J. Clin. Immunol.* **2018**, *38*, 300–306. [CrossRef] [PubMed]
42. Macher, H.C.; Suárez-Artacho, G.; Jiménez-Arriscado, P.; Álvarez-Gómez, S.; García-Fernández, N.; Guerrero, J.M.; Molinero, P.; Trujillo-Arribas, E.; Gómez-Bravo, M.A.; Rubio, A. Evaluation of the State of Transplanted Liver Health by Monitoring of Organ-Specific Genomic Marker in Circulating DNA from Receptor. *Adv. Exp. Med. Biol.* **2016**, *924*, 113–116. [PubMed]
43. Zou, J.; Duffy, B.; Slade, M.; Young, A.L.; Steward, N.; Hachem, R.; Mohanakumar, T. Rapid detection of donor cell free DNA in lung transplant recipients with rejections using donor–recipient HLA mismatch. *Hum. Immunol.* **2017**, *78*, 342–349. [CrossRef] [PubMed]
44. Bach, C.; Tomova, E.; Goldmann, K.; Weisbach, V.; Roesler, W.; Mackensen, A.; Winkler, J.; Spriewald, B.M. Monitoring of Hematopoietic Chimerism by Real-Time Quantitative PCR of Micro Insertions/Deletions in Samples with Low DNA Quantities. *Transfus. Med. Hemotherapy* **2015**, *42*, 38–45. [CrossRef] [PubMed]
45. Heid, C.A.; Stevens, J.; Livak, K.J.; Williams, P.M. Real time quantitative PCR. *Genome Res* **1996**, *6*, 986–994. [CrossRef] [PubMed]
46. George, D.; Czech, J.; John, B.; Yu, M.; Jennings, L.J. Detection and quantification of chimerism by droplet digital PCR. *Chimerism* **2013**, *4*, 102–108. [CrossRef]
47. Basu, A.S. Digital Assays Part I: Partitioning Statistics and Digital PCR. *SLAS Technol.* **2017**, *22*, 369–386. [CrossRef]
48. Pinheiro, L.B.; Coleman, V.A.; Hindson, C.M.; Herrmann, J.; Hindson, B.J.; Bhat, S.; Emslie, K.R. Evaluation of a Droplet Digital Polymerase Chain Reaction Format for DNA Copy Number Quantification. *Anal. Chem.* **2011**, *84*, 1003–1011. [CrossRef]
49. Mika, T.; Baraniskin, A.; Ladigan, S.; Wulf, G.; Dierks, S.; Haase, D.; Schork, K.; Turewicz, M.; Eisenacher, M.; Schmiegel, W.; et al. Digital droplet PCR-based chimerism analysis for monitoring of hematopoietic engraftment after allogeneic stem cell transplantation. *Int. J. Lab. Hematol.* **2019**. [CrossRef]
50. Goh, S.K.; Do, H.; Testro, A.; Pavlovic, J.; Vago, A.; Lokan, J.; Jones, R.M.; Christophi, C.; Dobrovic, A.; Muralidharan, V. The Measurement of Donor-Specific Cell-Free DNA Identifies Recipients With Biopsy-Proven Acute Rejection Requiring Treatment After Liver Transplantation. *Transplant. Direct* **2019**, *5*, e462. [CrossRef]
51. Whitlam, J.B.; Ling, L.; Skene, A.; Kanellis, J.; Ierino, F.L.; Slater, H.R.; Bruno, D.L.; Power, D.A. Diagnostic application of kidney allograft-derived absolute cell-free DNA levels during transplant dysfunction. *Am. J. Transplant.* **2018**, *19*, 1037–1049. [CrossRef] [PubMed]
52. Yang, J.; Verleden, S.; Zarinsefat, A.; Vanaudenaerde, B.; Vos, R.; Verleden, G.; Sarwal, R.; Sigdel, T.; Liberto, J.; Damm, I.; et al. Cell-Free DNA and CXCL10 Derived from Bronchoalveolar Lavage Predict Lung Transplant Survival. *J. Clin. Med.* **2019**, *8*, 241. [CrossRef] [PubMed]
53. Behjati, S.; Tarpey, P.S. What is next generation sequencing? *Arch. Dis. Child. Educ. Pract. Ed.* **2013**, *98*, 236–238. [CrossRef] [PubMed]
54. Aloisio, M.; Bortot, B.; Gandin, I.; Severini, G.M.; Athanasakis, E. A semi-nested real-time PCR method to detect low chimerism percentage in small quantity of hematopoietic stem cell transplant DNA samples. *Genome* **2017**, *60*, 183–192. [CrossRef] [PubMed]
55. Kim, J.; Hwang, I.S.; Shin, S.; Choi, J.R.; Lee, S.T. SNP-based next-generation sequencing reveals low-level mixed chimerism after allogeneic hematopoietic stem cell transplantation. *Ann. Hematol.* **2018**, *97*, 1731–1734. [CrossRef] [PubMed]

56. Bloom, R.D.; Bromberg, J.S.; Poggio, E.D.; Bunnapradist, S.; Langone, A.J.; Sood, P.; Matas, A.J.; Mehta, S.; Mannon, R.B.; Sharfuddin, A.; et al. Cell-Free DNA and Active Rejection in Kidney Allografts. *J. Am. Soc. Nephrol.* **2017**, *28*, 2221–2232. [CrossRef]
57. Huang, E.; Sethi, S.; Peng, A.; Najjar, R.; Mirocha, J.; Haas, M.; Vo, A.; Jordan, S.C. Early clinical experience using donor-derived cell-free DNA to detect rejection in kidney transplant recipients. *Am. J. Transplant.* **2019**, *19*, 1663–1670. [CrossRef]
58. Jordan, S.C.; Bunnapradist, S.; Bromberg, J.S.; Langone, A.J.; Hiller, D.; Yee, J.P.; Sninsky, J.J.; Woodward, R.N.; Matas, A.J. Donor-derived Cell-free DNA Identifies Antibody-mediated Rejection in Donor Specific Antibody Positive Kidney Transplant Recipients. *Transplant. Direct* **2018**, *4*, e379. [CrossRef]
59. Holdenrieder, S.; Burges, A.; Reich, O.; Spelsberg, F.W.; Stieber, P. DNA Integrity in Plasma and Serum of Patients with Malignant and Benign Diseases. *Ann. N. Y. Acad. Sci.* **2008**, *1137*, 162–170. [CrossRef]
60. Lam, N.Y.L. EDTA Is a Better Anticoagulant than Heparin or Citrate for Delayed Blood Processing for Plasma DNA Analysis. *Clin. Chem.* **2004**, *50*, 256–257. [CrossRef]
61. Wong, D.; Moturi, S.; Angkachatchai, V.; Mueller, R.; DeSantis, G.; Van den Boom, D.; Ehrich, M. Optimizing blood collection, transport and storage conditions for cell free DNA increases access to prenatal testing. *Clin. Biochem.* **2013**, *46*, 1099–1104. [CrossRef] [PubMed]
62. Norton, S.E.; Lechner, J.M.; Williams, T.; Fernando, M.R. A stabilizing reagent prevents cell-free DNA contamination by cellular DNA in plasma during blood sample storage and shipping as determined by digital PCR. *Clin. Biochem.* **2013**, *46*, 1561–1565. [CrossRef] [PubMed]
63. Hidestrand, M.; Stokowski, R.; Song, K.; Oliphant, A.; Deavers, J.; Goetsch, M.; Simpson, P.; Kuhlman, R.; Ames, M.; Mitchell, M.; et al. Influence of Temperature during Transportation on Cell-Free DNA Analysis. *Fetal Diagn. Ther.* **2012**, *31*, 122–128. [CrossRef] [PubMed]
64. Ward Gahlawat, A.; Lenhardt, J.; Witte, T.; Keitel, D.; Kaufhold, A.; Maass, K.K.; Pajtler, K.W.; Sohn, C.; Schott, S. Evaluation of Storage Tubes for Combined Analysis of Circulating Nucleic Acids in Liquid Biopsies. *Int. J. Mol. Sci.* **2019**, *20*, 704. [CrossRef] [PubMed]
65. Deans, Z.C.; Butler, R.; Cheetham, M.; Dequeker, E.M.C.; Fairley, J.A.; Fenizia, F.; Hall, J.A.; Keppens, C.; Normanno, N.; Schuuring, E.; et al. IQN path ASBL report from the first European cfDNA consensus meeting: Expert opinion on the minimal requirements for clinical ctDNA testing. *Virchows Archiv.* **2019**, *474*, 681–689. [CrossRef] [PubMed]
66. Swinkels, D.W. Effects of Blood-Processing Protocols on Cell-free DNA Quantification in Plasma. *Clin. Chem.* **2003**, *49*, 525–526. [CrossRef] [PubMed]
67. Kumar, M.; Choudhury, Y.; Ghosh, S.K.; Mondal, R. Application and optimization of minimally invasive cell-free DNA techniques in oncogenomics. *Tumor Biol.* **2018**, *40*, 101042831876034. [CrossRef] [PubMed]
68. El Messaoudi, S.; Rolet, F.; Mouliere, F.; Thierry, A.R. Circulating cell free DNA: Preanalytical considerations. *Clin. Chim. Acta* **2013**, *424*, 222–230. [CrossRef]
69. Mauger, F.; Dulary, C.; Daviaud, C.; Deleuze, J.-F.; Tost, J. Comprehensive evaluation of methods to isolate, quantify, and characterize circulating cell-free DNA from small volumes of plasma. *Anal. Bioanal. Chem.* **2015**, *407*, 6873–6878. [CrossRef]
70. Pérez-Barrios, C.; Nieto-Alcolado, I.; Torrente, M.; Jiménez-Sánchez, C.; Calvo, V.; Gutierrez-Sanz, L.; Palka, M.; Donoso-Navarro, E.; Provencio, M.; Romero, A. Comparison of methods for circulating cell-free DNA isolation using blood from cancer patients: Impact on biomarker testing. *Transl. Lung Cancer Res.* **2016**, *5*, 665–672. [CrossRef]
71. Markus, H.; Contente-Cuomo, T.; Farooq, M.; Liang, W.S.; Borad, M.J.; Sivakumar, S.; Gollins, S.; Tran, N.L.; Dhruv, H.D.; Berens, M.E.; et al. Evaluation of pre-analytical factors affecting plasma DNA analysis. *Sci. Rep.* **2018**, *8*, 7375. [CrossRef] [PubMed]
72. Eikmans, M.; Gielis, E.M.; Ledeganck, K.J.; Yang, J.; Abramowicz, D.; Claas, F.F.J. Non-invasive Biomarkers of Acute Rejection in Kidney Transplantation: Novel Targets and Strategies. *Front. Med.* **2019**, *5*, 358. [CrossRef] [PubMed]
73. Lefaucheur, C.; Viglietti, D.; Bentlejewski, C.; Duong van Huyen, J.-P.; Vernerey, D.; Aubert, O.; Verine, J.; Jouven, X.; Legendre, C.; Glotz, D.; et al. IgG Donor-Specific Anti-Human HLA Antibody Subclasses and Kidney Allograft Antibody-Mediated Injury. *J. Am. Soc. Nephrol.* **2015**, *27*, 293–304. [CrossRef] [PubMed]
74. Nasr, M.; Sigdel, T.; Sarwal, M. Advances in diagnostics for transplant rejection. *Expert Rev. Mol. Diagn.* **2016**, *16*, 1121–1132. [CrossRef] [PubMed]

75. Lo, Y.M.D.; Tein, M.S.C.; Pang, C.C.P.; Yeung, C.K.; Tong, K.-L.; Hjelm, N.M. Presence of donor-specific DNA in plasma of kidney and liver-transplant recipients. *Lancet* **1998**, *351*, 1329–1330. [CrossRef]
76. Garcia Moreira, V.; Prieto Garcia, B.; Baltar Martin, J.M.; Ortega Suarez, F.; Alvarez, F.V. Cell-Free DNA as a Noninvasive Acute Rejection Marker in Renal Transplantation. *Clin. Chem.* **2009**, *55*, 1958–1966. [CrossRef] [PubMed]
77. Lui, Y.Y.N. Origin of Plasma Cell-free DNA after Solid Organ Transplantation. *Clin. Chem.* **2003**, *49*, 495–496. [CrossRef]
78. Snyder, T.M.; Khush, K.K.; Valantine, H.A.; Quake, S.R. Universal noninvasive detection of solid organ transplant rejection. *Proc. Natl. Acad. Sci. USA* **2011**, *108*, 6229–6234. [CrossRef]
79. Zheng, Y.W.L.; Chan, K.C.A.; Sun, H.; Jiang, P.; Su, X.; Chen, E.Z.; Lun, F.M.F.; Hung, E.C.W.; Lee, V.; Wong, J.; et al. Nonhematopoietically Derived DNA Is Shorter than Hematopoietically Derived DNA in Plasma: A Transplantation Model. *Clin. Chem.* **2011**, *58*, 549–558. [CrossRef]
80. Burnham, P.; Khush, K.; De Vlaminck, I. Myriad Applications of Circulating Cell-Free DNA in Precision Organ Transplant Monitoring. *Ann. Am. Thorac. Soc.* **2017**, *14*, S237–S241. [CrossRef]
81. Choi, J.; Bano, A.; Azzi, J. Biomarkers in Solid Organ Transplantation. *Clin. Lab. Med.* **2019**, *39*, 73–85. [CrossRef] [PubMed]
82. Khush, K.K. Personalized treatment in heart transplantation. *Curr. Opin. Organ Transplant.* **2017**, *22*, 215–220. [CrossRef] [PubMed]
83. Menon, M.C.; Murphy, B.; Heeger, P.S. Moving Biomarkers toward Clinical Implementation in Kidney Transplantation. *J. Am. Soc. Nephrol.* **2017**, *28*, 735–747. [CrossRef] [PubMed]
84. Knight, S.R.; Thorne, A.; Lo Faro, M.L. Donor-specific Cell-free DNA as a Biomarker in Solid Organ Transplantation. A Systematic Review. *Transplantation* **2019**, *103*, 273–283. [CrossRef] [PubMed]
85. De Vlaminck, I.; Valantine, H.A.; Snyder, T.M.; Strehl, C.; Cohen, G.; Luikart, H.; Neff, N.F.; Okamoto, J.; Bernstein, D.; Weisshaar, D.; et al. Circulating cell-free DNA enables noninvasive diagnosis of heart transplant rejection. *Sci. Transl. Med.* **2014**, *6*, 241ra277. [CrossRef]
86. Ragalie, W.S.; Stamm, K.; Mahnke, D.; Liang, H.L.; Simpson, P.; Katz, R.; Tomita-Mitchell, A.; Kindel, S.J.; Zangwill, S.; Mitchell, M.E. Noninvasive Assay for Donor Fraction of Cell-Free DNA in Pediatric Heart Transplant Recipients. *J. Am. Coll. Cardiol.* **2018**, *71*, 2982–2983. [CrossRef] [PubMed]
87. Schütz, E.; Fischer, A.; Beck, J.; Harden, M.; Koch, M.; Wuensch, T.; Stockmann, M.; Nashan, B.; Kollmar, O.; Matthaei, J.; et al. Graft-derived cell-free DNA, a noninvasive early rejection and graft damage marker in liver transplantation: A prospective, observational, multicenter cohort study. *PLOS Med.* **2017**, *14*, e1002286. [CrossRef]
88. De Vlaminck, I.; Martin, L.; Kertesz, M.; Patel, K.; Kowarsky, M.; Strehl, C.; Cohen, G.; Luikart, H.; Neff, N.F.; Okamoto, J.; et al. Noninvasive monitoring of infection and rejection after lung transplantation. *Proc. Natl. Acad. Sci. USA* **2015**, *112*, 13336–13341. [CrossRef]
89. Agbor-Enoh, S.; Jackson, A.M.; Tunc, I.; Berry, G.J.; Cochrane, A.; Grimm, D.; Davis, A.; Shah, P.; Brown, A.W.; Wang, Y.; et al. Late manifestation of alloantibody-associated injury and clinical pulmonary antibody-mediated rejection: Evidence from cell-free DNA analysis. *J. Heart Lung Transplant.* **2018**, *37*, 925–932. [CrossRef]
90. Agbor-Enoh, S.; Wang, Y.; Tunc, I.; Jang, M.K.; Davis, A.; De Vlaminck, I.; Luikart, H.; Shah, P.D.; Timofte, I.; Brown, A.W.; et al. Donor-derived cell-free DNA predicts allograft failure and mortality after lung transplantation. *EBioMedicine* **2019**, *40*, 541–553. [CrossRef]
91. Kanzow, P.; Kollmar, O.; Schütz, E.; Oellerich, M.; Schmitz, J.; Beck, J.; Walson, P.D.; Slotta, J.E. Graft-Derived Cell-Free DNA as an Early Organ Integrity Biomarker After Transplantation of a Marginal HELLP Syndrome Donor Liver. *Transplantation* **2014**, *98*, e43–e45. [CrossRef] [PubMed]
92. Ng, H.I.; Zhu, X.; Xuan, L.; Long, Y.; Mao, Y.; Shi, Y.; Sun, L.; Liang, B.; Scaglia, F.; Choy, K.W.; et al. Analysis of fragment size distribution of cell-free DNA: A potential non-invasive marker to monitor graft damage in living-related liver transplantation for inborn errors of metabolism. *Mol. Genet. Metab.* **2019**, *127*, 45–50. [CrossRef] [PubMed]
93. Sigdel, T.K.; Vitalone, M.J.; Tran, T.Q.; Dai, H.; Hsieh, S.-c.; Salvatierra, O.; Sarwal, M.M. A Rapid Noninvasive Assay for the Detection of Renal Transplant Injury. *Transplant. J.* **2013**, *96*, 97–101. [CrossRef] [PubMed]
94. Cechova, H.; Leontovycova, M.; Pavlatova, L. Chimerism as an important marker in post-transplant monitoring chimerism monitoring. *HLA* **2018**, *92*, 60–63. [CrossRef] [PubMed]

95. Bader, P.; Niethammer, D.; Willasch, A.; Kreyenberg, H.; Klingebiel, T. How and when should we monitor chimerism after allogeneic stem cell transplantation? *Bone Marrow Transplant.* **2004**, *35*, 107–119. [CrossRef] [PubMed]
96. Horky, O.; Mayer, J.; Kablaskova, L.; Razga, F.; Krejci, M.; Kissova, J.; Borsky, M.; Jeziskova, I.; Dvorakova, D. Increasing hematopoietic microchimerism is a reliable indicator of incipient AML relapse. *Int. J. Lab. Hematol.* **2010**, *33*, 57–66. [CrossRef]
97. Huisman, C.; De Weger, R.A.; De Vries, L.; Tilanus, M.G.J.; Verdonck, L.F. Chimerism analysis within 6 months of allogeneic stem cell transplantation predicts relapse in acute myeloid leukemia. *Bone Marrow Transplant.* **2007**, *39*, 285–291. [CrossRef]
98. Qin, X.Y.; Li, G.X.; Qin, Y.Z.; Wang, Y.; Wang, F.R.; Liu, D.H.; Xu, L.P.; Chen, H.; Han, W.; Wang, J.Z.; et al. Quantitative chimerism: An independent acute leukemia prognosis indicator following allogeneic hematopoietic SCT. *Bone Marrow Transplant.* **2014**, *49*, 1269–1277. [CrossRef]
99. Schuurhuis, G.J.; Heuser, M.; Freeman, S.; Béné, M.-C.; Buccisano, F.; Cloos, J.; Grimwade, D.; Haferlach, T.; Hills, R.K.; Hourigan, C.S.; et al. Minimal/measurable residual disease in AML: A consensus document from the European LeukemiaNet MRD Working Party. *Blood* **2018**, *131*, 1275–1291. [CrossRef]
100. Tsirigotis, P.; Byrne, M.; Schmid, C.; Baron, F.; Ciceri, F.; Esteve, J.; Gorin, N.C.; Giebel, S.; Mohty, M.; Savani, B.N.; et al. Relapse of AML after hematopoietic stem cell transplantation: Methods of monitoring and preventive strategies. A review from the ALWP of the EBMT. *Bone Marrow Transplant.* **2016**, *51*, 1431–1438. [CrossRef]
101. Araki, D.; Wood, B.L.; Othus, M.; Radich, J.P.; Halpern, A.B.; Zhou, Y.; Mielcarek, M.; Estey, E.H.; Appelbaum, F.R.; Walter, R.B. Allogeneic Hematopoietic Cell Transplantation for AML: Is it Time to Move Toward a Minimal Residual Disease-Based Definition of Complete Remission? *Clin. Lymphoma Myeloma Leuk.* **2015**, *15*, S8. [CrossRef]
102. Antin, J.H.; Childs, R.; Filipovich, A.H.; Giralt, S.; Mackinnon, S.; Spitzer, T.; Weisdorf, D. Establishment of complete and mixed donor chimerism after allogeneic lymphohematopoietic transplantation: Recommendations from a workshop at the 2001 Tandem Meetings of the International Bone Marrow Transplant Registry and the American Society of Blood and Marrow Transplantation. *Biol. Blood Marrow Transplant.* **2001**, *7*, 473–485. [PubMed]
103. Sellmann, L.; Rabe, K.; Bünting, I.; Dammann, E.; Göhring, G.; Ganser, A.; Stadler, M.; Weissinger, E.M.; Hambach, L. Diagnostic value of highly-sensitive chimerism analysis after allogeneic stem cell transplantation. *Bone Marrow Transplant.* **2018**, *53*, 1457–1465. [CrossRef] [PubMed]
104. Kim, S.Y.; Jeong, M.H.; Park, N.; Ra, E.; Park, H.; Seo, S.H.; Kim, J.Y.; Seong, M.W.; Park, S.S. Chimerism monitoring after allogeneic hematopoietic stem cell transplantation using quantitative real-time PCR of biallelic insertion/deletion polymorphisms. *J. Mol. Diagn.* **2014**, *16*, 679–688. [CrossRef] [PubMed]
105. Koldehoff, M.; Steckel, N.K.; Hlinka, M.; Beelen, D.W.; Elmaagacli, A.H. Quantitative analysis of chimerism after allogeneic stem cell transplantation by real-time polymerase chain reaction with single nucleotide polymorphisms, standard tandem repeats, and Y-chromosome-specific sequences. *Am. J. Hematol.* **2006**, *81*, 735–746. [CrossRef]
106. Vicente, D.C.; Laranjeira, A.B.; Miranda, E.C.; Yunes, J.A.; De Souza, C.A. Chimerism interpretation with a highly sensitive quantitative PCR method: 6 months median latency before chimerism drop below 0.1. *Bone Marrow Transplant.* **2016**, *51*, 874–875. [CrossRef] [PubMed]
107. Wiedemann, B.; Klyuchnikov, E.; Kroger, N.; Zabelina, T.; Stahl, T.; Zeschke, S.; Badbaran, A.; Ayuk, F.; Alchalby, H.; Wolschke, C.; et al. Chimerism studies with quantitative real-time PCR in stem cell recipients with acute myeloid leukemia. *Exp. Hematol.* **2010**, *38*, 1261–1271. [CrossRef]
108. Waterhouse, M.; Pfeifer, D.; Follo, M.; Duyster, J.; Schäfer, H.; Bertz, H.; Finke, J. Early mixed hematopoietic chimerism detection by digital droplet PCR in patients undergoing gender-mismatched hematopoietic stem cell transplantation. *Clin. Chem. Lab. Med.* **2017**, *55*, 1115–1121. [CrossRef]
109. David, K.A.; Cooper, D.; Strair, R. Clinical Studies in Hematologic Microtransplantation. *Curr. Hematol. Malig. Rep.* **2017**, *12*, 51–60. [CrossRef]
110. Guo, M.; Chao, N.J.; Li, J.-Y.; Rizzieri, D.A.; Sun, Q.-Y.; Mohrbacher, A.; Krakow, E.F.; Sun, W.-J.; Shen, X.-L.; Zhan, X.-R.; et al. HLA-Mismatched Microtransplant in Older Patients Newly Diagnosed With Acute Myeloid Leukemia. *JAMA. Oncol.* **2018**, *4*, 54. [CrossRef]

111. Liu, L.; Zhang, X.; Qiu, H.; Tang, X.; Han, Y.; Fu, C.; Jin, Z.; Zhu, M.; Miao, M.; Wu, D. HLA-mismatched stem cell microtransplantation compared to matched-sibling donor transplantation for intermediate/high-risk acute myeloid leukemia. *Ann. Hematol.* **2019**, *98*, 1249–1257. [CrossRef] [PubMed]
112. Punwani, N.; Merin, N.; Mohrbacher, A.; Yaghmour, G.; Sano, A.; Ramezani, L.; Chaudhary, P.M.; Ramsingh, G. Unrelated HLA mismatched microtransplantation in a patient with refractory secondary acute myeloid leukemia. *Leuk. Res. Rep.* **2018**, *9*, 18–20. [CrossRef] [PubMed]
113. Rubnitz, J.E.; Inaba, H.; Kang, G.; Gan, K.; Hartford, C.; Triplett, B.M.; Dallas, M.; Shook, D.; Gruber, T.; Pui, C.-H.; et al. Natural killer cell therapy in children with relapsed leukemia. *Pediatric Blood Cancer* **2015**, *62*, 1468–1472. [CrossRef] [PubMed]
114. Kaeuferle, T.; Krauss, R.; Blaeschke, F.; Willier, S.; Feuchtinger, T. Strategies of adoptive T -cell transfer to treat refractory viral infections post allogeneic stem cell transplantation. *J. Hematol. Oncol.* **2019**, *12*, 13. [CrossRef]
115. Aljurf, M.; Abalkhail, H.; Alseraihy, A.; Mohamed, S.Y.; Ayas, M.; Alsharif, F.; Alzahrani, H.; Al-Jefri, A.; Aldawsari, G.; Al-Ahmari, A.; et al. Chimerism Analysis of Cell-Free DNA in Patients Treated with Hematopoietic Stem Cell Transplantation May Predict Early Relapse in Patients with Hematologic Malignancies. *Biotechnol. Res. Int.* **2016**, *2016*, 1–6. [CrossRef] [PubMed]
116. Duque-Afonso, J.; Waterhouse, M.; Pfeifer, D.; Follo, M.; Duyster, J.; Bertz, H.; Finke, J. Cell-free DNA characteristics and chimerism analysis in patients after allogeneic cell transplantation. *Clin. Biochem.* **2018**, *52*, 137–141. [CrossRef] [PubMed]
117. Rozmus, J.; Ivison, S.; Kariminia, A.; Leung, V.M.; Sung, S.; Subrt, P.; Lee, S.J.; Boilard, E.; Walker, I.; Foley, R.; et al. Higher levels of free plasma mitochondrial DNA are associated with the onset of chronic GvHD. *Bone Marrow Transplant.* **2018**, *53*, 1263–1269. [CrossRef] [PubMed]

4

Identification of Structural Variation from NGS-Based Non-Invasive Prenatal Testing

Ondrej Pös [1,2,*], Jaroslav Budis [2,3,4], Zuzana Kubiritova [1,5], Marcel Kucharik [2], Frantisek Duris [2,4], Jan Radvanszky [1,5] and Tomas Szemes [1,2,3]

1 Faculty of Natural Sciences, Comenius University, 841 04 Bratislava, Slovakia
2 Geneton Ltd., 841 04 Bratislava, Slovakia
3 Comenius University Science Park, 841 04 Bratislava, Slovakia
4 Slovak Center of Scientific and Technical Information, 811 04 Bratislava, Slovakia
5 Institute for Clinical and Translational Research, Biomedical Research Center, Slovak Academy of Sciences, 845 05 Bratislava, Slovakia
* Correspondence: pos1@uniba.sk

Abstract: Copy number variants (CNVs) are an important type of human genome variation, which play a significant role in evolution contribute to population diversity and human genetic diseases. In recent years, next generation sequencing has become a valuable tool for clinical diagnostics and to provide sensitive and accurate approaches for detecting CNVs. In our previous work, we described a non-invasive prenatal test (NIPT) based on low-coverage massively parallel whole-genome sequencing of total plasma DNA for detection of CNV aberrations ≥600 kbp. We reanalyzed NIPT genomic data from 5018 patients to evaluate CNV aberrations in the Slovak population. Our analysis of autosomal chromosomes identified 225 maternal CNVs (47 deletions; 178 duplications) ranging from 600 to 7820 kbp. According to the ClinVar database, 137 CNVs (60.89%) were fully overlapping with previously annotated variants, 66 CNVs (29.33%) were in partial overlap, and 22 CNVs (9.78%) did not overlap with any previously described variant. Identified variants were further classified with the AnnotSV method. In summary, we identified 129 likely benign variants, 13 variants of uncertain significance, and 83 likely pathogenic variants. In this study, we use NIPT as a valuable source of population specific data. Our results suggest the utility of genomic data from commercial CNV analysis test as background for a population study.

Keywords: copy number variants; next generation sequencing; non-invasive prenatal testing; population study

---

1. Introduction

Copy number variation (CNV) is a segment of DNA with length ≥1 kbp which is presented at a variable copy number in comparison to the reference genome. CNVs include insertions, deletions and duplications, which result in copy number gain or copy number loss [1]. It was shown that CNVs are important cause of structural variations in the human genome [2]. Research of the past decades revealed that these variations are functionally and evolutionary significant and contribute to the population diversity and human genetic diseases [3,4].

Various methods for CNV detection have been developed, from the conventional cytogenetic analysis (e.g., G-banded karyotype) through microarray-based methods (e.g., comparative genomic hybridization) to next-generation sequencing (NGS) [5]. Genomic microarrays provide a genome-wide coverage at a much higher resolution than a conventional cytogenetic analysis. This is the reason why microarray-based methods have been standard for CNV detection [6,7]. However, this method has limited resolution, accuracy, and several other limitations are noted in the literature [8]. In recent years, NGS has become a valuable tool for clinical diagnostics and to provide sensitive and accurate approaches for detecting genomic variations, e.g., CNVs. With the reducing cost of this method, numbers of NGS based CNV detection tests is increasing [9,10].

In our previous study, we described non-invasive prenatal test (NIPT) based on analysis of plasma DNA from pregnant women [11–13]. This test uses low-coverage massively parallel sequencing of whole-genome for detection of CNV aberrations [14]. With the informed consent of these patients we generated an amount of credible genomic data from thousands of pregnant women. Since these patients represent a relatively standard sample of local female population, we hypothesized this data could be used not only for primary purpose as prenatal screening but also as a valuable source of data for population study. The objective of the present study is based on our previous work which suggests the use of NIPT as a valuable source of population specific allelic frequencies [15].

2. Results

We obtained CNV profile for 22 autosomes from 5018 pregnant women (Figure 1). Together, we identified 225 CNVs ranging from 600 kbp to 7820 kbp with median size 820 kbp (Table S1). These variants include 178 duplications (79.11%) and 47 deletions (20.89%) with median size 830 kbp for duplications and 800 kbp for deletions. As can be seen, the majority of identified CNVs were approximately 600–700 kbp long (Figure 2a). Most variants (28) were found on the chromosome 2, while on the chromosome 15 we detected only one variant. We did not identify any deletions on chromosomes 11, 15, 20, and 22 (Figure 2b). The identified CNVs came from 212 individuals, corresponding to frequency 4.2% of CNV ≥ 600 kbp in our cohort. The vast majority of individuals (95.28%) displayed a single CNV; only 4.72% exhibited more than one variant. The most frequently detected variant was the CNV duplication in chromosome location 2p22 with a total of 11 detection events; however, the frequency of every CNV was calculated as less than 1%, thus all variants were considered to be rare. The largest CNV was duplication spanning 7820 kbp in chromosome location 10q21.1.

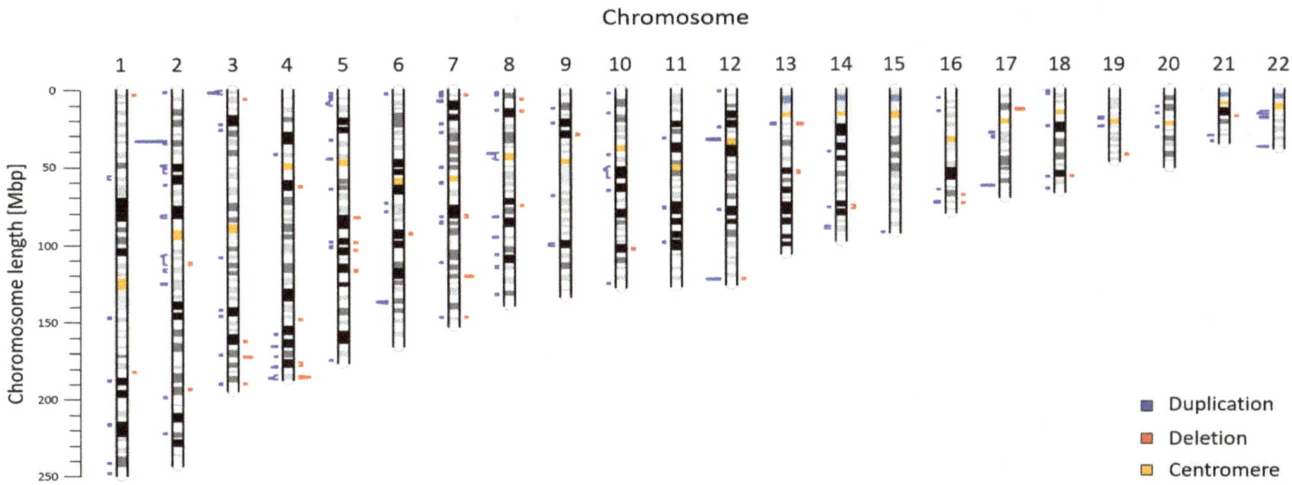

Figure 1. Chromosomal location of maternal CNVs identified by NIPT. The length of blue (duplication) and red (deletion) bars corresponds to the frequency of CNV ranging from minimum of 1 to maximum of 11 detections.

Variants were compared with ClinVar database records and following results were obtained. Together, 137 CNVs (60.89%) were overlapping with previously described variants in full extent, 66 CNVs (29.33%) were partially overlapping and 22 CNVs (9.78%) did not overlap with any previously described variant according to ClinVar. Some of our CNVs overlap with variants previously observed among patients with pathogenic phenotypes, e.g., developmental delay, intellectual disability, etc. (Table 1).

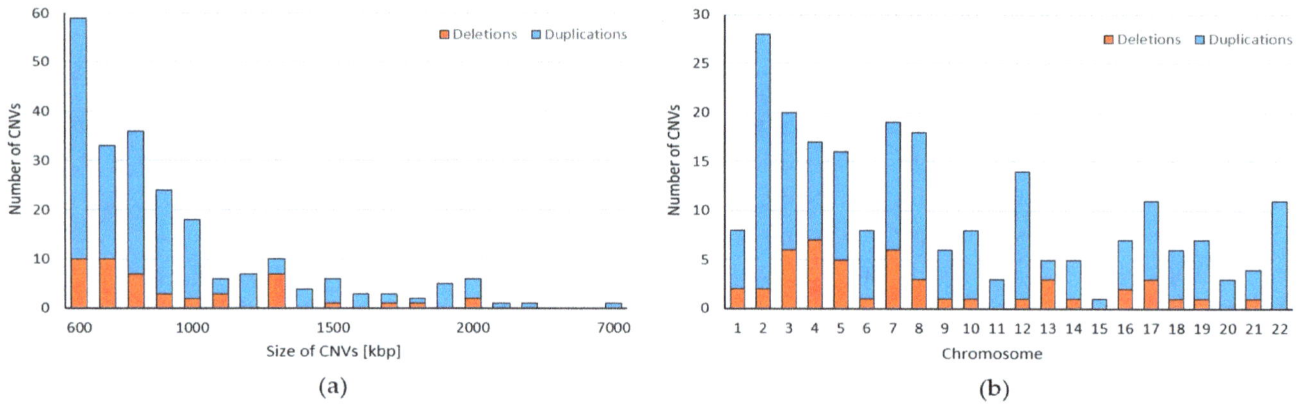

Figure 2. Characteristics of maternal CNVs identified from the NIPT. (**a**) Size distribution of detected CNVs ranging from ≥600 kbp to ≥7000 kbp. (**b**) Genomic distribution of CNV deletions (red) and duplications (blue) ≥600 kbp in Slovak population within the chromosomes.

Table 1. Variants overlapping with CNVs that were previously observed among patients with pathogenic phenotypes. Data acquired from the ClinVar database.

Variant Type	Location	Identifier	Phenotype	Events	Reference
Duplication	1q21.1-21.2	dbVar: nsv531885	Developmental delay AND/OR other significant developmental or morphological phenotypes, Global developmental delay	1	[16,17]
Duplication	2q33.1	OMIM: 609728.0002	Autosomal Recessive Spastic Ataxia with Leukoencephalopathy	1	[18]
Duplication	7q11.23	dbVar: nsv532240	Encephalopathy, Global developmental delay, Muscular hypotonia	1	[17]
Deletion	13q12.12	dbVar: nsv491643	Developmental delay AND/OR other significant developmental or morphological phenotypes, Seizures, Intellectual disability, Intrauterine growth retardation	2	[16]
Duplication	17q12	dbVar: nsv2775541	Developmental delay AND/OR other significant developmental or morphological phenotypes, Behavioral abnormality	1	[16,17]
Duplication	22q11.21	dbVar: nssv577068 nsv530653	Global developmental delay	3	[16,17]
Duplication	22q11.21	dbVar: nssv578923 nsv531796	Developmental delay AND/OR other significant developmental or morphological phenotypes	1	[16,17]
Duplication	22q11.23	dbVar: nssv13653977 nsv2769497	Short stature, Macrocephalus, Abnormality of the face, Intellectual disability	2	[16]

The identified variants were classified based on criteria in AnnotSV database [19]. In summary, we identified 129 likely benign variants, 13 variants of uncertain significance and 83 likely pathogenic variants. According to AnnotSV, 207 CNVs overlap with known genes and only 18 CNVs were localized in non-coding areas. Regarding the type of CNV, we identified approximately 3.8 times more CNV gains than CNV losses. These variants were more frequently present in non-coding regions; however, duplications overlap coding regions nearly 6.4 times more frequently than the deletions (Table 2).

Table 2. Data shows number of identified CNVs sorted by the type of variant and number of Mega base pairs (Mbp) attributed to specific genomic location.

Type of Variant	Number of CNVs	Total Sequence (Mbp)	Coding Regions (Mbp)	Non-Coding Regions (Mbp)
CNV gain	178	191.54	3.27	188.27
CNV loss	47	46.98	0.44	46.54
Sum	225	238.52	3.71	234.81

3. Discussion

Knowledge of population genetic studies, e.g., Human Genome Project, has changed genomics and had tremendous impact on current medicine [20,21]. Detection of CNVs within and between populations is important to understand the plasticity of our genome and to elucidate its possible contribution to disease management [22]. Based on these statements, we are suggesting the additional utility of genomic data generated through routine NIPT screening based on low-coverage massively parallel whole-genome sequencing of total plasma DNA from pregnant women. This test provides a lot of credible genomic data that can be used as background for population studies. Our results show that 4.2% of individuals carry CNV ≥ 600 kbp, suggesting a relatively high frequency of large CNVs in the Slovak population. These findings are consistent with results from Cooper et al., which presented one of the largest studies investigating the role of rare CNVs in intellectual disability and developmental delay, analyzing data from 15,767 affected individuals and 8329 controls. They showed that 25.7% of affected individuals and 11.5% of controls harbor CNVs > 400 kbp [23].

Overall, there were approximately four times higher frequency of duplications compared to deletions (Table 2). The underrepresentation of deletions is consistent with previous reports, where large deletions were less common than large duplications when considering CNVs > 500 kbp [24,25]. These results are concordant with the hypothesis that CNV losses are more deleterious [26]. All variants together span 238.52 Mbp; however, only 3.71 Mbp (1.56%) were identified in coding regions. These 3.71 Mbp were spread through 207 CNVs (92%) overlapping with coding sequences. Since the gene density is calculated at 5–23 genes per Mbp [27], there is a low probability that a CNV ≥ 600 kbp will occur exactly in the non-coding region. Therefore, we expected most CNVs of such length to be at least partially overlapping the coding regions. We have shown that duplications affect coding regions approximately two times more frequently than deletions (1.71% vs 0.93% for duplications and deletions, respectively). Sudmant et al. also found that duplications and deletions exhibit fundamentally different population-genetic properties. Duplications are subjected to weaker selective constraint, hence affect genes four times more likely than deletions, indicating that they provide a larger target for adaptive selection [3].

Clinically relevant CNVs can be found in databases such as ClinVar, DECIPHER, ECARUCA and the International Standards for Cytogenomic Arrays Database. When we compared our results with ClinVar database, we found at least 22 variants (17 CNV gains; 5 CNV losses) in regions without any previous record (Figure S1). For example, we have identified a CNV loss on the chromosome location 3q26.3 that is present consistently in three of our samples, but it was not previously described in the database. This deletion encompassing approximately one half of sequence from the 3′ end of a gene N-acetylated alpha-linked acidic dipeptidase-like 2 (*NAALADL2*). It has been shown that deletions involving *NAALADL2* are found in the general population [28]. On a closer view, we found that our largest duplication in chromosome 10q21.1 overlaps the complete sequence of gene Protocadherin Related 15 (*PCDH15*). Duplications in this gene have been shown to be associated with Usher syndrome type 1 (OMIM: # 601067), which is characterized by deafness, vestibular areflexia, and prepubertal onset of retinitis pigmentosa [29,30]. Although the NIPT enables the detection of maternal CNVs, current analyses do not interpret these findings. Maternal aberrations can be clinically actionable or potentially harmful for the fetus. Brison et al. suggest reporting these variants if clinically relevant because it can improve pregnancy management and promote the health of the fetus or the mother or both [31]. On the other hand, the identification and reporting of such CNVs represent a big challenge for genetic counselors; thus, further guidelines to improve patient counseling are needed [32]. It is also known that performing NIPT may incidentally lead to the diagnosis of maternal malignancy. Giles et al. showed, 80% of genetic counselors recognized it would be beneficial in the future to use NIPT for neoplasm screening, however, more than 90% affirmed that guidelines are necessary to better prepare for these cases [33].

Performing large numbers of parental samples is expensive, but the need for parental testing will diminish by accumulating data about benign CNVs [16]. Recently, an updated, higher resolution map of CNVs that are not associated with adverse phenotypes, based on 55 studies, was developed. Zarrei et al. estimated that up to 9.5% of the genome contributes to CNV. Additionally, they have found approximately 100 genes that can be homozygously deleted without producing apparent phenotypic consequences. This map is a great contribution to the interpretation of new CNV findings, for clinical and research applications [34]. As clinical laboratories adopt CNV analysis, these resources will become invaluable for the clinician to discriminate pathogenic from non-disease associated CNVs [8]. However, there is still a need for appropriate recommendations or guidelines related to evaluation of CNV findings and for their classifications. The main limitation of our study remains the size of detected CNVs; however, with improving laboratory and computational methods, as well as lowering the cost of sequencing, this limit should decrease. Currently, our method was validated to CNVs with minimal length 600 kbp, while the vast majority of CNVs are smaller than 500 kbp [35]. On the other hand, CNVs larger than 500 kbp are strongly associated with morbid consequences such as developmental disorders and cancer [22]. Despite mentioned limitation, we showed, NIPT may be utilized for the identification of common structural variations in population, and it could contribute to the interpretation of CNV findings in clinical research.

4. Materials and Methods

In our previous work we described non-invasive prenatal test (NIPT) based on low-coverage (0.3×) massively parallel whole-genome sequencing of total plasma DNA for detection of CNV aberrations longer than 600 kbp [14]. This test generates amount of credible genomic data, from thousands of pregnant women which represent a relatively standard sample of local population. We reanalyzed NIPT genomic data from 5018 patients to calculate frequencies of CNV aberrations in the Slovak population. All subjects gave their informed consent for inclusion before they participated in the study. Informed consent includes permission to process the sample for further analysis maintaining the anonymity but does not include a statement for contacting the patient again in case of a clinically significant maternal finding. Therefore, it was possible to use samples processed in the past, but due to anonymization we were not able to contact the patients and associate the finding with the phenotype. The study has been approved by the Ethical Committee of the Bratislava Self-Governing Region (Sabinovska ul.16, 820 05 Bratislava) on 30 April 2015 under the decision ID 03899_2015.

4.1. Sample Preparation and Sequencing

Blood from pregnant women was collected into EDTA tubes and kept at 4 °C temperature until plasma separation. Blood plasma was separated within 36 h after collection and stored at −20 °C until DNA isolation. DNA was isolated using Qiagen DNA Blood Mini kit (QIAGEN, Hilden, Germany). Standard fragment libraries for massively parallel sequencing were prepared from isolated DNA using an Illumina TruSeq Nano kit (Illumina, San Diego, CA, USA) and a modified protocol described previously [11]. Briefly, to decrease laboratory costs, we used reduced volumes of reagents, which was compensated by nine cycles of PCR instead of eight as per protocol. Physical size selection of cfDNA fragments was performed using specific volumes of magnetic beads in order to enrich fetal fraction. Illumina NextSeq 500/550 High Output Kit v2 (75 cycles) (Illumina, San Diego, CA, USA) was used for massively parallel sequencing of prepared libraries using pair-end sequencing with read length of 2 × 35 bp on an Illumina NextSeq 500 platform.

4.2. Mapping and Read Count Correction

Sequencing reads were aligned to the human reference genome (hg19) using Bowtie 2 algorithm [36]. NextSeq-produced fastq files (two per sample; R1 and R2) were directly mapped using the Bowtie 2 algorithm with very-sensitive option. Next, for each sample, the unique reads were processed to eliminate the GC bias according to [37] with the exclusion of intrarun normalization. Briefly, for each sample the number of unique reads from each 20 kbp bin on each chromosome was counted. With empty bins filtered out, the locally weighted scatterplot smoothing (LOESS) regression was used to predict the expected read count for each bin based on its GC content. The LOESS-corrected read count for a particular bin was then calculated as RC= RC − ||RC− RC||, where RC is the global average of read counts through all bins; RC is the fitted read count of that bin, and RC is its observed read count. PCA normalization has been further carried out to remove higher-order population artifacts on autosomal chromosomes [38,39]. At first, bin counts are transformed into a principal space. The first component represents the highest variability across individuals in the control set. To normalize the sample, bin counts corresponding to predefined number of top components are removed to reduce common noise in euploid samples. Bins without sufficient coverage that correspond to the low complexity genomic regions were excluded from the analysis.

4.3. Segment Identification and CNV Calling

Normalized bin counts were analyzed by circular binary segmentation (CBS) algorithm provided by the R package DNAcopy (Seshan VE, Olshen A. DNAcopy: DNA copy number data analysis. R package version 1.48.0. 2016.) to identify same-coverage segments. CBS partitions a chromosome into regions with equal copy numbers. Segments longer than 600 kbp with abnormal copy number (at least 60% gain or loss of a single chromosomal segment) were marked as maternal and annotated using AnnotSV tool [40] and ClinVar database [41].

4.4. Data Processing

All computational steps were executed using Snakemake workflow engine [42]. Evaluation of maternal calls and generation of plots were performed using in-house Python scripts.

5. Conclusions

CNVs represent an important source of variations in the human genome. They are functionally and evolutionary significant and contribute to the population diversity and human genetic diseases. As NGS has become a valuable tool in research and in clinical settings, the number of NGS based tests has increased. Among them, CNV detection tests are also increasing. In this study, we confirmed our hypothesis and demonstrated that NIPT can be used also for the identification of common structural variations in population.

Author Contributions: Conceptualization, F.D., J.R., and T.S.; methodology, M.K. and J.B.; software, M.K.; Validation, O.P., Z.K., and J.B.; formal analysis, O.P., Z.K., and J.B.; investigation, O.P. and Z.K.; resources, T.S.; data curation, J.B. and M.K.; writing—original draft preparation, O.P.; writing—review and editing, O.P. and Z.K.; visualization, J.B.; supervision, J.B., J.R., and T.S.; project administration, T.S.; funding acquisition, T.S.

References

1. Feuk, L.; Carson, A.R.; Scherer, S.W. Structural variation in the human genome. *Nat. Rev. Genet.* **2006**, *7*, 85–97. [CrossRef] [PubMed]
2. Escaramís, G.; Docampo, E.; Rabionet, R. A decade of structural variants: Description, history and methods to detect structural variation. *Brief. Funct. Genom.* **2015**, *14*, 305–314. [CrossRef] [PubMed]
3. Sudmant, P.H.; Mallick, S.; Nelson, B.J.; Hormozdiari, F.; Krumm, N.; Huddleston, J.; Coe, B.P.; Baker, C.; Nordenfelt, S.; Bamshad, M.; et al. Global diversity, population stratification, and selection of human copy-number variation. *Science* **2015**, *349*, aab3761. [CrossRef] [PubMed]
4. Mikhail, F.M. Copy number variations and human genetic disease. *Curr. Opin. Pediatr.* **2014**, *26*, 646–652. [CrossRef] [PubMed]
5. Martin, C.L.; Kirkpatrick, B.E.; Ledbetter, D.H. Copy number variants, aneuploidies, and human disease. *Clin. Perinatol.* **2015**, *42*, 227–242. [CrossRef] [PubMed]
6. Kearney, H.M.; Thorland, E.C.; Brown, K.K.; Quintero-Rivera, F.; South, S.T. Working group of the American College of Medical Genetics Laboratory quality assurance committee American College of Medical Genetics standards and guidelines for interpretation and reporting of postnatal constitutional copy number variants. *Genet. Med.* **2011**, *13*, 680–685. [CrossRef] [PubMed]

7. Zhao, M.; Wang, Q.; Wang, Q.; Jia, P.; Zhao, Z. Computational tools for copy number variation (CNV) detection using next-generation sequencing data: Features and perspectives. *BMC Bioinf.* **2013**, *14* (Suppl. 11), S1. [CrossRef] [PubMed]
8. Coughlin, C.R.; Scharer, G.H.; Shaikh, T.H. Clinical impact of copy number variation analysis using high-resolution microarray technologies: Advantages, limitations and concerns. *Genome Med.* **2012**, *4*, 80. [CrossRef]
9. Russo, C.D.; Di Giacomo, G.; Cignini, P.; Padula, F.; Mangiafico, L.; Mesoraca, A.; D'Emidio, L.; McCluskey, M.R.; Paganelli, A.; Giorlandino, C. Comparative study of aCGH and next generation sequencing (NGS) for chromosomal microdeletion and microduplication screening. *J. Prenat. Med.* **2014**, *8*, 57–69.
10. Wang, H.; Nettleton, D.; Ying, K. Copy number variation detection using next generation sequencing read counts. *BMC Bioinf.* **2014**, *15*, 109. [CrossRef]
11. Minarik, G.; Repiska, G.; Hyblova, M.; Nagyova, E.; Soltys, K.; Budis, J.; Duris, F.; Sysak, R.; Gerykova Bujalkova, M.; Vlkova-Izrael, B.; et al. Utilization of benchtop next generation sequencing platforms ion torrent PGM and MiSeq in noninvasive prenatal testing for chromosome 21 trisomy and testing of impact of in silico and physical size selection on its analytical performance. *PLoS ONE* **2015**, *10*, e0144811. [CrossRef] [PubMed]
12. Budis, J.; Gazdarica, J.; Radvanszky, J.; Szucs, G.; Kucharik, M.; Strieskova, L.; Gazdaricova, I.; Harsanyova, M.; Duris, F.; Minarik, G.; et al. Combining count- and length-based z-scores leads to improved predictions in non-invasive prenatal testing. *Bioinformatics* **2018**, *35*, 1284–1291. [CrossRef] [PubMed]
13. Pös, O.; Budiš, J.; Szemes, T. Recent trends in prenatal genetic screening and testing. *F1000 Res.* **2019**, *8*. [CrossRef] [PubMed]
14. Kucharik, M.; Gnip, A.; Hyblova, M.; Budis, J.; Strieskova, L.; Harsanyova, M.; Duris, F.; Radvanszky, J.; Minarik, G.; Szemes, T. Non-invasive prenatal testing by low coverage genomic sequencing: Detection limits of screened chromosomal microdeletions. *BioRxiv* **2019**, 686345. [CrossRef]
15. Jaroslav, B.; Juraj, G.; Jan, R.; Maria, H.; Iveta, G.; Lucia, S.; Richard, F.; Frantisek, D.; Gabriel, M.; Martina, S.; et al. Non-invasive prenatal testing as a valuable source of population specific allelic frequencies. *J. Biotechnol.* **2019**, *299*, 72–78.
16. Miller, D.T.; Adam, M.P.; Aradhya, S.; Biesecker, L.G.; Brothman, A.R.; Carter, N.P.; Church, D.M.; Crolla, J.A.; Eichler, E.E.; Epstein, C.J.; et al. Consensus statement: Chromosomal microarray is a first-tier clinical diagnostic test for individuals with developmental disabilities or congenital anomalies. *Am. J. Hum. Genet.* **2010**, *86*, 749–764. [CrossRef] [PubMed]
17. Kaminsky, E.B.; Kaul, V.; Paschall, J.; Church, D.M.; Bunke, B.; Kunig, D.; Moreno-De-Luca, D.; Moreno-De-Luca, A.; Mulle, J.G.; Warren, S.T.; et al. An evidence-based approach to establish the functional and clinical significance of copy number variants in intellectual and developmental disabilities. *Genet. Med.* **2011**, *13*, 777–784. [CrossRef]
18. Bayat, V.; Thiffault, I.; Jaiswal, M.; Tétreault, M.; Donti, T.; Sasarman, F.; Bernard, G.; Demers-Lamarche, J.; Dicaire, M.-J.; Mathieu, J.; et al. Mutations in the mitochondrial methionyl-tRNA synthetase cause a neurodegenerative phenotype in flies and a recessive ataxia (ARSAL) in humans. *PLoS Biol.* **2012**, *10*, e1001288. [CrossRef]
19. Geoffroy, V. AnnotSV. Available online: https://lbgi.fr/AnnotSV/ (accessed on 7 February 2019).
20. Carrasco-Ramiro, F.; Peiró-Pastor, R.; Aguado, B. Human genomics projects and precision medicine. *Gene Ther.* **2017**, *24*, 551–561. [CrossRef]
21. Beyene, J.; Pare, G. Statistical genetics with application to population-based study design: A primer for clinicians. *Eur. Heart, J.* **2014**, *35*, 495–500. [CrossRef]
22. Valsesia, A.; Macé, A.; Jacquemont, S.; Beckmann, J.S.; Kutalik, Z. The growing importance of CNVs: New insights for detection and clinical interpretation. *Front. Genet.* **2013**, *4*, 92. [CrossRef] [PubMed]
23. Cooper, G.M.; Coe, B.P.; Girirajan, S.; Rosenfeld, J.A.; Vu, T.H.; Baker, C.; Williams, C.; Stalker, H.; Hamid, R.; Hannig, V.; et al. A copy number variation morbidity map of developmental delay. *Nat. Genet.* **2011**, *43*, 838–846. [CrossRef] [PubMed]
24. Pietiläinen, O.P.H.; Rehnström, K.; Jakkula, E.; Service, S.K.; Congdon, E.; Tilgmann, C.; Hartikainen, A.-L.; Taanila, A.; Heikura, U.; Paunio, T.; et al. Phenotype mining in CNV carriers from a population cohort. *Hum. Mol. Genet.* **2011**, *20*, 2686–2695. [CrossRef] [PubMed]
25. Guyatt, A.L.; Stergiakouli, E.; Martin, J.; Walters, J.; O'Donovan, M.; Owen, M.; Thapar, A.; Kirov, G.; Rodriguez, S.; Rai, D.; et al. Association of copy number variation across the genome with neuropsychiatric

traits in the general population. *Am. J. Med. Genet. B Neuropsychiatr. Genet.* **2018**, *177*, 489–502. [CrossRef] [PubMed]

26. Männik, K.; Mägi, R.; Macé, A.; Cole, B.; Guyatt, A.L.; Shihab, H.A.; Maillard, A.M.; Alavere, H.; Kolk, A.; Reigo, A.; et al. Copy number variations and cognitive phenotypes in unselected populations. *JAMA* **2015**, *313*, 2044–2054. [CrossRef]
27. Venter, J.C.; Adams, M.D.; Myers, E.W.; Li, P.W.; Mural, R.J.; Sutton, G.G.; Smith, H.O.; Yandell, M.; Evans, C.A.; Holt, R.A.; et al. The sequence of the human genome. *Science* **2001**, *291*, 1304–1351. [CrossRef] [PubMed]
28. Millson, A.; Lagrave, D.; Willis, M.J.H.; Rowe, L.R.; Lyon, E.; South, S.T. Chromosomal loss of 3q26.3-3q26.32, involving a partial neuroligin 1 deletion, identified by genomic microarray in a child with microcephaly, seizure disorder, and severe intellectual disability. *Am. J. Med. Genet. A* **2012**, *158*, 159–165. [CrossRef]
29. Aller, E.; Jaijo, T.; García-García, G.; Aparisi, M.J.; Blesa, D.; Díaz-Llopis, M.; Ayuso, C.; Millán, J.M. Identification of large rearrangements of the PCDH15 gene by combined MLPA and a CGH: Large duplications are responsible for Usher syndrome. *Invest. Ophthalmol. Vis. Sci.* **2010**, *51*, 5480–5485. [CrossRef]
30. OMIM ENTRY—# 601067—USHER SYNDROME, TYPE ID.; USH1D. Available online: https://www.omim.org/entry/601067?search=usher%20syndrome%20pcdh15&highlight=%28syndrome%7Csyndromic%29 (accessed on 31 July 2019).
31. Brison, N.; Van Den Bogaert, K.; Dehaspe, L.; van den Oever, J.M.E.; Janssens, K.; Blaumeiser, B.; Peeters, H.; Van Esch, H.; Van Buggenhout, G.; Vogels, A.; et al. Accuracy and clinical value of maternal incidental findings during noninvasive prenatal testing for fetal aneuploidies. *Genet. Med.* **2017**, *19*, 306–313. [CrossRef]
32. Imbert-Bouteille, M.; Chiesa, J.; Gaillard, J.-B.; Dorvaux, V.; Altounian, L.; Gatinois, V.; Mousty, E.; Finge, S.; Bourquard, P.; Vermeesch, J.R.; et al. An incidental finding of maternal multiple myeloma by non invasive prenatal testing. *Prenat. Diagn.* **2017**, *37*, 1257–1260. [CrossRef]
33. Giles, M.E.; Murphy, L.; Krstić, N.; Sullivan, C.; Hashmi, S.S.; Stevens, B. Prenatal cfDNA screening results indicative of maternal neoplasm: Survey of current practice and management needs. *Prenatal. Diagnosis* **2017**, *37*, 126–132. [CrossRef] [PubMed]
34. Zarrei, M.; MacDonald, J.R.; Merico, D.; Scherer, S.W. A copy number variation map of the human genome. *Nat. Rev. Genet.* **2015**, *16*, 172–183. [CrossRef] [PubMed]
35. Database of Genomic Variants. Available online: http://dgv.tcag.ca/v103_20131106/app/statistics (accessed on 14 May 2019).
36. Langmead, B.; Salzberg, S.L. Fast gapped-read alignment with Bowtie 2. *Nat. Methods* **2012**, *9*, 357–359. [CrossRef] [PubMed]
37. Liao, C.; Yin, A.-H.; Peng, C.-F.; Fu, F.; Yang, J.-X.; Li, R.; Chen, Y.-Y.; Luo, D.-H.; Zhang, Y.-L.; Ou, Y.-M.; et al. Noninvasive prenatal diagnosis of common aneuploidies by semiconductor sequencing. *Proc. Natl. Acad. Sci. USA* **2014**, *111*, 7415–7420. [CrossRef] [PubMed]
38. Zhao, C.; Tynan, J.; Ehrich, M.; Hannum, G.; McCullough, R.; Saldivar, J.-S.; Oeth, P.; van den Boom, D.; Deciu, C. Detection of fetal subchromosomal abnormalities by sequencing circulating cell-free DNA from maternal plasma. *Clin. Chem.* **2015**, *61*, 608–616. [CrossRef]
39. Price, A.L.; Patterson, N.J.; Plenge, R.M.; Weinblatt, M.E.; Shadick, N.A.; Reich, D. Principal components analysis corrects for stratification in genome-wide association studies. *Nat. Genet.* **2006**, *38*, 904–909. [CrossRef] [PubMed]
40. Geoffroy, V.; Herenger, Y.; Kress, A.; Stoetzel, C.; Piton, A.; Dollfus, H.; Muller, J. AnnotSV: An integrated tool for structural variations annotation. *Bioinformatics* **2018**, *34*, 3572–3574. [CrossRef]
41. Landrum, M.J.; Lee, J.M.; Riley, G.R.; Jang, W.; Rubinstein, W.S.; Church, D.M.; Maglott, D.R. ClinVar: Public archive of relationships among sequence variation and human phenotype. *Nucleic Acids Res.* **2014**, *42*, D980–D985. [CrossRef]
42. Koster, J.; Rahmann, S. Snakemake—A scalable bioinformatics workflow engine. *Bioinformatics* **2012**, *28*, 2520–2522. [CrossRef]

5

Cell-Free, Embryo-Specific sncRNA as a Molecular Biological Bridge between Patient Fertility and IVF Efficiency

Angelika V. Timofeeva *, Vitaliy V. Chagovets, Yulia S. Drapkina, Nataliya P. Makarova, Elena A. Kalinina and Gennady T. Sukhikh

Kulakov National Medical Research Center of Obstetrics, Gynecology, and Perinatology, Ministry of Health of Russia, Ac. Oparina 4, Moscow 117997, Russia; vvchagovets@gmail.com (V.V.C.); julia.drapkina@gmail.com (Y.S.D.); npmakarova@gmail.com (N.P.M.); e_kalinina@oparina4.ru (E.A.K.); gtsukhikh@mail.ru (G.T.S.)

* Correspondence: avtimofeeva28@gmail.com or v_timofeeva@oparina4.ru

Abstract: Small noncoding RNAs (sncRNAs) are key regulators of the majority of human reproduction events. Understanding their function in the context of gametogenesis and embryogenesis will allow insight into the possible causes of in vitro fertilization (IVF) implantation failure. The aim of this study was to analyze the sncRNA expression profile of the spent culture media on day 4 after fertilization and to reveal a relationship with the morphofunctional characteristics of gametes and resultant embryos, in particular, with the embryo development and implantation potential. Thereto, cell-free, embryo-specific sncRNAs were identified by next generation sequencing (NGS) and quantified by reverse transcription coupled with polymerase chain reaction (RT-PCR) in real-time. Significant differences in the expression level of let-7b-5p, let-7i-5p, piR020401, piR16735, piR19675, piR20326, and piR17716 were revealed between embryo groups of various morphological gradings. Statistically significant correlations were found between the expression profiles of piR16735 and piR020401 with the oocyte-cumulus complex number, let-7b-5p and piR020401 with metaphase II oocyte and two pronuclei embryo numbers, let-7i-5p and piR20497 with the spermatozoid count per milliliter of ejaculate, piR19675 with the percentage of linearly motile spermatozoids, let-7b-5p with the embryo development grade, and let-7i-5p with embryo implantation. According to partial least squares discriminant analysis (PLS-DA), the expression levels of let-7i-5p (Variable Importance in Projection score (VIP) = 1.6262), piR020401 (VIP = 1.45281), and piR20497 (VIP = 1.42765) have the strongest influences on the implantation outcome.

Keywords: miRNA; piRNA; NGS; RT-PCR; embryo culture medium

1. Introduction

Infertility affects almost 15% of couples globally. One of the most effective methods to treat infertility is assisted reproductive technology (ART), which covers a wide spectrum of treatments. Although ART has the highest pregnancy and live birth rates, it still has limited efficacy. Only 32.3% of ART cycles performed for the first time in women under 37 years old result in the birth of a healthy child. Patients aged more than 37 years old who commence in vitro fertilization (IVF) treatment have only a 12.3% success rate. A successful result in ART programs depends on a balanced interaction of a top-quality embryo with an optimally receptive endometrium.

With the worldwide move towards single embryo transfer, there is a renewed focus on developing a reliable method of assessment of embryo viability. The ability to select an embryo with the highest developmental and implantation potential is paramount to maximizing the probability of attaining a successful pregnancy.

Numerous systems have been devised to grade and rank embryos [1]. One of the most common grading systems is based on morphological criteria developed by Gardner et al. [2]. The grading usually considers the morphology of an inner cell mass (ICM) and trophectoderm (TE) for embryos at the blastocyst stage as well as the degree of expansion of the blastocyst cavity. However, sometimes, it is difficult to interpret data on morphological characteristics and morphological traits as they do not fully reflect the physiology of the embryo and its chromosomal complement. Moreover, even good quality embryos do not always implant.

Different biomarkers of embryo quality and implantation potential have been proposed in addition to morphological criteria. They include noninvasive metabolomic and proteomic profiling of embryo culture media [3–5]. However, these biomarkers have their own disadvantages. The metabolomic profile of embryo culture media requires the use of very sensitive instruments which can detect even slight changes in media composition. This is crucial because embryo metabolomic activity depends on the media. Embryo proteomic profile assessment remains inaccurate, because one must enrich media with exogenous major proteins to assure normal embryo development.

Cells and tissues that are involved in human reproduction processes are unique because they constantly undergo dramatic reorganization during gametogenesis and embryogenesis at both transcriptomic and proteomic levels. Recently, scientists have started to focus on cell proteomic profile changes which are controlled by non-coding RNAs. MicroRNAs (miRNAs), small endogenous interfering RNAs (siRNAs), and piwi-interacting RNAs (piRNAs) have the most crucial roles among these molecules [6]. Most miRNAs are transcribed by RNA polymerase II [7]. The result of this process is the formation of a primary transcript (pri-miRNAs) which is 2000–4000 nucleotides in length, has a 5′-7-methyl-guanosine cap structure, a 3′-poly (A) tail, and a complex secondary structure. Further processing of pri-miRNAs continues in the cell nucleus under the control of conserved RNase III Drosha and the DiGeorge Syndrome Critical Region Gene 8 (DGCR8) cofactor with the formation of a hairpin structure, 70–100 nucleotides in length, termed precursor miRNA (pre-miRNA). The resulting pre-miRNA is transported to the cytoplasm via a process that involves Exportin-5 and RAN (RAs-related Nuclear protein) GTPase. The pre-miRNA is further cleaved by RNase III Dicer and Tar RNA-binding protein (TRBP) to generate a short, partially double-stranded (ds) RNA with approximately 22 nucleotides in which one strand is the mature miRNA. Afterwards, the mature miRNA strand is taken up by a multi-protein complex that is identical to the RNA Induced Silencing Complex (RISC) that supports RNA interference (RNAi) and miRNA-bound complex functions to regulate translation. Meanwhile, another strand of miRNA undergoes degradation.

The biogenesis of piRNAs is less studied in comparison with that of miRNAs and siRNAs. It does not require Dicer, because the precursor piRNAs are single-stranded [8]. There are several classes of piRNAs according to next generation sequencing (NGS) data [9]. The class I piRNAs are produced by clustered genetic loci. The piwi-protein-mediated cleavage of the transposon RNA target produces the class II piRNAs. The last group of piRNAs originates from different genomic regions, including three prime untranslated regions of some mRNAs. This fact demonstrates that piRNAs might regulate the level of gene expression in addition to chromosomal rearrangement maintenance by repressing transposable element expression. The piRNA amplification model and transposons neutralization occur via the ping-pong mechanism in germ cells [10].

miRNAs are synthesized in the granulosa and cumulus cells and secreted into the follicular fluid. The influence of these molecules on oocyte maturation and embryo quality has been already shown [11–13]. The effect of female age on the miRNA profile of an oocyte has also been clearly demonstrated [14]. Moreover, miRNA expression levels depend on the degree of oocyte maturation [15]. Elke F. Roovers et al. showed dramatic expression of piRNAs in human, bovine, and macaque oocytes [16]. The piRNA profile in oocytes was similar to that in sperm at the pachytene stage. The principal role of this stage during meiosis is the regulation of transposons expression and, consequently, it controls genome instability [16]. The role of piRNAs and miRNAs during spermatogenesis is described in detail in a review article [17]. This article focuses on the epigenetic regulatory functions of DNA modification, mRNA translation and stability, and the maintenance and self-renewal of germ stem cells. The crucial role of sperm small non-coding RNAs (sncRNAs) produced after fertilization in embryogenesis, implantation, and subsequent embryo development has been recently proven [18]. It has also been shown that successful embryonic development immediately after fertilization depends on coordinated maternal mRNA degradation and zygotic genome activation, resulting in embryonic mRNA production and protein translation. It has been found that piRNAs, miRNAs, and siRNAs are essential for the maternal–zygotic transition [19–21].

In the light of the above, the development of non-invasive methods for predicting the onset of pregnancy and its further development is a highly topical problem at present. An example of such a method is the evaluation of miRNAs specific for the villus and decidual tissues in the peripheral blood plasma of women in both the first trimester of pregnancy and the day of embryo transfer (ET) in ART programs [22]. It was demonstrated that MiR (miR-23a-3p, 27a-3p, 29a-3p, 100-5p, 127-3p, and 486-5p) expression levels significantly changed in the group of women with recurrent miscarriage compared to the group with normal pregnancy, but the clinical pregnancy outcome could not be predicted in the IVF/ET (in vitro fertilization/embryo transfer) patients.

Another diagnostic method for predicting pregnancy is to evaluate the expression profile of sncRNAs secreted by embryo cells into their culture media. To the best of our knowledge, very few studies have considered the interactions that exist between sncRNA expression levels in culture medium and embryo implantation potential [23–25]. To create more effective treatment for infertile couples in the IVF program, the implication of sncRNAs from spent culture medium in the formation of the morphofunctional characteristics of gametes and resultant embryos, as well as in embryo development and implantation potential, should be investigated. Therefore, the present study was aimed at revealing and analyzing such interactions.

2. Results and Discussion

2.1. *Identification of Embryo-Specific sncRNAs*

Deep sequencing was performed to identify embryo-specific sncRNAs. A blastocyst of excellent quality (4AA) from one couple was enrolled in the study. The fraction of RNA secreted both into the blastocoele and into its culture medium on day 5 after fertilization was analyzed. Sequence reads aligned to miRBase v21 and piRNABase with a count of at least 10 were compared by plotting Venn diagrams of the miRNA expression pattern (Figure 1A) and piRNA expression pattern (Figure 1B). Figure 1 demonstrates a wider spectrum of piRNAs (132 and 128 subtypes in blastocoele liquid and embryo culture medium, respectively, Figure 1B) in comparison with miRNAs (49 and 36 subtypes in blastocoele liquid and embryo culture medium, respectively, Figure 1A). Among them, 73.5% miRNAs and 34.7% piRNAs are secreted by the embryo into both the blastocoele liquid and the embryo culture medium. All of the miRNAs secreted by the embryo into the culture medium were detected in the blastocoele liquid. As for the piRNAs (Figure 1B), 33.7% were uniquely expressed in the blastocoele liquid and 31.6% in the embryo culture medium.

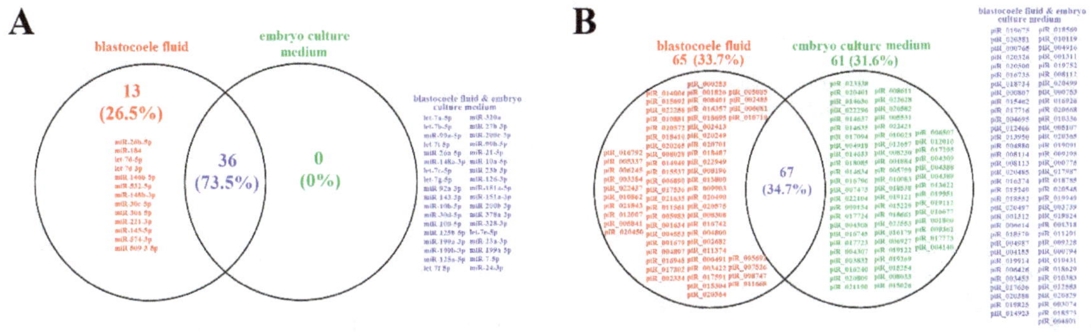

Figure 1. Venn diagrams for microRNA (miRNA) expression pattern (**A**) and piwi-interacting RNA (piRNA) expression pattern (**B**). The sncRNAs (miRNA and piRNA) unique to the blastocoele fluid are written in red; sncRNAs written in green are unique to the blastocyst spent culture medium by the fifth day after fertilization; sncRNAs are written in blue provided that they are detected both in the blastocoele fluid and in the blastocyst spent culture medium.

2.2. Real Time PCR Analysis of sncRNA Revealed by NGS

miRNAs and piRNAs characterized by detectable PCR signals with Ct < 35 cycles (Table 1) and specific PCR products (a single peak of the amplification product melting curve) were selected for subsequent analysis by RT-PCR in 87 RNA samples from embryo culture medium on day 4 after fertilization.

Table 1. Small non-coding RNA (sncRNA) identification by next generation sequencing (NGS) and real time polymerase chain reaction (PCR) (if Ct > 35 cycles in most of the studied samples, these data were excluded from the analysis, "+"—data suitable for analysis).

sncRNA	NGS Read Count (Blastocoele Liquid/Embryo Culture Medium)	5′–3′ Sequence of Sense Primer, Tm (Melting Temperature)	PCR (Embryo Culture Medium)
hsa-let-7b-5p	527/282	Hs_let-7b_1 miScript Primer Assay, Cat.No. MS00003122, Tm = 55 °C	+
hsa-let-7a-5p	537/258	Hs_let-7a_2 miScript Primer Assay, Cat.No. MS00031220, Tm = 55 °C	Ct > 35
hsa-miR-99a-5p	351/145	Hs_miR-99a_2 miScript Primer Assay, Cat.No. MS00032158, Tm = 55 °C	Ct > 35
hsa-miR-148a-3p	194/136	Hs_miR-148a_1 miScript Primer Assay, Cat.No. MS00003556, Tm = 55 °C	Ct > 35
hsa-let-7i-5p	239/114	Hs_let-7i_1 miScript Primer Assay, Cat.No. MS00003157, Tm = 55 °C	+
hsa-miR-26a-5p	209/111	Hs_miR-26a_2 miScript Primer Assay, Cat.No. MS00029239, Tm = 55 °C	Ct > 35
hsa-let-7c-5p	192/110	Hs_let-7c_1 miScript Primer Assay, Cat.No. MS00003129, Tm = 55 °C	+
hsa-let-7g-5p	145/89	Hs_let-7g_2 miScript Primer Assay, Cat.No. MS00008337, Tm = 55 °C	Ct > 35
hsa-miR-92a-3p	144/91	Hs_miR-92_1 miScript Primer Assay, Cat.No. MS00006594, Tm = 55 °C	+
hsa-miR-143-3p	135/81	Hs_miR-143_1 miScript Primer Assay, Cat.No. MS00003514, Tm = 55 °C	+
hsa-miR-125b-5p	69/57	Hs_miR-125b_1 miScript Primer Assay, Cat.No. MS00006629, Tm = 55 °C	Ct > 35
hsa-miR-100-5p	77/50	Hs_miR-100_2 miScript Primer Assay, Cat.No. MS00031234, Tm = 55 °C	Ct > 35
hsa-miR-125a-5p	63/36	Hs_miR-125a_1 miScript Primer Assay, Cat.No. MS00003423, Tm = 55 °C	Ct > 35

Table 1. *Cont.*

sncRNA	NGS Read Count (Blastocoele Liquid/Embryo Culture Medium)	5′–3′ Sequence of Sense Primer, Tm (Melting Temperature)	PCR (Embryo Culture Medium)
hsa-miR-320a	57/30	Hs_miR-320a_1 miScript Primer Assay, Cat.No. MS00014707, Tm = 55 °C	Ct > 35
hsa-miR-27b-3p	56/28	Hs_miR-27b_2 miScript Primer Assay, Cat.No. MS00031668, Tm = 55 °C	Ct > 35
hsa-miR-200c-3p	47/28	Hs_miR-200c_1 miScript Primer Assay, Cat.No. MS00003752, Tm = 55 °C	Ct > 35
Hsa-piR020401, DQ598029	0/23,808	GGCTGGTCTCGAACTCCTGACCTCAGGT, Tm = 45 °C	+
Hsa-piR023338, DQ601914	0/27,727	TAGTCCCAGCTACTTGGGAGGCTGAGGCA, Tm = 45 °C	+
Hsa-piR019675, DQ596992	2708/2645	GCAATAACAGGTCTGTGATGCCCTTAGA, Tm = 53 °C	+
Hsa-piR016735, DQ593039	1676/1240	CCTGGGAATACCGGGTGCTGTAGGCTTA, Tm = 50 °C	+
Hsa-piR017716, DQ594453	620/680	TTCCCTGGTGGTCTAGTGGTTAGGATTCGGC, Tm = 45 °C	+
Hsa-piR020326, DQ597916	1810/2540	GGCATTGGTGGTTCAGTGGTAGAATTCTCGC, Tm = 60 °C	+
Hsa-piR020497, DQ598177	291/350	TGTAGCTCAGTGGTAGAGCGCGTGCT, Tm = 45 °C	+
Hsa-piR022296, DQ600515	0/10,609	TACTCAGGAGGCTGAGGCAGGAGAATGGC, Tm = 45 °C	+

Two-dimensional hierarchical clustering of real-time RT-PCR data was performed to classify 87 embryo cultivation medium samples according to the degree of expression profile similarity of the studied sncRNAs. The Pearson and Euclidean distance correlations were used to calculate the difference between dendrogram nodes. The heat map for the RT-PCR data clustering is presented in Figure 2. The sncRNA clustering patterns did not clearly separate culture media samples from excellent, good, fair, and poor-quality embryos according to Gardner grading scale, or from morulas, cavernous morulas, and 3–10 cell embryos.

This may be explained by undulating changes in sncRNA expression which depend on the embryo development stage. Certain patterns of the studied sncRNAs expression throughout the development stages and inside subgroups of blastocysts of different quality can be observed in the boxplot in Figure 3. For example, the trends of the piR19675 and piR020401 expression level medians are similar for the "morula", "good quality blastocyst", and "excellent quality blastocyst" groups but are dramatically different in other groups. Meanwhile, the median trend in the let-7b-5p expression level is similar in the "3–10 cell embryo", "cavernous morula", and "poor quality blastocyst" groups.

The level of let-7b-5p expression increases in "morula" and "good and excellent quality blastocyst" groups and decreases notably in the "fair quality blastocyst" group. The Mann–Whitney U-test revealed statistically significant differences in the expression levels of let-7b-5p, let-7i-5p, piR020401, piR16735, piR19675, and piR20326 in the culture medium from embryos of various development grades (Table 2). Moreover, blastocysts of different quality according to Gardner's grading scale were shown to differ significantly from each other in expression levels of let-7i-5p, piR020401, and piR17716. The obtained data are in good agreement with those of Abd El Naby et al., who demonstrated miRNA expression dynamics during the preimplantation stage period from bovine zygote to blastocyst [15].

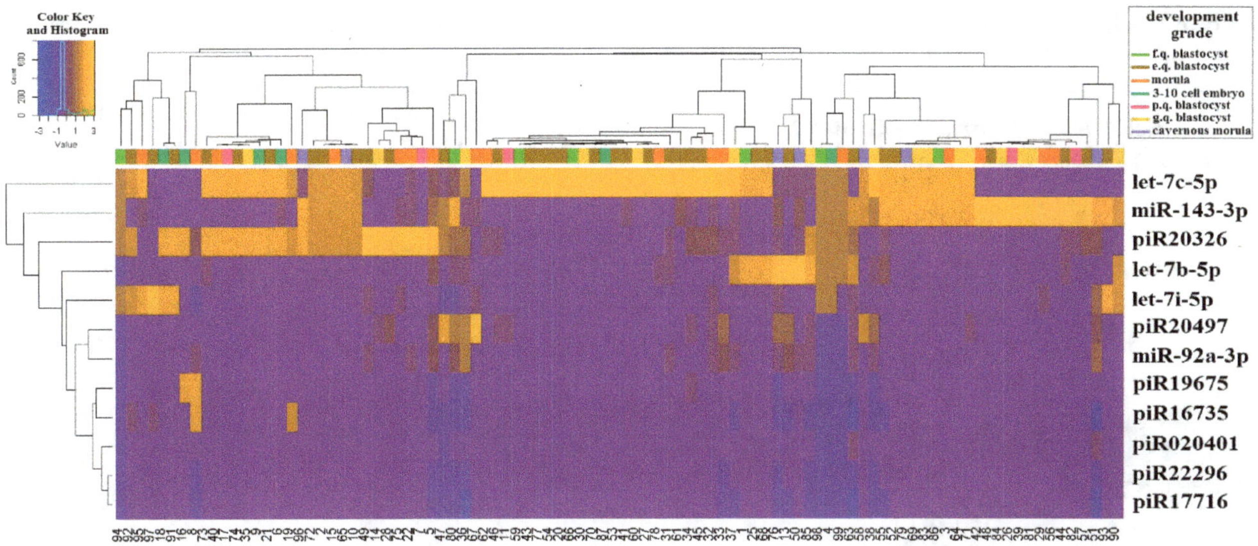

Figure 2. Two-dimensional hierarchical clustering of real-time RT-PCR data on sncRNAs from 87 embryo cultivation medium samples. E.q.—excellent quality, g.q.—good quality, f.q.—fair quality, p.q.—poor quality.

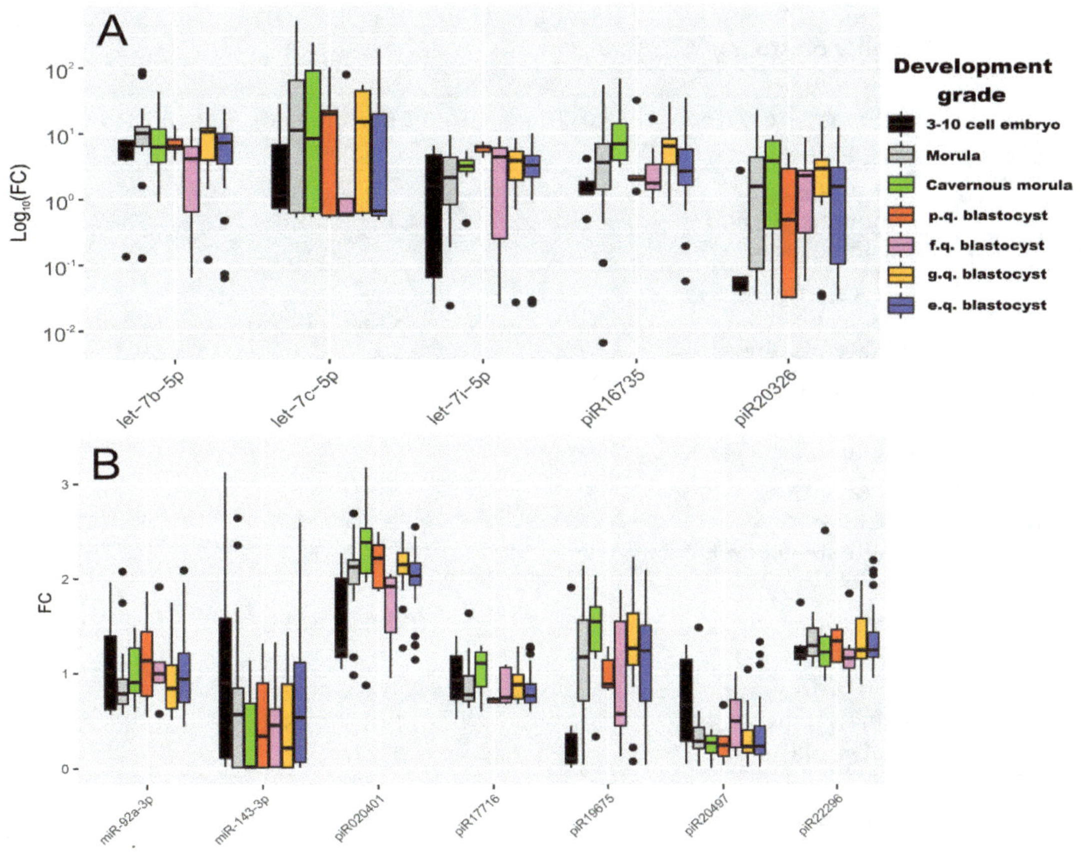

Figure 3. sncRNA expression level fold change in the culture medium of embryos of different development grade. (**A**) sncRNA expression level fold change (FC) with values varying from 0.01 to 400 (data are presented on a logarithmic scale). (**B**) sncRNA expression level FC with values varying from 0.1 to 3.5 (data are presented on a linear scale). E.q.—excellent quality, g.q.—good quality, f.q.—fair quality, p.q.—poor quality.

Table 2. Pairwise comparison of the fold changes in the expression level of sncRNAs in the sample groups shown in Figure 3.

sncRNA	Group 1–Group 2	*p*-Value	Group 1, Me(Q1, Q3) *	Group 2, Me(Q1, Q3) *
let-7b-5p	"3–10 cell embryon"–"morula"	0.04	6.16 (4.08, 7.86)	10.1 (6.42, 12.86)
	"morula"–"fair quality blastocyst"	0.02	10.1 (6.42, 12.86)	4.18 (1.94, 6.71)
	"morula"–"excellent quality blastocyst"	0.05	10.1 (6.42, 12.86)	7.24 (3.48, 10.07)
let-7i-5p	"morula"–"poor quality blastocyst"	0.02	2.2 (0.87, 4.33)	6.2 (5.19, 6.34)
	"cavernous morula"–"poor quality blastocyst"	0.01	3.17 (2.67, 3.89)	6.2 (5.19, 6.34)
	"poor quality blastocyst"–"excellent quality blastocyst"	0.01	6.2 (5.19, 6.34)	3.26 (2.18, 4.52)
piR020401	"morula"–"fair quality blastocyst"	0.02	2.12 (1.94, 2.2)	1.92 (1.43, 2.01)
	"fair quality blastocyst"–"good quality blastocyst"	<0.001	1.92 (1.43, 2.01)	2.14 (2.05, 2.27)
piR16735	"3–10 cell embryon"–"cavernous morula"	<0.001	1.21 (1.21, 1.88)	6.78 (3.97, 14.92)
	"3–10 cell embryon"–"good quality blastocyst"	0.01	1.21 (1.21, 1.88)	6.3 (3.51, 8.2)
	"cavernous morula"–"fair quality blastocyst"	0.04	6.78 (3.97, 14.92)	1.74 (1.41, 3.85)
	"cavernous morula"–"excellent quality blastocyst"	0.02	6.78 (3.97, 14.92)	2.61 (1.62, 5.62)
piR17716	"poor quality blastocyst"–"good quality blastocyst"	0.04	0.72 (0.69, 0.73)	0.88 (0.73, 0.98)
piR19675	"3–10 cell embryo"–"good quality blastocyst"	0.02	0.1 (0.05, 0.36)	1.26 (1.08, 1.63)
	"3–10 cell embryo"–"excellent quality blastocyst"	0.03	0.1 (0.05, 0.36)	1.23 (0.7, 1.5)
piR20326	"3–10 cell embryo"–"good quality blastocyst"	0.02	0.06 (0.04, 0.06)	2.88 (1.09, 3.92)

* The data are presented as medians (Me) and quartiles Q1 and Q3 in the format Me (Q1; Q3).

Since embryo sncRNA expression patterns which contribute to the embryo quality and development potential depend on the degree of oocyte maturation, concomitant diseases of the female reproductive and endocrine systems, and also, sperm characteristics, a correlation matrix was calculated and plotted to reveal these interactions (Figure 4). The analyzed samples ranged, according to the embryo development grading scale, in the following way: "3–10 cell embryo" < "morula" < "cavernous morula" < "poor quality blastocyst" < "fair quality blastocyst" < "good quality blastocyst" < "excellent quality blastocyst". Considering all 87 samples (Figure 4A), a reliable correlation of sncRNA expression profiles was found both within one class and between classes of these molecules. The expression levels of piR17716 and piR20497 negatively correlated with the embryo development grade, positively correlated with each other and with miR-92a-3p and let-7c-5p, and negatively correlated with piR16735, piR20326, let-7b-5p, and let-7i-5p. In contrast, the expression level of let-7c-5p negatively correlated with the expression of other Let7 family members (let-7b-5p and let-7i-5p) and piR020401. Such a complex interrelation of sncRNA expression patterns in the culture media of embryos of various development grades is probably a manifestation of the fine system of signaling pathway regulation necessary for the implementation of the embryogenesis program. The crucial role of miRNAs and piRNAs in the post-transcriptional gene regulation is already well known [12–21].

A correlation matrix with 48 samples of culture media from embryos which were transferred to the uterine cavity was calculated (Figure 4B) to analyze the interaction between pregnancy, the quality of oocytes retrieved, sperm quality, the number of blastocysts obtained, and the expression profile of sncRNAs. The correlation matrix revealed that the expression level of the Let7 family in embryo culture media can be a potential biomarker for IVF efficiency prognosis. For example, the expression level of let-7b-5p negatively correlated with both the embryo development grade and the number of M2 oocytes and 2PN cells. Moreover, the M2 oocyte number as a percentage of oocyte–cumulus complexes (OCC) and the 2PN embryos number as a percentage of the M2 oocyte number had positive correlations with fallopian tube presence. Many experts have assumed the effects of an epithelium in the fallopian tubes on oocyte maturation and the ability to fertilize [26]. In turn, the let-7b-5p expression level was positively correlated with the piR020401 expression profile, which had a negative correlation with the number of OCC, M2-oocytes, and 2PN-cells. In addition, this piRNA was also positively correlated with the expression patterns of piR16735, piR19675, and piR20326. As for the piR16735 and piR1967 expression patterns, the former was negatively correlated with the OCC number, and the latter was negatively correlated with sperm motility. Meanwhile, the sperm concentration was positively correlated with sperm progressive motility and normal sperm morphology.

A promising interaction between let-7i-5p, piR17716, and piR20497 was found. The expression pattern of let-7i-5p had a negative correlation with the piR17716 and piR20497 expression profiles. Along with this, the piR17716 expression pattern was positively correlated with implantation, while the piR20497 expression profile had a negative relationship with sperm concentration. However, for let-7i-5p, an inverse correlation was obtained. A negative correlation of its expression with the implantation rate and a positive correlation with the sperm concentration were found.

As for let-7c-5p, a positive correlation of its expression with piR16735 and piR19675 was observed. piR16735 expression was negatively correlated with OCC, whereas piR19675 expression had a negative relationship with the number of motile sperm. It was difficult to identify whether the analyzed sncRNA in the embryo culture medium originated from gametes involved in fertilization or from the activated embryo genome after the maternal–zygotic transition, but there is strong evidence for the contribution of sperm RNA to the embryo developmental potential and its implantation ability according to the literature data [18].

The number of OCCs collected in one stimulated cycle correlated negatively with the number of blastocysts as a percentage of OCCs or 2PN (Figure 4B). The obtained results were consistent with literature data, indicating that the number of retrieved oocytes in a stimulated cycle has a negative relationship with the number of blastocysts obtained [27,28].

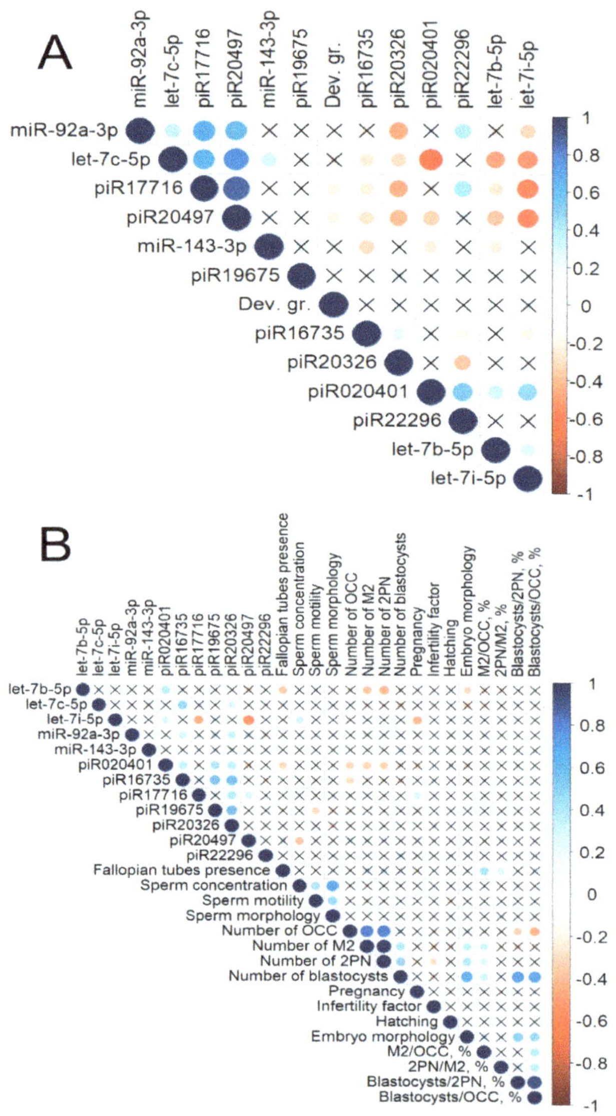

Figure 4. Correlation matrix based on the non-parametric Spearman rank correlation method. Significant ($p < 0.05$) correlations are indicated by a dot, non-significant correlations are indicated by a cross, positive correlations are marked in blue, and negative correlations in red—the more significant the correlation, the larger the dot size. (**A**) Correlation analysis of the sncRNA expression level in 87 embryo cultivation medium samples and the quality of embryos. (**B**) Correlation analysis of the sncRNA expression level in 48 cultivation medium samples of embryos transferred to the uterus and the following parameters: fallopian tube presence—the presence of fallopian tubes in the female of each couple; sperm concentration—spermatozoid count per milliliter of ejaculate from the male of each couple; sperm motility—percentage of linearly motile spermatozoids from the male of each couple; sperm morphology—percentage of morphologically normal spermatozoids from the male of each couple; number of OCC—the number of oocyte–cumulus complexes from the female of each couple; number of M2—metaphase II oocyte number from the female of each couple; number of 2PN—two pronuclei embryo numbers from each couple; number of blastocysts—blastocyst number of each couple; pregnancy—the development of pregnancy after embryo transfer into the uterus; infertility factor—primary or secondary infertility in the female of each couple; hatching—embryo hatching before transfer to the uterine cavity; dev. gr.—embryo development stage and quality according to the Gardner grading scale; embryo morphology—morphological parameters of embryos according to the Gardner grading scale; M2/OCC, %—metaphase II oocyte number as a percentage of OCC; 2PN/M2, %—number of two pronuclei embryos as a percentage of M2 oocyte; blastocysts/2PN, %—blastocyst number as a percentage of 2PN embryos; blastocysts/OCC, %—blastocyst number as a percentage of OCC.

The relationship of the individual ovarian response to gonadotropins in IVF programs and post-transcriptional regulation of genes responsible for oocyte maturation was shown by Cengiz Karakaya et al. [11]. They found that a poor ovarian response to IVF is associated with up-regulation of 16 miRNAs and down-regulation of 88 miRNAs in cumulus cells. It is possible that the spectrum of sncRNAs secreted by a fertilized oocyte after four days of in vitro cultivation and correlated with the number of OCC and M2-oocytes may be a consequence of the post-transcriptional regulation of genes through cumulus–oocyte communication.

The PLS-DA model based on RT-PCR data was developed to study differences between the culture medium from embryos which implanted or did not implant. The fold change of the sncRNA expression level in each of the 48 samples was used for the model (Figure 5). Figure 5A represents the score plot of the developed PLS-DA model. Three clusters of data points can be distinguished. The first one has an abscise of less than −0.75 and represents the embryos which failed to implant despite there being an appropriate quality. The second cluster lies between −0.75 and 0.5 of the abscise and contains data points corresponding to embryos with similar morphological properties (excellent, good, and fair quality according to Gardner's grading scale) and molecular biomarkers (sncRNA expression profile), but some of them implanted (highlighted in gray), and some of them failed to implant (highlighted in red). This phenomenon can be explained by independent factors irrelevant to the embryo quality, e.g., endometrial receptivity, chronic inflammation, or this might have happened because of the other sncRNAs which were not analyzed in the current study. The third cluster is characterized by an abscise greater than 0.5. This cluster corresponds to embryos with high implantation potential, according to morphological grading scale and molecular biomarkers, that successfully implanted. The contribution of sncRNA to the distribution of the data points on the score plot can be estimated by the Variable Importance in Projection (VIP) score (Figure 5B). The following sncRNA have VIP > 1 and those with the highest impact are let-7i-5p (VIP = 1.6262), piR020401 (VIP = 1.45281), piR20497 (VIP = 1.42765), and piR17716 (VIP = 1.14438). Notably, the expression profile of these sncRNAs has a strong correlation with sperm quality, the number of OCCs, oocyte maturity, and oocyte fertilization ability (Figure 4B). Thus, one can suppose that these molecules contribute to the embryo implantation potential. In addition, the Mann–Whitney test revealed significant differences in let-7i-5p and piR020401 expression levels in the culture media from embryos of clusters II and III which succeeded in implanting (highlighted in gray in Figure 5A) in comparison to embryos from cluster I which failed to implant (highlighted in red in Figure 5A): 2.38 ± 2.25 vs. 5.20 ± 1.84, $p < 0.001$ for let-7i-5p; 1.85 ± 0.46 vs. 2.28 ± 0.35, p = 0.005 for piR020401. The culture media from the embryos with implantation failure (highlighted in red in Figure 5A) differed significantly in let-7i-5p expression when comparing the embryos of cluster I and cluster II: 5.20 ± 1.84 vs. 3.41 ± 1.79, p = 0.018. Consequently, let-7i-5p can be proposed as a marker of the embryo's implantation potential on the 4th day after fertilization.

Thus, the PLS-DA model developed by us reflects the implantation potential of an embryo according to the expression profile of sncRNAs in the culture medium on the 4th day after fertilization but cannot be used to predict the onset of pregnancy and its development. The fact remains that the embryo transferred into the uterine cavity, and endometrial cells are able to reciprocally exchange signals, in particular, in the form of secreted extracellular vesicles containing small non-coding RNA. This topic is discussed in detail in a review article by Ferlita A. et al. [29]. Even if the molecular-biological profile of the embryo is normal and the spectrum of secreted sncRNAs corresponds to an embryo with high implantation potential, abnormal secretion of RNA molecules and proteins by the endometrium can adversely affect the embryo implantation process. Thus, to improve the performance of IVF programs before embryo transfer to the uterus, it is necessary not only to evaluate the expression profile of small non-coding embryonic RNA, as proposed in this work, but also to assess endometrial receptivity, in particular, focusing on the extracellular sncRNAs present in endometrial fluid during the window of implantation.

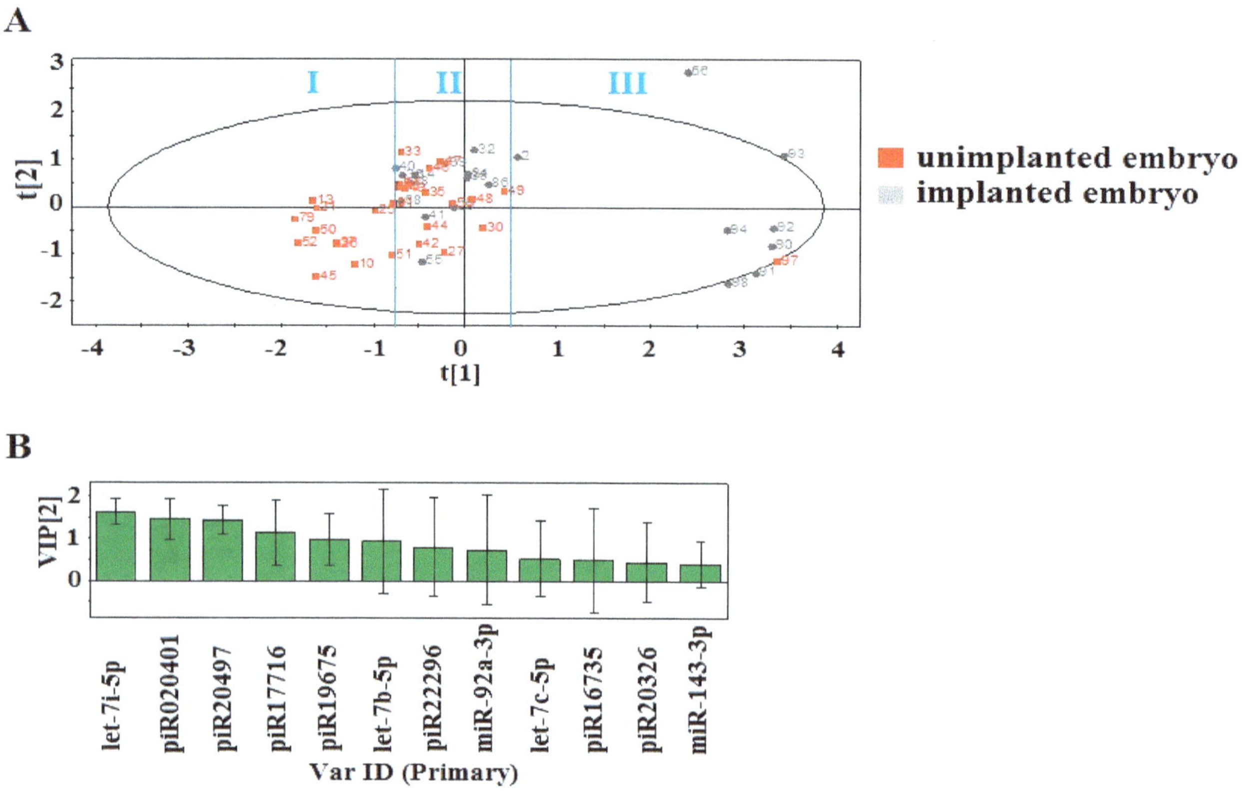

Figure 5. Partial least squares discriminant analysis (PLS-DA) of 2 $^{\Delta\Delta Ct}$ RT-PCR data on the expression of sncRNAs in the 48-embryo culture medium samples. Arabic numerals denote the sample number. Roman numerals (I, II, III) denote the cluster of samples depending on the result of embryo transfer into the uterine cavity and the profile of RNA expression in the cultivation medium. (**A**) score plot with the imposition of information of the sncRNA expression level on the results of embryo transfer into the uterine cavity, (**B**) Variable Importance in Projection (VIP) score.

2.3. Identification of piRNA Targets

piRNA sequences were annotated using the piRBase database (Available online: http://www.regulatoryrna.org), which contains information on the piRNA genome location (hg38) (Table 3). Studied piRNAs are mapped to the transposable elements (TE) such as long interspersed elements (LINEs) and short interspersed elements (SINEs), including Alu elements. These data are in good agreement with published information confirming that most mammalian genomes are dominated by LINE and SINE retrotransposons [30], where LINE and SINE are the most abundantly represented TE classes in the bovine testis, oocyte, and zygote pilRNAs, representing over 50% of all TE mapped reads [31]. Some piRNAs, analyzed in the current study, were mapped to tRNA (tRNA-GluGAG, tRNA-AlaGCY, tRNA-Gly-GGG, tRNA-GlyGGY, tRNA-GlyGCC and tRNAValAAC) and rRNA species, which are specifically processed into piRNAs. Recent studies have also revealed that some tRNA-derived small RNAs associate with the Argonaute (AGO) proteins or P-element induced wimpy testis proteins (PIWI) [32], and the accumulation of small tRNAs and rRNAs and their association with the RNA interference machinery seems to be characteristic of highly proliferative cells and tightly controlled to avoid apoptosis [33]. Additionally, we found out from piRBase that piRNAs are mapped not only to the repeat elements but also to the region of protein coding genes, and these piRNAs are referred to as gene-derived, in particular, *LY6G5C-*, *EFCAB11-*, *VAC14-*, *COLQ-*, and *PTGES3L*-derived piRNAs (Table 3). These findings are in accordance with the published literature and confirm that some mRNAs in flies and vertebrates are known to be processed into piRNAs [21,34–36].

Table 3. Genomic localization and potential piwiRNA target genes.

Genbank ID	PiRBase ID for piRNAs	Location in hg38 *	RefSeq Gene	Repeat Information	Potential Gene Targets
hsa_piR_ 017716|gb|DQ594453|Homo	piR-hsa-24672, aliases piR-60565, PIR55564	chr6_GL000255v2_ alt:247775-247806: +		Name: tRNA-Glu-GAG, Family: tRNA, Class: tRNA	
hsa_piR_ 020401|gb|DQ598029|Homo	piR-hsa-28244, aliases piR-36095 PIR59140	chr6_GL000253v2_ alt:2984874-2984902: −	LY6G5C NM_025262	Name: AluSx, Family: Alu, Class: SINE	*PAK3,CXorf38,FRRS1L, C9orf85,EEF1D,CACNA2D1, RBM28,POLH,TRIM52, FER,NAF1,NSUN7, LIAS,PHC3,RABL2B, KIAA1755,ZNF13,PHF20, ZHX3,TCFL5,CARF*
hsa_piR_ 016735|gb|DQ593039|Homo	piR-hsa-23317, aliases piR-33151 PIR54150	chr6:4428052-4428083: +		Name: 5S, Family: rRNA, Class: rRNA	
hsa_piR_ 019675|gb|DQ596992|Homo	piR-hsa-27282, aliases piR-35058 PIR58103	chr14:89875053-89875081: −	EFCAB11 NM_001284269;NM_145231; NM_001284267	Name: SSU-rRNA_Hsa, Family: rRNA, Class: rRNA	*MDN1,MLKL,METTL22, ARHGEF10L*
		chr21:8211110-8211138: +	RNA18SN2 NR_146146; RNA18SN4 NR_146119; RNA18SN3 NR_146152; RNA18SN5 NR_003286; RNA45SN2 NR_146144; LOC100507412 NR_038958; RNA18SN1 NR_145820	Name: SSU-rRNA_Hsa, Family: rRNA, Class: rRNA	
		chr21:8255319-8255347: +	RNA28SN5 NR_003287	Name: SSU-rRNA_Hsa, Family: rRNA, Class: rRNA	
		chr21:8394145-8394173: +	LOC100507412 NR_038958; RNA28SN5 NR_003287; RNA18SN2 NR_146146; RNA18SN5 NR_003286; RNA45SN3 NR_146151; RNA18SN3 NR_146152; RNA18SN4 NR_146119; RNA18SN1 NR_145820;	Name: SSU-rRNA_Hsa, Family: rRNA, Class: rRNA	
		chr21:8438355-8438383: +	RNA45SN1 NR_145819; RNA18SN2 NR_146146; RNA45SN5 NR_046235; RNA18SN5 NR_003286; RNA18SN3 NR_146152; RNA18SN4 NR_146119; RNA18SN1 NR_145820;	Name: SSU-rRNA_Hsa, Family: rRNA, Class: rRNA	
		chr22_KI270733v1_ random:127410-127438: +	RNA45SN1 NR_145819; RNA45SN5 NR_046235; RNA45SN2 NR_146144; RNA45SN4 NR_146117; RNA18SN5 NR_003286; RNA18SN3 NR_146152; RNA18SN4 NR_146119; RNA18SN2 NR_146146; RNA18SN1 NR_145820;	Name: SSU-rRNA_Hsa, Family: rRNA, Class: rRNA	
		chr22_KI270733v1_ random:172491-172519: +	RNA18SN3 NR_146152; RNA18SN5 NR_003286; RNA18SN4 NR_146119; RNA18SN2 NR_146146; RNA18SN1 NR_145820;	Name: SSU-rRNA_Hsa, Family: rRNA, Class: rRNA	
hsa_piR_ 020497|gb|DQ598177|Homo	piR-hsa-28392, aliases piR-36243 PIR59288	chr6:28863725-28863757: −, chr6_GL000250v2_ alt:129309-129341: −, chr6_GL000251v2_ alt:354265-354297: −, chr6_GL000252v2_ alt:129327-129359: −, chr6_GL000253v2_ alt:129283-129315: −, chr6_GL000254v2_ alt:129319-129351: −, chr6_GL000255v2_ alt:129307-129339: −, chr6_GL000256v2_ alt:172970-173002: −		Name: tRNA-Ala-GCY, Family: tRNA, Class: tRNA	

Table 3. *Cont.*

Genbank ID	PiRBase ID for piRNAs	Location in hg38 *	RefSeq Gene	Repeat Information	Potential Gene Targets
hsa_piR_ 020326|gb| DQ597916|Homo	piR-hsa-28131, aliases piR-35982 PIR59027	chr1:16545979-16546010: −, chr1:16861919-16861950: +		Name: tRNA-Gly-GGG, Family: tRNA, Class: tRNA	
		chr16:707895 05-70789536: +	VAC14 NM_001351157, NM_018052	Name: tRNA-Gly-GGY, Family: tRNA, Class: tRNA	
		chr3:155079 90-15508021: +	COLQ NM_080539, NM_005677	Name: tRNA-Gly-GGG, Family: tRNA, Class: tRNA	
		chr6:27902948- 27902979: −		Name: tRNA-Gly-GGY, Family: tRNA, Class: tRNA	
hsa_piR_ 022296| gb|DQ600515|Homo	piR-hsa-30715, aliases piR-38581 PIR61626	chr18:13769187-13769216: +		Name: AluY, Family: Alu, Class: SINE	*NR6A1,RELL1,IL17RB, POLA2,PLPP4,PRPF3, C3orf70,CDH18,ZNF726, UGCG,IL17RD,MCF2L2, TBC1D24,AMN1,KIN, PLPP4,ZNF124,ACTR8, LPP,CPT1A,ZNF320*
hsa_piR_ 023338| gb|DQ601914|Homo	piR-hsa-30937, aliases piR-39980 PIR63025	chr17:42974237-42974266: +	PTGES3L NM_001142654, NM_001142653, NM_001261430; PTGES3L-AARSD1 NM_025267, NM_001136042;	Name: AluSq, Family: Alu, Class: SINE	

* "+" denotes sense strand which contains the exact nucleotide sequence to the messenger RNA (Mrna); "−" denotes antisense strand, serves as the template for the transcription and contains complementary nucleotide sequence to the transcribed mRNA.

Recent studies have suggested that piRNAs have the potential to target mRNAs in addition to their traditional transposon-derived transcripts [37–39]. piRNA target data have been collected from literature in piRBase and related to mice and fruitflies but not to human piRNA until now. Therefore, BLAST (Available online: https://blast.ncbi.nlm.nih.gov) was used to identify human piRNA overlap with protein coding genes. mRNA was considered a potential target for piRNA if the mapping direction for the piRNA–mRNA pair was opposite. This approach was proposed by S. Russell for Bowtie application [31]. Lists of potential gene-targets were obtained for hsa-piR020401, hsa-piR019675, and hsa-piR022296 (Table 3). We focused on the 25 gene targets of hsa-piR020401 and hsa-piR019675. The expression profiles of these molecules correlated with the OCC, M2, 2PN number, and the percentage of sperm with linear motility. A Gene Ontology (GO) analysis of these genes in the PANTHER Classification System (Available online: http://pantherdb.org/) showed that within the molecular function category, 28% of genes are related to binding processes (including transcription factor like 5 (*TCFL5*), Rho Guanine Nucleotide Exchange Factor 10 Like (*ARHGEF10L*), p21-activated kinase 3 (*PAK3*), Eukaryotic Translation Elongation Factor 1 Delta (*EEF1D*), Polyhomeotic-like protein 3 (*PHC3*), RNA Binding Motif Protein 28 (*RBM28*), Feline Encephalitis Virus-Related Kinase (*FER*)), 24% of genes are involved in catalytic activity (Methyltransferase Like 22 (*METTL22*), DNA polymerase eta (*POLH*), *PAK3*, member of RAS oncogene family like 2B (*RABL2B*), Lipoic acid synthetase (*LIAS*), *FER*), *ARHGEF10L* is considered a molecular function regulator, Zinc fingers and homeoboxes protein 3 (*ZHX3*) and *PHC3* are implicated in transcription regulator activity, Eukaryotic Translation Elongation Factor 1 Delta (*EEF1D*) is implicated in translation regulator activity, and Calcium Voltage-Gated Channel Auxiliary Subunit Alpha2delta 1 (*CACNA2D1*) is involved in transporter activity. Therefore, it seemed important to focus on several genes whose function may be associated with reproduction. For instance, Lawrence M. Roth et al. revealed the role of *TCFL5* in normal human spermatogenesis [40]. *ARHGEF10L* is involved in the positive regulation of cytoskeleton organization, and thereby is implicated in microtubule dynamics, signal transduction, gene expression, and enzymatic regulation [41]. *PHC3* is a ubiquitously expressed member of the polycomb gene family, encoding a diverse set of regulatory proteins that are involved in the maintenance of the expression patterns that control development [42,43] and are responsible for long-term silencing of genes by altering chromatin structure through the deacetylation of histone tails and by inhibiting adenosine triphosphate (ATP)-dependent chromatin remodeling [44,45]. The loss of *PHC3* may favor tumorigenesis by potentially disrupting the ability of cells to remain in G0 [46]. *POLH* is associated with the replication of damaged DNA by synthesis across a lesion in the template strand, which allows DNA synthesis to continue beyond the lesion. Observations of Ohkumo T. et al. in the Caenorhabditis elegans suggest that *POLH* contributes to damage tolerance against UV irradiation to ensure the successful completion of embryogenesis; this provides important insights into its role in DNA damage tolerance in germ and embryonic cells [47]. *ZHX3* belong to the homeodomain transcription factor family, which is crucial for the development from embryogenesis to cell differentiation, including neuronal differentiation [48].

2.4. Functional Annotation of miRNA Target Genes

Potential and experimentally verified target mRNAs were determined using four separate webtools to explore the biological significance of the studied miRNAs. These webtools were DianaTools microT-CDS, DianaTools_TargetScan, DianaTools_Tarbase, and miRtargetlink. DianaTools microT-CDS and DianaTools_TargetScan were used to predict the target genes, whereas DianaTools_Tarbase and miRtargetlink allowed identification of the experimentally validated targets. Kyoto Encyclopedia of Genes and Genomes (KEGG) pathways with p-values of < 0.05 were united with DianaTools (Table 4) to identify signaling pathways regulated by several miRNAs at the same time. It is important to focus attention on the regulatory effects of miR-92a-3p, let-7b-5p, let-7c-5p, and let-7i-5p on signaling pathways.

Their effector molecules might be involved in gameto- and embryogenesis. It is essential to mention signaling pathways regulating the pluripotency of stem cells, the extracellular matrix (ECM)–receptor interaction; adherens junctions; RNA transport; protein processing in the endoplasmic reticulum; protein digestion and absorption; ubiquitin mediated proteolysis; the Phosphoinositide 3-kinase (PI3K)-Akt serine/threonine kinase signaling pathway; the cell cycle; the Wingless-type Mouse mammary tumor virus integration site family member (Wnt)-signaling pathway; the Hippo signaling pathway; the FoxO signaling pathway; the mitogen-activated protein kinase (MAPK) signaling pathway; the transforming growth factor beta (TGFβ) signaling pathway; the p53 signaling pathway; the estrogen signaling pathway; oocyte meiosis; and valine, leucine, and isoleucine biosynthesis or degradation. For instance, the PI3K-Akt, MAPK, Hippo, and Wnt signaling pathways participate in protein synthesis, cell survival, migration, invasion, cell cycle progression, and cellular proliferation and differentiation [49,50].

The first cell differentiation event in embryogenesis occurs when the outer blastomeres of the embryo form a trophectoderm, and the remaining blastomeres form the inner cell mass (ICM), giving rise to embryonic stem cells which have the potential to self-renew and differentiate into different cell types and tissues. The balance between differentiation and self-renewal in embryonic stem cells is maintained, among others, by the Hippo pathway [50,51].

Moreover, it has been reported that the Hippo pathway can interact with other pathways to promote and maintain pluripotency. For example, the transcriptional co-activator with PDZ-binding motif (TAZ), the major mediator of the Hippo pathway, associates with Smad2/3 (directly phosphorylated by TGFβ receptors) and maintains the nuclear accumulation of Smad complexes, thereby promoting the expression of pluripotency markers (Oct4, Nanog) in response to TGFβ stimulation [51]. In turn, the TGFβ signaling pathway is modulated by deubiquitinating enzymes by regulating TGFBR1, TGFBR2, R-SMADs, co-SMAD, and I-SMAD [52]. The most crucial role in embryological events has canonical Wnt signaling, which is implicated in cell fate decisions, stem cell maintenance, body-axis determination in vertebrate embryos, and gastrulation [53].

Table 4. Predicted and experimentally supported miRNA targets.

Kyoto Encyclopedia of Genes and Genomes (KEGG) Pathway	p-Value	Number of Genes	miRNAs
DianaTools microT-CDS Algorithm for the Prediction of miRNA Targets in Both the three prime untranslated regions (3′UTRs) and Coding Sequence (CDS)			
ECM–receptor interaction	$< 1 \times 10^{-325}$	15	hsa-let-7b-5p, hsa-let-7c-5p, hsa-let-7i-5p, hsa-miR-92a-3p
Mucin type O-Glycan biosynthesis	4.21×10^{-11}	5	hsa-let-7b-5p, hsa-let-7c-5p, hsa-let-7i-5p
Glycosaminoglycan biosynthesis—chondroitin sulfate/dermatan sulfate	8.44×10^{-8}	4	hsa-let-7b-5p, hsa-let-7c-5p, hsa-let-7i-5p
Amoebiasis	2.48×10^{-7}	11	hsa-let-7b-5p, hsa-let-7c-5p, hsa-let-7i-5p
Signaling pathways regulating stem cell pluripotency	1.19×10^{-5}	16	hsa-let-7b-5p, hsa-let-7c-5p, hsa-let-7i-5p
Protein digestion and absorption	0.006789	11	hsa-let-7b-5p, hsa-let-7c-5p, hsa-let-7i-5p
PI3K-Akt signaling pathway	0.009286	27	hsa-let-7b-5p, hsa-let-7c-5p
Wnt signaling pathway	0.017555	11	hsa-let-7c-5p
DianaTools_TargetScan—Predicts Biological Targets of miRNAs Considering Matches to 3′ UTRs			
Mucin type O-Glycan biosynthesis (hsa00512)	4.43×10^{-12}	1	hsa-let-7b-5p, hsa-let-7c-5p, hsa-let-7i-5p
Valine, leucine, and isoleucine biosynthesis (hsa00290)	1.79×10^{-8}	1	hsa-let-7b-5p, hsa-let-7c-5p, hsa-let-7i-5p
Signaling pathways regulating stem cell pluripotency (hsa04550)	7.77×10^{-6}	6	hsa-let-7b-5p, hsa-let-7c-5p, hsa-let-7i-5p
Biosynthesis of amino acids (hsa01230)	0.000132	2	hsa-let-7b-5p, hsa-let-7c-5p, hsa-let-7i-5p
2-Oxocarboxylic acid metabolism (hsa01210)	0.000166	1	hsa-let-7b-5p, hsa-let-7c-5p, hsa-let-7i-5p
Oocyte meiosis (hsa04114)	0.008317	3	hsa-let-7b-5p, hsa-let-7c-5p
Valine, leucine, and isoleucine degradation (hsa00280)	0.026219	1	hsa-let-7c-5p
Arginine and proline metabolism (hsa00330)	0.026219	2	hsa-let-7c-5p
DianaTools_Tarbase—Database of Experimentally Supported miRNA Targets			
Lysine degradation (hsa00310)	6.06959×10^{-13}	20	hsa-let-7b-5p, hsa-let-7i-5p, hsa-miR-92a-3p
Cell cycle (hsa04110)	1.13021×10^{-11}	52	hsa-let-7b-5p, hsa-let-7i-5p, hsa-miR-92a-3p
Viral carcinogenesis (hsa05203)	5.62537×10^{-10}	60	hsa-let-7b-5p, hsa-let-7i-5p, hsa-miR-92a-3p
Hepatitis B (hsa05161)	5.35744×10^{-08}	54	hsa-let-7b-5p, hsa-let-7i-5p, hsa-miR-92a-3p
Oocyte meiosis (hsa04114)	1.75682×10^{-07}	41	hsa-let-7b-5p, hsa-let-7i-5p, hsa-miR-92a-3p
Chronic myeloid leukemia (hsa05220)	2.5576×10^{-7}	27	hsa-let-7b-5p, hsa-let-7i-5p, hsa-miR-92a-3p
Hippo signaling pathway (hsa04390)	4.05654×10^{-7}	50	hsa-let-7b-5p, hsa-let-7i-5p, hsa-miR-92a-3p
Proteoglycans in cancer (hsa05205)	1.70389×10^{-6}	64	hsa-let-7b-5p, hsa-let-7i-5p, hsa-miR-92a-3p
Thyroid hormone signaling pathway (hsa04919)	2.7327×10^{-6}	39	hsa-let-7b-5p, hsa-let-7i-5p, hsa-miR-92a-3p
Adherens junctions (hsa04520)	3.42243×10^{-6}	30	hsa-let-7i-5p, hsa-miR-92a-3p
FoxO signaling pathway (hsa04068)	2.15597×10^{-5}	50	hsa-let-7b-5p, hsa-let-7i-5p, hsa-miR-92a-3p
ECM–receptor interaction (hsa04512)	2.47672×10^{-5}	12	hsa-let-7i-5p

Table 4. *Cont.*

DianaTools_Tarbase—Database of Experimentally Supported miRNA Targets			
Bacterial invasion of epithelial cells (hsa05100)	4.61176×10^{-5}	28	hsa-let-7b-5p, hsa-let-7i-5p, hsa-miR-92a-3p
Glioma (hsa05214)	7.08143×10^{-5}	20	hsa-let-7b-5p, hsa-let-7i-5p
Pathways in cancer (hsa05200)	0.000154251	99	hsa-let-7i-5p, hsa-miR-92a-3p
Colorectal cancer (hsa05210)	0.000634129	25	hsa-let-7b-5p, hsa-let-7i-5p, hsa-miR-92a-3p
Thyroid cancer (hsa05216)	0.000693712	12	hsa-let-7b-5p, hsa-let-7i-5p, hsa-miR-92a-3p
Prostate cancer (hsa05215)	0.001016403	35	hsa-let-7b-5p, hsa-let-7i-5p, hsa-miR-92a-3p
Epstein–Barr virus infection (hsa05169)	0.002168481	51	hsa-let-7b-5p, hsa-let-7i-5p
MAPK signaling pathway (hsa04010)	0.003378721	59	hsa-let-7b-5p, hsa-let-7i-5p, hsa-miR-92a-3p
Endocytosis (hsa04144)	0.003828609	62	hsa-let-7b-5p, hsa-let-7i-5p, hsa-miR-92a-3p
Huntington's disease (hsa05016)	0.003922408	50	hsa-let-7b-5p, hsa-miR-92a-3p
Small cell lung cancer (hsa05222)	0.004508148	24	hsa-let-7b-5p, hsa-let-7i-5p
TGF-beta signaling pathway (hsa04350)	0.006182085	29	hsa-let-7b-5p, hsa-let-7i-5p, hsa-miR-92a-3p
p53 signaling pathway (hsa04115)	0.007856274	20	hsa-let-7b-5p, hsa-let-7i-5p
Melanoma (hsa05218)	0.008148048	18	hsa-let-7b-5p, hsa-let-7i-5p
Shigellosis (hsa05131)	0.009612655	13	hsa-miR-92a-3p
Bladder cancer (hsa05219)	0.009747186	14	hsa-let-7b-5p, hsa-let-7i-5p
Signaling pathways regulating pluripotency of stem cells (hsa04550)	0.01296196	26	hsa-miR-92a-3p
Estrogen signaling pathway (hsa04915)	0.01345837	17	hsa-let-7i-5p
Transcriptional misregulation in cancer (hsa05202)	0.01729337	45	hsa-let-7i-5p, hsa-miR-92a-3p
Ubiquitin mediated proteolysis (hsa04120)	0.01776272	34	hsa-let-7b-5p
Protein processing in endoplasmic reticulum (hsa04141)	0.02539808	38	hsa-let-7b-5p, hsa-let-7i-5p
RNA transport (hsa03013)	0.02748059	41	hsa-let-7b-5p
Valine, leucine and isoleucine biosynthesis (hsa00290)	0.03746544	2	hsa-miR-92a-3p

In our study, statistically significant correlations were found between the expression profile of let-7b-5p and the metaphase II oocyte number, the two pronuclei embryos number, and the embryo development grade, while the expression profile of let-7i-5p correlated with the sperm count per milliliter of ejaculate and with embryo implantation. let-7b-5p and let-7i-5p regulate the FoxO signaling pathway, controlling the expression of 35 genes (Table 4), among which *FOXO1* plays the most important role. Kuscu N et al. showed that mouse FoxO1, FoxO3, and FoxO4 proteins are regulated by the PI3K/Akt signaling pathway and differentially expressed in prophase I, metaphase I, and metaphase II oocytes, as well as in fertilized oocytes, 2-cell embryos, 4-cell embryos, 8-cell embryos, morula, and blastocysts [54]. Therefore, they are implicated in oocyte maturation and preimplantation embryo development.

According to Targetscan and Tarbase, the let-7 family regulates valine, leucine, and isoleucine biosynthesis or degradation (Table 4). Perkel, K. J., et al. demonstrated that the individually cultured embryos growing at different developmental rates consume pyruvate, lactate, acetate, isoleucine, leucine, valine, threonine, alanine, methionine, lysine, glycine, arginine, phenylalanine, histidine tryptophan, and tyrosine in varying amounts from spent culture medium [55]. For instance, significantly higher levels of valine, leucine, and isoleucine consumption by 16-cell fast growing embryos compared with their slow growing counterparts (developmentally delayed by 12–24 h) were found using the proton nuclear magnetic resonance method. Decreased leucine levels in the embryo culture medium correlated with the pregnancy rate in a study by Brison et al. [56]. These data show that it is important to characterize the growing embryo not only by morphological criteria, but by metabolomic and transcriptomic profiles, as well to assess an embryo's developmental and implantation potential.

The miRtargetlink database was used for hsa-let-7c-5p, since no target genes were detected for it by DianaTools_Tarbase. The miRtargetlink contains information on "miRNA/gene target" interactions confirmed by reporter analysis as strong interactions. The list of gene targets of miR-92a-3p, let-7b-5p, let-7c-5p, and let-7i-5p were subjected to ontology and pathway analysis using PANTHER Classification System (Available online: http://pantherdb.org/) and were subsequently classified based on their biological process (Figure 6). Among them, the common target genes of let-7c-5p and let-7i-5p are *GPS1*, *COPS6*, and *COPS8*. *GPS1* is a G Protein Pathway Suppressor 1 known to suppress mitogen-activated signal transduction in mammalian cells. *COPS6* and *COPS8* are subunits 6 and 8 of COP9 Signalosome and are involved in various cellular and developmental processes, for instance, in the regulation of the ubiquitin (Ubl) conjugation pathway. Let-7b-5p and let-7c-5p regulate the expression level of High Mobility Group AT-Hook 2 (*HMGA2*), which is an essential component of the enhancesome and acts as a transcriptional regulating factor; Neuroblastoma RAS proto-oncogene (*NRAS*), which has intrinsic GTPase activity and controls cell proliferation and anti-apoptosis pathways [57]; *AGO1* (Argonaute 1, RISC catalytic component), which degrades and represses the translation of mRNA bound to miRNA as well as performing transcriptional gene silencing of promoter regions bound to short antigene RNAs; *IGF1R* (Insulin Like Growth Factor 1 Receptor, implicated in cell growth and survival control); *TGFBR1* (Transforming Growth Factor Beta Receptor 1, involved in the regulation of cellular processes, including division, differentiation, motility, adhesion, and death); and *TNFRSF10B* (tumor necrosis factor Receptor Superfamily Member 10b, causes cell apoptosis through adapter molecule Fas Associated Via Death Domain (FADD) and effector caspases). The only common experimentally proven target gene of let-7b-5p and let-7i-5p is *TLR4* (Toll Like Receptor 4, transmembrane cell-surface receptor), which was initially discovered in D. Melanogaster as a gene controlling body patterning during embryonic development [58] and plays a key role in the innate immune system [59].

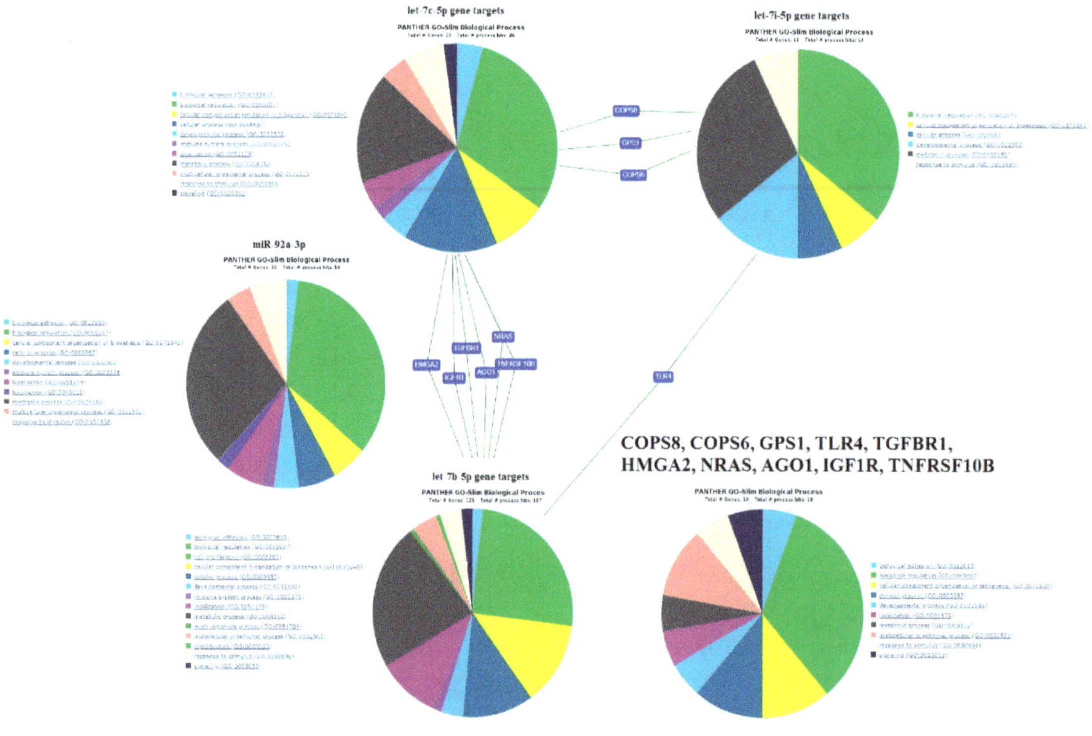

Figure 6. Functional classification of genes targeted by let-7b-5p, let-7c-5p, let-7i-5p, and miR-92a-3p from miRtargetlink and the PANTHER databases.

The essential roles of the let-7 and miR-92 family in determining the blastocyst developmental and implantation potential were confirmed by data from Kim J et al. [60]. They showed a significant increase in the expression level of let-7b-5p and miR-92a-3p in outgrowth embryos compared with blastocysts and non-outgrowth embryos. Regulation of the «Mucin type O-Glycan biosynthesis signaling pathway (hsa00512)» under the control of let-7b-5p, let-7c-5p, and let-7i-5p according to the DianaTools database (Table 4) as well as the regulation of the expression level of IGF1R under the influence of let-7b-5p according to the miRtargetlink database (see above) emphasizes the important roles of these miRNA subtypes in implantation processes, since Mucin 1, being an integral transmembrane mucin glycoprotein expressed on the apical surface of the endometrium, acts as an inhibitor of embryo attachment, whereas IGF1R increases on the surfaces of the endometrium during the receptive stage and contributes to adhesive interaction with the embryo [29].

3. Materials and Methods

3.1. Patients

Forty couples aged between 27 and 40 years with reproductive disorders and IVF indications were enrolled in the study, and 87 samples of spent embryo culture medium (Irvine OneStep) were obtained on day 4 after fertilization during IVF cycles. The morphological characteristics of studied embryos were evaluated on day 5 and were as follows: there were blastocysts of excellent (>3AA, $n = 32$), good (3–6 AB, 3–6 BA, 1–2 AA, $n = 16$), fair (3–6 BB, 3–6 AC, 3–6 CA, 1–2 AB, 1–2 BA, $n = 11$), and poor quality (1–6 BC, 1–6 CB, 1–6 CC, 1–2 BB, $n = 6$) according to the Gardner grading scale [2], as well as morula embryos ($n = 14$) and 3–10 cell embryos ($n = 8$). Out of the 87 embryos, 7 morulas, 3 cavernous morulas, and 38 blastocysts were transferred to the uterus at 5 or 6 days after fertilization. When ovarian hyperstimulation syndrome risk was present, the freeze-all strategy was applied. Before initiation of the IVF program, the following information from a couple was analyzed: medical history, hormonal profile, rhesus blood group system, blood coagulability, infection status, Papanicolaou test, pelvic ultrasound data, and spermogram. There were no statistically significant age or anthropometric measurement differences between patients enrolled in the current study. Patients with normal ovarian reserve, a regular menstrual cycle, and without any extragenital diseases were selected to minimize the influence of maternal factors on embryo implantation. The inclusion criteria were as follows: patients aged 20–37 years old with normal ovarian reserve, tubal factor infertility, and a regular menstrual cycle. The exclusion criteria were patients with contraindications for IVF treatment including extragenital diseases, stage 3–4 moderate and severe endometriosis, polycystic ovary syndrome, endometrial pathology, intramural and subserosal myomas of more than 4 cm, submucosal myomas distorting the uterine cavity, genetic disorders, congenital genitourinary anomalies, a history of invasive surgery on the ovaries, or severe male infertility.

Forty patients with tubal infertility factor were enrolled in the current study. All patients underwent a short antagonist protocol. Gonadotropin stimulation commenced on day 3–4 of the menstrual cycle. Subcutaneous administration of the gonadotropin-releasing hormone (GnRH) antagonist began when the follicular size was more than or equal to 14 mm. All patients had routine monitoring via transvaginal sonography and hormonal profiling of follicle-stimulating hormone (FSH), luteinizing hormone (LH), estrogen, and progesterone levels. The mature oocytes were retrieved after 34–36 h of human chorionic gonadotropin (HCG) injection. The collected matured oocytes were fertilized in vitro using IVF and the intracytoplasmic sperm injection (ICSI) method. Progesterone vaginal suppository or dydrogesterone enhancesome urred daily after oocyte retrieval and continued until the result of the first blood pregnancy test. All embryos were cultivated in microdrops in the incubator. The assisted hatching procedure was done in individual cases. Single embryo transfer (SET) was performed on day 5 after oocyte retrieval. If the pregnancy test was positive, the patient continued progesterone supplementation until 8 weeks of pregnancy. In the case of a negative result, the medication was discontinued.

All patients were divided into three groups depending on the IVF program result (Table 5).

Group I: 22 patients who had ovarian stimulation and SET in a stimulated cycle with a negative pregnancy blood test.
Group II: 18 patients who had ovarian stimulation and SET in a stimulated cycle with a positive pregnancy blood test.
Group III: 3 patients from group I who had implantation failure in the previous stimulated cycle and frozen-thawed (FT) embryo transfer with a positive result. All embryos were thawed at the blastocyst stage. The endometrium was prepared with the exogenous administration of oral micronized estradiol forms and progesterone. FT embryo transfer was performed on day 19–20 of the menstrual cycle. The endometrium thickness at the day of embryo transfer was 9–12 mm.

Table 5. Comparison of the clinical and demographic characteristics of patients enrolled in the current study.

Clinical and Demographic Characteristics	Group I (Implantation Failure)—25 Patients	Group II (Successful Implantation)—14 Patients	Group III (Successful Implantation in A Frozen Embryo Transfer Protocol)—4 Patients	*p*
Body mass index (BMI) *	22.1 (2.0)	22.2 (1.4)	22.0 (1.8)	>0.05
Menstrual cycle length (days) *	29.9 (1.6)	28.4 (0.9)	29.0 (2.7)	>0.05
Average age (years) *	32.3 (3.5)	32.0 (3.1)	30.0 (2.4)	>0,05
Follicle-stimulating hormone level on day 2–3 of menstrual cycle (mIU/mL) *	7.8 (1.4)	7.2 (1.4)	5.8 (0.5)	>0.05
Anti-mullerian hormone level (ng/mL) *	2.4 (1.0)	2.4 (0.7)	2.8 (0.5)	>0.05
Antral follicle count on day 2–3 of the menstrual cycle *	8.2 (1.6)	7.8 (1.5)	8.3 (1.0)	>0.05
Male factor infertility **	18 patients (64%) had male factor infertility	14 patients (78%) had male factor infertility	2 patients (50%) had male factor infertility	>0.05
Primary infertility **	13 patients (46%) had primary infertility	8 patients (44%) had primary infertility	2 patients (50%) had primary infertility	>0.05
Assisted hatching **	In 20 (71%) patients, assisted hatching was performed	In 10 (56%) patients, assisted hatching was performed	In 2 (50%) patients, assisted hatching was performed	>0.05
Number of oocytes retrieved *	7.8 (3.7)	7.4 (4.5)	8.0 (3.7)	>0.05
Number of blastocysts *	1.4 (1.3)	1.4 (1.2)	2.8 (2.0)	>0.05

* Data are presented as an arithmetic mean (M) and standard deviation (SD) in the format of M(SD), specifying the significant differences when using ANOVA test; ** data are presented as absolute N numbers and percentages of the total number of patients in group P in the format of N(P%), specifying the significant differences when using the $\chi 2$ test. Comment: Group III (successful implantation in a frozen embryo transfer protocol) coincides with Group I (implantation failure), since patients from Group I were transferred to Group III after their previous failed IVF.

The ethics committee of the National Medical Research Center for Obstetrics, Gynecology, and Perinatology, named after Academician V.I. Kulakov of Ministry of Healthcare of the Russian Federation, approved this study (Ethic's committee approval protocol No13, approval date: 6 December 2013).

3.2. RNA Isolation from Embryo Culture Medium and Blastocoele Fluid

Twenty-five microliters of embryo culture medium or several nanoliters of blastocoele fluid adjusted to 200 µL with 0.9% NaCl were treated with 1000 µL of QIAzol Lysis Reagent (Qiagen, Hilden, Germany), followed by mixing with 200 µL of chloroform, centrifugation for 15 min at 12,000 g (4 °C), collection of 600 µL aqua phase, and RNA isolation using the miRNeasy Serum/Plasma Kit (Qiagen, Hilden, Germany).

3.3. sncRNA Deep Sequencing

cDNA libraries were synthesized using 7 µL of the 14 µL total RNA column eluate (miRNeasy Serum/Plasma Kit, Qiagen, Hilden, Germany), extracted from embryo culture medium and blastocoele fluid using the NEBNext® Multiplex Small RNA Library Prep Set for Illumina® (Set11 and Set2, New England Biolab®, Frankfurt am Main, Germany), amplified for 30 PCR cycles, and sequenced on the NextSeq 500 platform (Illumina, San Diego, CA, USA). The adapters were removed with Cutadapt. All trimmed reads shorter than 16 bp and longer than 35 bp were filtered, and only reads with a mean quality higher than 15 were retained. The remaining reads were mapped to the GRCh38.p15 human genomes miRBase v21 and piRNABase with the bowtie aligner [61]. Aligned reads were counted with the featureCount tool from the Subread package [62] and with the fracOverlap 0.9 option, so the whole read was forced to have a 90% intersection with sncRNA features. Differential expression analysis of the sncRNA count data was performed with the DESeq2 package [63].

3.4. Reverse Transcription and Quantitative Real-Time PCR

Five microliters of the 14 μL total RNA column eluate (miRNeasy Serum/Plasma Kit, Qiagen, Hilden, Germany) extracted from the embryo culture medium was converted into cDNA in a reaction mixture (20 μL) containing 1x Hispec buffer, 1x Nucleics mix, and miScript RT, according to the miScript® II RT Kit protocol (Qiagen, Hilden, Germany); then, the sample volume was adjusted with deionized water to 200 μL. The synthesized cDNA (3 μL) was used as a template for real-time PCR using a forward primer specific for the studied sncRNA (Table 1) and the miScript SYBR Green PCR Kit (Qiagen, Hilden, Germany). The following PCR conditions were used: (1) 15 min at 95 °C and (2) 40 cycles at 94 °C for 15 s, an optimized annealing temperature (45–60 °C) for 30 s, and 70 °C for 30 s, followed by heating the reaction mixture from 65 to 95 °C by 0.1 °C to plot the melting curve of the PCR product in a StepOnePlus™ thermocycler (Applied Biosystems, Foster City, CA, USA). The relative expression of sncRNA in the embryo culture medium was determined by the ΔΔCt method using hsa-piR023338 (DQ601914, GenBank, available online: https://www.ncbi.nlm.nih.gov/genbank/) as the reference RNA and culture medium without any embryo incubated for 4 days at 37 °C as a reference sample. hsa-piR023338 was chosen as the reference RNA due to its identical expression level in all 87 analyzed samples.

3.5. Data Processing

Scripts written in R were used for the resulting data processing [62,64]. The Shapiro–Wilk test was used to test whether the analyzed parameters were normally distributed. The $\chi 2$ test was used for comparing categorical variables. The ANOVA test was used for the analysis of the three groups of normally distributed parameters. Finally, the Mann–Whitney U-test was used for the pairwise comparison of the non-normally distributed parameters. Absolute numbers (N) and percentages of the total number of patients in a group (P) in the N (P%) format were used to describe categorical binary data. The arithmetic mean (M) and standard deviation (SD) in M (SD) format were used to evaluate the normally distributed quantitative data. Non-normally distributed parameters were described as medians (Me) and quartiles (Q1 and Q3) in the Me (Q1; Q3) format. The Spearman correlation analysis was performed, since both quantitative and qualitative data were analyzed. A 95% confidence interval for the correlation coefficients was determined using the Fisher transform strategy. The threshold for the statistical significance was $p \leq 0.05$. The p-value was specified in the $p < 0.001$ format if it was less than 0.001. In addition, the morphological characteristics of the embryos and the obtained experimental data were analyzed using PLS-DA [65].

4. Conclusions

In recent years, the attention of scientists has been drawn to the study of the role of sncRNA in embryogenesis. In the course of the present research work, a complex interrelation of sncRNA expression patterns in the culture media of embryos at various development grades on day 4 after fertilization was revealed. These findings probably reflect the manifestation of the fine regulation system of signaling pathways which is necessary for implementation of the embryogenesis program. The pathway analysis of miRNA and piRNA target genes emphasizes the role of sncRNA described in this article in the control of chromatin structure, genome stability, DNA replication, gene transcription, protein synthesis, cell survival, migration and invasion, cell cycle progression, and cellular proliferation and differentiation, i.e., the processes that determine the normal course of gametogenesis and embryogenesis. Correlations between the sncRNA expression patterns and the number of oocyte–cumulus complexes, metaphase II oocytes, and two pronuclei embryos as well as the spermatozoid count, the percentage of linearly motile spermatozoids, the embryo development grade, and embryo implantation provide evidence of the roles of these molecules in human reproductive system regulation.

Author Contributions: Concept and design of the study, A.V.T. and E.A.K.; collection of materials and creation of a clinical database of samples, Y.S.D. and N.P.M.; generation of experimental data, A.V.T.; statistical data processing, V.V.C.; data interpretation and manuscript writing, A.V.T., Y.S.D. and V.V.C.; data curation and project administration, E.A.K. and G.T.S. All authors approved the submitted version of the manuscript.

References

1. Gardner, D.K.; Balaban, B. Assessment of human embryo development using morphological criteria in an era of time-lapse, algorithms and 'OMICS': Is looking good still important? *Mol. Hum. Reprod.* **2016**, *22*, 704–718. [CrossRef] [PubMed]
2. Gardner, D.K.; Schoolcraft, W.B. In-vitro culture of human blastocysts. In *Towards Reproductive Certainty: Infertility and Genetics Beyond 1999: The Plenary Proceedings of the 11th World Congress on In Vitro Fertilization and Human Reproductive Genetics*; Parthenon Press: New York, NY, USA, 1999; pp. 378–388.
3. Gardner, D.K.; Lane, M.; Stevens, J.; Schoolcraft, W.B. Noninvasive assessment of human embryo nutrient consumption as a measure of developmental potential. *Fertil. Steril.* **2001**, *76*, 1175–1180. [CrossRef]
4. Katz-Jaffe, M.G.; Linck, D.W.; Schoolcraft, W.B.; Gardner, D.K. A proteomic analysis of mammalian preimplantation embryonic development. *Reproduction* **2005**, *130*, 899–905. [CrossRef] [PubMed]
5. Poli, M.; Ori, A.; Child, T.; Jaroudi, S.; Spath, K.; Beck, M.; Wells, D. Characterization and quantification of proteins secreted by single human embryos prior to implantation. *EMBO Mol. Med.* **2015**, *7*, 1465–1479. [CrossRef] [PubMed]
6. Hale, B.J.; Yang, C.-X.; Ross, J.W. Small RNA regulation of reproductive function. *Mol. Reprod. Dev.* **2014**, *81*, 148–159. [CrossRef] [PubMed]
7. Chua, J.H.; Armugam, A.; Jeyaseelan, K. MicroRNAs: Biogenesis, function and applications. *Curr. Opin. Mol. Ther.* **2009**, *11*, 189–199.
8. Houwing, S.; Kamminga, L.M.; Berezikov, E.; Cronembold, D.; Girard, A.; van den Elst, H.; Filippov, D.V.; Blaser, H.; Raz, E.; Moens, C.B.; et al. A role for Piwi and piRNAs in germ cell maintenance and transposon silencing in Zebrafish. *Cell* **2007**, *129*, 69–82. [CrossRef]
9. Girard, A.; Sachidanandam, R.; Hannon, G.J.; Carmell, M.A. A germline-specific class of small RNAs binds mammalian Piwi proteins. *Nature* **2006**, *442*, 199–202. [CrossRef]
10. Hirakata, S.; Siomi, M.C. piRNA biogenesis in the germline: From transcription of piRNA genomic sources to piRNA maturation. *Biochim. Biophys. Acta Gene Regul. Mech.* **2016**, *1859*, 82–92. [CrossRef]
11. Karakaya, C.; Guzeloglu-Kayisli, O.; Uyar, A.; Kallen, A.N.; Babayev, E.; Bozkurt, N.; Unsal, E.; Karabacak, O.; Seli, E. Poor ovarian response in women undergoing in vitro fertilization is associated with altered microRNA expression in cumulus cells. *Fertil. Steril.* **2015**, *103*, 1469–1476.e3. [CrossRef]
12. Machtinger, R.; Rodosthenous, R.S.; Adir, M.; Mansour, A.; Racowsky, C.; Baccarelli, A.A.; Hauser, R. Extracellular microRNAs in follicular fluid and their potential association with oocyte fertilization and embryo quality: An exploratory study. *J. Assist. Reprod. Genet.* **2017**, *34*, 525–533. [CrossRef] [PubMed]
13. Sang, Q.; Yao, Z.; Wang, H.; Feng, R.; Wang, H.; Zhao, X.; Xing, Q.; Jin, L.; He, L.; Wu, L.; et al. Identification of MicroRNAs in Human Follicular Fluid: Characterization of MicroRNAs That Govern Steroidogenesis in Vitro and Are Associated With Polycystic Ovary Syndrome in Vivo. *J. Clin. Endocrinol. Metab.* **2013**, *98*, 3068–3079. [CrossRef] [PubMed]
14. Battaglia, R.; Vento, M.E.; Ragusa, M.; Barbagallo, D.; La Ferlita, A.; Di Emidio, G.; Borzi, P.; Artini, P.G.; Scollo, P.; Tatone, C.; et al. MicroRNAs Are Stored in Human MII Oocyte and Their Expression Profile Changes in Reproductive Aging. *Biol. Reprod.* **2016**, *95*, 131. [CrossRef] [PubMed]
15. Abd El Naby, W.S.; Hagos, T.H.; Hossain, M.M.; Salilew-Wondim, D.; Gad, A.Y.; Rings, F.; Cinar, M.U.;

Tholen, E.; Looft, C.; Schellander, K.; et al. Expression analysis of regulatory microRNAs in bovine cumulus oocyte complex and preimplantation embryos. *Zygote* **2013**, *21*, 31–51. [CrossRef] [PubMed]

16. Roovers, E.F.; Rosenkranz, D.; Mahdipour, M.; Han, C.-T.; He, N.; Chuva de Sousa Lopes, S.M.; van der Westerlaken, L.A.J.; Zischler, H.; Butter, F.; Roelen, B.A.J.; et al. Piwi Proteins and piRNAs in Mammalian Oocytes and Early Embryos. *Cell Rep.* **2015**, *10*, 2069–2082. [CrossRef] [PubMed]
17. Luo, L.-F.; Hou, C.-C.; Yang, W.-X. Small non-coding RNAs and their associated proteins in spermatogenesis. *Gene* **2016**, *578*, 141–157. [CrossRef] [PubMed]
18. Yuan, S.; Schuster, A.; Tang, C.; Yu, T.; Ortogero, N.; Bao, J.; Zheng, H.; Yan, W. Sperm-borne miRNAs and endo-siRNAs are important for fertilization and preimplantation embryonic development. *Development* **2016**, *143*, 635–647. [CrossRef]
19. Svoboda, P.; Flemr, M. The role of miRNAs and endogenous siRNAs in maternal-to-zygotic reprogramming and the establishment of pluripotency. *EMBO Rep.* **2010**, *11*, 590–597. [CrossRef]
20. Giraldez, A.J.; Mishima, Y.; Rihel, J.; Grocock, R.J.; Van Dongen, S.; Inoue, K.; Enright, A.J.; Schier, A.F. Zebrafish MiR-430 promotes deadenylation and clearance of maternal mRNAs. *Science* **2006**, *312*, 75–79. [CrossRef]
21. Siomi, M.C.; Sato, K.; Pezic, D.; Aravin, A.A. PIWI-interacting small RNAs: The vanguard of genome defence. *Nat. Rev. Mol. Cell Biol.* **2011**, *12*, 246–258. [CrossRef]
22. Yang, Q.; Gu, W.-W.; Gu, Y.; Yan, N.-N.; Mao, Y.-Y.; Zhen, X.-X.; Wang, J.-M.; Yang, J.; Shi, H.-J.; Zhang, X.; et al. Association of the peripheral blood levels of circulating microRNAs with both recurrent miscarriage and the outcomes of embryo transfer in an in vitro fertilization process. *J. Transl. Med.* **2018**, *16*, 186. [CrossRef] [PubMed]
23. Capalbo, A.; Ubaldi, F.M.; Cimadomo, D.; Noli, L.; Khalaf, Y.; Farcomeni, A.; Ilic, D.; Rienzi, L. MicroRNAs in spent blastocyst culture medium are derived from trophectoderm cells and can be explored for human embryo reproductive competence assessment. *Fertil. Steril.* **2016**, *105*, 225–235.e3. [CrossRef] [PubMed]
24. Noli, L.; Capalbo, A.; Dajani, Y.; Cimadomo, D.; Bvumbe, J.; Rienzi, L.; Ubaldi, F.M.; Ogilvie, C.; Khalaf, Y.; Ilic, D. Human Embryos Created by Embryo Splitting Secrete Significantly Lower Levels of miRNA-30c. *Stem Cells Dev.* **2016**, *25*, 1853–1862. [CrossRef] [PubMed]
25. Rosenbluth, E.M.; Shelton, D.N.; Sparks, A.E.T.; Devor, E.; Christenson, L.; Van Voorhis, B.J. MicroRNA expression in the human blastocyst. *Fertil. Steril.* **2013**, *99*, 855–861.e3. [CrossRef] [PubMed]
26. No, J.; Zhao, M.; Lee, S.; Ock, S.A.; Nam, Y.; Hur, T.-Y. Enhanced in vitro maturation of canine oocytes by oviduct epithelial cell co-culture. *Theriogenology* **2018**, *105*, 66–74. [CrossRef] [PubMed]
27. Ji, J.; Liu, Y.; Tong, X.H.; Luo, L.; Ma, J.; Chen, Z. The optimum number of oocytes in IVF treatment: An analysis of 2455 cycles in China. *Hum. Reprod.* **2013**, *28*, 2728–2734. [CrossRef]
28. Chen, Y.; Xu, X.; Wang, Q.; Zhang, S.; Jiang, L.; Zhang, C.; Ge, Z. Optimum oocyte retrieved and transfer strategy in young women with normal ovarian reserve undergoing a long treatment protocol: A retrospective cohort study. *J. Assist. Reprod. Genet.* **2015**, *32*, 1459–1467. [CrossRef]
29. La Ferlita, A.; Battaglia, R.; Andronico, F.; Caruso, S.; Cianci, A.; Purrello, M.; Di Pietro, C. Non-Coding RNAs in Endometrial Physiopathology. *Int. J. Mol. Sci.* **2018**, *19*, 2120. [CrossRef]
30. Platt, R.N.; Vandewege, M.W.; Ray, D.A. Mammalian transposable elements and their impacts on genome evolution. *Chromosom. Res.* **2018**, *26*, 25–43. [CrossRef]
31. Russell, S.; Patel, M.; Gilchrist, G.; Stalker, L.; Gillis, D.; Rosenkranz, D.; LaMarre, J. Bovine piRNA-like RNAs are associated with both transposable elements and mRNAs. *Reproduction* **2017**, *153*, 305–318. [CrossRef]
32. Sobala, A.; Hutvagner, G. Transfer RNA-derived fragments: Origins, processing, and functions. *Wiley Interdiscip. Rev. RNA* **2011**, *2*, 853–862. [CrossRef] [PubMed]
33. Bühler, M.; Spies, N.; Bartel, D.P.; Moazed, D. TRAMP-mediated RNA surveillance prevents spurious entry of RNAs into the Schizosaccharomyces pombe siRNA pathway. *Nat. Struct. Mol. Biol.* **2008**, *15*, 1015–1023. [CrossRef] [PubMed]
34. Saito, K.; Inagaki, S.; Mituyama, T.; Kawamura, Y.; Ono, Y.; Sakota, E.; Kotani, H.; Asai, K.; Siomi, H.; Siomi, M.C. A regulatory circuit for piwi by the large Maf gene traffic jam in Drosophila. *Nature* **2009**, *461*, 1296–1299. [CrossRef] [PubMed]
35. Robine, N.; Lau, N.C.; Balla, S.; Jin, Z.; Okamura, K.; Kuramochi-Miyagawa, S.; Blower, M.D.; Lai, E.C. A Broadly Conserved Pathway Generates 3′UTR-Directed Primary piRNAs. *Curr. Biol.* **2009**, *19*, 2066–2076. [CrossRef] [PubMed]

36. Hirano, T.; Iwasaki, Y.W.; Lin, Z.Y.-C.; Imamura, M.; Seki, N.M.; Sasaki, E.; Saito, K.; Okano, H.; Siomi, M.C.; Siomi, H. Small RNA profiling and characterization of piRNA clusters in the adult testes of the common marmoset, a model primate. *RNA* **2014**, *20*, 1223–1237. [CrossRef] [PubMed]
37. Zhang, P.; Kang, J.-Y.; Gou, L.-T.; Wang, J.; Xue, Y.; Skogerboe, G.; Dai, P.; Huang, D.-W.; Chen, R.; Fu, X.-D.; et al. MIWI and piRNA-mediated cleavage of messenger RNAs in mouse testes. *Cell Res.* **2015**, *25*, 193–207. [CrossRef] [PubMed]
38. Gou, L.-T.; Dai, P.; Yang, J.-H.; Xue, Y.; Hu, Y.-P.; Zhou, Y.; Kang, J.-Y.; Wang, X.; Li, H.; Hua, M.-M.; et al. Pachytene piRNAs instruct massive mRNA elimination during late spermiogenesis. *Cell Res.* **2014**, *24*, 680–700. [CrossRef] [PubMed]
39. Rouget, C.; Papin, C.; Boureux, A.; Meunier, A.-C.; Franco, B.; Robine, N.; Lai, E.C.; Pelisson, A.; Simonelig, M. Maternal mRNA deadenylation and decay by the piRNA pathway in the early Drosophila embryo. *Nature* **2010**, *467*, 1128–1132. [CrossRef]
40. Roth, L.M.; Michal, M.; Michal, M.; Cheng, L. Protein expression of the transcription factors DMRT1, TCLF5, and OCT4 in selected germ cell neoplasms of the testis. *Hum. Pathol.* **2018**, *82*, 68–75. [CrossRef]
41. Duquette, P.M.; Lamarche-Vane, N. Rho GTPases in embryonic development. *Small GTPases* **2014**, *5*, e972857. [CrossRef]
42. Ringrose, L.; Paro, R. Epigenetic Regulation of Cellular Memory by the Polycomb and Trithorax Group Proteins. *Annu. Rev. Genet.* **2004**, *38*, 413–443. [CrossRef] [PubMed]
43. Lund, A.H.; van Lohuizen, M. Polycomb complexes and silencing mechanisms. *Curr. Opin. Cell Biol.* **2004**, *16*, 239–246. [CrossRef] [PubMed]
44. Van der Vlag, J.; Otte, A.P. Transcriptional repression mediated by the human polycomb-group protein EED involves histone deacetylation. *Nat. Genet.* **1999**, *23*, 474–478. [CrossRef] [PubMed]
45. Shao, Z.; Raible, F.; Mollaaghababa, R.; Guyon, J.R.; Wu, C.; Bender, W.; Kingston, R.E. Stabilization of Chromatin Structure by PRC1, a Polycomb Complex. *Cell* **1999**, *98*, 37–46. [CrossRef]
46. Deshpande, A.M.; Akunowicz, J.D.; Reveles, X.T.; Patel, B.B.; Saria, E.A.; Gorlick, R.G.; Naylor, S.L.; Leach, R.J.; Hansen, M.F. PHC3, a component of the hPRC-H complex, associates with 2A7E during G0 and is lost in osteosarcoma tumors. *Oncogene* **2007**, *26*, 1714–1722. [CrossRef] [PubMed]
47. Ohkumo, T.; Masutani, C.; Eki, T.; Hanaoka, F. Deficiency of the Caenorhabditis elegans DNA polymerase eta homologue increases sensitivity to UV radiation during germ-line development. *Cell Struct. Funct.* **2006**, *31*, 29–37. [CrossRef]
48. Bürglin, T.R. *Homeodomain Subtypes and Functional Diversity*; Springer: Dordrecht, The Netherlands, 2011; pp. 95–122. [CrossRef]
49. Mo, J.-S.; Park, H.W.; Guan, K.-L. The Hippo signaling pathway in stem cell biology and cancer. *EMBO Rep.* **2014**, *15*, 642–656. [CrossRef]
50. Hers, I.; Vincent, E.E.; Tavaré, J.M. Akt signalling in health and disease. *Cell. Signal.* **2011**, *23*, 1515–1527. [CrossRef]
51. Varelas, X.; Sakuma, R.; Samavarchi-Tehrani, P.; Peerani, R.; Rao, B.M.; Dembowy, J.; Yaffe, M.B.; Zandstra, P.W.; Wrana, J.L. TAZ controls Smad nucleocytoplasmic shuttling and regulates human embryonic stem-cell self-renewal. *Nat. Cell Biol.* **2008**, *10*, 837–848. [CrossRef]
52. Kim, S.-Y.; Baek, K.-H. TGF-β signaling pathway mediated by deubiquitinating enzymes. *Cell. Mol. Life Sci.* **2019**, *76*, 653–665. [CrossRef]
53. Song, J.L.; Nigam, P.; Tektas, S.S.; Selva, E. microRNA regulation of Wnt signaling pathways in development and disease. *Cell. Signal.* **2015**, *27*, 1380–1391. [CrossRef] [PubMed]
54. Kuscu, N.; Celik-Ozenci, C. FOXO1, FOXO3, AND FOXO4 are differently expressed during mouse oocyte maturation and preimplantation embryo development. *Gene Expr. Patterns* **2015**, *18*, 16–20. [CrossRef] [PubMed]
55. Perkel, K.J.; Madan, P. Spent culture medium analysis from individually cultured bovine embryos demonstrates metabolomic differences. *Zygote* **2017**, *25*, 662–674. [CrossRef] [PubMed]
56. Brison, D.R.; Houghton, F.D.; Falconer, D.; Roberts, S.A.; Hawkhead, J.; Humpherson, P.G.; Lieberman, B.A.; Leese, H.J. Identification of viable embryos in IVF by non-invasive measurement of amino acid turnover. *Hum. Reprod.* **2004**, *19*, 2319–2324. [CrossRef] [PubMed]
57. Stephen, A.G.; Esposito, D.; Bagni, R.K.; McCormick, F. Dragging Ras Back in the Ring. *Cancer Cell* **2014**, *25*, 272–281. [CrossRef] [PubMed]

58. Gerttula, S.; Jin, Y.S.; Anderson, K. V Zygotic expression and activity of the Drosophila Toll gene, a gene required maternally for embryonic dorsal-ventral pattern formation. *Genetics* **1988**, *119*.
59. Vijay, K. Toll-like receptors in immunity and inflammatory diseases: Past, present, and future. *Int. Immunopharmacol.* **2018**, *59*, 391–412. [CrossRef]
60. Kim, J.; Lee, J.; Jun, J.H. Identification of differentially expressed microRNAs in outgrowth embryos compared with blastocysts and non-outgrowth embryos in mice. *Reprod. Fertil. Dev.* **2019**, *31*, 645. [CrossRef]
61. Langmead, B.; Trapnell, C.; Pop, M.; Salzberg, S.L. Ultrafast and memory-efficient alignment of short DNA sequences to the human genome. *Genome Biol.* **2009**, *10*, R25. [CrossRef]
62. R: The R Project for Statistical Computing. Available online: https://www.r-project.org/ (accessed on 5 June 2019).
63. Love, M.I.; Huber, W.; Anders, S. Moderated estimation of fold change and dispersion for RNA-seq data with DESeq2. *Genome Biol.* **2014**, *15*, 550. [CrossRef]
64. Dayal, V. *An Introduction to R for Quantitative Economics*; SpringerBriefs in Economics; Springer India: New Delhi, India, 2015; ISBN 978-81-322-2339-9. [CrossRef]
65. Wold, S.; Sjöström, M.; Eriksson, L. PLS-regression: A basic tool of chemometrics. *Chemom. Intell. Lab. Syst.* **2001**, *58*, 109–130. [CrossRef]

6

Why the Gold Standard Approach by Mammography Demands Extension by Multiomics? Application of Liquid Biopsy miRNA Profiles to Breast Cancer Disease Management

Pavol Zubor [1,2,*,†], Peter Kubatka [2,3,*,†], Karol Kajo [4,5], Zuzana Dankova [2], Hubert Polacek [6,7], Tibor Bielik [1], Erik Kudela [1], Marek Samec [1,2], Alena Liskova [1], Dominika Vlcakova [1], Tatiana Kulkovska [1], Igor Stastny [1,2], Veronika Holubekova [2], Jan Bujnak [8,9], Zuzana Laucekova [1], Dietrich Büsselberg [10], Mariusz Adamek [11], Walther Kuhn [12], Jan Danko [1] and Olga Golubnitschaja [13,14,15]

1 Department of Obstetrics and Gynaecology, Jessenius Faculty of Medicine, Comenius University in Bratislava, Martin University Hospital, 03659 Martin, Slovak Republic; tbielik57@gmail.com (T.B.); Erik.Kudela@jfmed.uniba.sk (E.K.); marek.samec@gmail.com (M.S.); alenka.liskova@gmail.com (A.L.); Dominika.Vlcak@gmail.com (D.V.); tatiana.kulkovska@gmail.com (T.K.); igor.stastny@uniba.sk (I.S.); zuzana_laucekova@yahoo.com (Z.L.); danko@jfmed.uniba.sk (J.D.)

2 Division of Oncology, Biomedical Center Martin, Jessenius Faculty of Medicine, Comenius University in Bratislava, 03601 Martin, Slovak Republic; zuzana.dankova@jfmed.uniba.sk (Z.D.); holubekova@jfmed.uniba.sk (V.H.)

3 Department of Medical Biology, Jessenius Faculty of Medicine, Comenius University in Bratislava, 03601 Martin, Slovak Republic

4 Department of Pathology, St. Elizabeth Cancer Institute Hospital, 81250 Bratislava, Slovak Republic; kkajo@ousa.sk

5 Biomedical Research Centre, Slovak Academy of Sciences, 81439 Bratislava, Slovak Republic

6 Center for Cancer Prevention, 03659 Martin, Slovak Republic; polacek@jfmed.uniba.sk

7 Department of Radiology, Jessenius Faculty of Medicine, Comenius University in Bratislava, 03659 Martin, Slovak Republic

8 Department of Obstetrics and Gynaecology, Kukuras Michalovce Hospital, 07101 Michalovce, Slovak Republic; janbujnak@hotmail.com

9 Oncogynecology Unit, Penta Hospitals International, Svet Zdravia, Michalovce 07101, Slovak Republic

10 Weill Cornell Medicine in Qatar, Qatar Foundation-Education City, Doha 24144, Qatar; dib2015@qatar-med.cornell.edu

11 Department of Thoracic Surgery, Faculty of Medicine and Dentistry, Medical University of Silesia, 40055 Katowice, Poland; m.adamek@e.pl

12 Centre of Obstetrics, Gynaecology and Gynaecologic Oncology, DonauIsar Klinikum Deggendorf-Dingolfing-Landau, 94469 Deggendorf, Germany; walther.kuhn@donau-isar-klinikum.de

13 Radiological Hospital, Rheinische Friedrich-Wilhelms-University of Bonn, 53105 Bonn, Germany; Olga.Golubnitschaja@ukbonn.de

14 Breast Cancer Research Centre, Rheinische Friedrich-Wilhelms-University of Bonn, 53105 Bonn, Germany

15 Centre for Integrated Oncology, Cologne-Bonn, Rheinische Friedrich-Wilhelms-University of Bonn, 53105 Bonn, Germany

* Correspondence: pavol.zubor@jfmed.uniba.sk (P.Z.); kubatka@jfmed.uniba.sk (P.K.)

† These authors contributed equally to this work.

Abstract: In the global context, the epidemic of breast cancer (BC) is evident for the early 21st century. Evidence shows that national mammography screening programs have sufficiently reduced BC related mortality. Therefore, the great utility of the mammography-based screening is not an issue. However, both false positive and false negative BC diagnosis, excessive biopsies, and irradiation linked to mammography application, as well as sub-optimal mammography-based screening, such as in the case of high-dense breast tissue in young females, altogether increase awareness among the experts regarding the limitations of mammography-based screening. Severe concerns regarding the mammography as the "golden standard" approach demanding complementary tools to cover the evident deficits led the authors to present innovative strategies, which would sufficiently improve the quality of the BC management and services to the patient. Contextually, this article provides insights into mammography deficits and current clinical data demonstrating the great potential of non-invasive diagnostic tools utilizing circulating miRNA profiles as an adjunct to conventional mammography for the population screening and personalization of BC management.

Keywords: breast cancer; screening; liquid biopsy; omics; multi-level diagnostics; individualized patient profile; miRNA; mammography; predictive and preventive approach; personalized medicine

1. Introduction

Cancer is one of the leading healthcare burdens worldwide. In 2018 18.1 million (95% UI: 17.5–18.7 million) new cases of cancer (17 million excluding non-melanoma skin cancer) and 9.6 million (95% UI: 9.3–9.8 million) cancer related deaths (9.5 million excluding non-melanoma skin cancer) have been estimated worldwide. Figure 1 summarizes most frequent cancer types [1]. To this end, a big portion of cancer-related deaths can be avoided at the level of primary prevention: innovative screening programs and targeted preventive measures are essential tools to identify and mitigate modifiable risks individually and in a timely manner [2–4].

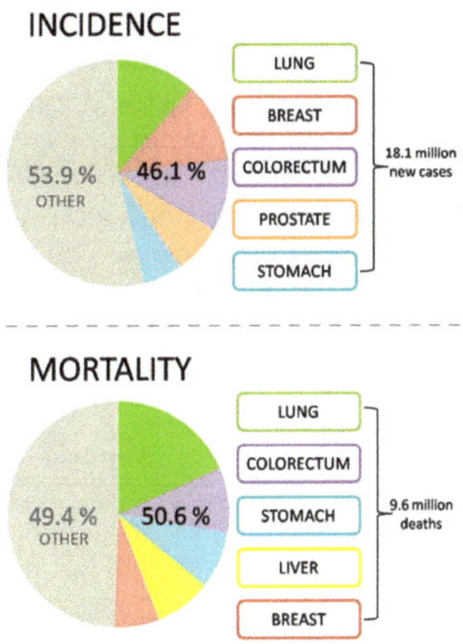

Figure 1. Global cancer statistics demonstrating the most prevalent cancer types; the image is based on the data published by 2019 [1].

Further, at the level of secondary prevention, cancer-diagnosed patients could demonstrate longer survival rates and have much better quality of life if the disease-management process would adapt treatment algorithms that are tailored exactly to the individualized patient profiles [5,6]. For the population screening, early and predictive diagnosis, as well as prognosis and disease monitoring, a multi-level diagnostics (multi-omics, sub-cellular and medical imaging) method utilizing the great information potential of liquid biopsy is considered to be the most appropriate tool [7] and is thoroughly analyzed in the current paper using the example of breast cancer (BC) management.

2. Breast Cancer in the Context of Global Cancer Mortality

The 28 Member States of the European Union (EU-28) with a population of 504.6 million had 5.0 million deaths in 2012 with more than one fourth attributable to cancer [8,9]. Detailed analyses revealed 29.2% of deaths among men and 22.5% of deaths among women were caused by cancer alone in 2012. Next to the disorders of the circulatory system (cerebrovascular and heart diseases), cancer is the second most common cause of deaths in the EU, being one of the major public health burdens in the European Union (EU). According to the International Agency for Research on Cancer (IARC, 2012), out of 1.26 million cancer-related deaths in the EU-28, breast cancer (BC) alone is responsible for 91,500 deaths annually [10]. In the global context, the epidemic of BC is associated with a number of external and internal risk factors attributed to the early 21st century [11].

BC is the most frequent tumor in female populations worldwide, with an incidence rate of 43.1 per 100,000 world age-standardized rate (ASR-W), a mortality rate of 12.9 per 100,000 ASR-W, and a 5-year prevalence of 239.9. In low-, middle- and high-income countries the incidence rates are persistently increasing [12]. To this end, the contribution to the BC incidence by the European Region is higher than the global average [13]. Specifically for the EU-28, the incidence and mortality rates are as high as 80.3 and 14.4 per 100,000 ASR-W, respectively [14]. Most of EU-28, including the biggest sufferers such as the UK [15], France [16], Italy [17], Germany [18], and Belgium [19], have established national programs for BC screening by mammography as the golden standard for reducing mortality from BC. Evidence shows that national mammography screening programs have sufficiently reduced BC related mortality [20,21]. Therefore, the adequacy and usefulness of the mammography-based screening for women aged by 50 to 74 years is generally well-accepted [22]. Therefore, the great utility of the mammography-based screening is not an issue. However, it is important to decide for which population the mammography-based screening should be considered to be optimal i.e., the "golden standard" approach (see Table 1). Both false positive and false negative BC diagnosis [23], excessive biopsies and irradiation linked to mammography application, as well as sub-optimal mammography-based screening, e.g., in case of high-dense breast tissue in young females [24], altogether increases awareness among the experts regarding the mammography-based screening limitations. Severe concerns regarding the mammography as the "golden standard" approach demanding complementary tools to cover the evident deficits [25] led us to present innovative strategies, which would sufficiently improve the quality of the BC management and services to the patient.

Table 1. Categories of women and mammography screening applicability.

Categories of Women	Applicability of Mammography
Postmenopausal with fatty breasts	The breast density gradually decreases after menopause, applicable every two years
Young with very dense breast parenchyma	Low diagnostic sensitivity
Pregnancy	Mammography is not contraindicated in the first and second trimesters (sufficient shading of the uterus is necessary); USG and MRI are predominant diagnostic methods
Family history of BC	An annual mammography, supplemental imaging (MRI, USG) in dense breast tissue
Genetic predisposition to BC	An annual mammogram starting at age 25
Diagnosis of atypical hyperplasia or lobular carcinoma in situ	An annual mammogram beginning at the time of diagnosis
Average BC risk with no symptoms	An annual mammogram combined with USG in dense breast tissue

BC, breast cancer; MRI, magnetic resonance imagining; USG, ultrasonography.

Contextually, this article is aiming to provide insights into mammography deficits and current clinical data demonstrating the great potential of non-invasive diagnostic tool utilizing circulating miRNA profiles as an adjunct to conventional mammography for the population screening and personalization of BC management [6]. To this end, innovative screening strategies should consider primary [11] and secondary [26] levels of predictive and preventive medical approach, including both non-modifiable and modifiable risk factors [27] based on comprehensive individual patient profiles that means application of multi-omics [28,29], big data processing [30] and artificial intelligence such as machine learning approach [5]. Table 1 summarizes the categories of women with the mammography screening applicability.

3. Breast Cancer Screening by Mammography: An Evolution

Radiography is the oldest and most common form of medical imaging. A mammography is an x-ray medical method examining the breast using lower doses of radiation. Because these x-rays do not pass through breast tissue easily, the mammogram machine has two plates that compress the breast to spread the tissue apart, resulting in a more accurate image with less radiation [31]. Over the past decades, technology changed over from the analog to digital picturing.

The analog mammography used film as both a receptor and a display for the image to produce static, fixed images. The advantage of analog mammography linked with computed radiographic systems included much less costs than digital mammography, however the main disadvantage, that the image is far inferior to digital mammography and also the storage in protective sleeve required large amounts of space, was the reason for leaving it worldwide. Thus, the digital technology fluently re-placed the analog mammograms.

Digital mammography uses detectors that change x-rays into electrical signals (pixels), which are transferred into a digital receptor and converts x-rays energy into numbers. It produces an image displayed on a monitor or printed on high-resolution printer. In comparison with analogue film, digital mammography provides images with more contrast, allows image manipulation, and archive films which reduces the risk of misplacement or damage. Moreover, the digital detector provides a crisp image with no limitations on breast size and detects cancer cells earlier than analog mammography. This makes this method superior to film mammography [32,33]. Digital breast tomosynthesis mammography (DBT) is relatively new technology being developed to improve detection and characterization of breast lesions, especially in women with non-fatty breasts. The x-ray dose for a tomosynthesis image is similar to that of a regular mammogram. DBT creates a 3-dimensional (3D) picture of the breast using several low dose x-rays obtained at different angles. The breast is positioned and compressed in the same way as in a regular mammography, but unlike for regular mammography, the x-ray tube moves in a circular arc around the breast in DBT [34,35]. DBT provides an advantage in detection of breast masses compared to 2-dimensional (2D) mammography, since it allows the separation of the tissue layers and the noticeable reduction of occlusions caused by overlapping anatomical structures. While DBT slice images provide advantages for detecting mass lesions, it is more difficult to get an overview and evaluate the distribution of microcalcifications compared to 2D mammography images. Therefore, the parallel using of 2D mammography and DBT slice images seems to be necessary in clinical diagnostic practice. Due to the cumulative patient dose, using both 2D mammography and DBT at one session is not an acceptable screening method. However, the DBT method offers a possibility of reprocessing the tomosynthesis data to create a 2D mammography-like image (synthetic mammography, SMMG) from DBT image data at the dose level of a single DBT screening. The use of SMMG with DBT provides significant benefit of increased diagnostic accuracy compared with regular mammography [36]. Here, the initial evidence suggests that SMMG may reduce recall rates and increase cancer detection rates when added to digital MMG screening [37]. Figure 2 describes the advantages and disadvantages of analog versus digital mammography.

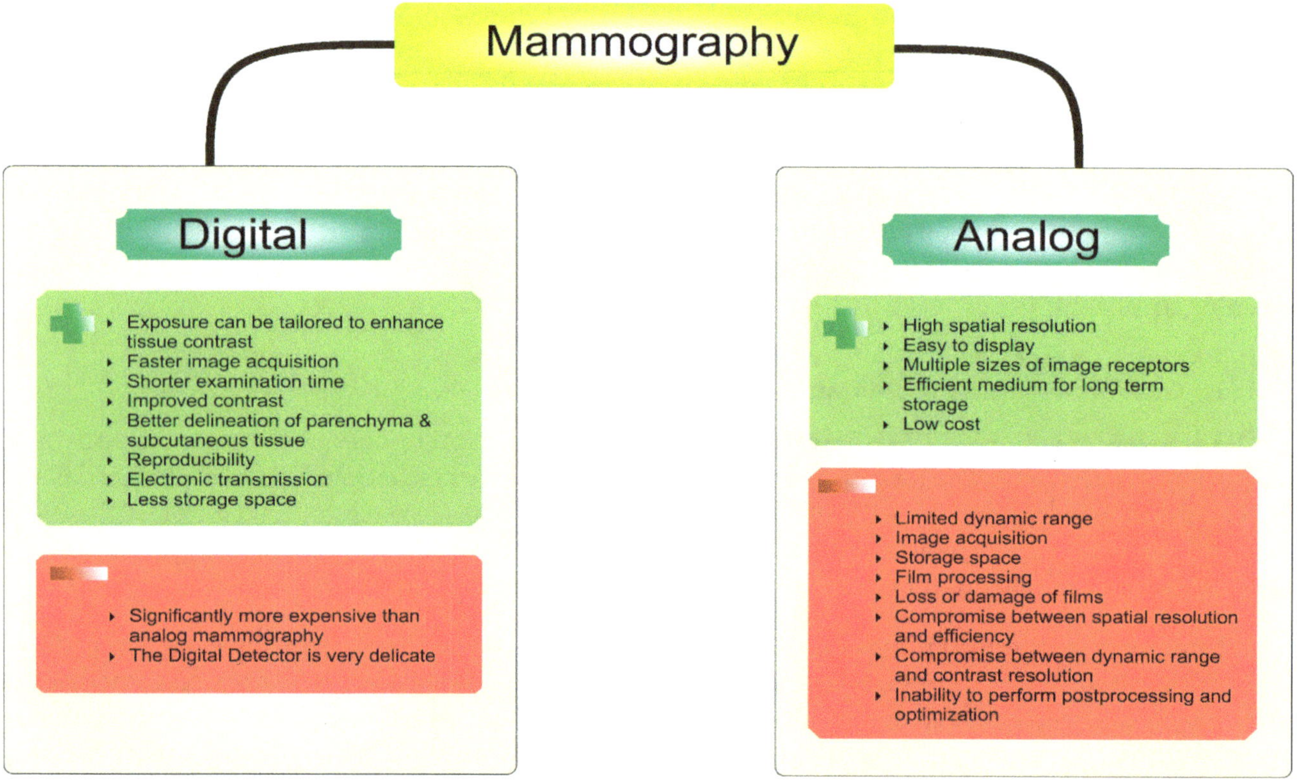

Figure 2. Advantages and disadvantages of digital versus analog mammography.

Regular screening for BC with mammography (along with other examinations such as breast self-examinations) is widely recommended with the aim to reduce mortality of BC. Although controversy remains over the best screening programs and to whom it should be offered (e.g., for the primary cancer free population, screening of secondary cancer after previous BC surgery, screening of population with different genetic background or different age-groups of screened population), these methods are regularly used in clinical practice, following local or national guidelines. Reflecting this, there are multiple approaches for BC screening. E.g., according to the Canadian Task Force on Preventive Health Care, the use of mammography for women at average risk of BC aged 40–74 years includes following recommendations: for women aged 40–49 years, it is recommended to not routinely screen with mammography; for women aged 50–69 years and also 70–74 years, it is recommended to routinely screen with mammography every two to three years [38].

Although the results of mammographic screening in women aged 50–70 years are sometimes disputed [39], there is a consensus among clinicians that BC screening of women in this age group is effective. On the other hand, there is no consensus about the value of BC mammographic screening among women who are aged 40–49 years having denser breasts compared to postmenopausal women [40]. Regarding women under 40 years of age, in which BC is rare and typically presents symptomatically, the best imaging modality is controversial. Routine screening of women in this age group in the absence of significant BC risk factors is not recommended. Some authors showed the superior sensitivity of ultrasound screening for BC in women under the age of 40 years; however, they noted that mammography and/or MRI remain essential adjuncts, particularly in the identification of multifocal disease [41]. The large body of analysis comparing ultrasound and mammography to evaluate women aged 30–39 with symptoms of possible BC has demonstrated that ultrasound screening is a superior diagnostic tool [42], which may impact on the consideration of current clinical practice guidelines that nowadays recommend mammography as the first evaluation in these women.

A special group for screening is women under 50 years who underwent breast conservation therapy, as those women may benefit from breast screening as an adjunct method to MMG. In this regard, the addition of MRI to annual MMG screening improves the detection of early-stage but biologically aggressive BC [43].

The last group, where much of the variability in BC screening programs exists, represents the high-risk population. The annual mammography in women with one or two first degree relatives with invasive BC starts 5 to 10 years younger than the youngest case in the family, but no earlier than age 25 and no later than age 40. Women with a breast biopsy showing atypical hyperplasia or lobular carcinoma in situ and following surgical management to rule out invasive carcinoma have annual mammography. Women with a history of chest wall radiation (i.e., mantle radiation for treatment of Hodgkin's lymphoma) at age 30 or younger have an annual mammography and breast screening MRI starting 5 to 10 years after radiation given, but starting no earlier than age 25 and no later than age 40. For women with *BRCA1* or *BRCA2* genes mutations, current guidelines recommend annual screening by clinical breast examination and mammography starting at age 30 [44]. However, these approaches may vary from country to country.

4. Breast Cancer Screening by Mammography and Profiling of Genetic Risk

Routine BC screening is recommended for women from the age of 50 years; however, high-risk individuals (with a strong family history of the disease) may be included for screening at earlier age. About twenty percent of all BCs occur in women under 50 years old, and the vast majority of these women do not have any family history of the disease. Most of these tumors have poorer prognosis, therefore early diagnosis by mammography screening, irrespective of known family history, can be clinically beneficial due to reduced BC mortality [45]. Covering the genetic risk assessment into mammography screening programs (by modification of screening frequency or using alternative modalities such as MRI and USG) has been supposed as the way which maximize benefits and minimize harms [46]. Therefore risk-stratified mammography screening based on genetic risk seems to be more effective compared to prevailing age-stratified approaches [47].

4.1. Low- and Intermediate-Risk Women

Intermediate-risk women include cases with a breast biopsy that shows changes such as atypical ductal or lobular hyperplasia, or lobular carcinoma in situ. A calculated risk of BC in these women is ranged from 20% to 29% based upon family history, personal health history, or certain genetic markers. Average-risk women (low-risk women) with none of the above risk factors have a 10–13% lifetime risk of BC [48]. Mammography screening in low-or intermediate-risk women aged less than 50 years is intensively discussed. Arguments for the lower age of mammographic screening include the individual and societal gains linked with increased survival rates, greater work life participation, and lower treatment costs due to early detection. On the other hand, arguments against the lower age of mammography screening include the possible harms and higher costs of full population screening. In this regard, screening in this specific age group of women is accompanied by more cases of false-positive results and unnecessary biopsies because of lower screening specificity. A recent review by Nelson et al. [49] assessed the studies of screening in intermediate-risk women, including mammography screening. Results demonstrated that false-positive results are common and are higher for annual screening, younger women, and women with dense breasts. It seems that the absolute benefits (e.g., number of deaths prevented) are smaller than for older women, because of general lower BC incidence and lower sensitivity of mammography in women aged 40–49 years [33,50].

Several older clinical studies did not demonstrate a significant reductions in BC mortality resulting from screening low-risk women aged 40–50 years [51,52]. A more recent study of Moss et al. [53] enrolled women aged 39–41 years from 23 UK NHS Breast Screening Programme units. Participants were randomly assigned to intervention groups with annual screening by mammography up to an age of 48 years, or to a control group receiving usual medical care (invitation for screening at age 50 years and every 3 years thereafter) respectively. Results showed a significant reduction in BC mortality in the intervention group compared with the control group in the first 10 years after diagnosis but not thereafter from tumors diagnosed during the intervention phase. The overall BC incidence during 17-year follow-up was similar between the groups. A meta-analysis of eight trials revealed that mammography screening reduces BC mortality by 15% for intermediate-risk women aged 39 to 49 years [54].

4.2. High-Risk Women

Validated risk assessment models demonstrated that high-risk women are considered to be those with lifetime risk of BC that is greater than 20% and very high-risk women with a 30% or greater risk for the disease [55]. High-risk individuals include women with a known *BRCA1* or *BRCA2* mutation and their first-degree relatives, women with a personal history of invasive BC or ductal carcinoma in situ and lobular carcinoma in situ or atypical hyperplasia, Li-Fraumeni, Cowden/*PTEN* or Bannayan-Riley-Ruvalcaba syndrome (and first-degree relatives), mutation in specific genes (*ATM, CDH1, CHEK2, NBN, NF1, PALB2, PTEN, STK11* or *TP53*), and a history of chest irradiation between the age of 10 and 30 [56]. A germline gene mutation in *BRCA1* or *BRCA2* results in a significantly elevated lifetime risk of developing breast and ovarian cancer estimated at up to 7 and 25 times, respectively, compared to average risk population. It is supposed that more than 90% of hereditary cases of BC (and also ovarian cancer) are a result of a mutation in *BRCA1/2* [57]. The estimated prevalence of *BRCA1* and *BRCA2* mutations is dependent on the population and can vary between 1 in 300 and 1 in 800, respectively [58].

Meta-analysis of the three studies that compared MRI plus mammography versus mammography alone in screening of young women at high BC risk revealed the sensitivity of MRI plus mammography to be 94% (95%CI 86–98%) and the incremental sensitivity of MRI to be 58% (95% CI 47–70%) [59]. Regarding the high-risk healthy women over the age of 50, there are no clear-cut guidelines for how to continue screening them. Most clinicians continue to screen these women with annual MRI, moreover, in some older women, mammary gland tissue becomes less dense, making it easier to recognize lesions using mammography [60].

Management of high-risk women for the development of BC is debatable, mainly in women carrying a *BRCA1/BRCA2* or *p53* genes mutation because they can develop cancer at an earlier age [61]. This is because mammography alone has limitations in screening younger women with a specifically denser mammary gland tissue or with special tumor phenotypes [60]. Therefore, magnetic resonance imaging can be used along with mammography in these women to increase sensitivity of the screening program. Regarding the high-risk individuals with *BRCA1* or *BRCA2* mutations, current guidelines suggest to begin annual MRI imaging at age 25 and to add mammography at age 30 [44]. Recent meta-analysis of Phi et al. [62] showed that additional screening sensitivity from mammography above that from MRI is limited in *BRCA1* mutation carriers. On the other hand, mammography contributes to screening sensitivity in *BRCA2* mutation carriers, especially those over 40 years. Authors summarized that a differential screening schedule by BRCA status is worth considering [62]. The results of a prospective multicenter trial enrolling 296 carriers of the *BRCA1/2* mutation showed that carriers of the *BRCA* mutation younger than 40 years may not benefit from full-field digital mammography surveillance in addition to dynamic contrast agent-enhanced MR imaging [63].

Based on above-mentioned clinical data, we can conclude than women at high-risk of BC require a close breast surveillance. On the other hand, there is no evidence that more frequent mammography screening or screening with other modalities actually reduces the risk of BC mortality in women with an intermediate or low BC risk (including women with extremely dense breast at mammography) [64]. In addition, mammographic screening has several weaknesses: (a) the risk of false positives; (b) the risk of false negatives; (c) X-ray radiation exposition may trigger BC in high-risk women; (d) mammography performance is operator dependent [65–67].

Moreover, clinical practice demonstrates common diagnostic problems in the distinguishing the pure atypical ductal hyperplasia from advanced lesions, such as DCIS and/or invasive ductal carcinoma following a mammography, and even combined with follow-up core needle biopsy. In this regard, accumulating evidence from oncological research confirming the role of miRNAs in BC progression should be helpful. Therefore, it seems logical to use the potential of miRNA molecules as biomarkers in early BC detection as a follow up of mammography and core needle biopsy [68].

For women who are at high risk for BC and are unable to undergo an MRI evaluation (or are pregnant), ultrasonography of the breast is considered as a useful diagnostic tool. Moreover, ultrasound has been suggested as an adjunct screening method that can detect BC that is missed when using mammography. In this regard, Health Quality Ontario [69] investigated the benefits of ultrasound as an adjunct to mammography compared with mammography alone in women at average and high BC risk. After including five prospective studies, authors concluded that there is low-quality evidence that screening with mammography and adjunct ultrasound detects additional cases of disease, with improved sensitivity compared to mammography alone. Moreover, the results did not show that the use of ultrasound as an adjunct to mammography might reduce BC-related mortality in high-risk women. Due to certain limitations of mammography, particularly in women with dense breasts, ultrasound in combination with contrast-enhanced magnetic resonance imaging, are suggested to supplement mammography for the early detection of BC [70].

4.3. Genetic Profiling as a Tool for the Risk Assessment

Another method focused on the more specific and early detection of BC itself and its risk assessment involves genetic signature profiling. It includes either only genetic variants, or gynecological characteristics or it combines all these factors together. The first mentioned genetic model is mainly the BRCAPRO with several modifications [71] and BOADICEA [72]. Both are able to predict the risk of the disease by mutations analyses in highly penetrant genes, such as *BRCA1* and *BRCA2*. These variants have high individual benefit, but due to rare incidence, they are not suitable for general screening. Therefore, other genetic models include several single nucleotide polymorphisms (SNPs) in low penetrant genes, as they are more frequent in the general population and thus are preferable in the primary screening programs. Some models analyze 7 [73], 12 [74], 51 [72], 77 [75], 88 [76] or 153 SNPs [77]. However, the predictive ability of these genetic models, explained by an area under the ROC curve (AUC) is individually low, ranging from 0.53 to 0.68 [75,77]. When combined multiplicatively with other risk models (i.e., BRCAT / Gail model), a substantial improvement in specificity and sensitivity was observed [72,73,76,78]. Addition of a genetic risk model (12 SNPs) to the BRCAT had a greater effect among African Americans than in whites as it reclassifies the high-risk status of several women undergoing screening mammography [74].

Comparative analyses focusing on the evaluation of the best model's discriminative ability explained by the area under the ROC curve, including genetic, Gail/demographic, and mammography models, revealed the domination of single mammography model above other models. Better identification of women with elevated risk for BC could be attained by the combination of these models as it increases the AUC values by a statistically significant amount [79–81]. It seems that BC risk-stratification based on the combination of mammography screening, genomics and classical risk factors could augment comprehensive risk prediction, provide many benefits in further treatment or monitoring and facilitate tailored preventive intervention.

4.4. Proteomic Profiling as a Tool for Screening Guidelines

The clinical complications due to increased breast density lead to confusion for physicians in the management of women with dense breasts and their follow-up. In this regard, the discussion between patients and health-care providers regarding the need for supplemental screening is necessary. A biochemical clinical approach not affected by density of mammary gland or risk profile of the women would provide an important tool in the management of women with dense breasts or other risks and doubtful imaging results. With the discovery of key biomarkers and protein signatures for BC, proteomic technologies are fully available to provide an ideal diagnostic adjunct to imaging. Research studies have demonstrated that breast tumors are linked with complex changes in the levels of both serum protein biomarkers (SPB) and tumor associated autoantibodies (TAAb) [82].

Recently, Videssa® Breast as a combinatorial proteomic biomarker assay has been comprised of SPB and TAAb integrated with patient-specific clinical data to produce a diagnostic score that reliably detects BC as an adjunctive tool to imaging. Certain blood-based biomarkers are associated with higher mammographic density [83], therefore it is unknown whether the biomarkers included in Videssa® Breast might be impacted as well. Reese et al. [84] aimed to assess the performance of Videssa® Breast in women with dense and non-dense breasts and determine whether this test could help as an additional tool to clinicians in managing women with dense breasts and questionable imaging results. Results of this study demonstrated that Videssa® Breast has high sensitivity and specificity in detecting BC, irrespective of density status. Moreover, a negative Videssa® Breast test gives an assurance to women with dense breasts that they likely do not have BC.

In addition to above mentioned study, Lourenco et al. [85] conducted two prospective clinical trials with the aim to assess a blood-based Videssa Breast test for accurately detection of BC and reduce false positives imaging results. Moreover, they used the Videsa test to detect BC for use in conjunction with imaging to aid healthcare providers in making informed decisions on treating young women (under 50 years old) with difficult-to-assess imaging findings. Authors showed that Videssa Breast can effectively detect BC when used in combination with imaging, improves the management of BC in individuals under 50 years old with challenging or absent imaging findings, and can apparently decrease undesirable clinical procedures. The aforementioned results pointed to the benefit of the integration of SPB and TAAb data in BC diagnosis. Moreover, these data sustain the further progress of combinatorial proteomic approaches for detecting BC.

5. Liquid Biopsy as Marker for Breast Cancer Control and Management

Traditional BC diagnostic tools include clinical and physical examinations, imaging mammography, ultrasound, and/or magnetic resonance imaging, followed by histopathology. Ultrasound as a non-invasive and safe tool is very helpful, but since it is unable to screen the general population for cancer, it cannot replace mammograms, especially in women above 40. Nevertheless, once a suspect lesion in the breast is diagnosed, the bioptic verification is necessary.

Histopathology as an invasive approach to examining cancerous tissues once the disease is installed has for decades been a golden standard for assessment of the tumor biology and if available can also facilitate assessment of ipsilateral lymph node status and serve as a decision-making tool in disease management. Currently, the histological and partial genetic profile of solid tumors is achieved from biopsy or surgical excisional specimens, but these invasive techniques cannot always be performed routinely. It is well-known that tumors consist of subpopulations of cells and needle biopsy takes only a small amount of tumor tissue that does not reflect its full heterogeneity, making the capturing of aggressive clones problematic. Moreover, neither tumor cells show heterogeneity, nor their metastases, which carries different genomic aberrations. As core biopsy of tumor tissue reveals only the portion of this heterogeneity, especially in patients with metastases and in overall assessment it does not seems to be fully representative [86,87], except of full excisional biopsy.

Knowledge, that cancer tissue is associated with mutations in genes, specific genetic alterations and protein expressions, together with those needle biopsies in BC diagnostics have some disadvantages leading to false decisions, sets the identification of other tumor biology markers as useful tools for diagnostic, prognostic and therapeutic purposes. In line with this, other information indicates that surgically resected primary tumor alone does not provide sufficient information about the future diseases biology and seeding of metastases, which can be dissociated at different sites and can harbor unique genomic characteristics that are not detectable in the corresponding primary tumor of the same patient. Thus, the international oncology community is in active pursuit of non-invasive methods for the diagnosis and monitoring of BC patients, which could be introduced in clinical practice. Nowadays efforts are focusing on monitoring specific bodily fluid biomarkers for early and minimally invasive detection [88]. The background for this is the natural behavior of the malign disease, leading to its spread. As disease advances, tumor cells are released from primary tumors (e.g., circulating tumor cells – CTC) and/or metastases or tumor cells release their own nucleic acids (DNA, RNA, miRNA, etc.) into the circulation (circulating free DNA – cfDNA). Analysis of these particles with tumor origin has led to a new diagnostic procedure known as the Liquid Biopsy [89,90].

Early detection of BC disease, treatment of BC, and metastasis monitoring are of eminent importance to ensure favorable prognosis in an individual. Although conventional diagnostic methods, i.e., breast X-ray mammography is precise ("gold standard") clinical method, they may bring about radioactive/invasive harms in patients. In this regard, liquid biopsy as a noninvasive approach is convenient for repeated sampling in clinical oncology practice. Emerging interests in "liquid biopsies" have encouraged researchers to recognize and develop clinically-valid noninvasive genomic and epigenomic signatures that can be exploited as biomarkers capable of detecting premalignant and early-stage tumors, or as biomarkers for prognostic and metastatic evaluation, including cancer relapse monitoring [91]. Importantly, these genomic and epigenomic signatures that are frequently deregulated in cancers, have great potential to serve as promising entities for multifarious purposes within clinical oncology [92].

The term "liquid biopsy" refers to the use of circulating (cell-free) tumor DNA (ctDNA), circulating tumor cells (CTCs), and other non-invasive biomarkers such as long non-coding RNAs (lncRNA), messenger and microRNAs (mRNAs and miRNAs), proteins (soluble or membrane-associated proteins and glycoproteins) and exosomes for the early diagnosis, prognosis, monitoring of clinical progression and response to treatment [93] as was demonstrated in various cancer types including HPV-Associated oropharyngeal cancer [94], and BC [95–97]. Thus, liquid biopsy can be used as an additional diagnostic tool to core or excisional biopsy of primary tumor or its metastasis, or indirect diagnostic tool in case of technically non-performable, non-achievable localization of tumor/metastasis. The high importance for wide clinical application of liquid biopsy supports the results from the studies analyzing temporal and spatial heterogeneity of the tumor tissue. Several studies have described the important role of CTCs in the clinical management of BC disease, notably the ones in association with primary metastases [98,99]. On the other hand, Mansouri et al. [100] evaluated whether CTCs may serve as a clinical prognostic marker for survival in primary BC. Their meta-analysis pointed to CTCs as valid prognostic marker in primary BC prior to any systemic therapy mainly when it is studied through CellSearch® using, concluding that the more the CTCs are linked with increased death and relapse rates in patients. In another study, the prognostic value of CTCs with an epithelial-mesenchymal transition (EMT) phenotype (expression of *TWIST1*, *SNAIL1*, *SLUG*, *ZEB1* transcription factors was analyzed) in primary BC patients were assessed [101]. CTC EMT was determined in 77 from 427 (18.0%) patients. Considering all subgroups of patients, individuals without detectable CTC EMT in peripheral blood manifested longer disease-free survival compared to patients with detectable CTC EMT. Likewise, plasma DNA mutations in ER + MBC seems very promising markers in the early prediction of therapeutic response. In this regard, Kumar et al. [102] used digital PCR-based target enrichment, which was followed by next-generation sequencing to analyze plasma DNA mutations in *ESR1*, *PIK3CA*, and *TP53* in a prospective cohort of 58 patients with ER + MBC. This assay found *ESR1*, *PIK3CA*, and *TP53* plasma ctDNA mutations in 55%, 32%, and 32% of individuals and revealed ctDNA mutant allele fractions that were frequently discordant among the analyzed genes.

Despite the initial optimism and expectancy due to identification of CTCs and ctDNA from liquid biopsies in cancer patients, most recent data indicate that although these markers provide a high grade of cancer specificity, both groups of clinical indicators are rare in body fluids. Thus, these markers may be insufficient as clinically valid diagnostic markers. In general, ctDNA represents only less than 1% of the total cfDNA detected in body fluids. In this regard, the ratio of CTCs to white blood cells consists approximately 1:1 million [103]. Thus, a study that assessed the ability of ctDNA to recognize specific mutations in patients with primary tumors demonstrated positive result in only 73% of colorectal, 57% of gastroesophageal, and 48% of pancreatic carcinomas [104]. These data may be considered rather disappointing contemplating the fact that each of these mutations were known apriori before screening [105]. Importantly, other molecules derived from tumor mass, such as non-coding RNAs (including miRNA) that are far more plentiful than ctDNA or CTCs in body fluids, are relatively stable in biofluids. These RNA molecules are often deregulated, even in the initial stages of carcinogenesis. These features favor RNA markers (when compared to CTCs and cfDNA) for further methodical development as noninvasive liquid biopsy diagnostic and prognostic biomarkers for cancer disease, including BC. In addition, liquid biopsy miRNA biomarkers are applicable only for cancer patients but for also healthy individuals with benign diseases. Thus, cancer screening, staging, and response to treatment may be more effectively assessed by evaluating specific miRNA expression levels in body fluids [106]. The role of liquid biopsy analyzing miRNA signatures as adjunct to conventional screening of BC is summarized in Figure 3.

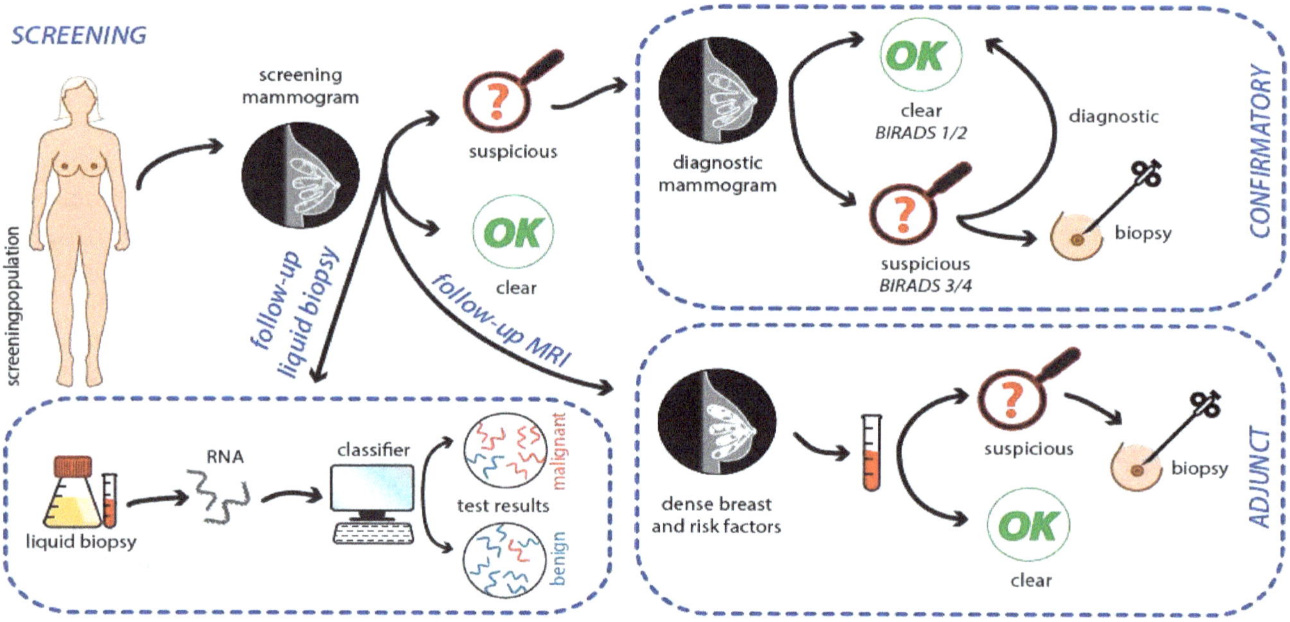

Figure 3. Liquid biopsy miRNA adjunct to conventional screening of BC within personalized diagnosis and improved management of the disease.

Extracellular miRNA Molecules as an Important Tool of Liquid Biopsy in BC Screening

Circulatory tumor cells (CTCs) come either from primary or metastatic cancer tissue (Figure 4). In addition to previously studied ctDNA and CTCs there are also other circulating nucleic acids. The presence of circulating cell-free miRNA (cfmiRNA) molecules in plasma is the latest knowledge in the liquid biopsy era studied in BC patients.

MicroRNAs (miRNAs) are short, non-coding RNAs of typically 22 nucleotides in length, which regulate gene expression at the post-transcriptional level and thus are responsible for proteome shaping [107,108], regulating post-transcriptional gene expression by binding to the 3′ untranslated regions of mRNA [109]. The sequence that is crucial for this binding is known as the 'seed sequence', situated mostly at positions 2–7 of the miRNA 5′-end [110]. MiRNAs are encoded by genomic DNA and are located mostly in intergenic regions (about 52%), intronic regions of genes (40%) and within exons (8%) [111,112]. They represent the human genome information, previously considered to be junk DNA, nowadays believed to be the hidden treasure regarding their potential relevance in diagnosis, prognosis, treatment and follow-up of cancer [113], including BC [6]. The miRNAs were initially discovered in 1993 [114]. A few years later, human studies were launched when its role in cancer was described [115,116]. Since then, an enormous number of studies of miRNA role in various disease's ethiopathogenesis was conducted, describing their roles in or connection to them, including women's cancers. Nowadays, the miRBase, a searchable database of published miRNAs contains more than 1800 human miRNAs sequences [117,118].

The first reports describing the existence of a miRNA signature characterizing human BC were published in 2005, suggesting the involvement of miRNAs in the pathogenesis of this human neoplasm [119,120]. The following years of research showed that miRNAs play a vital role in tumor initiation, progression, drug resistance and disease metastasis. Moreover, the tissue and cancer specificity of several miRNAs enabled us to generate miRNA fingerprints for several cancer types in women, reflecting their reproductive organs [110,121–123], and this specific miRNA expression profile can better classify tumors as compared with the mRNAs. MiRNAs thus have not only important diagnostic purposes, but also high prognostic value, while opening new possibilities in the cancer management, treatment stratification, and in the designing of personalized therapy [108,110,124], most of all in BC [6].

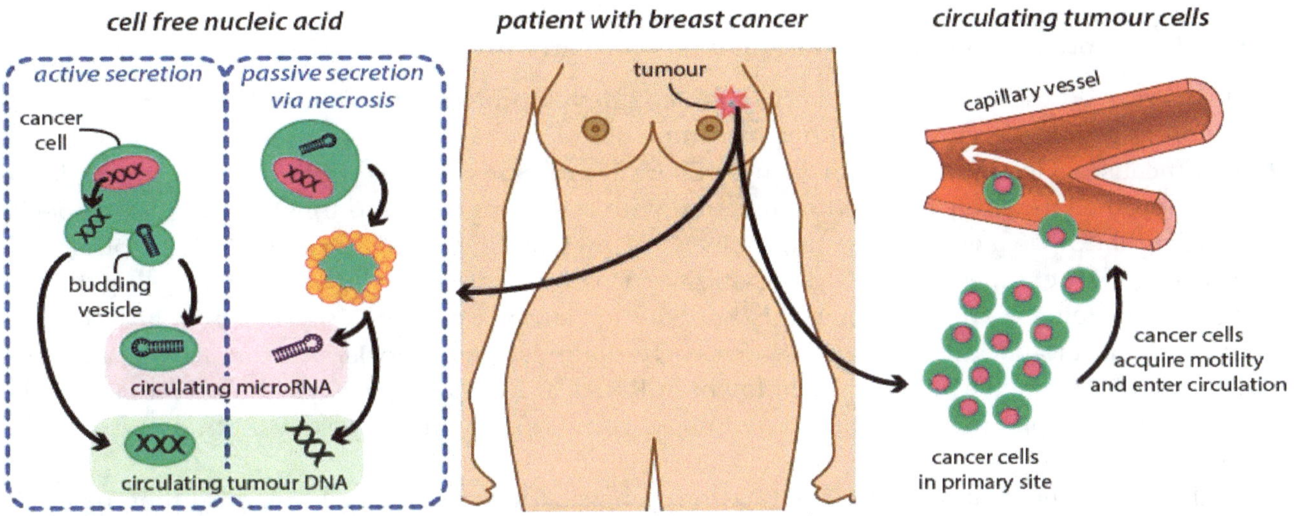

Figure 4. The background of the natural behavior of the malign BC disease leading to its spread through the circulating tumor cells and consequent secretion of cell free nucleic acids.

6. Circulating miRNA

Apart from the tumor microenvironment, miRNAs can be found and isolated from various body fluids including serum, plasma, saliva, urine, breast milk, seminal fluid, cerebrospinal fluids and others [125–127]. The miRNA molecules found in the circulation are derived from tumor tissue cells, cells with short half-life as platelets, broken cells after tissue injury, apoptotic or necrotic cells and chronic inflammation [128,129]. These miRNAs, called circulating miRNAs or extracellular miRNAs (ECmiRNAs), are typically contained within exosomes and vesicles or protein bound complexes, which are shed from tumor cells into the circulation. Packaging complexes protect RNA from degradation, making them remarkably stable [130] and resistant to RNases, fluctuations in pH, long storage periods and to multiple freeze/thaw cycles [125,131]. The stability is most probably caused by the transport mechanisms as miRNAs are conjugated in complexes, providing them the protection. ECmiRNAs can be released and transported a) in the membrane-derived vesicles - either in microparticles (microvesicles) or in smaller exosomes b) in HDL or LDL lipoprotein complexes c) in AGO protein complex or d) wrapped in large apoptotic bodies [132–134]. Such exported extracellular miRNAs can be taken up by variety of recipient even distant cells, where they can alter target gene expression. ECmiRNAs thus represent cell-cell communication, which contributes in carcinogenesis to tumor progression, metastasis and therapy resistance [107,108,125,135,136]. The export of miRNAs most likely has selective patterns and is not only passive. Cells secrete specific miRNAs due to cellular signals or environmental cues and load them into specific vesicles. Moreover, some miRNAs seem to be expressed only to be exported, as they were not detected in the parent cell, only in the extracellular vesicles [132,137].

Due to the above mentioned characteristics of the ECmiRNAs and non-invasive method of acquisition, and despite it only being several years since its discovery, extracellular miRNAs represent excellent detectable biomarkers and seem to be very promising useful biomarkers in non-invasive monitoring and management of BC [135,138,139].

Considering this, to evaluate extracellular miRNAs as a biomarker in BC, the sensitive detection method is essential. This is possible with several methods, e.g., high-throughput sequencing, quantitative real-time PCR (qPCR), digital PCR (dPCR) or microarrays on chip. All these techniques are now fully available, differing only in some requirements. For example, in qPCR experiments it is necessary to find a stable extracellular miRNA reference in plasma or serum of BC patients. To achieve this, a few miRNAs as miR-10b, -16, -30a, -103, -148b, -191, -192 or RNU6 are usually used as endogenous reference markers [140,141].

7. miRNAs as Potential Blood-Based Biomarkers for Early Breast Cancer Detection

It was proved that miRNA has a crucial role in development of breast tumors and that miRNA expression is highly deregulated either in tumor tissue, metastatic tissue, or in plasma from BC patients. It causes a loss of control for many biological processes such as proliferation, differentiation, apoptosis, epithelial-mesenchymal transposition with cell migration, and miRNA can play a specific role as a regulator of metastasis in many levels of metastatic cascade as well. On the other side, except oncogenic activities, miRNA also exhibits oncosupressor characteristics by targeting miRNA coding oncoproteins [142]. Reflecting this knowledge, circulating miRNAs can serve as one of the most promising biomarkers in oncology for early diagnosis, prognosis and therapeutic response prediction [143]. Thus, the biological employ of miRNAs in BC management as liquid biopsy marker recently obtained major interest due its advantages compared to other markers, both in translational and clinical research.

Malignant tumors can stay clinically asymptomatic for quite a long time, until they reach the size to be clinically detected or spread to the distant organs forming metastases. There has been a worldwide effort to identify BC in its early stages for decades. This involved self-examination for a palpable mass and mammography screening (MMG), enriched by ultrasound. The advance in primary screening, mainly thanks to modification of approaches (MMG + genetic risk subdivided populations), enabled us to detect BC quite quickly, however, it is still necessary to undergo core-cut biopsy or fine needle aspiration to set the diagnosis and biological profile of the cancer [144]. Nevertheless, as was shown recently, mammography could not be an early screening tool exclusively, and there exist other possibilities, overcoming its limited specificity and sensitivity, e.g., liquid biopsy technology [68]. Thus, an ideal approach should be a combination of MMG and some sensitive based "liquid biopsy" biomarker.

The definition of biomarker states that it is "a biological molecule found in blood, other body fluids, or tissues that is a sign of a normal or abnormal process or of a condition or disease" [145]. Such a marker should be readily accessible, sensitive enough to detect all types of tumors, and specific enough to not give false positive results [109]. Based on their usage, we can classify the BC biomarker as including risk screening, prognostic, predictive and diagnostic and disease monitoring biomarkers. Thanks to developments in molecular biology, new circulating biomarkers (ctDNAs, mRNAs, cell surface receptors, transcription factors, and secreted proteins) have been discovered and they have been proved to be extremely valuable tools for establishing reliable and early BC diagnosis in a minimally invasive way [146], and this potential of peripheral blood based liquid biopsy can be also fully used in the follow-up testing after anti-cancer treatment [147].The question "why this race is favoring miRNAs" is answered by their biology. The miRNAs have many properties and characteristics, such as low complexity of their molecules, tissue-specificity, stability and easy quantification and amplification, making them excellent potential biomarkers for many pathological and physiological processes. Therefore, miRNAs are becoming a point of interest in cancer detection, since they have been proven to be selectively secreted from malignant cells in the mammary gland, and are expressed differently in the blood of healthy individuals, and patients affected with BC [133]. Their onset and progress in BC detection, profiling and management thus consequently shows great promise for new options in screening, diagnosis and therapeutic interventions [148].

8. Tumor vs Serum miRNA Profile as a Background for miRNAs Based Screening

Cookson and al. [149] tried to answer the question of whether the circulating miRNA profiles resemble those miRNAs inside the tumor. The study analyzed plasma and tissue from patients before and after tumor resection. There were 210 miRNAs overall. Those miRNAs that have been overexpressed in plasma could be matched with those in tumors. This fact suggests that the presence of circulating miRNA can represent the solid tumor [149]. This finding has become the basis for subsequent studies focusing on BC tissue and circulating miRNAs in clinically oriented research. Many characteristics used for their detection have been identified using RNA-seq to generate profiles of miRNA expression in paired samples tumor-serum from patients with carcinoma. This resulted in a set of differently expressed miRNA between the tumor and the corresponding serum, suggesting that only a small amount of miRNA is released from primary tumor into circulation [150].

The first effort for clinical utilization of miRNAs was therefore oriented on BC screening, subsequently modified according to molecular subgroups, and later, adding the predictive potential to determine tumor biological features and aggressiveness. Initially, a wide panel of miRNAs was profiled, followed by spectra of selected miRNAs in validating studies. First of all, it was necessary to find the difference in miRNAs profiles between BC tissue and normal breast healthy tissue, e.g., miR-21, miR-125b, miR-145, mir-155 [119], followed by the finding of positive linkage between miRNAs present in the primary BC and patients plasma. Here, matching miRNAs have been subsequently validated for screening purposes, describing their expression profile or signaling function [150]. Contuinuing this approach, Heneghan et al. [148] analyzed miRNA from tumor tissues and blood samples by qRT-PCR. They quantified the level of 7 candidate miRNA of 148 patients with BC and 44 health controls. Overexpressed levels of miR-195 and let-7a reflected the presence of tumors and when these 2 miRNAs were evaluated two weeks after surgery, their expression was very low [148]. Others have conducted similar studies analyzing circulating miRNAs in patients with BC compared to healthy controls. They concluded that a plasmatic level of miRNAs could be a valid distinguishing biomarker for BC, finding a significantly higher level of several miRNAs for this purpose in their studies [151].

Nowadays, circulating miRNAs associated with neoplasia have the potential to detect cancer even in its earliest stages. It was proved that miRNAs can be used as screening tool for BC (e.g., panel of miR-127-3p, miR-148b, miR-376a, miR-376c, miR-409-3p, miR-652 and miR-801), having high distinguish ability between healthy women, and those with benign, as well as malign breast tumors, especially with better discriminatory power in younger women [148,152]. Moreover, circulating miRNAs are also to differentiate between various tumors, e.g., in compliance with their histological features such as hormone receptors or the state of lymph nodes in BC patients (expression of miR-10b, miR-373, miR299-5p, miR-411, miR-215 and miR-452 in nodal positive patients) [153], showing high specificity and sensitivity indicating metastatic disease [154]. Furthermore, a panel including miR-200a, miR-200b, miR-200c, miR-210, miR-215 and miR-486-5p can predict the onset of metastasis for up to 2 years prior to clinical diagnosis in BC patients [155].

The extremely valuable predicting role of circulating miRNAs showed that some of them, e.g., overexpression of miR-302b and miR-425, can be associated with early BC stage [151], or miR-182 [156], miR-155 [157], and miR-21 [158]. In particular, miR-155 and multifunctional miR-21 have recently been of high scientific interest. Meta-analyses based on relevant articles collected from several scientific databases showed that high levels of their expression correlate with detection of early stages of the disease in screening approach, and also with creating distant metastases [159–161]. For interest, meta-analysis involving 3 studies with 184 patients showed a screening biomarker diagnostic value with sensitivity of 79% (95%CI: 72–84%) and a specificity of 85% (95%CI: 75–92%) for mi-R155 [162]. In addition, plasmatic miR-21 and miR-155 correlate with tumor receptor status.

Another important feature of miRNAs presented in extracellular liquids is fact that they can play a role in cross-talk of cancer cells with cells of surrounding tissue, potentiating their use as biomarkers in BC [163,164]. From the known circulating miRNAs and miRNAs expressed in tissues, the evidence of abnormal activation in BC patients extends to the circulating miR-16, miR-18a, miR-21, miR-145, let-151a, miR-155 and the tissue-specific miR-7, miR-21, miR-145, miR-155/154, miR-182, miR-203, miR-213 suggesting their values as non-invasive marker, and in addition as a potential approach to overcome chemo-resistance [165].

Others correlate with treatment response and may have significant utility as predictive markers or may serve as non-invasive predictors for tumor relapse and overall survival, e.g., in triple-negative BC patients (miR-18b, miR-103, miR-107 and miR-652). Furthermore, this 4-miRNAs signature is capable of distinguishing tumors from patients with early relapse to those without recurrence [166]. This therefore makes this blood-based signature a potential risk predictor for distinguishing metastatic disease from recurrence in the early disease stage, serving as blood-based screening tool for BC relapse. This concept is nor predictive in general, nor cancer type specific. E.g. high levels of serum miR-19a may represent a biomarker for favorable clinical outcome in patients with metastatic HER2-positive BC [167].

Singling out the current knowledge in these molecules, provides blood-based miRNA liquid biopsy also the most valuable opportunity to forgo invasive methods such as tissue biopsy and associated complications in BC diagnosis, and the screening of relapse in clinical praxis [168]. Circulating miRNAs associated with BC screening approach are summarized in Table 2.

Table 2. Circulating miRNAs associated with BC screening approach.

miRNA	Expression (BC vs. Normal)	Sample Type	References
miR-15a	Upregulated	serum	[169]
miR-18a	Upregulated	serum	[169–171]
miR-107	Upregulated	serum	[169]
miR-425	Upregulated	serum	[169]
miR-139-5p	Downregulated	serum	[169]
miR-143	Downregulated	serum	[169]
miR-145	Downregulated	serum	[169,172]
miR-365	Downregulated	serum	[169]
miR-155	Upregulated	serum	[157,162,172–177]
miR-1	Upregulated	serum	[178]
miR-133a	Upregulated	serum	[178,179]
miR-133b	Upregulated	serum	[178]
miR-92a	Upregulated	serum	[178]
miR-148b	Upregulated	plasma	[152,179,180]
miR-376c	Upregulated	plasma	[180]
miR-409-3p	Upregulated	plasma	[152,179,180]
miR-801	Upregulated	plasma	[180]
miR-16	Upregulated	plasma	[181,182]
miR-21	Upregulated	plasma/serum	[133,175,176,181,183–185]
miR-451	Upregulated	plasma	[184]
miR-145	Downregulated	plasma	[184]

Table 2. *Cont.*

miRNA	Expression (BC vs. Normal)	Sample Type	References
miR-222	Upregulated	serum	[170,182]
miR-127	Upregulated	plasma	[152]
miR-376a	Upregulated	plasma	[152]
miR-652	Upregulated	plasma	[152]
miR-801	Upregulated	plasma	[152]
miR-484	Upregulated	serum	[186]
miR-1246	Upregulated	serum	[187]
miR-1307	Upregulated	serum	[187]
miR-6861	Upregulated	serum	[187]
miR-4634	Downregulated	serum	[187]
miR-6875	Downregulated	serum	[187]
miR-181b	Upregulated	serum	[173]
miR-24	Upregulated	serum	[173]
miR-505	Upregulated	plasma	[133]
miR-125	Upregulated	plasma	[133]
miR-96	Upregulated	plasma	[133]
miR-195	Upregulated	serum	[148]
miR-199a	Upregulated	serum	[188]
Let-7a	Upregulated	serum	[148]
miR-106a	Upregulated	serum	[175]
miR-126	Downregulated	serum	[175]
miR-335	Downregulated	serum	[175]
Let-7c	Downregulated	serum	[189]
miR-182	Upregulated	serum	[156]
miR-25	Upregulated	serum	[182]
miR-324	Upregulated	serum	[182]

9. The Role of Circulating miRNAs in Profiling of BC at the Time of Sampling for Screening

As positivity of hormone receptors and estrogen signaling pathway play an important role in cancer development, progression and therapeutic response, neither tissue specific nor circulatory miRNAs reflect the endocrine tumor status. It was proved that miRNA serum profile is dependent on tumor endocrine status and may be differentially expressed (e.g., miR-21, miR-155) in the serum of women with hormone sensitive compared to women with hormone insensitive BC. Its serum concentration was found to be lower in PR tumor positivity [190]. Contrary to that, concentration of miR-182 levels was significantly increased in patients with PR + BC. Considering this fact, miR-155 and miR-182 were suggested as valued plasma biomarker for Luminal type BC diagnosis [156]. A validating study providing deeper insight into the underlying molecular portrait of Luminal A-like BC subtype selected from initial 76 deregulated miRNAs for further analysis 10 miRNAs (miR-19b, miR-29a, miR-93, miR-181a, miR-182, miR-223, miR-301a, miR-423-5p, miR-486-5 and miR-652). The biomarker potential was confirmed by RQ-PCR for four miRNAs (miR-29a, miR-181a, miR-223, and miR-652) and by binary logistic regression for three miRNAs (miR-29a, miR-181, and miR-652). A combination of these three miRNAs could reliably differentiate between cancers and controls. The expression profiles of these three miRNAs in plasma in combination with mammography, could have potential to facilitate accurate subtype-specific BC detection [191]. As with the luminal type, plasmatic miRNAs (e.g., miR-130a, miR-146a, miR-373) also differed between HER2-positive and -negative tumors [192–194], and the miRNAs distinguishing potential for plasma-based screening was also showed for TNBC. Mishra et al. [195] proclaimed miR-195-5p and miR-495 as prospective circulating surrogate molecular markers for early detection of either Luminal or TNBC [195]. In addition, the other circulating miRNAs (miR-16, miR-21 and miR-199a-5p) have proved this concept, finding them to be underexpressed when compared with non-TNBC. Moreover, plasma miR-199a-5p expression in TNBC significantly differed in pre and postoperative levels, and the expression levels were associated with disease stage. These results suggest that the miR-199a-5p is a TNBC-specific marker with diagnostic value and strong insight into targeted therapy during the treatment of TNBC [196]. All above mentioned just confirmed the strong leading role of miRNAs in non-invasive approach for BC screening and management. Moreover, Zhu et al. [190] highlighted that the stability of miRNAs as such screening biomarkers, by examining differential expression in the samples of patients, is safe even for samples have been conserved for 10 years [190], and if Zang et al. [197] showed the sensitivity and specificity for BC diagnosis for miRNA-30a at 74.0 and 65.6%, respectively, overweighting the sensitivities of conventional circulating tumor markers CEA and CA153 being 12.0 and 14.0%, respectively [197]. miRNAs based screening and prediction of BC is very valuable for clinical management and also in patients with a genetically increased risk for disease development. To identify a prognostic marker among asymptomatic women without a BC diagnosis, however with high risk predictive factors for developing the tumor, was an aim of a study by Taslim et al. [198], who performed genome analysis of 41-miRNA model expression in breast tissue in women without tumor with high and low BC risk (based upon Gail risk model), they have revealed miRNAs correlating with high risk for BC developing. Moreover, it was reported that altered or disrupted serum concentration of selected miRNAs, led to development of BC among these women within next 18 months [198]. All these results serve as proof-of-principle that miRNAs in women without BC may be useful for predicting BC risk and/or as an adjunct biomarker for BC early detection in screening programs among those who already developed cancer. The miRNAs identified herein may be involved in breast carcinogenic pathways because they were first identified in the breast tissues of healthy women. Circulating serum miRNAs predicting BC profile (ER, PR, TNBC, Her2+ status, stage, nodal affection) are shown in Table 3.

Table 3. Circulating serum miRNAs predicting BC profile.

miRNA	Expression	ER+/ER−	PR+/PR−	HER2+/HER2−	TNBC+/−	Nodal Affection	Stage of BC	References
miR-10b	up	−	−	−	+	yes	early	[153,199,200]
miR-18a	up	−	−	−	+			[199,201]
miR-18b	up	−	−	−	+			[199,202]
miR-20a	up			+	−	yes	early	[200,203,204]
miR-21	up	−	−	+	−	yes	early/advanced	[184,200]
miR-29a	down	+	+	−	−		early	[191]
miR-34a	up	−	−	+	−	yes		[200,205]
miR-103	down	+	+	−	−	yes		[154,199]
miR-107	down	+	+		−	yes		[154,199]
miR-125a	down	−	−	+	−		early	[119,199]
miR-125b	down	−	−	+	−		early	[119,199]
miR-138	up	+				yes	early	[206]
miR-143	down	−	−	−	+		early	[196,202]
miR-153	up	−	−	−	+			[199]
miR-155	up	−	−	−	+	yes	early	[119,199,200]
miR-181a	down	+	+	−	−		early	[191]
miR-193b	up	−	−	+	−	yes		[155,205]
miR-200a	up			+	−	yes		[200,205]

Table 3. *Cont.*

miRNA	Expression	ER+/ER−	PR+/PR−	HER2+/HER2−	TNBC+/−	Nodal Affection	Stage of BC	References
miR-200b	up			+	−	yes		[200,205]
miR-200c	up			+	−	yes		[200,205]
miR-342	up	+	+	+	−		early	[109,199]
miR-373	up			+	−	yes		[193,200]
miR-375	up	−	−	+	−	yes		[154,205]
miR-429	up	−	−	+	−	yes		[155,199]
miR-484	up	+				yes	early	[206]
miR-486-5p	up	+	+	−	−	yes		[155,191]
miR-642-3p	up						early	[207]
miR-652	down	+	+	−	−	yes	early	[152,154,191]
miR-801	up	+				yes	early	[152,160,206]
miR-1202-5p	up						early	[207]
miR-1207-5p	up						early	[207]

10. Pros and Cons for miRNAs in BC Screening and Management

In the context of personalized medicine, the achievement of CTC cultures and cell free nucleic acids via the liquid biopsy provides outstanding potential to noninvasively diagnostics, prediction, and monitoring of the changing patterns of drug susceptibility in individual patients as their cancer cells acquire new mutations [104,200] Analysis of cfDNA in BC individuals provides an opportunity for non-invasive sampling of tumor DNA and supports its clinical validity as a promising 'liquid biopsy' tumor biomarker [201]. Despite that the quantification of miRNAs and cfDNAs within BC screening in dual use mammography does not exist in literature, the clinical diagnostic may serve as a useful tool in BC diagnosis. Specific attention is placed on short miRNAs, therefore miRNA profiling in individuals may be a promising biomarker and prediction tool that could be utilized in all phases of carcinogenesis within personalized management of breast carcinoma [6,208]. However, miRNA detection as a part of liquid biopsy biomarkers still needs to be validated. There are various limitations of circulating microRNAs as biomarkers of BC. Up to now, a number of studies have been focusing on selected miRNAs that could be applicated as prognostic or predictive biomarkers. Interestingly, certain levels of potential miRNAs occur in healthy subjects as well as in patients' blood or plasma samples. Therefore, alterations of miRNAs expression levels between controls and patients are generally quite low [207]. Importantly, the origin of miRNAs as a part liquid biopsy can influence the effectiveness of this non-invasive diagnostic method. It is well-known that the majority of miRNAs in blood are packed in extracellular vesicles such as macrovesicles or exosomes [209]. Numerous studies analyzed different miRNA expression profiles in plasma, serum and peripheral blood exosomes in BC patients in comparison with healthy individuals [207,210]. These findings suggest the fundamental importance of selecting proper sampling methods for the quantification of circulating miRNAs. Furthermore, single miRNA as a diagnostic and prognostic biomarker has limitations in attributes, including specificity and sensitivity. Moreover, levels of individual miRNA could be overlapped between patients and healthy controls and lead to generation of false positive or false negative results [211]. Nowadays, it is an active field of cancer research because it is necessary to clarify its biological context in body fluids. Moreover, even though several miRNAs have been identified as biomarkers solely or in signatures in multiple studies, many reports differ in their opinion on the detected miRNA. The reason for this is most likely the variability in the design of the studies, cohort characteristics, isolation and detection methodologies or data analysis [212]. Thus, here is an urgent need for to standardize detection and quantification assays with appropriate normalization controls. Actually, for determination of circulating BC biomarkers, there are currently extensively used genomic and proteomic methods. However, they have limited multiplexing capabilities and involve multi-step, high-cost, and time-consuming processes demanding skilled people, which limits significantly their applicability for point-of-case diagnosis [146]. Therefore, there is an urgent need to develop portable, easy handling, time-efficient, cost-effective and quantitative tools for reliable determination of circulating biomarkers at different molecular levels. These analytical tools should be afterwards implemented in decentralized and resource-limited settings. Afterwards, a panel of cancer-specific circulating miRNAs should be created with corresponding tumor grades, responses to treatment, recurrence, and patient survival, which in combination with other biomarkers detectable in liquid biopsies could increase the sensitivity and specificity of cancer detection. Equally, more research studies are necessary for the establishment of feasibility of these applications between patients' subgroups.

11. Conclusions and Expert Recommendations

Despite the extensive use of mammography as the gold standard for breast cancer (BC) screening, the occurrences of both – false-positive and false-negative diagnosis, as well as over diagnosis and the high expenditures, is an issue in gynecological oncology. Consequently, BC management demands new strategies to for compensate the existing deficits. New strategies should consider:

- unmet needs of young populations such as innovative screening programs for early and predictive diagnosis, for example in case of planned pregnancies to avoid pregnancy associated BC [213]
- new diagnostic tests with more predictive power for both – primary BC prevention (by risk assessment to mitigate modifiable risks) and secondary prevention to mitigate the risk of metastatic disease [26,29,214]
- the great potential of multi-level diagnostics by phenotyping and multiomics, in order to adapt the treatment algorithms to the individualized patient profiles [3,28,215,216].

Contextually, the overall concepts of predictive, preventive and personalized medicine are strongly recommended to advance the overall BC management [217].

Early and predictive diagnostic approaches (specifically in premenopausal women), extended and innovative screening programs focused on young female populations (with dense breast parenchyma), targeted prevention in high-risk groups, and optimized treatment concepts are necessary for better controlling of BC. A multiomics clinical approach using liquid biopsy and based on the utilization of the circulating biomarkers has a great potential to improve and enrich the cancer screening and its later management. Promising candidate biomarkers include proteins, RNA, DNA, also autoantibodies, metabolites, and lipids, which can be applied in the detection (involving the pre-invasive and early stages of the disease), diagnosis, and treatment monitoring of BC. While protein-based cancer biomarkers have been introduced in routine pathological practice for many years, nucleic acids-based biomarkers such as miRNAs are relatively new. Blood-based biomarkers for BC screening are still at the early phases of development, and many clinical/preclinical issues including cost effectiveness need to be resolved before their standard introduction into clinical practice. However, despite this, a novel approach for BC screening based on the combination of mammography with liquid biopsy based methods (miRNAs or other) is very promising for specific clinical settings, which may refine the diagnosis and lead to more personalized cancer treatment.

Acknowledgments: This work has been supported by European Association for Predictive, Preventive and Personalized Medicine, EPMA, Brussels, Belgium.

References

1. Ferlay, J.; Colombet, M.; Soerjomataram, I.; Mathers, C.; Parkin, D.M.; Piñeros, M.; Znaor, A.; Bray, F. Estimating the global cancer incidence and mortality in 2018: GLOBOCAN sources and methods. *Int. J. Cancer* **2018**, *144*, 1941–1953. [CrossRef] [PubMed]
2. Golubnitschaja, O.; Sridhar, K.C. Liver metastatic disease: New concepts and biomarker panels to improve individual outcomes. *Clin. Exp. Metastasis* **2016**, *33*, 743–755. [CrossRef] [PubMed]
3. Golubnitschaja, O. Feeling cold and other underestimated symptoms in breast cancer: Anecdotes or individual profiles for advanced patient stratification? *EPMA J.* **2017**, *8*, 17–22. [CrossRef]

4. Zubor, P.; Gondova, A.; Polivka, J.; Kasajova, P.; Konieczka, K.; Danko, J.; Golubnitschaja, O. Breast cancer and Flammer syndrome: Any symptoms in common for prediction, prevention and personalised medical approach? *EPMA J.* **2017**, *8*, 129–140. [CrossRef] [PubMed]
5. Fröhlich, H.; Patjoshi, S.; Yeghiazaryan, K.; Kehrer, C.; Kuhn, W.; Golubnitschaja, O. Premenopausal breast cancer: Potential clinical utility of a multi-omics based machine learning approach for patient stratification. *EPMA J.* **2018**, *9*, 175–186. [CrossRef]
6. Zubor, P.; Kubatka, P.; Dankova, Z.; Gondova, A.; Kajo, K.; Hatok, J.; Samec, M.; Jagelkova, M.; Krivus, S.; Holubekova, V.; et al. miRNA in a multiomic context for diagnosis, treatment monitoring and personalized management of metastatic breast cancer. *Future Oncol.* **2018**, *14*, 1847–1867. [CrossRef]
7. Golubnitschaja, O.; Polivka, J.; Yeghiazaryan, K.; Berliner, L. Liquid biopsy and multiparametric analysis in management of liver malignancies: New concepts of the patient stratification and prognostic approach. *EPMA J.* **2018**, *9*, 271–285. [CrossRef]
8. Statistical Office of the European Communities; Kotzeva, M.; Brandmüller, T.; Önnerfors, Å. *Eurostat Regional Yearbook 2014*; Publications Office of the European Union: Luxembourg, 2014; ISBN 978-92-79-38906-1.
9. Ferlay, J.; Steliarova-Foucher, E.; Lortet-Tieulent, J.; Rosso, S.; Coebergh, J.W.W.; Comber, H.; Forman, D.; Bray, F. Cancer incidence and mortality patterns in Europe: Estimates for 40 countries in 2012. *Eur. J. Cancer* **2013**, *49*, 1374–1403. [CrossRef]
10. Ferlay, J.; Soerjomataram, I.; Dikshit, R.; Eser, S.; Mathers, C.; Rebelo, M.; Parkin, D.M.; Forman, D.; Bray, F. Cancer incidence and mortality worldwide: Sources, methods and major patterns in GLOBOCAN 2012: Globocan 2012. *Int. J. Cancer* **2015**, *136*, E359–E386. [CrossRef]
11. Golubnitschaja, O.; Debald, M.; Yeghiazaryan, K.; Kuhn, W.; Pešta, M.; Costigliola, V.; Grech, G. Breast cancer epidemic in the early twenty-first century: Evaluation of risk factors, cumulative questionnaires and recommendations for preventive measures. *Tumor Biol.* **2016**, *37*, 12941–12957. [CrossRef]
12. DeSantis, C.E.; Ma, J.; Goding Sauer, A.; Newman, L.A.; Jemal, A. Breast cancer statistics, 2017, racial disparity in mortality by state: Breast Cancer Statistics, 2017. *CA Cancer J. Clin.* **2017**, *67*, 439–448. [CrossRef] [PubMed]
13. Malvezzi, M.; Carioli, G.; Bertuccio, P.; Boffetta, P.; Levi, F.; La Vecchia, C.; Negri, E. European cancer mortality predictions for the year 2019 with focus on breast cancer. *Ann. Oncol.* **2019**, *30*, 781–787. [CrossRef] [PubMed]
14. Altobelli, E.; Lattanzi, A. Breast cancer in European Union: An update of screening programmes as of March 2014 (Review). *Int. J. Oncol.* **2014**, *45*, 1785–1792. [CrossRef]
15. Breast Cancer Breast Screening Programme, England Statistics for 2014–2015, Health and Social Care Information Centre. Available online: https://digital.nhs.uk/data-and-information/publications/statistical/breast-screening-programme/breast-screening-programme-england-2014-15 (accessed on 4 June 2019).
16. *Programme de Dépistage du Cancer du sein en France: Résultats 2006*; Institut de Veille Sanitaire: Saint-Maurice, France, 2009; Available online: http://invs.santepubliquefrance.fr/publications/2009/plaquette_depistage_cancer_sein_2006/depistage_cancer_sein_2006.pdf (accessed on 4 June 2019).
17. Giordano, L.; Castagno, R.; Giorgi, D.; Piccinelli, C.; Ventura, L.; Segnan, N.; Zappa, M. Breast cancer screening in Italy: Evaluating key performance indicators for time trends and activity volumes. *Epidemiol. Prev.* **2015**, *39*, 30–39. [PubMed]
18. Marnach-Kopp, B. Mammographie-Screening in Deutschland, Informationen und Adressen. Kooperationsgemeinschaft Mammographie 2008. Available online: http://www.zentralestelle-bayern.de/downloads/info_adressen.pdf (accessed on 4 June 2019).
19. Laat je Borsten Zien: Doen of Niet? Resultaten Screening. Available online: https://www.zol.be/sites/default/files/deelsites/medische-beeldvorming/verwijzers/sympoisa/20140920/dr.-van-steen-laat-je-borsten-zien-doen-of-niet.pdf (accessed on 4 June 2019).
20. Skol, A.D.; Sasaki, M.M.; Onel, K. The genetics of breast cancer risk in the post-genome era: Thoughts on study design to move past BRCA and towards clinical relevance. *Breast Cancer Res.* **2016**, *18*, 99. [CrossRef] [PubMed]
21. Chetlen, A.; Mack, J.; Chan, T. Breast cancer screening controversies: Who, when, why, and how? *Clin. Imaging* **2016**, *40*, 279–282. [CrossRef] [PubMed]
22. Pérez-Solis, M.A.; Maya-Nuñez, G.; Casas-González, P.; Olivares, A.; Aguilar-Rojas, A. Effects of the lifestyle habits in breast cancer transcriptional regulation. *Cancer Cell Int.* **2016**, *16*, 7. [CrossRef] [PubMed]
23. Drukteinis, J.S.; Mooney, B.P.; Flowers, C.I.; Gatenby, R.A. Beyond Mammography: New Frontiers in Breast Cancer Screening. *Am. J. Med.* **2013**, *126*, 472–479. [CrossRef] [PubMed]

24. Wilczek, B.; Wilczek, H.E.; Rasouliyan, L.; Leifland, K. Adding 3D automated breast ultrasound to mammography screening in women with heterogeneously and extremely dense breasts: Report from a hospital-based, high-volume, single-center breast cancer screening program. *Eur. J. Radiol.* **2016**, *85*, 1554–1563. [CrossRef] [PubMed]
25. Lebron-Zapata, L.; Jochelson, M.S. Overview of Breast Cancer Screening and Diagnosis. *PET Clin.* **2018**, *13*, 301–323. [CrossRef] [PubMed]
26. Polivka, J.; Kralickova, M.; Polivka, J.; Kaiser, C.; Kuhn, W.; Golubnitschaja, O. Mystery of the brain metastatic disease in breast cancer patients: Improved patient stratification, disease prediction and targeted prevention on the horizon? *EPMA J.* **2017**, *8*, 119–127. [CrossRef] [PubMed]
27. Yung, R.L.; Ligibel, J.A. Obesity and breast cancer: Risk, outcomes, and future considerations. *Clin. Adv. Hematol. Oncol.* **2016**, *14*, 790–797. [PubMed]
28. Golubnitschaja, O.; Filep, N.; Yeghiazaryan, K.; Blom, H.J.; Hofmann-Apitius, M.; Kuhn, W. Multi-omic approach decodes paradoxes of the triple-negative breast cancer: Lessons for predictive, preventive and personalised medicine. *Amino Acids* **2018**, *50*, 383–395. [CrossRef] [PubMed]
29. Golubnitschaja, O.; Yeghiazaryan, K.; Abraham, J.-A.; Schild, H.H.; Costigliola, V.; Debald, M.; Kuhn, W. Breast cancer risk assessment: A non-invasive multiparametric approach to stratify patients by MMP-9 serum activity and RhoA expression patterns in circulating leucocytes. *Amino Acids* **2017**, *49*, 273–281. [CrossRef] [PubMed]
30. Hinkson, I.V.; Davidsen, T.M.; Klemm, J.D.; Chandramouliswaran, I.; Kerlavage, A.R.; Kibbe, W.A. A Comprehensive Infrastructure for Big Data in Cancer Research: Accelerating Cancer Research and Precision Medicine. *Front. Cell Dev. Biol.* **2017**, *5*, 83. [CrossRef] [PubMed]
31. Koch, H. Mammography as a method for diagnosing breast cancer. *Radiol. Bras.* **2016**, *49*, VII. [CrossRef] [PubMed]
32. Gay, H.; Pietrosanu, R.; George, S.; Tzias, D.; Mehta, R.; Patel, C.; Heller, S.; Wilkinson, L. PB.13: Comparison between analogue and digital mammography: A reader's perspective. *Breast Cancer Res.* **2013**, *15*, P13. [CrossRef]
33. Van Ravesteyn, N.T. Tipping the Balance of Benefits and Harms to Favor Screening Mammography Starting at Age 40 Years: A Comparative Modeling Study of Risk. *Ann. Intern. Med.* **2012**, *156*, 609. [CrossRef] [PubMed]
34. Helvie, M.A. Digital Mammography Imaging: Breast Tomosynthesis and Advanced Applications. *Radiol. Clin. N. Am.* **2010**, *48*, 917–929. [CrossRef] [PubMed]
35. Niklason, L.T.; Christian, B.T.; Niklason, L.E.; Kopans, D.B.; Castleberry, D.E.; Opsahl-Ong, B.H.; Landberg, C.E.; Slanetz, P.J.; Giardino, A.A.; Moore, R.; et al. Digital tomosynthesis in breast imaging. *Radiology* **1997**, *205*, 399–406. [CrossRef] [PubMed]
36. Uchiyama, N.; Kikuchi, M.; Machida, M.; Arai, Y.; Murakami, R.; Otsuka, K.; Jerebko, A.; Kelm, M.; Mertelmeier, T. Diagnostic Usefulness of Synthetic MMG (SMMG) with DBT (Digital Breast Tomosynthesis) for Clinical Setting in Breast Cancer Screening. In *Breast Imaging*; Tingberg, A., Lång, K., Timberg, P., Eds.; Springer International Publishing: Cham, Switzerland, 2016; Volume 9699, pp. 59–67, ISBN 978-3-319-41545-1.
37. Takahashi, T.A.; Lee, C.I.; Johnson, K.M. Breast cancer screening: Does tomosynthesis augment mammography? *Cleve Clin. J. Med.* **2017**, *84*, 522–527. [CrossRef] [PubMed]
38. Tonelli, M.; Connor Gorber, S.; Joffres, M.; Dickinson, J.; Singh, H.; Lewin, G.; Birtwhistle, R.; Fitzpatrick-Lewis, D.; Hodgson, N.; Ciliska, D.; et al. Recommendations on screening for breast cancer in average-risk women aged 40–74 years. *CMAJ Can. Med. Assoc. J. J. Assoc. Medicale Can.* **2011**, *183*, 1991–2001.
39. Olsen, O.; Gøtzsche, P.C. Cochrane review on screening for breast cancer with mammography. *Lancet* **2001**, *358*, 1340–1342. [CrossRef]
40. Miller, A.B. The role of screening mammography in the era of modern breast cancer treatment. *Climacteric* **2018**, *21*, 204–208. [CrossRef] [PubMed]
41. Redmond, C.E.; Healy, G.M.; Murphy, C.F.; O'Doherty, A.; Foster, A. The use of ultrasonography and digital mammography in women under 40 years with symptomatic breast cancer: A 7-year Irish experience. *Ir. J. Med. Sci. (1971-)* **2017**, *186*, 63–67. [CrossRef] [PubMed]
42. Lehman, C.D.; Lee, C.I.; Loving, V.A.; Portillo, M.S.; Peacock, S.; DeMartini, W.B. Accuracy and Value of Breast Ultrasound for Primary Imaging Evaluation of Symptomatic Women 30–39 Years of Age. *Am. J. Roentgenol.* **2012**, *199*, 1169–1177. [CrossRef] [PubMed]

43. Cho, N.; Han, W.; Han, B.-K.; Bae, M.S.; Ko, E.S.; Nam, S.J.; Chae, E.Y.; Lee, J.W.; Kim, S.H.; Kang, B.J.; et al. Breast Cancer Screening with Mammography Plus Ultrasonography or Magnetic Resonance Imaging in Women 50 Years or Younger at Diagnosis and Treated with Breast Conservation Therapy. *JAMA Oncol.* **2017**, *3*, 1495. [CrossRef]
44. Wellings, E.; Vassiliades, L.; Abdalla, R. Breast Cancer Screening for High-Risk Patients of Different Ages and Risk—Which Modality Is Most Effective? *Cureus* **2016**. [CrossRef]
45. Anders, C.K.; Hsu, D.S.; Broadwater, G.; Acharya, C.R.; Foekens, J.A.; Zhang, Y.; Wang, Y.; Marcom, P.K.; Marks, J.R.; Febbo, P.G.; et al. Young Age at Diagnosis Correlates with Worse Prognosis and Defines a Subset of Breast Cancers with Shared Patterns of Gene Expression. *J. Clin. Oncol.* **2008**, *26*, 3324–3330. [CrossRef]
46. Meisel, S.F.; Pashayan, N.; Rahman, B.; Side, L.; Fraser, L.; Gessler, S.; Lanceley, A.; Wardle, J. Adjusting the frequency of mammography screening on the basis of genetic risk: Attitudes among women in the UK. *Breast* **2015**, *24*, 237–241. [CrossRef]
47. Pashayan, N.; Duffy, S.W.; Chowdhury, S.; Dent, T.; Burton, H.; Neal, D.E.; Easton, D.F.; Eeles, R.; Pharoah, P. Polygenic susceptibility to prostate and breast cancer: Implications for personalised screening. *Br. J. Cancer* **2011**, *104*, 1656–1663. [CrossRef] [PubMed]
48. Gagnon, J.; Lévesque, E.; Borduas, F.; Chiquette, J.; Diorio, C.; Duchesne, N.; Dumais, M.; Eloy, L.; Foulkes, W.; Gervais, N.; et al. Recommendations on breast cancer screening and prevention in the context of implementing risk stratification: Impending changes to current policies. *Curr. Oncol.* **2016**, *23*, 615. [CrossRef] [PubMed]
49. Nelson, H.D.; Pappas, M.; Cantor, A.; Griffin, J.; Daeges, M.; Humphrey, L. Harms of Breast Cancer Screening: Systematic Review to Update the 2009 U.S. Preventive Services Task Force Recommendation. *Ann. Intern. Med.* **2016**, *164*, 256. [CrossRef] [PubMed]
50. Hellquist, B.N.; Czene, K.; Hjälm, A.; Nyström, L.; Jonsson, H. Effectiveness of population-based service screening with mammography for women ages 40 to 49 years with a high or low risk of breast cancer: Socioeconomic status, parity, and age at birth of first child: Mammography Effectiveness in Risk Groups. *Cancer* **2015**, *121*, 251–258. [CrossRef] [PubMed]
51. Moss, S.M.; Cuckle, H.; Evans, A.; Johns, L.; Waller, M.; Bobrow, L. Effect of mammographic screening from age 40 years on breast cancer mortality at 10 years' follow-up: A randomised controlled trial. *Lancet* **2006**, *368*, 2053–2060. [CrossRef]
52. The Canadian National Breast Screening Study-1: Breast Cancer Mortality after 11 to 16 Years of Follow-Up. A Randomized Screening Trial of Mammography in Women Age 40 to 49 Years. Available online: https://www.ncbi.nlm.nih.gov/pubmed/12204013 (accessed on 4 June 2019).
53. Moss, S.M.; Wale, C.; Smith, R.; Evans, A.; Cuckle, H.; Duffy, S.W. Effect of mammographic screening from age 40 years on breast cancer mortality in the UK Age trial at 17 years' follow-up: A randomised controlled trial. *Lancet Oncol.* **2015**, *16*, 1123–1132. [CrossRef]
54. Nelson, H.D. Screening for Breast Cancer: An Update for the U.S. Preventive Services Task Force. *Ann. Intern. Med.* **2009**, *151*, 727. [CrossRef]
55. Sutton, T.; Reilly, P.; Johnson, N.; Garreau, J.R. Breast cancer in women under 50: Most are not high risk. *Am. J. Surg.* **2018**, *215*, 848–851. [CrossRef]
56. Sung, J.S.; Stamler, S.; Brooks, J.; Kaplan, J.; Huang, T.; Dershaw, D.D.; Lee, C.H.; Morris, E.A.; Comstock, C.E. Breast Cancers Detected at Screening MR Imaging and Mammography in Patients at High Risk: Method of Detection Reflects Tumor Histopathologic Results. *Radiology* **2016**, *280*, 716–722. [CrossRef]
57. Ford, D.; Easton, D.F.; Stratton, M.; Narod, S.; Goldgar, D.; Devilee, P.; Bishop, D.T.; Weber, B.; Lenoir, G.; Chang-Claude, J.; et al. Genetic heterogeneity and penetrance analysis of the BRCA1 and BRCA2 genes in breast cancer families. The Breast Cancer Linkage Consortium. *Am. J. Hum. Genet.* **1998**, *62*, 676–689. [CrossRef]
58. Roa, B.B.; Boyd, A.A.; Volcik, K.; Richards, C.S. Ashkenazi Jewish population frequencies for common mutations in BRCA1 and BRCA2. *Nat. Genet.* **1996**, *14*, 185–187. [CrossRef] [PubMed]
59. Lord, S.J.; Lei, W.; Craft, P.; Cawson, J.N.; Morris, I.; Walleser, S.; Griffiths, A.; Parker, S.; Houssami, N. A systematic review of the effectiveness of magnetic resonance imaging (MRI) as an addition to mammography and ultrasound in screening young women at high risk of breast cancer. *Eur. J. Cancer* **2007**, *43*, 1905–1917. [CrossRef] [PubMed]
60. Hartman, A.-R.; Daniel, B.L.; Kurian, A.W.; Mills, M.A.; Nowels, K.W.; Dirbas, F.M.; Kingham, K.E.;

Chun, N.M.; Herfkens, R.J.; Ford, J.M.; et al. Breast magnetic resonance image screening and ductal lavage in women at high genetic risk for breast carcinoma. *Cancer* **2004**, *100*, 479–489. [CrossRef] [PubMed]

61. Lowry, K.P.; Lee, J.M.; Kong, C.Y.; McMahon, P.M.; Gilmore, M.E.; Cott Chubiz, J.E.; Pisano, E.D.; Gatsonis, C.; Ryan, P.D.; Ozanne, E.M.; et al. Annual screening strategies in BRCA1 and BRCA2 gene mutation carriers: A comparative effectiveness analysis. *Cancer* **2012**, *118*, 2021–2030. [CrossRef] [PubMed]
62. Phi, X.-A.; Saadatmand, S.; De Bock, G.H.; Warner, E.; Sardanelli, F.; Leach, M.O.; Riedl, C.C.; Trop, I.; Hooning, M.J.; Mandel, R.; et al. Contribution of mammography to MRI screening in BRCA mutation carriers by BRCA status and age: Individual patient data meta-analysis. *Br. J. Cancer* **2016**, *114*, 631–637. [CrossRef] [PubMed]
63. Van Zelst, J.C.M.; Mus, R.D.M.; Woldringh, G.; Rutten, M.J.C.M.; Bult, P.; Vreemann, S.; de Jong, M.; Karssemeijer, N.; Hoogerbrugge, N.; Mann, R.M. Surveillance of Women with the *BRCA* 1 or *BRCA* 2 Mutation by Using Biannual Automated Breast US, MR Imaging, and Mammography. *Radiology* **2017**, *285*, 376–388. [CrossRef]
64. Autier, P.; Boniol, M. Mammography screening: A major issue in medicine. *Eur. J. Cancer* **2018**, *90*, 34–62. [CrossRef]
65. The Independent UK Panel on Breast Cancer Screening; Marmot, M.G.; Altman, D.G.; Cameron, D.A.; Dewar, J.A.; Thompson, S.G.; Wilcox, M. The benefits and harms of breast cancer screening: An independent review: A report jointly commissioned by Cancer Research UK and the Department of Health (England) October 2012. *Br. J. Cancer* **2013**, *108*, 2205–2240. [CrossRef]
66. Théberge, I.; Chang, S.-L.; Vandal, N.; Daigle, J.-M.; Guertin, M.-H.; Pelletier, É.; Brisson, J. Radiologist Interpretive Volume and Breast Cancer Screening Accuracy in a Canadian Organized Screening Program. *JNCI J. Natl. Cancer Inst.* **2014**, *106*. [CrossRef]
67. Frères, P.; Wenric, S.; Boukerroucha, M.; Fasquelle, C.; Thiry, J.; Bovy, N.; Struman, I.; Geurts, P.; Collignon, J.; Schroeder, H.; et al. Circulating microRNA-based screening tool for breast cancer. *Oncotarget* **2016**, *7*, 5416. [CrossRef]
68. Fu, S.W.; Lee, W.; Coffey, C.; Lean, A.; Wu, X.; Tan, X.; Man, Y.; Brem, R.F. miRNAs as potential biomarkers in early breast cancer detection following mammography. *Cell Biosci.* **2016**, *6*, 6. [CrossRef] [PubMed]
69. Health Quality Ontario Ultrasound as an Adjunct to Mammography for Breast Cancer Screening: A Health Technology Assessment. *Ont. Health Technol. Assess. Ser.* **2016**, *16*, 1–71.
70. Hooley, R.J.; Andrejeva, L.; Scoutt, L.M. Breast Cancer Screening and Problem Solving Using Mammography, Ultrasound, and Magnetic Resonance Imaging. *Ultrasound Q.* **2011**, *27*, 23–47. [CrossRef] [PubMed]
71. Biswas, S.; Atienza, P.; Chipman, J.; Hughes, K.; Barrera, A.M.G.; Amos, C.I.; Arun, B.; Parmigiani, G. Simplifying clinical use of the genetic risk prediction model BRCAPRO. *Breast Cancer Res. Treat.* **2013**, *139*, 571–579. [CrossRef] [PubMed]
72. The Consortium of Investigators of Modifiers of BRCA1/2; The Breast Cancer Association Consortium; Lee, A.J.; Cunningham, A.P.; Kuchenbaecker, K.B.; Mavaddat, N.; Easton, D.F.; Antoniou, A.C. BOADICEA breast cancer risk prediction model: Updates to cancer incidences, tumour pathology and web interface. *Br. J. Cancer* **2014**, *110*, 535–545. [CrossRef] [PubMed]
73. Mealiffe, M.E.; Stokowski, R.P.; Rhees, B.K.; Prentice, R.L.; Pettinger, M.; Hinds, D.A. Assessment of Clinical Validity of a Breast Cancer Risk Model Combining Genetic and Clinical Information. *JNCI J. Natl. Cancer Inst.* **2010**, *102*, 1618–1627. [CrossRef] [PubMed]
74. McCarthy, A.M.; Armstrong, K.; Handorf, E.; Boghossian, L.; Jones, M.; Chen, J.; Demeter, M.B.; McGuire, E.; Conant, E.F.; Domchek, S.M. Incremental impact of breast cancer SNP panel on risk classification in a screening population of white and African American women. *Breast Cancer Res. Treat.* **2013**, *138*, 889–898. [CrossRef] [PubMed]
75. Dite, G.S.; Mahmoodi, M.; Bickerstaffe, A.; Hammet, F.; Macinnis, R.J.; Tsimiklis, H.; Dowty, J.G.; Apicella, C.; Phillips, K.-A.; Giles, G.G.; et al. Using SNP genotypes to improve the discrimination of a simple breast cancer risk prediction model. *Breast Cancer Res. Treat.* **2013**, *139*, 887–896. [CrossRef]
76. Cuzick, J.; Brentnall, A.R.; Segal, C.; Byers, H.; Reuter, C.; Detre, S.; Lopez-Knowles, E.; Sestak, I.; Howell, A.; Powles, T.J.; et al. Impact of a Panel of 88 Single Nucleotide Polymorphisms on the Risk of Breast Cancer in High-Risk Women: Results from Two Randomized Tamoxifen Prevention Trials. *J. Clin. Oncol.* **2017**, *35*, 743–750. [CrossRef]

77. Wu, Y.; Abbey, C.K.; Liu, J.; Ong, I.; Peissig, P.; Onitilo, A.A.; Fan, J.; Yuan, M.; Burnside, E.S. Discriminatory Power of Common Genetic Variants in Personalized Breast Cancer Diagnosis. *Proc. SPIE Int. Soc. Opt. Eng.* **2016**, *9787*, 978706.
78. Wacholder, S.; Hartge, P.; Prentice, R.; Garcia-Closas, M.; Feigelson, H.S.; Diver, W.R.; Thun, M.J.; Cox, D.G.; Hankinson, S.E.; Kraft, P.; et al. Performance of Common Genetic Variants in Breast-Cancer Risk Models. *N. Engl. J. Med.* **2010**, *362*, 986–993. [CrossRef] [PubMed]
79. Liu, J.; Page, D.; Nassif, H.; Shavlik, J.; Peissig, P.; McCarty, C.; Onitilo, A.A.; Burnside, E. Genetic variants improve breast cancer risk prediction on mammograms. *AMIA Annu. Symp. Proc.* **2013**, *2013*, 876–885. [PubMed]
80. Lee, C.P.L.; Choi, H.; Soo, K.C.; Tan, M.-H.; Chay, W.Y.; Chia, K.S.; Liu, J.; Li, J.; Hartman, M. Mammographic Breast Density and Common Genetic Variants in Breast Cancer Risk Prediction. *PLoS ONE* **2015**, *10*, e0136650. [CrossRef] [PubMed]
81. Burnside, E.S.; Liu, J.; Wu, Y.; Onitilo, A.A.; McCarty, C.A.; Page, C.D.; Peissig, P.L.; Trentham-Dietz, A.; Kitchner, T.; Fan, J.; et al. Comparing Mammography Abnormality Features to Genetic Variants in the Prediction of Breast Cancer in Women Recommended for Breast Biopsy. *Acad. Radiol.* **2016**, *23*, 62–69. [CrossRef] [PubMed]
82. Henderson, M.C.; Hollingsworth, A.B.; Gordon, K.; Silver, M.; Mulpuri, R.; Letsios, E.; Reese, D.E. Integration of Serum Protein Biomarker and Tumor Associated Autoantibody Expression Data Increases the Ability of a Blood-Based Proteomic Assay to Identify Breast Cancer. *PLoS ONE* **2016**, *11*, e0157692. [CrossRef] [PubMed]
83. Horne, H.N.; Sherman, M.E.; Pfeiffer, R.M.; Figueroa, J.D.; Khodr, Z.G.; Falk, R.T.; Pollak, M.; Patel, D.A.; Palakal, M.M.; Linville, L.; et al. Circulating insulin-like growth factor-I, insulin-like growth factor binding protein-3 and terminal duct lobular unit involution of the breast: A cross-sectional study of women with benign breast disease. *Breast Cancer Res.* **2016**, *18*, 24. [CrossRef]
84. Reese, D.E.; Henderson, M.C.; Silver, M.; Mulpuri, R.; Letsios, E.; Tran, Q.; Wolf, J.K. Breast density does not impact the ability of Videssa® Breast to detect breast cancer in women under age 50. *PLoS ONE* **2017**, *12*, e0186198. [CrossRef] [PubMed]
85. Lourenco, A.P.; Benson, K.L.; Henderson, M.C.; Silver, M.; Letsios, E.; Tran, Q.; Gordon, K.J.; Borman, S.; Corn, C.; Mulpuri, R.; et al. A Noninvasive Blood-based Combinatorial Proteomic Biomarker Assay to Detect Breast Cancer in Women Under the Age of 50 Years. *Clin. Breast Cancer* **2017**, *17*, 516–525.e6. [CrossRef]
86. Liquid Biopsy: The Future Work for Clinical Pathologist. Available online: https://www.austinpublishinggroup.com/clinical-pathology/fulltext/ajcp-v2-id1034.php (accessed on 4 June 2019).
87. Siravegna, G.; Bardelli, A. Genotyping cell-free tumor DNA in the blood to detect residual disease and drug resistance. *Genome Biol.* **2014**, *15*, 449. [CrossRef]
88. Cardoso, A.R.; Moreira, F.T.C.; Fernandes, R.; Sales, M.G.F. Novel and simple electrochemical biosensor monitoring attomolar levels of miRNA-155 in breast cancer. *Biosens. Bioelectron.* **2016**, *80*, 621–630. [CrossRef]
89. Crowley, E.; Di Nicolantonio, F.; Loupakis, F.; Bardelli, A. Liquid biopsy: Monitoring cancer-genetics in the blood. *Nat. Rev. Clin. Oncol.* **2013**, *10*, 472–484. [CrossRef] [PubMed]
90. Pantel, K.; Alix-Panabières, C. Liquid biopsy: Potential and challenges. *Mol. Oncol.* **2016**, *10*, 371–373. [CrossRef] [PubMed]
91. Zhai, L.-Y.; Li, M.-X.; Pan, W.-L.; Chen, Y.; Li, M.-M.; Pang, J.-X.; Zheng, L.; Chen, J.-X.; Duan, W.-J. In Situ Detection of Plasma Exosomal MicroRNA-1246 for Breast Cancer Diagnostics by a Au Nanoflare Probe. *ACS Appl. Mater. Interfaces* **2018**, *10*, 39478–39486. [CrossRef] [PubMed]
92. Kim, H.-Y.; Choi, H.-J.; Lee, J.-Y.; Kong, G. Cancer Target Gene Screening: A web application for breast cancer target gene screening using multi-omics data analysis. *Brief. Bioinform.* **2019**. [CrossRef] [PubMed]
93. Beaver, J.A.; Jelovac, D.; Balukrishna, S.; Cochran, R.L.; Croessmann, S.; Zabransky, D.J.; Wong, H.Y.; Valda Toro, P.; Cidado, J.; Blair, B.G.; et al. Detection of Cancer DNA in Plasma of Patients with Early-Stage Breast Cancer. *Clin. Cancer Res.* **2014**, *20*, 2643–2650. [CrossRef] [PubMed]
94. Chera, B.S.; Kumar, S.; Beaty, B.T.; Marron, D.; Jefferys, S.R.; Green, R.L.; Goldman, E.C.; Amdur, R.; Sheets, N.; Dagan, R.; et al. Rapid Clearance Profile of Plasma Circulating Tumor HPV Type 16 DNA during Chemoradiotherapy Correlates with Disease Control in HPV-Associated Oropharyngeal Cancer. *Clin. Cancer Res. Off. J. Am. Assoc. Cancer Res.* **2019**. [CrossRef] [PubMed]

95. Ignatiadis, M.; Rack, B.; Rothé, F.; Riethdorf, S.; Decraene, C.; Bonnefoi, H.; Dittrich, C.; Messina, C.; Beauvois, M.; Trapp, E.; et al. Liquid biopsy-based clinical research in early breast cancer: The EORTC 90091-10093 Treat CTC trial. *Eur. J. Cancer Oxf. Engl. 1990* **2016**, *63*, 97–104. [CrossRef] [PubMed]
96. Coombes, R.C.; Page, K.; Salari, R.; Hastings, R.K.; Armstrong, A.; Ahmed, S.; Ali, S.; Cleator, S.; Kenny, L.; Stebbing, J.; et al. Personalized Detection of Circulating Tumor DNA Antedates Breast Cancer Metastatic Recurrence. *Clin. Cancer Res. Off. J. Am. Assoc. Cancer Res.* **2019**. [CrossRef] [PubMed]
97. Li, X.; Chen, N.; Zhou, L.; Wang, C.; Wen, X.; Jia, L.; Cui, J.; Hoffman, A.R.; Hu, J.-F.; Li, W. Genome-wide target interactome profiling reveals a novel EEF1A1 epigenetic pathway for oncogenic lncRNA MALAT1 in breast cancer. *Am. J. Cancer Res.* **2019**, *9*, 714–729.
98. Bidard, F.-C.; Proudhon, C.; Pierga, J.-Y. Circulating tumor cells in breast cancer. *Mol. Oncol.* **2016**, *10*, 418–430. [CrossRef]
99. Hall, C.; Valad, L.; Lucci, A. Circulating Tumor Cells in Breast Cancer Patients. *Crit. Rev. Oncog.* **2016**, *21*, 125–139. [CrossRef] [PubMed]
100. Mansouri, S.; Hesari, P.M.; Naghavi-al-Hosseini, F.; Majidzadeh-A, K.; Farahmand, L. The Prognostic Value of Circulating Tumor Cells in Primary Breast Cancer Prior to any Systematic Therapy: A Systematic Review. *Curr. Stem Cell Res. Ther.* **2019**, *14*. [CrossRef] [PubMed]
101. Mego, M.; Karaba, M.; Minarik, G.; Benca, J.; Silvia, J.; Sedlackova, T.; Manasova, D.; Kalavska, K.; Pindak, D.; Cristofanilli, M.; et al. Circulating Tumor Cells with Epithelial–to–mesenchymal Transition Phenotypes Associated with Inferior Outcomes in Primary Breast Cancer. *Anticancer Res.* **2019**, *39*, 1829–1837. [CrossRef]
102. Kumar, S.; Lindsay, D.; Chen, Q.B.; Garrett, A.L.; Tan, X.M.; Anders, C.K.; Carey, L.A.; Gupta, G.P. Tracking plasma DNA mutation dynamics in estrogen receptor positive metastatic breast cancer with dPCR-SEQ. *NPJ Breast Cancer* **2018**, *4*, 39. [CrossRef] [PubMed]
103. Diaz, L.A.; Bardelli, A. Liquid Biopsies: Genotyping Circulating Tumor DNA. *J. Clin. Oncol.* **2014**, *32*, 579–586. [CrossRef]
104. Bettegowda, C.; Sausen, M.; Leary, R.J.; Kinde, I.; Wang, Y.; Agrawal, N.; Bartlett, B.R.; Wang, H.; Luber, B.; Alani, R.M.; et al. Detection of Circulating Tumor DNA in Early- and Late-Stage Human Malignancies. *Sci. Transl. Med.* **2014**, *6*, 224ra24. [CrossRef] [PubMed]
105. Diehl, F.; Schmidt, K.; Choti, M.A.; Romans, K.; Goodman, S.; Li, M.; Thornton, K.; Agrawal, N.; Sokoll, L.; Szabo, S.A.; et al. Circulating mutant DNA to assess tumor dynamics. *Nat. Med.* **2008**, *14*, 985–990. [CrossRef] [PubMed]
106. Shigeyasu, K.; Toden, S.; Zumwalt, T.J.; Okugawa, Y.; Goel, A. Emerging Role of MicroRNAs as Liquid Biopsy Biomarkers in Gastrointestinal Cancers. *Clin. Cancer Res.* **2017**, *23*, 2391–2399. [CrossRef] [PubMed]
107. Shah, M.Y.; Ferrajoli, A.; Sood, A.K.; Lopez-Berestein, G.; Calin, G.A. microRNA Therapeutics in Cancer—An Emerging Concept. *EBioMedicine* **2016**, *12*, 34–42. [CrossRef] [PubMed]
108. Takahashi, R.; Miyazaki, H.; Ochiya, T. The Roles of MicroRNAs in Breast Cancer. *Cancers* **2015**, *7*, 598–616. [CrossRef] [PubMed]
109. Van Schooneveld, E.; Wildiers, H.; Vergote, I.; Vermeulen, P.B.; Dirix, L.Y.; Van Laere, S.J. Dysregulation of microRNAs in breast cancer and their potential role as prognostic and predictive biomarkers in patient management. *Breast Cancer Res.* **2015**, *17*, 21. [CrossRef] [PubMed]
110. Abba, M.L.; Patil, N.; Leupold, J.H.; Moniuszko, M.; Utikal, J.; Niklinski, J.; Allgayer, H. MicroRNAs as novel targets and tools in cancer therapy. *Cancer Lett.* **2017**, *387*, 84–94. [CrossRef] [PubMed]
111. Mohr, A.; Mott, J. Overview of MicroRNA Biology. *Semin. Liver Dis.* **2015**, *35*, 003–011. [CrossRef] [PubMed]
112. Hsu, P.W.C. miRNAMap: Genomic maps of microRNA genes and their target genes in mammalian genomes. *Nucleic Acids Res.* **2006**, *34*, D135–D139. [CrossRef] [PubMed]
113. Fiorucci, G.; Chiantore, M.V.; Mangino, G.; Percario, Z.A.; Affabris, E.; Romeo, G. Cancer regulator microRNA: Potential relevance in diagnosis, prognosis and treatment of cancer. *Curr. Med. Chem.* **2012**, *19*, 461–474. [CrossRef] [PubMed]
114. Lee, R.C.; Feinbaum, R.L.; Ambros, V. The C. elegans heterochronic gene lin-4 encodes small RNAs with antisense complementarity to lin-14. *Cell* **1993**, *75*, 843–854. [CrossRef]
115. Iorio, M.V.; Croce, C.M. Commentary on microRNA Fingerprint in Human Epithelial Ovarian Cancer. *Cancer Res.* **2016**, *76*, 6143–6145. [CrossRef] [PubMed]
116. Calin, G.A.; Dumitru, C.D.; Shimizu, M.; Bichi, R.; Zupo, S.; Noch, E.; Aldler, H.; Rattan, S.; Keating, M.; Rai, K.; et al. Nonlinear partial differential equations and applications: Frequent deletions

and down-regulation of micro-RNA genes miR15 and miR16 at 13q14 in chronic lymphocytic leukemia. *Proc. Natl. Acad. Sci. USA* **2002**, *99*, 15524–15529. [CrossRef]

117. Homo Sapiens miRNAs (1917 Sequences) [GRCh38]. Available online: http://www.mirbase.org/cgi-bin/mirna_summary.pl?org=hsa (accessed on 4 June 2019).
118. Griffiths-Jones, S. miRBase: microRNA sequences, targets and gene nomenclature. *Nucleic Acids Res.* **2006**, *34*, D140–D144. [CrossRef]
119. Iorio, M.V.; Ferracin, M.; Liu, C.-G.; Veronese, A.; Spizzo, R.; Sabbioni, S.; Magri, E.; Pedriali, M.; Fabbri, M.; Campiglio, M.; et al. MicroRNA Gene Expression Deregulation in Human Breast Cancer. *Cancer Res.* **2005**, *65*, 7065–7070. [CrossRef]
120. Lu, J.; Getz, G.; Miska, E.A.; Alvarez-Saavedra, E.; Lamb, J.; Peck, D.; Sweet-Cordero, A.; Ebert, B.L.; Mak, R.H.; Ferrando, A.A.; et al. MicroRNA expression profiles classify human cancers. *Nature* **2005**, *435*, 834–838. [CrossRef] [PubMed]
121. Jia, W.; Wu, Y.; Zhang, Q.; Gao, G.; Zhang, C.; Xiang, Y. Expression profile of circulating microRNAs as a promising fingerprint for cervical cancer diagnosis and monitoring. *Mol. Clin. Oncol.* **2015**, *3*, 851–858. [CrossRef]
122. Keller, A.; Leidinger, P.; Borries, A.; Wendschlag, A.; Wucherpfennig, F.; Scheffler, M.; Huwer, H.; Lenhof, H.-P.; Meese, E. miRNAs in lung cancer—Studying complex fingerprints in patient's blood cells by microarray experiments. *BMC Cancer* **2009**, *9*, 353. [CrossRef] [PubMed]
123. Iorio, M.V.; Visone, R.; Di Leva, G.; Donati, V.; Petrocca, F.; Casalini, P.; Taccioli, C.; Volinia, S.; Liu, C.-G.; Alder, H.; et al. MicroRNA Signatures in Human Ovarian Cancer. *Cancer Res.* **2007**, *67*, 8699–8707. [CrossRef] [PubMed]
124. Raychaudhuri, M.; Bronger, H.; Buchner, T.; Kiechle, M.; Weichert, W.; Avril, S. MicroRNAs miR-7 and miR-340 predict response to neoadjuvant chemotherapy in breast cancer. *Breast Cancer Res. Treat.* **2017**, *162*, 511–521. [CrossRef] [PubMed]
125. Sohel, M.H. Extracellular/Circulating MicroRNAs: Release Mechanisms, Functions and Challenges. *Achiev. Life Sci.* **2016**, *10*, 175–186. [CrossRef]
126. Turchinovich, A.; Tonevitsky, A.G.; Burwinkel, B. Extracellular miRNA: A Collision of Two Paradigms. *Trends Biochem. Sci.* **2016**, *41*, 883–892. [CrossRef] [PubMed]
127. Weber, J.A.; Baxter, D.H.; Zhang, S.; Huang, D.Y.; How Huang, K.; Jen Lee, M.; Galas, D.J.; Wang, K. The MicroRNA Spectrum in 12 Body Fluids. *Clin. Chem.* **2010**, *56*, 1733–1741. [CrossRef] [PubMed]
128. Mo, M.-H.; Chen, L.; Fu, Y.; Wang, W.; Fu, S.W. Cell-free Circulating miRNA Biomarkers in Cancer. *J. Cancer* **2012**, *3*, 432–448. [CrossRef]
129. Shah, M.Y.; Calin, G.A. The Mix of Two Worlds: Non-Coding RNAs and Hormones. *Nucleic Acid Ther.* **2013**, *23*, 2–8. [CrossRef] [PubMed]
130. Cocucci, E.; Racchetti, G.; Meldolesi, J. Shedding microvesicles: Artefacts no more. *Trends Cell Biol.* **2009**, *19*, 43–51. [CrossRef] [PubMed]
131. Sourvinou, I.S.; Markou, A.; Lianidou, E.S. Quantification of Circulating miRNAs in Plasma. *J. Mol. Diagn.* **2013**, *15*, 827–834. [CrossRef]
132. Boon, R.A.; Vickers, K.C. Intercellular Transport of MicroRNAs. *Arterioscler. Thromb. Vasc. Biol.* **2013**, *33*, 186–192. [CrossRef] [PubMed]
133. Matamala, N.; Vargas, M.T.; Gonzalez-Campora, R.; Minambres, R.; Arias, J.I.; Menendez, P.; Andres-Leon, E.; Gomez-Lopez, G.; Yanowsky, K.; Calvete-Candenas, J.; et al. Tumor MicroRNA Expression Profiling Identifies Circulating MicroRNAs for Early Breast Cancer Detection. *Clin. Chem.* **2015**, *61*, 1098–1106. [CrossRef] [PubMed]
134. Arroyo, J.D.; Chevillet, J.R.; Kroh, E.M.; Ruf, I.K.; Pritchard, C.C.; Gibson, D.F.; Mitchell, P.S.; Bennett, C.F.; Pogosova-Agadjanyan, E.L.; Stirewalt, D.L.; et al. Argonaute2 complexes carry a population of circulating microRNAs independent of vesicles in human plasma. *Proc. Natl. Acad. Sci. USA* **2011**, *108*, 5003–5008. [CrossRef] [PubMed]
135. Théry, C. Exosomes: Secreted vesicles and intercellular communications. *F1000 Biol. Rep.* **2011**, *3*, 15. [CrossRef]
136. Kahlert, C.; Kalluri, R. Exosomes in tumor microenvironment influence cancer progression and metastasis. *J. Mol. Med.* **2013**, *91*, 431–437. [CrossRef]

137. Pigati, L.; Yaddanapudi, S.C.S.; Iyengar, R.; Kim, D.-J.; Hearn, S.A.; Danforth, D.; Hastings, M.L.; Duelli, D.M. Selective Release of MicroRNA Species from Normal and Malignant Mammary Epithelial Cells. *PLoS ONE* **2010**, *5*, e13515. [CrossRef]
138. Chen, X.; Ba, Y.; Ma, L.; Cai, X.; Yin, Y.; Wang, K.; Guo, J.; Zhang, Y.; Chen, J.; Guo, X.; et al. Characterization of microRNAs in serum: A novel class of biomarkers for diagnosis of cancer and other diseases. *Cell Res.* **2008**, *18*, 997–1006. [CrossRef]
139. Schrauder, M.G.; Strick, R.; Schulz-Wendtland, R.; Strissel, P.L.; Kahmann, L.; Loehberg, C.R.; Lux, M.P.; Jud, S.M.; Hartmann, A.; Hein, A.; et al. Circulating Micro-RNAs as Potential Blood-Based Markers for Early Stage Breast Cancer Detection. *PLoS ONE* **2012**, *7*, e29770. [CrossRef]
140. Witwer, K.W. Circulating MicroRNA Biomarker Studies: Pitfalls and Potential Solutions. *Clin. Chem.* **2015**, *61*, 56–63. [CrossRef] [PubMed]
141. Witwer, K.W.; Buzás, E.I.; Bemis, L.T.; Bora, A.; Lässer, C.; Lötvall, J.; Nolte-'t Hoen, E.N.; Piper, M.G.; Sivaraman, S.; Skog, J.; et al. Standardization of sample collection, isolation and analysis methods in extracellular vesicle research. *J. Extracell. Vesicles* **2013**, *2*, 20360. [CrossRef] [PubMed]
142. Piva, R.; Spandidos, D.A.; Gambari, R. From microRNA functions to microRNA therapeutics: Novel targets and novel drugs in breast cancer research and treatment. *Int. J. Oncol.* **2013**, *43*, 985–994. [CrossRef] [PubMed]
143. Armand-Labit, V.; Pradines, A. Circulating cell-free microRNAs as clinical cancer biomarkers. *Biomol. Concepts* **2017**, *8*, 61–81. [CrossRef] [PubMed]
144. Khoury, S.; Tran, N. Circulating microRNAs: Potential Biomarkers for Common Malignancies. *Biomark. Med.* **2015**, *9*, 131–151. [CrossRef] [PubMed]
145. NCI Dictionary of Cancer Terms. Available online: https://www.cancer.gov/publications/dictionaries/cancer-terms (accessed on 4 June 2019).
146. Susana Campuzano; María Pedrero; José Pingarrón Non-Invasive Breast Cancer Diagnosis through Electrochemical Biosensing at Different Molecular Levels. *Sensors* **2017**, *17*, 1993. [CrossRef] [PubMed]
147. Hamam, R.; Ali, A.M.; Alsaleh, K.A.; Kassem, M.; Alfayez, M.; Aldahmash, A.; Alajez, N.M. microRNA expression profiling on individual breast cancer patients identifies novel panel of circulating microRNA for early detection. *Sci. Rep.* **2016**, *6*, 25997. [CrossRef]
148. Heneghan, H.M.; Miller, N.; Lowery, A.J.; Sweeney, K.J.; Newell, J.; Kerin, M.J. Circulating microRNAs as Novel Minimally Invasive Biomarkers for Breast Cancer. *Ann. Surg.* **2010**, *251*, 499–505. [CrossRef]
149. Cookson, V.J.; Bentley, M.A.; Hogan, B.V.; Horgan, K.; Hayward, B.E.; Hazelwood, L.D.; Hughes, T.A. Circulating microRNA profiles reflect the presence of breast tumours but not the profiles of microRNAs within the tumours. *Cell. Oncol.* **2012**, *35*, 301–308. [CrossRef]
150. Zhu, J.; Zheng, Z.; Wang, J.; Sun, J.; Wang, P.; Cheng, X.; Fu, L.; Zhang, L.; Wang, Z.; Li, Z. Different miRNA expression profiles between human breast cancer tumors and serum. *Front. Genet.* **2014**, *5*, 149. [CrossRef]
151. Zhao, H.; Shen, J.; Medico, L.; Wang, D.; Ambrosone, C.B.; Liu, S. A Pilot Study of Circulating miRNAs as Potential Biomarkers of Early Stage Breast Cancer. *PLoS ONE* **2010**, *5*, e13735. [CrossRef]
152. Cuk, K.; Zucknick, M.; Madhavan, D.; Schott, S.; Golatta, M.; Heil, J.; Marmé, F.; Turchinovich, A.; Sinn, P.; Sohn, C.; et al. Plasma MicroRNA Panel for Minimally Invasive Detection of Breast Cancer. *PLoS ONE* **2013**, *8*, e76729. [CrossRef] [PubMed]
153. Chen, W.; Cai, F.; Zhang, B.; Barekati, Z.; Zhong, X.Y. The level of circulating miRNA-10b and miRNA-373 in detecting lymph node metastasis of breast cancer: Potential biomarkers. *Tumor Biol.* **2013**, *34*, 455–462. [CrossRef] [PubMed]
154. Inns, J.; James, V. Circulating microRNAs for the prediction of metastasis in breast cancer patients diagnosed with early stage disease. *Breast* **2015**, *24*, 364–369. [CrossRef] [PubMed]
155. Madhavan, D.; Peng, C.; Wallwiener, M.; Zucknick, M.; Nees, J.; Schott, S.; Rudolph, A.; Riethdorf, S.; Trumpp, A.; Pantel, K.; et al. Circulating miRNAs with prognostic value in metastatic breast cancer and for early detection of metastasis. *Carcinogenesis* **2016**, *37*, 461–470. [CrossRef]
156. Wang, P.-Y.; Gong, H.-T.; Li, B.-F.; Lv, C.-L.; Wang, H.-T.; Zhou, H.-H.; Li, X.-X.; Xie, S.-Y.; Jiang, B.-F. Higher expression of circulating miR-182 as a novel biomarker for breast cancer. *Oncol. Lett.* **2013**, *6*, 1681–1686. [CrossRef] [PubMed]
157. Sun, Y.; Wang, M.; Lin, G.; Sun, S.; Li, X.; Qi, J.; Li, J. Serum MicroRNA-155 as a Potential Biomarker to Track Disease in Breast Cancer. *PLoS ONE* **2012**, *7*, e47003. [CrossRef] [PubMed]

158. Fang, R.; Zhu, Y.; Hu, L.; Khadka, V.S.; Ai, J.; Zou, H.; Ju, D.; Jiang, B.; Deng, Y.; Hu, X. Plasma MicroRNA Pair Panels as Novel Biomarkers for Detection of Early Stage Breast Cancer. *Front. Physiol.* **2019**, *9*, 1879. [CrossRef]
159. Li, S.; Yang, X.; Yang, J.; Zhen, J.; Zhang, D. Serum microRNA-21 as a potential diagnostic biomarker for breast cancer: A systematic review and meta-analysis. *Clin. Exp. Med.* **2016**, *16*, 29–35. [CrossRef]
160. Schwarzenbach, H. Circulating nucleic acids as biomarkers in breast cancer. *Breast Cancer Res.* **2013**, *15*, 211. [CrossRef]
161. Markou, A.; Zavridou, M.; Sourvinou, I.; Yousef, G.; Kounelis, S.; Malamos, N.; Georgoulias, V.; Lianidou, E. Direct Comparison of Metastasis-Related miRNAs Expression Levels in Circulating Tumor Cells, Corresponding Plasma, and Primary Tumors of Breast Cancer Patients. *Clin. Chem.* **2016**, *62*, 1002–1011. [CrossRef] [PubMed]
162. Wang, F.; Hou, J.; Jin, W.; Li, J.; Yue, Y.; Jin, H.; Wang, X. Increased Circulating MicroRNA-155 as a Potential Biomarker for Breast Cancer Screening: A Meta-Analysis. *Molecules* **2014**, *19*, 6282–6293. [CrossRef] [PubMed]
163. Moldovan, L.; Batte, K.; Wang, Y.; Wisler, J.; Piper, M. Analyzing the Circulating MicroRNAs in Exosomes/Extracellular Vesicles from Serum or Plasma by qRT-PCR. In *Circulating MicroRNAs*; Kosaka, N., Ed.; Humana Press: Totowa, NJ, USA, 2013; Volume 1024, pp. 129–145, ISBN 978-1-62703-452-4.
164. Chen, X.; Liang, H.; Zhang, J.; Zen, K.; Zhang, C.-Y. Horizontal transfer of microRNAs: Molecular mechanisms and clinical applications. *Protein Cell* **2012**, *3*, 28–37. [CrossRef] [PubMed]
165. Bahrami, A.; Aledavood, A.; Anvari, K.; Hassanian, S.M.; Maftouh, M.; Yaghobzade, A.; Salarzaee, O.; ShahidSales, S.; Avan, A. The prognostic and therapeutic application of microRNAs in breast cancer: Tissue and circulating microRNAs. *J. Cell. Physiol.* **2018**, *233*, 774–786. [CrossRef] [PubMed]
166. Kleivi Sahlberg, K.; Bottai, G.; Naume, B.; Burwinkel, B.; Calin, G.A.; Borresen-Dale, A.-L.; Santarpia, L. A Serum MicroRNA Signature Predicts Tumor Relapse and Survival in Triple-Negative Breast Cancer Patients. *Clin. Cancer Res.* **2015**, *21*, 1207–1214. [CrossRef] [PubMed]
167. Anfossi, S.; Giordano, A.; Gao, H.; Cohen, E.N.; Tin, S.; Wu, Q.; Garza, R.J.; Debeb, B.G.; Alvarez, R.H.; Valero, V.; et al. High Serum miR-19a Levels Are Associated with Inflammatory Breast Cancer and Are Predictive of Favorable Clinical Outcome in Patients with Metastatic HER2+ Inflammatory Breast Cancer. *PLoS ONE* **2014**, *9*, e83113. [CrossRef] [PubMed]
168. Bertoli, G.; Cava, C.; Castiglioni, I. MicroRNAs: New Biomarkers for Diagnosis, Prognosis, Therapy Prediction and Therapeutic Tools for Breast Cancer. *Theranostics* **2015**, *5*, 1122–1143. [CrossRef] [PubMed]
169. Kodahl, A.R.; Lyng, M.B.; Binder, H.; Cold, S.; Gravgaard, K.; Knoop, A.S.; Ditzel, H.J. Novel circulating microRNA signature as a potential non-invasive multi-marker test in ER-positive early-stage breast cancer: A case control study. *Mol. Oncol.* **2014**, *8*, 874–883. [CrossRef]
170. Godfrey, A.C.; Xu, Z.; Weinberg, C.R.; Getts, R.C.; Wade, P.A.; DeRoo, L.A.; Sandler, D.P.; Taylor, J.A. Serum microRNA expression as an early marker for breast cancer risk in prospectively collected samples from the Sister Study cohort. *Breast Cancer Res.* **2013**, *15*, R42. [CrossRef]
171. Guo, L.-J.; Zhang, Q.-Y. Decreased serum miR-181a is a potential new tool for breast cancer screening. *Int. J. Mol. Med.* **2012**, *30*, 680–686. [CrossRef] [PubMed]
172. Mar-Aguilar, F.; Mendoza-Ramírez, J.A.; Malagón-Santiago, I.; Espino-Silva, P.K.; Santuario-Facio, S.K.; Ruiz-Flores, P.; Rodríguez-Padilla, C.; Reséndez-Pérez, D. Serum Circulating microRNA Profiling for Identification of Potential Breast Cancer Biomarkers. *Dis. Markers* **2013**, *34*, 163–169. [CrossRef] [PubMed]
173. Sochor, M.; Basova, P.; Pesta, M.; Dusilkova, N.; Bartos, J.; Burda, P.; Pospisil, V.; Stopka, T. Oncogenic MicroRNAs: miR-155, miR-19a, miR-181b, and miR-24 enable monitoring of early breast cancer in serum. *BMC Cancer* **2014**, *14*, 448. [CrossRef] [PubMed]
174. Roth, C.; Rack, B.; Müller, V.; Janni, W.; Pantel, K.; Schwarzenbach, H. Circulating microRNAs as blood-based markers for patients with primary and metastatic breast cancer. *Breast Cancer Res.* **2010**, *12*, R90. [CrossRef] [PubMed]
175. Wang, F.; Zheng, Z.; Guo, J.; Ding, X. Correlation and quantitation of microRNA aberrant expression in tissues and sera from patients with breast tumor. *Gynecol. Oncol.* **2010**, *119*, 586–593. [CrossRef] [PubMed]
176. Zhang, L.; Dong, B.; Ren, P.; Ye, H.; Shi, J.; Qin, J.; Wang, K.; Wang, P.; Zhang, J. Circulating plasma microRNAs in the detection of esophageal squamous cell carcinoma. *Oncol. Lett.* **2018**, *16*, 3303–3318. [CrossRef]

177. Zhao, S.; Wu, Q.; Gao, F.; Zhang, C.; Yang, X. Serum microRNA-155 as a potential biomarker for breast cancer screening. *Chin. Sci. Bull.* **2012**, *57*, 3466–3468. [CrossRef]
178. Chan, M.; Liaw, C.S.; Ji, S.M.; Tan, H.H.; Wong, C.Y.; Thike, A.A.; Tan, P.H.; Ho, G.H.; Lee, A.S.-G. Identification of Circulating MicroRNA Signatures for Breast Cancer Detection. *Clin. Cancer Res.* **2013**, *19*, 4477–4487. [CrossRef]
179. Shen, J.; Hu, Q.; Schrauder, M.; Yan, L.; Wang, D.; Medico, L.; Guo, Y.; Yao, S.; Zhu, Q.; Liu, B.; et al. Circulating miR-148b and miR-133a as biomarkers for breast cancer detection. *Oncotarget* **2014**, *5*, 5284. [CrossRef]
180. Cuk, K.; Zucknick, M.; Heil, J.; Madhavan, D.; Schott, S.; Turchinovich, A.; Arlt, D.; Rath, M.; Sohn, C.; Benner, A.; et al. Circulating microRNAs in plasma as early detection markers for breast cancer. *Int. J. Cancer* **2013**, *132*, 1602–1612. [CrossRef]
181. Ng, E.K.O.; Li, R.; Shin, V.Y.; Jin, H.C.; Leung, C.P.H.; Ma, E.S.K.; Pang, R.; Chua, D.; Chu, K.-M.; Law, W.L.; et al. Circulating microRNAs as Specific Biomarkers for Breast Cancer Detection. *PLoS ONE* **2013**, *8*, e53141. [CrossRef]
182. Hu, Z.; Dong, J.; Wang, L.-E.; Ma, H.; Liu, J.; Zhao, Y.; Tang, J.; Chen, X.; Dai, J.; Wei, Q.; et al. Serum microRNA profiling and breast cancer risk: The use of miR-484/191 as endogenous controls. *Carcinogenesis* **2012**, *33*, 828–834. [CrossRef] [PubMed]
183. Gao, J.; Zhang, Q.; Xu, J.; Guo, L.; Li, X. Clinical significance of serum miR-21 in breast cancer compared with CA153 and CEA. *Chin. J. Cancer Res.* **2013**, *25*, 743–748.
184. Asaga, S.; Kuo, C.; Nguyen, T.; Terpenning, M.; Giuliano, A.E.; Hoon, D.S.B. Direct Serum Assay for MicroRNA-21 Concentrations in Early and Advanced Breast Cancer. *Clin. Chem.* **2011**, *57*, 84–91. [CrossRef] [PubMed]
185. Chen, H.; Liu, H.; Zou, H.; Chen, R.; Dou, Y.; Sheng, S.; Dai, S.; Ai, J.; Melson, J.; Kittles, R.A.; et al. Evaluation of Plasma miR-21 and miR-152 as Diagnostic Biomarkers for Common Types of Human Cancers. *J. Cancer* **2016**, *7*, 490–499. [CrossRef] [PubMed]
186. Zearo, S.; Kim, E.; Zhu, Y.; Zhao, J.T.; Sidhu, S.B.; Robinson, B.G.; Soon, P.S. MicroRNA-484 is more highly expressed in serum of early breast cancer patients compared to healthy volunteers. *BMC Cancer* **2014**, *14*, 200. [CrossRef] [PubMed]
187. Shimomura, A.; Shiino, S.; Kawauchi, J.; Takizawa, S.; Sakamoto, H.; Matsuzaki, J.; Ono, M.; Takeshita, F.; Niida, S.; Shimizu, C.; et al. Novel combination of serum microRNA for detecting breast cancer in the early stage. *Cancer Sci.* **2016**, *107*, 326–334. [CrossRef] [PubMed]
188. Zhang, L.; Xu, Y.; Jin, X.; Wang, Z.; Wu, Y.; Zhao, D.; Chen, G.; Li, D.; Wang, X.; Cao, H.; et al. A circulating miRNA signature as a diagnostic biomarker for non-invasive early detection of breast cancer. *Breast Cancer Res. Treat.* **2015**, *154*, 423–434. [CrossRef] [PubMed]
189. Li, X.-X.; Gao, S.-Y.; Wang, P.-Y.; Zhou, X.; Li, Y.-J.; Yu, Y.; Yan, Y.-F.; Zhang, H.-H.; Lv, C.-J.; Zhou, H.-H.; et al. Reduced expression levels of let-7c in human breast cancer patients. *Oncol. Lett.* **2015**, *9*, 1207–1212. [CrossRef] [PubMed]
190. Zhu, W.; Qin, W.; Atasoy, U.; Sauter, E.R. Circulating microRNAs in breast cancer and healthy subjects. *BMC Res. Notes* **2009**, *2*, 89. [CrossRef] [PubMed]
191. McDermott, A.M.; Miller, N.; Wall, D.; Martyn, L.M.; Ball, G.; Sweeney, K.J.; Kerin, M.J. Identification and Validation of Oncologic miRNA Biomarkers for Luminal A-like Breast Cancer. *PLoS ONE* **2014**, *9*, e87032. [CrossRef]
192. Stückrath, I.; Rack, B.; Janni, W.; Jäger, B.; Pantel, K.; Schwarzenbach, H. Aberrant plasma levels of circulating *miR-16, miR-107, miR-130a* and *miR-146a* are associated with lymph node metastasis and receptor status of breast cancer patients. *Oncotarget* **2015**, *6*, 13387. [CrossRef] [PubMed]
193. Eichelser, C.; Flesch-Janys, D.; Chang-Claude, J.; Pantel, K.; Schwarzenbach, H. Deregulated Serum Concentrations of Circulating Cell-Free MicroRNAs miR-17, miR-34a, miR-155, and miR-373 in Human Breast Cancer Development and Progression. *Clin. Chem.* **2013**, *59*, 1489–1496. [CrossRef] [PubMed]
194. Swellam, M.; El Magdoub, H.M.; Hassan, N.M.; Hefny, M.M.; Sobeih, M.E. Potential diagnostic role of circulating MiRNAs in breast cancer: Implications on clinicopathological characters. *Clin. Biochem.* **2018**, *56*, 47–54. [CrossRef] [PubMed]
195. Mishra, S.; Srivastava, A.K.; Suman, S.; Kumar, V.; Shukla, Y. Circulating miRNAs revealed as surrogate molecular signatures for the early detection of breast cancer. *Cancer Lett.* **2015**, *369*, 67–75. [CrossRef] [PubMed]

196. Shin, V.Y.; Siu, J.M.; Cheuk, I.; Ng, E.K.O.; Kwong, A. Circulating cell-free miRNAs as biomarker for triple-negative breast cancer. *Br. J. Cancer* **2015**, *112*, 1751–1759. [CrossRef] [PubMed]
197. Zhang, J.; Jiang, C.; Shi, X.; Yu, H.; Lin, H.; Peng, Y. Diagnostic value of circulating miR-155, miR-21, and miR-10 b as promising biomarkers in human breast cancer. *Int. J. Clin. Exp. Pathol.* **2016**, *9*, 10258–10265.
198. Taslim, C.; Weng, D.Y.; Brasky, T.M.; Dumitrescu, R.G.; Huang, K.; Kallakury, B.V.S.; Krishnan, S.; Llanos, A.A.; Marian, C.; McElroy, J.; et al. Discovery and replication of microRNAs for breast cancer risk using genome-wide profiling. *Oncotarget* **2016**, *7*, 86457. [CrossRef]
199. Kurozumi, S.; Yamaguchi, Y.; Kurosumi, M.; Ohira, M.; Matsumoto, H.; Horiguchi, J. Recent trends in microRNA research into breast cancer with particular focus on the associations between microRNAs and intrinsic subtypes. *J. Hum. Genet.* **2017**, *62*, 15–24. [CrossRef]
200. McGuire, A.; Brown, J.A.L.; Kerin, M.J. Metastatic breast cancer: The potential of miRNA for diagnosis and treatment monitoring. *Cancer Metastasis Rev.* **2015**, *34*, 145–155. [CrossRef]
201. Komatsu, S.; Ichikawa, D.; Takeshita, H.; Morimura, R.; Hirajima, S.; Tsujiura, M.; Kawaguchi, T.; Miyamae, M.; Nagata, H.; Konishi, H.; et al. Circulating miR-18a: A sensitive cancer screening biomarker in human cancer. *Vivo Athens Greece* **2014**, *28*, 293–297.
202. Hamam, R.; Hamam, D.; Alsaleh, K.A.; Kassem, M.; Zaher, W.; Alfayez, M.; Aldahmash, A.; Alajez, N.M. Circulating microRNAs in breast cancer: Novel diagnostic and prognostic biomarkers. *Cell Death Dis.* **2017**, *8*, e3045. [CrossRef] [PubMed]
203. Luengo-Gil, G.; Gonzalez-Billalabeitia, E.; Perez-Henarejos, S.A.; Navarro Manzano, E.; Chaves-Benito, A.; Garcia-Martinez, E.; Garcia-Garre, E.; Vicente, V.; Ayala de la Peña, F. Angiogenic role of miR-20a in breast cancer. *PLoS ONE* **2018**, *13*, e0194638. [CrossRef] [PubMed]
204. Schwarzenbach, H.; Milde-Langosch, K.; Steinbach, B.; Müller, V.; Pantel, K. Diagnostic potential of PTEN-targeting miR-214 in the blood of breast cancer patients. *Breast Cancer Res. Treat.* **2012**, *134*, 933–941. [CrossRef] [PubMed]
205. Wang, S.E.; Lin, R.-J. MicroRNA and HER2-overexpressing cancer. *MicroRNA* **2013**, *2*, 137–147. [CrossRef] [PubMed]
206. Graveel, C.R.; Calderone, H.M.; Westerhuis, J.J.; Winn, M.E.; Sempere, L.F. Critical analysis of the potential for microRNA biomarkers in breast cancer management. *Breast Cancer* **2015**, *7*, 59–79.
207. Wang, H.; Peng, R.; Wang, J.; Qin, Z.; Xue, L. Circulating microRNAs as potential cancer biomarkers: The advantage and disadvantage. *Clin. Epigenetics* **2018**, *10*, 59. [CrossRef]
208. Castro-Giner, F.; Gkountela, S.; Donato, C.; Alborelli, I.; Quagliata, L.; Ng, C.K.Y.; Piscuoglio, S.; Aceto, N. Cancer Diagnosis Using a Liquid Biopsy: Challenges and Expectations. *Diagnostics* **2018**, *8*, 31. [CrossRef]
209. Liu, T.; Zhang, Q.; Zhang, J.; Li, C.; Miao, Y.-R.; Lei, Q.; Li, Q.; Guo, A.-Y. EVmiRNA: A database of miRNA profiling in extracellular vesicles. *Nucleic Acids Res.* **2019**, *47*, D89–D93. [CrossRef]
210. Eichelser, C.; Stückrath, I.; Müller, V.; Milde-Langosch, K.; Wikman, H.; Pantel, K.; Schwarzenbach, H. Increased serum levels of circulating exosomal microRNA-373 in receptor-negative breast cancer patients. *Oncotarget* **2014**, *5*, 9650. [CrossRef]
211. Filipów, S.; Łaczmański, Ł. Blood Circulating miRNAs as Cancer Biomarkers for Diagnosis and Surgical Treatment Response. *Front. Genet.* **2019**, *10*, 169. [CrossRef]
212. Nassar, F.J.; Nasr, R.; Talhouk, R. MicroRNAs as biomarkers for early breast cancer diagnosis, prognosis and therapy prediction. *Pharmacol. Ther.* **2017**, *172*, 34–49. [CrossRef] [PubMed]
213. Polivka, J.; Altun, I.; Golubnitschaja, O. Pregnancy-associated breast cancer: The risky status quo and new concepts of predictive medicine. *EPMA J.* **2018**, *9*, 1–13. [CrossRef] [PubMed]
214. Avishai, E.; Yeghiazaryan, K.; Golubnitschaja, O. Impaired wound healing: Facts and hypotheses for multi-professional considerations in predictive, preventive and personalised medicine. *EPMA J.* **2017**, *8*, 23–33. [CrossRef] [PubMed]
215. Kunin, A.; Polivka, J.; Moiseeva, N.; Golubnitschaja, O. "Dry mouth" and "Flammer" syndromes—neglected risks in adolescents and new concepts by predictive, preventive and personalised approach. *EPMA J.* **2018**, *9*, 307–317. [CrossRef] [PubMed]
216. Girotra, S.; Yeghiazaryan, K.; Golubnitschaja, O. Potential biomarker panels in overall breast cancer management: Advancements by multilevel diagnostics. *Pers. Med.* **2016**, *13*, 469–484. [CrossRef] [PubMed]
217. Golubnitschaja, O.; Baban, B.; Boniolo, G.; Wang, W.; Bubnov, R.; Kapalla, M.; Krapfenbauer, K.; Mozaffari, M.S.; Costigliola, V. Medicine in the early twenty-first century: Paradigm and anticipation—EPMA position paper 2016. *EPMA J.* **2016**, *7*, 23. [CrossRef]

7

Aberrant Methylation Status of Tumour Suppressor Genes in Ovarian Cancer Tissue and Paired Plasma Samples

Dana Dvorská [1], Dušan Braný [1,*], Bálint Nagy [2], Marián Grendár [3], Robert Poka [4], Beáta Soltész [2], Marianna Jagelková [5,6], Katarína Zelinová [5,6], Zora Lasabová [6], Pavol Zubor [5] and Zuzana Danková [6]

1. Division of Molecular Medicine, Biomedical Center Martin, Jessenius Faculty of Medicine in Martin, Comenius University in Bratislava, 036 01 Martin, Slovakia
2. Department of Human Genetics, Faculty of Medicine, University of Debrecen, H-4032 Debrecen, Hungary
3. Bioinformatic Unit, Biomedical Center Martin, Jessenius Faculty of Medicine in Martin, Comenius University in Bratislava, 036 01 Martin, Slovakia
4. Institute of Obstetrics and Gynecology, Faculty of Medicine, University of Debrecen, H-4032 Debrecen, Hungary
5. Department of Gynaecology and Obstetrics, Martin University Hospital, Jessenius Faculty of Medicine in Martin, Comenius University in Bratislava, 036 01 Martin, Slovakia
6. Division of Oncology, Biomedical Center Martin, Jessenius Faculty of Medicine in Martin, Comenius University in Bratislava, 036 01 Martin, Slovakia

* Correspondence: dusan.brany@uniba.sk

Abstract: Ovarian cancer is a highly heterogeneous disease and its formation is affected by many epidemiological factors. It has typical lack of early signs and symptoms, and almost 70% of ovarian cancers are diagnosed in advanced stages. Robust, early and non-invasive ovarian cancer diagnosis will certainly be beneficial. Herein we analysed the regulatory sequence methylation profiles of the *RASSF1, PTEN, CDH1* and *PAX1* tumour suppressor genes by pyrosequencing in healthy, benign and malignant ovarian tissues, and corresponding plasma samples. We recorded statistically significant higher methylation levels ($p < 0.05$) in the *CDH1* and *PAX1* genes in malignant tissues than in controls (39.06 ± 18.78 versus 24.22 ± 6.93; 13.55 ± 10.65 versus 5.73 ± 2.19). Higher values in the *CDH1* gene were also found in plasma samples (22.25 ± 14.13 versus 46.42 ± 20.91). A similar methylation pattern with positive correlation between plasma and benign lesions was noted in the *CDH1* gene ($r = 0.886$, $p = 0.019$) and malignant lesions in the *PAX1* gene ($r = 0.771$, $p < 0.001$). The random forest algorithm combining methylation indices of all four genes and age determined 0.932 AUC (area under the receiver operating characteristic (ROC) curve) prediction power in the model classifying malignant lesions and controls. Our study results indicate the effects of methylation changes in ovarian cancer development and suggest that the *CDH1* gene is a potential candidate for non-invasive diagnosis of ovarian cancer.

Keywords: liquid biopsy; pyrosequencing; ovarian cancer; *CDH1*; *PTEN*; *PAX1*; *RASSF1*; cfDNA

---

1. Introduction

Ovarian cancer (OC) is one of the most lethal gynaecological cancerous diseases and the fifth leading cause of cancer-related death in women. Although it is globally diagnosed in almost 240,000 women annually and responsible for over 150,000 deaths each year [1], the incidence varies regionally and it is generally higher in both developed countries and caucasian women. The following OC rates in 100,000 women have been established: (1) Eastern Europe has 11.4 women diagnosed with OC in every 100,000, (2) the Central Europe and North America have over 8, (3) Central and South America have up to 6 and (4) the lowest rates are observed in Asia and Africa, at less than 3, usually [1,2]. It is also important that OC incidence has been slightly decreasing in developed countries since 1990 but increasing in developing countries [3–5]. OC is also very heterogeneous, with a variety of benign, borderline and malignant variants, almost all of which arise from transformation of epithelial, stromal and germ cell types [6]. Of these, the most common malignant neoplasms have their origin in the epithelium in up to 90% OC while the germ cell type are very rare with only 2–3% [6].

Ovarian tumours (OT) were traditionally divided into five groups: Serous, mucinous, endometrioid, clear cell and transitional [7]. However, each of those also comprised variable forms of benign, borderline and malignant lesions, with a total of over 40 tumour types. This classification primarily focused on the ovarian mesothelial surface as the point of origin for epithelial OT [7], but the more recent 2014 WHO classification, established in parallel with new FIGO staging implementation, excluded transitional cell tumours, but added the new sero-mucinous tumour group [7]. This latest classification is considered more consistent and it also comprises only 28 tumour types. While Meinhold-Heerlein et al. [7] have compared the traditional and new classification in detail many analyses have been based on older classification, and the results from those studies quoted herein are in their original form. However, it is essential to remember that OC is a very variable group of cancerous diseases and their formation is affected by many epidemiological factors [2]. It is especially typical in its lack of early signs and symptoms, and almost 70% of these cancers are diagnosed in advanced stages [8].

Diagnosed OC is primarily treated surgically with subsequent application of platine and taxane based adjuvant chemotherapy [9], but the efficacy of surgical treatment followed by chemotherapy rapidly decreases with the identified stage of the tumour [8]. The statistics for variable OC histotypes highlight that only 20% of women in advanced stages have five-year survival after diagnosis, while 89% diagnosed in stage I and 71% in stage II survive five years [8]. Moreover, OC metastases spread very quickly using two strategies: (1) Transcoelomic passive dissemination of tumour cell spheroids in the peritoneal cavity through ascites and (2) haematogenous metastasis of OC cells in the systemic circulation followed by the preferred seeding of the omentum [10–12]. It is therefore essential to use the robust, sensitive and non-invasive liquid biopsy (LB) approach for early OC diagnosis and differentiation. This methodology overcomes the typical limitations of solid tumour biopsies, such as invasiveness connected with unexpected complications during surgical procedures, difficulties in operating on organs that lie deep within the body, false negativity from sampling bias and the impossibility of repeating the same intervention [13]. Most importantly, solid biopsies provide no possibility for early tumour diagnosis and differentiation; they give tumour information only at particular points in time and they preclude monitoring the response to medication during treatment and follow up. In contrast, these are all possible with the LB approaches focused on detection and characterisation of abnormalities in circulating tumour cells (CTC), circulating tumour DNA (ctDNA), circulating cell-free microRNAs (cfmiRNA) and released exosome vesicles [10].

This cross-sectional study focuses on epigenetic changes, particularly alterations in methylation levels in primary tumours and the paired plasma samples. It then analyses the model's predictive power in distinguishing between diagnoses so that this can be used in clinically predicting the risk of OC. For this purpose, four tumour-suppressor genes: *Ras Association Domain Family Member 1* (*RASSF1*), *Paired Box 1* (*PAX1*), *Cadherin 1* (*CDH1*) and *Phosphatase and Tensin Homolog* (*PTEN*) were analysed.

RASSF1 inactivation is connected with the development and progression of many cancer types, and while this inactivation can be caused by deletions and point mutations, it most often arises from methylation changes [14,15]. *RASSF1* encodes a protein similar to the RAS effector proteins [16] and this product regulates cell-cycle progression and apoptosis pathways, especially the Ras/PI3K/AKT, Ras/RAF/MEK/ERK and Hippo apoptosis pathways [16]. It also regulates Bax-mediated cell death [17], interacts with the *XPA* DNA repair gene [18] and inhibits accumulation of cyclin D1 [19]. In addition, it inhibits anaphase-promoting complex (APC) activity and mitotic progression following interaction with CDC20 [20]. *PTEN* is a multifunctional tumour-suppressor gene encoding phosphatidylinositol-3,4,5-trisphosphate 3-phosphatase, which de-phosphorylates phosphoinositide substrates, and it is also an important regulator of insulin signalling, glucose metabolism and the PI3K/AKT/mTOR pathway [21]. The *CDH1* gene encodes the cadherin–calcium dependent cell adhesion protein which regulates the mobility and proliferation of epithelial cells [22], and abnormalities in this gene are connected with the development of colorectal [22], breast [23], gastric [24] and most likely also OC [25].

The *PAX1* gene is a member of the paired-box transcriptional factor gene family which regulates proper tissue development and cellular differentiation in embryos [26]. Although its role in cancer development and progression has not been sufficiently established [27], this gene is aberrantly hyper-methylated and down-regulated in different types of cancer [28], and is suggested to inhibit the phosphorylation of multiple kinases, especially after challenges with the *EGF* and *IL6* oncogenic growth factors [28]. In addition, *PAX1* has the ability to activate variable phosphatases, including DUSP-1, 5, and 6, and to inhibit EGF/MAPK signaling. Its interaction with *SET1B* increases histone H3K4 methylation and the DNA demethylation of numerous phosphatase-encoding genes [28].

Finally, changes in DNA methylation were repeatedly noted in OC and these affected the activity of variable genes, and therefore contributed to the tumorigenesis of variable histotypes. Furthermore, methylation changes in OC can be prognostic for shorter progression-free survival (PFS) and chemotherapy resistance [29–31]. Importantly for this study, all four selected genes have previously been observed to be abnormally methylated in primary OC tumours [25,32–40]. Moreover, lower mRNA expression in at least one OC histotype in the *RASSF1* and *PTEN* genes [41–43] and lower *CDH1* gene protein expression [44] in OC have been noted. Decreased *CDH1* gene expression also correlated with the methylation levels and all these genes can, therefore, be suitable candidates for methylation analysis of LB samples from women with OC.

2. Material and Methods

Our sample collection consists of 128 variable types of ovarian tissues and paired plasma samples from central European caucasian women who underwent surgical excision as part of their treatment. This includes healthy (mean age 63.83 ± 11.23 years), benign (64.57 ± 16.40 years) and malignant samples (57.20 ± 11.12 years), and also a few ovarian cancers subsequent to breast cancer (BC-OC) (59.25 ± 12.74 years) (Table 1). The histological characteristics of the OC samples are summarised by FIGO staging and subtype in Tables 2 and 3. Personal and gynaecological anamnesis information was obtained during medical examination and the study was approved by the ethical committee of Jessenius Faculty of Medicine in Martin Slovakia under number 1933/2016, and the research was performed in compliance with the Declaration of Helsinki.

Table 1. Number of samples according to the diagnosis.

Tissue		Diagnosis	Plasma		Paired Samples
n	%		*n*	%	*n*
17	13.3	Control	9	7.0	7
8	6.3	Benign	5	3.9	5
49	38.3	OC	33	25.8	33
4	3.1	BC-OC	3	2.3	3

Note: OC—ovarian cancer; BC-OC—ovarian cancer subsequent to breast cancer.

Table 2. FIGO classification of OC samples.

Stage	*n*	%
1A	5	10.2
1C	3	6.1
1C2	1	2.0
3b	1	2.0
3B	3	6.1
3C	17	34.7
4	7	14.3
4B	5	10.2
No data	7	14.3

Table 3. Histopathological subtypes of OC samples.

Subtype	*n*	%
Clear cell	2	4.1
Endometrioid	2	4.1
Mucinous	4	8.2
Serous	16	32.7
Serous papillary	25	51.0

2.1. DNA Isolation and Bisulfite Conversion

Immediately after section, the OT and healthy control tissues were stabilised in RNAlater solution and kept at −20 °C. DNA from tissue and plasma was isolated by DNeasy Blood and Tissue Kit (Qiagen GmbH, Hilden, Germany) and QIAamp® DSP Virus Kit (Qiagen GmbH, Hilden, Germany). The concentration was measured by Nanodrop™ 2000 (Thermo Fisher Inc, Wilmington, DE, USA) and QuBit (Thermo Fisher Inc, Wilmington, DE, USA). Finally, subsequent bisulfite DNA conversion was performed by the Epitect Bisulfite Kit (Qiagen Inc., Valencia, CA, USA) and the converted DNA was stored at −20 °C.

2.2. Methylation Analyses

Methylation of bisulfite modified DNA from control tissues, primary tumours and plasma samples were analysed by pyrosequencing on Pyromark Q96ID (Qiagen GmbH, Hilden, Germany). This is a very practical and reliable method and it provides quantitave methylation levels relatively quickly.

Sequencing was preceded by PCR amplification by Pyromark PCR Kit® (Qiagen GmbH). The total reaction volume was 25 μL and comprised 21 μL Pyromark PCR Kit mixture (12.5 μL MasterMix, 2.5 μL CoralLoad Concetrate, 1μL $MgCl_2$, 5μL Q-Solution), 2.5 μL primer pairs, 0.5 μL RNAse free water and 1 μL bisulfite modified DNA.

The PCR reaction steps were as follows: Activation of polymerase (95 °C, 15 min), 45 cycles (*RASSF1*) or 50 cycles (*PTEN, CDH1, PAX1*) of denaturation (94 °C, 15 s), annealing (56 °C (*RASSF1, PAX1*)/62 °C (*PTEN, CDH1*) for 30 s), extension (72 °C, 30 s) and final extension at 72 °C for 10 min. The qualitative and quantitative amplicon parameters were subsequently assessed by 1.5% agarose gel electrophoresis. The amplification process was the same for both DNA from tumorous tissue and circulating DNA. The PCR amplification primers were modified so that the reverse primer had a biotinylated 5′ end. This enabled separation of the strand for pyrosequencing analysis.

The PCR product was subsequently used for pyrosequencing. This was first mixed with streptavidin-coated sepharose beads (GE Healthcare Life Sciences, Chalfont, UK), binding buffer (Qiagen GmbH) and nuclease free water in a total volume of 80 μL in a Pyromark plate and shaken for 10 min at 2,500 rpm. The 5′-biotinylated strand for sequencing was immobilised by Pyromark working station, transferred to a mixture of 0.4 M sequencing primer and binding buffer (Qiagen GmbH) and incubated for 2 min at 80 °C. Finally, the pyrosequencing results were interpreted by

Pyromark Q96 software version 2.5.8 (Qiagen GmbH). The commercially available CpG methylation assays® (Qiagen GmbH) were used herein to analyse methylation levels in the following sequences (Table 4):

Table 4. Analysed sequences of genes *RASSF1*, *PTEN*, *CDH1* and *PAX*, and their location in human genome.

Gene	Sequence to Analyse	Location
RASSF1	5′-GGTAGCGCAGTCGCCGCGGGTCAAGTCGCGGC-3′	3:50,377,907–50,377,938
PTEN	5′-CGCGAGGCGAGGATAACGAGCTAAGCCTCGGC-3′	10:89,622,923–89,622,954
CDH1	5′-CGGCAGCGCGCCCTCACCTCTGCCCAGGACGCGGC-3′	16:68,772,300–68,772,331
PAX1	5′-CGGAATCTGCTAGCTTCGTCGGGCGCGA-3′	20:21,684,296–21,684,323

2.3. Biostatistical Analyses

Exploratory data analysis was performed by standard summary statistics; the methylation data was visualised by boxplot overlain with swarmplot; and the heat map provided methylation level mean values. Data normality was assessed by a quantile-quantile plot with 95% bootstrap confidence band, and robust ANOVA tested the hypothesis of equality of population methylation medians in the disease levels. Null hypothesis rejection was followed by the Tukey HSD post-hoc test. The discriminative ability of the methylation indices and age for various pairs of disease level was then assessed by the random forest learning algorithm. This quantified the importance of the individual predictors by minimal depth. The trained random forest discriminative ability was visualised by the ROC curve based on "out-of-bag" data and the AUC area under the ROC curve and the 95% AUC confidence interval quantified the discriminative ability of methylation indices and age. Finally, we determined sample size and performed power analysis for the one-way ANOVA test, assuming balanced design. All data analysis was performed in R, version 3.5.2 [45].

3. Results

3.1. CpG Sites Methylation Status of the RASSF1, PTEN, CDH1 and PAX1 Gene Regulatory Regions

DNA methylation profiling of 21 CpG sites in four genes, comprising six CpG sites in the *RASSF1* gene and five CpG sites in each of the *PTEN*, *CDH1* and *PAX1* genes was performed by pyrosequencing on the four different tissue types—healthy ovarian tissues, benign OT, OC and BC-OC, and also on the corresponding plasma samples. The mean value, median and standard deviation of each CpG site is provided in Supplementary Table S1 for tissues and Table S2 for plasma.

Figure 1 shows the methylation value of each CpG site in the tissue (A) and plasma samples (B) according to diagnosis. The general overview provides information on the hypomethylation of *PTEN* CpG sites for all diagnoses and tissue and plasma samples. Greater variability is obvious in the *RASSF1* and *PAX1* CpG sites, and the highest methylation was detected in the *CDH1* CpG sites. The comparison of tissue samples (Figure 1A) shows the increasing methylation trend from the control sample group to cancer samples in the *RASSF1*, *CDH1* and *PAX1* genes. The statistical comparison of means revealed the most significant differences between controls and OC samples in two *CDH1* and three *PAX1* gene CpG sites and between controls and BC-OC in three *PAX1* gene CpG sites. The increasing mean values for *RASSF1* were not statistically significant because of the wide variance in values (detailed statistical data is contained in Table S1). Figure 1B depicts variability in the plasma group methylation values, with no visible linear trends detected between the diagnostic groups. Finally, the statistically significant higher methylation levels were detected in OCs in four *CDH1* gene CpG sites compared to controls, and all detailed statistical data is listed in Table S2.

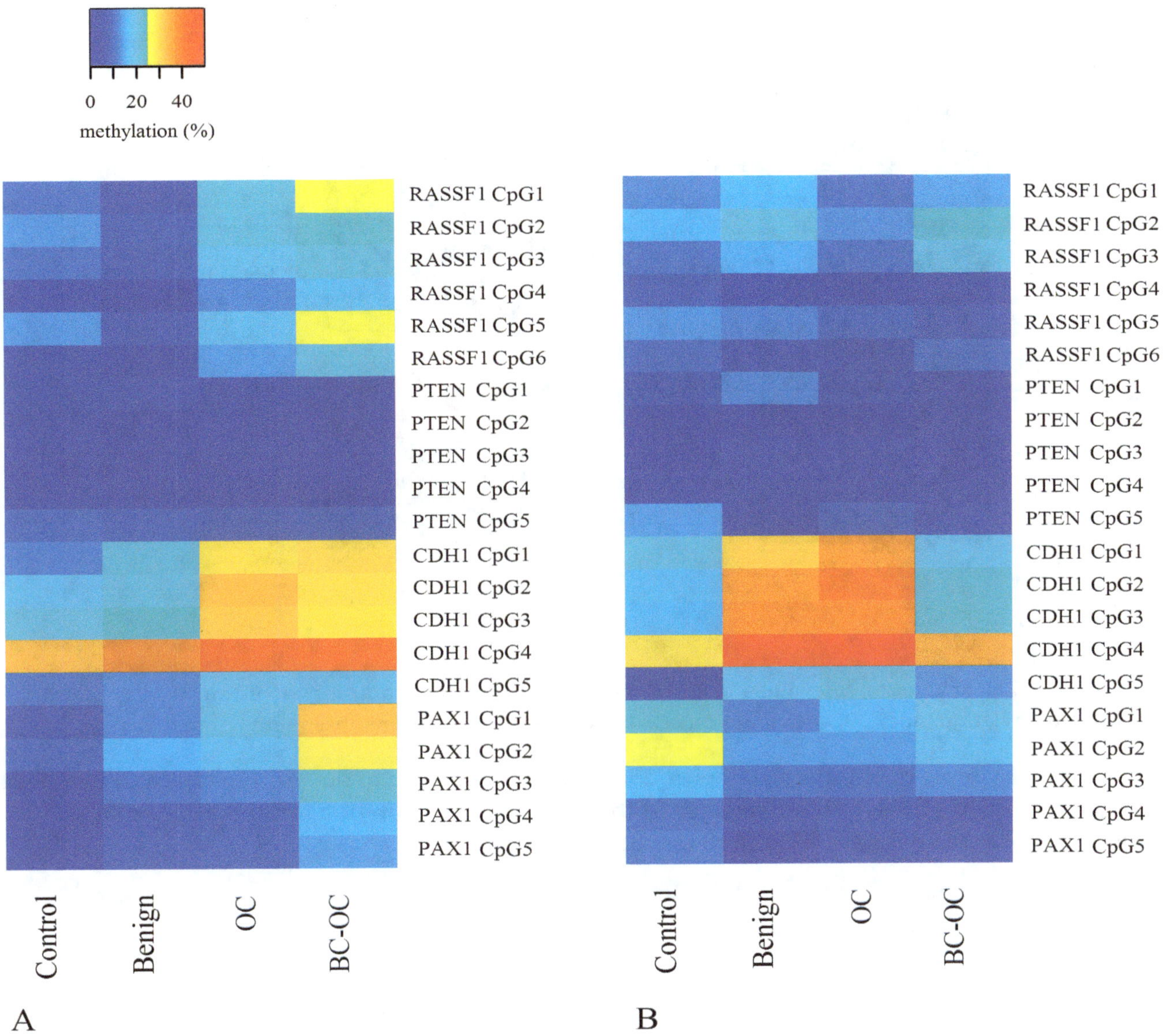

Figure 1. Heat map of each CpG site methylation values in tissue (**A**) and plasma samples (**B**) according to diagnosis. OC—ovarian cancer; BC-OC—ovarian cancer subsequent to breast cancer.

3.2. DNA Methylation Indices of the RASSF1, PTEN, CDH1 and PAX1 Gene Regulatory Regions

The methylation index (MI) is the mean percentage of methylation across all CpG sites analysed per gene, and that was calculated for each gene and diagnostic group in the tissue and plasma samples. The MI values of each diagnostic group are depicted by the heat maps in Figure 2A for tissue and 2B for plasma. Although *PTEN* gene methylation indices were very low, they were statistically significantly different between benign tumours (3.90 ± 4.10) and malignant OC (3.00 ± 2.60) in the tissue samples. No statistically significant differences were determined in *RASSF1* MI between diagnostic groups in either the tissue or plasma samples. In the tissue samples, the *CDH1* and *PAX1* gene methylation indices had increasing tendency toward OC, with statistically significant difference between controls (24.22 ± 6.93 for *CDH1* and 5.73 ± 2.19 for *PAX1*) and OCs (39.06 ± 18.78 and 13.55 ± 10.65) in both genes, and also between controls (5.73 ± 2.19) and BC-OC samples (22.90 ± 13.56) in the *PAX1* gene. In the plasma samples, there was only one statistically significant difference between controls (22.25 ± 14.13) and OC samples (46.42 ± 20.91), in the *CDH1* gene. Detailed data is provided in Supplementary Table S3.

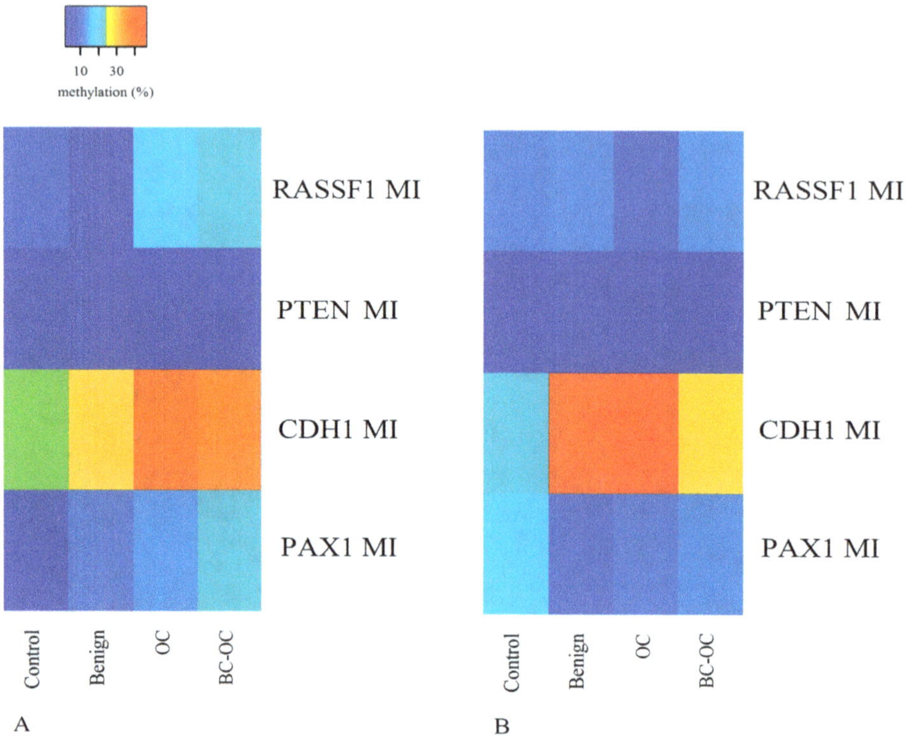

Figure 2. Heat map of MI values in tissue (**A**) and plasma (**B**) samples according to diagnosis. MI—methylation index, OC—ovarian cancer; BC-OC—ovarian cancer subsequent to breast cancer.

Analysed sample distribution is depicted in Figure 3 according to the sample type and diagnosis for all four studied genes. Wider variance in the mean values is obvious in the *PTEN* methylation index, and that explains the small differences between groups.

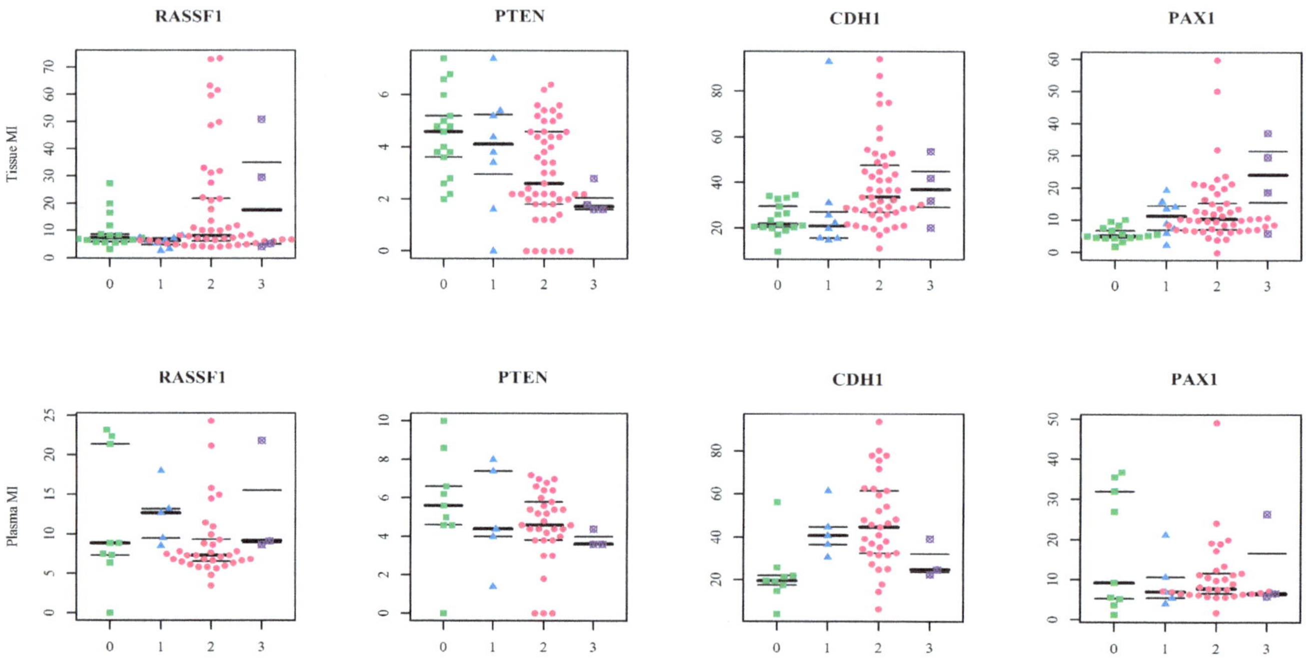

Figure 3. The swarmplots of methylation indices of *RASSF1*, *PTEN*, *CDH1* and *PAX1* genes, showing the mean values of all observations. MI—methylation index; 0—control tissues (green squares); 1—benign tumours (blue triangles); 2—ovarian cancers (pink circles); 3—ovarian cancers subsequent to breast cancer (purple crossed circles).

3.3. *Methylation Pattern by FIGO Staging and Histological Type*

The methylation levels of all CpG sites revealed little statistically significant differences between the compared FIGO stages (data not shown) and the MI data pattern is depicted in Table 5. Two following trends were observed: The *RASSF1* and *PAX1* methylation indices decreased with disease severity increase, and the *CDH1* and *PTEN* values had increasing tendency, thus followiing the stages. Statistical comparison of the mean values revealed significant differences only in the *PTEN* gene and between stages I and II, and stages I and IV in the *PTEN* gene.

Table 5. MI averages of CpG sites in regulatory regions of the *RASSF1*, *PTEN*, *CDH1* and *PAX1* genes in tissue samples according to FIGO stage.

	Stage I (n = 9)			Stage II (n = 21)			Stage IV (n = 12)			p-Value
	Mean	Median	SD	Mean	Median	SD	Mean	Median	SD	
MI_*RASSF1*	19.24	5.33	20.46	17.24	7.50	19.80	8.01	6.83	3.95	ns
MI_*PTEN*	2.09	2.00	1.37	3.47	3.60	1.49	3.63	3.60	1.37	a, b
MI_*CDH1*	32.89	29.00	16.81	39.98	30.50	20.72	48.38	44.25	17.76	ns
MI_*PAX1*	17.11	13.60	17.36	11.56	10.40	5.42	10.98	8.50	6.28	ns

Note: Due to number of cases in each category, we created only three groups, based on the FIGO stage numbers (for example 1A, 1B and 1C were merged into stage I); ns—non significant difference between groups; a—statistically significant difference ($p < 0.05$) between FIGO I and II; b—between FIGO I and IV.

Comparison of the methylation levels and indices between histological subtypes revealed only sporadic differences and indicated possible uniform methylation of the studied genes in this OC group.

3.4. *Comparison of Methylation Indices in Tissue and Plasma Paired Samples*

The MI values of paired tissue and plasma samples were compared to assess if the plasma methylation status reflects those observed in the primary tumours. Figure 4 shows similarities and differences in the methylation indices of all *RASSF1*, *PTEN*, *CDH1* and *PAX1* gene paired samples, and statistical analysis determined significant plasma and tissue correlation only in the *CDH1* and *PAX1* genes. The correlation coefficient of *CDH1* MI was negative in controls ($r = -0.812$, $p = 0.05$), but positive in benign samples ($r = 0.886$, $p = 0.019$) and OC samples ($r = 0.428$, $p = 0.015$). However, strong correlation was established between plasma and tissue in all *PAX1* CpG sites and between *PAX1* methylation indices, but only in the OC pairs ($r = 0.771$, $p < 0.001$).

The results for sample size determination, including the effect size and the number of observations in each group needed to maintain 80% power level, are included in Supplementary Table S4 for tissue samples and Table S5 for plasma samples.

3.5. *Evaluation of the Diagnostic Predictive Model Using ROC Analyses*

Finally, we used the random forest algorithm with ROC analysis to evaluate if the studied methylation indices combined with age could significantly distinguish between diagnoses. Table 6 provides the AUC values with 95% confidence intervals for tandem discriminations. The "Important variable" column denotes the most relevant variables based on importance measured by minimal graph depth. The best AUC values were obtained in the model which discriminate between controls and OC samples based on both the MI of the four genes in tissues (AUC = 0.932) and in plasma (AUC = 0.822). The first "tissue" model had perfect sensitivity (98%), but specificity was weak (34%). The same was observed in the "plasma" model, with 91% sensitivity and 56% specificity. Based on the AUC values and 95% CI, these models indicate strong predictive power, and can, therefore, be considered for diagnostic discrimination clinically.

Figure 4. Methylation indices of *RASSF1*, *PTEN*, *CDH1* and *PAX1* genes in paired tissue and plasma samples. MI—methylation index; C—control samples; B—benign; OC—ovarian cancer; BC-OC—ovarian cancer subsequent to breast cancer.

Table 6. Predictive performance of the model distinguishing diagnostic groups.

		AUC	95% CI	Overall Error Rate (%)	Important Variable
Tissue	Control vs. Benign	0.738	0.391–1.000	15.38	*PAX1*
	Control vs. OC	0.932	0.823–1.000	9.09	*PAX1*
	Benign vs. OC	0.499	0.195–0.802	12.5	*CDH1*
Plasma	Control vs. Benign	0.778	0.512–1.000	21.43	*CDH1*
	Control vs. OC	0.822	0.667–0.976	16.67	*CDH1*
	Benign vs. OC	0.630	0.309–0.951	13.16	*PTEN*

Note: OC—ovarian cancer, AUC—area under the curve, CI—confidence interval.

4. Discussion

OC is one of the most lethal cancerous diseases, and despite the progress in surgical techniques, better management, advanced chemotherapy and targeted therapy, the OC patient survival rate is still very low and almost 70% of all patients are diagnosed in advanced stages. This is mainly due to the lack of early symptoms in this disease [8]. Early OC screening approaches have been traditionally based on serum CA-125 concentration and trans-vaginal ultrasound, but both these methods lack sufficient sensitivity and specificity [46,47]. Therefore, non-invasive biomarkers able to predict early OC presence and monitor response to treatment and cancer progression are imperative. However, those with sufficient predictive power are still lacking and most possible candidates require further investigation [46,48,49]. Moreover, further use of the tumour markers is crucial in all gynaecological cancerous diseases [50–52]. For example, CTC, ctDNA, cfmiRNAs and released exosome vesicles can be analysed in the assessment of molecular markers. This assessment can be performed by a variety of methods, from basic PCR modifications to advanced multi-parallel sequencing approaches [10].

Herein, we focused on the epigenetic impact on OC formation by analysing the methylation levels of the *RASSF1*, *PTEN*, *CDH1* and *PAX1* tumour suppressor genes' regulatory sequences. We compared the methylation status of normal tissue, benign, OC and BC-OC samples. We then assessed the methylation levels in these genes in the corresponding plasma samples in order to estimate their potential for LB utilisation. Finally, we compared the methylation patterns in the tissue samples and the paired plasma samples to assess if these patterns were markedly different.

Previously published articles reported the importance of some genes' plasma methylation status in OC management. For example, the *C2 CD4 D*, *WNT6* and *COL23 A1* genes' methylation status is not only significantly altered in OC patients' ctDNA, but different responses to platinum-based therapy can also be associated with their methylation levels [53]. Moreover, Giannopolou et al. [54] discovered altered *ESR1* gene methylation levels in the ctDNA samples of high-grade serous OC patients, and these abnormalities are very similar to those in primary tumours [54].

The most notable methylation changes in solid OT are in the *BRCA1* [29,55], *OPCML* [56], *HOXA9* [33] and *P16 INK4α* genes [57]. *BRCA1* is methylated in up to 30% [29,55] of these tumours and *OPCML* is methylated in over 80% of OT [56]. Interestingly, *RASSF1 A* is the major *RASSF1* gene isoform and this has been repeatedly analysed for methylation changes in both OC tissues and plasma samples. For example, de Caceres et al. [32] noted methylation of the *RASSF1 A* gene in 50% of primary OT and detected an identical pattern of gene hyper-methylation in 41 of 50 (82%) matched plasma DNA. In addition, hyper-methylation of this gene was recorded in all OC histological types, grades and stages. Wu et al. [33] reported very similar results in solid tissues, with hyper-methylation in 49% of patients with various OC types and grades. In contrast, Agathanggelou et al. [58] found hyper-methylation in only 10% of solid OT.

Methylation changes in the *RASSF1 A* promoter have already been analysed for utilization in non-invasive differentiation of epithelial ovarian carcinoma from healthy controls. For example, 90% sensitivity and 86% specificity for cancer detection was determined when methylation levels of the *RASSF1 A*, *EP300* and *CALCA* genes were used in combination [59]. In addition, the multiplex methylation specific PCR (MSP) assay of seven different genes, including *RASSF1 A*, achieved 85.3% sensitivity and 90.5% specificity in differentiating epithelial OC from healthy controls in serum samples [60]. Variable *RASSF1 A* methylation has also been noted in tissue samples of serous and non-serous types and benign tumours [61].

Herein, the *RASSF1 A* MI differed in all four tissue types. It was higher in primary malignant tumours (17.98 ± 19.73) than in healthy tissue (9.50 ± 6.21), but the opposite was found in plasma ctDNA. Here, the controls have higher MI (11.74 ± 8.34) compared to malignant samples (8.93 ± 4.54). However, no statistically significant differences were established between particular CpGs or mean methylation indices. Importantly, although these differences proved insignificant, they can indicate wide methylation changes in *RASSF1* gene regulating sequences and the important role of methylation in silencing this gene in OC development as is generally suggested [29].

Although the methylation status of the *PTEN* CpG sites and methylation indices was generally low and homogenous, there were statistically significant differences between benign and malignant lesions. The decreasing tendency in methylation indices was observed in both tissue and plasma samples, but comparison between FIGO stages revealed a statistically significant increase in methylation indices only in the higher FIGO stages. This theoretically indicates its important role in cancer formation and spread. Herein, however, we compared only the main FIGO categories where subtypes were merged in one main category, because of the small sample size. In addition, the differences between the FIGO stages were not only low, but the methylation indices themselves also had low values. This leads to debate if such small increases in methylation levels can contribute to tumour progression from stage I to stage IV. This is especially pertinent when previous very low *PTEN* methylation levels have occurred without differences in OC FIGO stages [34].

Yang et al. [34] and Zuberi et al. [35] recorded relatively low methylation values in the *PTEN* gene with higher methylation in only 16% and 8.2% of the samples, respectively. Schondorf et al. [62] suggested that *PTEN* methylation should have only a subordinate role in OC progression and only the one study by Qi et al. [36] reported relatively higher methylation levels of this gene in OC. Overall, the low and homogeneous *PTEN* methylation concord with the most of previous studies, and the small increase in methylation should not reflect progression from stage I to stage IV, though statistical significance is established between the FIGO stages. Our results also indicate the potential effect of *PTEN* methylation changes in benign lesion formation. That suggests that further analysis of the methylation changes will be beneficial, because there is no rule that benign lesions must be less methylated than malignant lesions, and different mechanisms can lead to abnormal methylation in benign and malignant tumours [63].

In comparison to *RASSF1* and *PTEN*, the *PAX1* gene had greater variation in methylation levels with statistically significant differences between controls, OC and BC-OC. There was also wide variability in particular CpG islands, especially in the control and malignant samples. This gene's methylation provides a potential biomarker for cervical cancer screening [64–66], but important methylation changes in this gene were also noted in OC. In particular, Su et al. [37] reported 50% *PAX1* methylation rates in OC, but only 14.3% in borderline tumours and 4% in benign, and Hassan et al. [38] found hyper-methylation in 50%, 50%, 46.6% and 78% of patients depending on the FIGO stage of epithelial OC. In addition, the authors associated methylation levels with the presence of HPV16/18 infection. Neither of these studies can be quantified in the same manner as our quantitative pyrosequencing results because they used MSP in their assessment, but the significant differences we achieved between healthy tissues and cancer samples can be interpreted quite similarly—that dynamic methylation changes indicating OC's presence occur in the aforementioned gene's regulation sequences. Although our results established no statistical significance, they can certainly help move the debate on how this gene's methylation changes affect the formation of benign lesion because the methylation levels were as high they are in cancerous tissues.

Finally, the *CDH1* gene had statistically significant variability in methylation levels between control and malignant tissue samples, and this agrees with previous analyses [25,39,40]. Importantly, this gene also had statistically significant plasma samples differences in both mean methylation levels and those of particular CpG dinucleotides in both healthy and cancerous groups. Therefore, the *CDH1* gene could be considered a very promising candidate for non-invasive LB-based diagnosis of malignant OC, but its potential for non-invasive investigation of benign tumour presence requires further analysis.

It is important to realize that even non-significant, but observable differences in plasma samples methylation levels were revealed also in the *RASSF1*, *PTEN* and *PAX1* genes, and this could inspire additional analyses of these genes. The large deviation in the final methylation levels of these genes was very likely the main obstacle to achieving statistical significance. However, larger sample collection could help to solve this problem.

One of the issues in the analysis of plasma samples is the lack of definition and accepted consensus if changes, including those in ctDNA methylation, remain exactly the same as in the primary tumours that released them. While some authors consider that the aberrant patterns remain identical to those in the primary tumour [53,67–72], others support diverse patterns [73–75]. Herein, we conform to similarity because of the strong correlations between tissue and plasma samples in the *CDH1* and *PAX1* genes, and also in several *RASSF1* and *PTEN* gene samples. Discrepancies in tissue and plasma sample methylation levels could have also occurred herein because circulating cfDNA of cancer patients contains a mixture of DNA from both normal and cancerous cells. The ratio of ctDNA in the entire cfDNA fraction can vary because this depends on several tumour parameters [76]. Our methodology cannot estimate how plasma sample results are affected by the presence of non-tumorous cfDNA elements, and we cannot exclude that this is responsible for the variability obtained in the results. Although the ctDNA ratio is generally quite low it can vary widely, and some estimations of ctDNA abundance reach 90% [76,77]. Importantly, it has been demonstrated that the concentration of all circulating free molecules, and not only ctDNA, is associated with a tumour volume which gives shorter overall survival (OS) for OC patients [78].

Previously performed LB-based methylation analyses assessing the methylation pattern in OCs using non-NGS approaches have provided very interesting results. Those include (1) Teschendorff et al. [79], when they used a methylation-array based study and observed significant differences in the blood of healthy and cancerous patients, and (2) Flanagan et al. [80], who utilised the pyrosequencing approach and determined significant correlation of *SFN* gene methylation with PFS. In addition, their subsequent study [81] noted specific CpGs' alterations in blood DNA following relapse from platinum-based chemotherapy, and their results proved an independent significant association with survival. Moreover, both pyrosequencing and MSP have previously been used to assess methylation abnormalities in plasma samples from different cancer types, and some of these achieved significant results assessing the profiles of all circulating DNAs, not only ctDNA [82–86]. Therefore, even methodology which is unable to distinguish ctDNAs in the whole cfDNA fraction can be useful in assessing disease characteristics from LB samples. We also incline toward the consensus that changes in primary tumours should remain preserved in ctDNA elements and the changes noted in primary tumours could then be used as inspiration for initiating further experiments utilising LB samples. Therefore, known abnormalities in primary tumours can be very helpful because these can be searched and analysed in LB samples.

OC is widely recognised as a very heterogeneous group and it should, therefore, be important to analyse and interpret the results of each histotype. However, the rarity of some morphological subtypes causes study limitations, and the small numbers in some histotypes herein prevented performance of statistical analysis. Only the serous and serous papillary histotypes were sufficient for analysis, and these recorded no significant differences in their methylation levels and were even almost identical. While studies have confirmed that some differences are typical for all OC histotypes [25,32–34,56], it cannot be ignored that unintentional variance in our results is due to the heterogeneity of our collection and also the impossibility of separating and analysing particular histotypes.

The effect size and the number of observations in each group required to maintain 80% power were assessed by power analysis, and these are listed in supplementary Tables S4 and S5. Herein, the number of malignant tissues was mostly sufficient, but we occasionally noted a lack in the number of healthy controls and the number of benign lesions was quite often lower than proposed. Therefore, although the power analysis results may limit our study through having a sample size insufficient to assess methylation levels in all CpGs with sufficient statistical power, it can help design the same or similar experiments. The highest number of required observations, excluding three *PTEN* gene CpG sites in plasma samples, was 64, and it is quite possible to collect at least that number of samples for future analysis.

It has been noted that there are generally higher cfDNA levels in malignant OC patients than in healthy patients and in those with benign lesions [87,88]. The cfDNA levels are also lower in FIGO I–II stages than in the III–IV stages [78]. Although we assessed only four genes' methylation indices and not the entire cfDNA levels, and therefore cannot directly associate these parameters, analysis of the relationship between methylation changes, tumour stage and general cfDNA levels could most likely provide clinical implications. It would be beneficial, therefore, to design further studies which assess these relationships.

The cfDNA characteristics can also vary depending on age [89] and the presence of concomitant disease [90]. Furthermore, it is presumed that length of cfDNA fragments correlates with shorter PFS, and also OS [91]. Unfortunately, we can neither confirm nor preclude that our results are affected by other clinical attributes, such as the presence or absence of other patient disease. However, further studies should establish associations between a wide variety of clinical parameters, methylation and also cfDNA levels, as previously suggested.

Herein we evaluated the predictive performance of the combined model of DNA methylation signature (MI of the genes) and the age (in years) using ROC analyses computed by the random forest algorithm. We ran six models separated for sample type. The model which discriminated between control and OC samples had the highest predictive power—at 0.932 AUC in tissue samples and 0.822 in plasma. This initially appeared a perfect model with high sensitivity and clinical application potential, but 34% specificity for tissue samples and 56% for plasma samples means that too many healthy women would be incorrectly classified as cancer patients. The most likely reason for these results was the small number of controls, and this could have affected statistical analyses and hindered efforts to ensure a strong statistical model. Adequate AUC values were obtained in the model discriminating control samples and benign samples (tissue: AUC = 0.729, 85.7% sensitivity and 83.3% specificity; plasma: AUC = 0.778, 100% sensitivity and 66.7% specificity). Our last discrimination model of benign and OC samples had insufficient classifiers. In these models, *PAX1* MI and *CDH1* MI were selected as the most important variables for tissue and plasma respectively, and this indicates the previously described results about tissue and plasma non-conformity. Although these analyses promised a high-quality classifier, other possibilities should now be also considered. The most important would be the inclusion of additional control samples, as this would provide more accurate specificity. It is also worth considering the inclusion of more variables in the model, such as gynaecological anamnesis, life style risk factors and also CA-125 or other novel biomarkers with the potential to detect serous OC at earlier stages and predict patient survival prognosis [92–94]. Although this is beyond the scope of this pilot project, it should provide inspiration for further analyses focused on specific variables identified by a multinomial approach and offer personalised tailored therapy. Because of the great variability in this disease, the management of OC should be individualized, and performance status of the patient should be considered. This is best reflected in necessity for different treatment option for younger and older women [95,96]. In addition, future individual OC treatment should also be based on different patient physiological and health parameters.

In conclusion, our study confirmed higher methylation levels in several CpG sites and also higher methylation index in *RASSF1 A*, *PAX1* and *CDH1* genes in OC tissues compared to control tissues. However, no statistical significance was noted in the *RASSF1* gene because of high variance. Statistically significant higher MI and CpG sites methylation levels were detected in plasma samples only in the *CDH1* gene. *PTEN* gene was slightly and homogenously methylated. In addition, we recorded similar methylation patterns with strong correlations in the *CDH1* and *PAX1* genes in OC plasma and tissue, and the results suggest that the *CDH1* gene is a prospective candidate for non-invasive, LB-based differentiation in OC. Our diagnostic discrimination model combined the methylation indices of all four genes with age, and this revealed the highest AUC 0.932 predictive power in the model, comparing OC subjects with controls. Finally, it will prove highly beneficial to employ as many LB-based approaches as possible in cancer detection and treatment, especially in OC, which has such high risk and generally delayed diagnosis. LB is capable of providing early OC diagnosis which can help decrease the global expense of fighting this disease, but most importantly it can save lives, or at least prolong them with greater life quality for the thousands of women affected by this lethal cancer.

Author Contributions: Conceptualization: B.N., B.S. and Z.D.; data curation: M.G.; funding acquisition: P.Z. and Z.D.; investigation: D.D. and D.B.; methodology: Z.D., D.D., D.B., M.J. and K.Z.; project administration: P.Z., Z.L., and Z.D.; software: M.G. and Z.D.; supervision: B.N., P.Z. and Z.L.; writing—original draft: D.D., D.B. and Z.D.; writing—reviewing and editing: B.N., R.P. and Z.D.

Abbreviations

AUC	Area under the curve
BC-OC	Ovarian cancer subsequent to breast cancer
CDH1	Cadherin 1
cfmiRNA	Circulating cell-free microRNAs
CTC	Circulating tumour cells
ctDNA	Circulating tumour DNA
LB	Liquid biopsy
MI	Methylation index
MSP	Methylation specific PCR
OC	Ovarian cancer
OS	Overall survival
OT	Ovarian tumours
PAX1	Paired Box 1
PFS	Progression-free survival
PTEN	Phosphatase and Tensin Homolog
RASSF1	Ras Association Domain Family Member 1
ROC	Receiver operating characteristic

References

1. Ferlay, J.; Soerjomataram, I.; Ervik, M.; Dikshit, R.; Eser, S.; Mathers, C.; Rebelo, M.; Parkin, D.M.; Forman, D.; Bray, F. GLOBOCAN 2012 v1.0, Cancer Incidence and Mortality Worldwide: IARC CancerBase No. 11. Lyon, France: International Agency for Research on Cancer. 2013. Available online: http://globocan.iarc.fr (accessed on 28 February 2014).
2. Brett M., R.; Jennifer B., P.; Thomas A., S.; Brett M., R.; Jennifer B., P.; Thomas A., S. Epidemiology of ovarian cancer: A reviewepidemiology of ovarian cancer: A review. *Cancer Biol. Med.* **2017**, *14*, 9–32. [CrossRef] [PubMed]
3. Bray, F.; Loos, A.H.; Tognazzo, S.; La Vecchia, C. Ovarian cancer in europe: Cross-sectional trends in incidence and mortality in 28 countries, 1953–2000. *Int. J. Cancer* **2005**, *113*, 977–990. [CrossRef] [PubMed]
4. Malvezzi, M.; Carioli, G.; Rodriguez, T.; Negri, E.; La Vecchia, C. Global trends and predictions in ovarian cancer mortality. *Ann. Oncol.* **2016**, *27*, 2017–2025. [CrossRef] [PubMed]
5. Teng, Z.; Han, R.; Huang, X.; Zhou, J.; Yang, J.; Luo, P.; Wu, M. Increase of incidence and mortality of ovarian cancer during 2003–2012 in Jiangsu province, China. *Front. Public Heal.* **2016**, *4*, 146. [CrossRef] [PubMed]
6. Sankaranarayanan, R.; Ferlay, J. Worldwide Burden of gynaecological cancer: The size of the problem. *Best Pract. Res. Clin. Obstet. Gynaecol.* **2006**, *20*, 207–225. [CrossRef] [PubMed]
7. Meinhold-Heerlein, I.; Fotopoulou, C.; Harter, P.; Kurzeder, C.; Mustea, A.; Wimberger, P.; Hauptmann, S.; Sehouli, J. The new WHO classification of ovarian, fallopian tube, and primary peritoneal cancer and its clinical implications. *Arch. Gynecol. Obstet.* **2016**, *293*, 695–700. [CrossRef]
8. Torre, L.A.; Trabert, B.; DeSantis, C.E.; Miller, K.D.; Samimi, G.; Runowicz, C.D.; Gaudet, M.M.; Jemal, A.; Siegel, R.L. Ovarian cancer statistics, 2018. CA. *Cancer J. Clin.* **2018**, *68*, 284–296. [CrossRef]
9. Wimberger, P.; Wehling, M.; Lehmann, N.; Kimmig, R.; Schmalfeldt, B.; Burges, A.; Harter, P.; Pfisterer, J.; du Bois, A. Influence of residual tumor on outcome in ovarian cancer patients with FIGO stage IV disease. *Ann. Surg. Oncol.* **2010**, *17*, 1642–1648. [CrossRef]

10. Giannopoulou, L.; Zavridou, M.; Kasimir-Bauer, S.; Lianidou, E.S. Liquid biopsy in ovarian cancer: The potential of circulating MiRNAs and exosomes. *Transl. Res.* **2019**, *205*, 77–91. [CrossRef]
11. Tan, D.S.; Agarwal, R.; Kaye, S.B. Mechanisms of transcoelomic metastasis in ovarian cancer. *Lancet Oncol.* **2006**, *7*, 925–934. [CrossRef]
12. Yeung, T.L.; Leung, C.S.; Yip, K.P.; Au Yeung, C.L.; Wong, S.T.C.; Mok, S.C. Cellular and molecular processes in ovarian cancer metastasis. A review in the theme: Cell and molecular processes in cancer metastasis. *Am. J. Physiol. Physiol.* **2015**, *309*, C444–C456. [CrossRef] [PubMed]
13. Bedard, P.L.; Hansen, A.R.; Ratain, M.J.; Siu, L.L. Tumour heterogeneity in the clinic. *Nature.* **2013**, *501*, 355–364. [CrossRef] [PubMed]
14. Donninger, H.; Vos, M.D.; Clark, G.J. The RASSF1 a tumor suppressor. *J. Cell Sci.* **2007**, *120*, 3163–3172. [CrossRef] [PubMed]
15. Grawenda, A.M.; O'Neill, E. Clinical utility of RASSF1 a methylation in human malignancies. *Br. J. Cancer* **2015**, *113*, 372–381. [CrossRef] [PubMed]
16. Van der Weyden, L.; Adams, D.J. The ras-association domain family (RASSF) members and their role in human tumourigenesis. *Biochim. Biophys. Acta Rev. Cancer* **2007**, *1776*, 58–85. [CrossRef] [PubMed]
17. Law, J.; Yu, V.C.; Baksh, S. Modulator of apoptosis 1: A highly regulated RASSF1 a-interacting BH3-like protein. *Mol. Biol. Int.* **2012**, *2012*, 536802. [CrossRef] [PubMed]
18. Donninger, H.; Clark, J.; Rinaldo, F.; Nelson, N.; Barnoud, T.; Schmidt, M.L.; Hobbing, K.R.; Vos, M.D.; Sils, B.; Clark, G.J. The RASSF1 a tumor suppressor regulates XPA-mediated DNA repair. *Mol. Cell. Biol.* **2015**, *35*, 277–287. [CrossRef] [PubMed]
19. Shivakumar, L.; Minna, J.; Sakamaki, T.; Pestell, R.; White, M.A. The RASSF1 a tumor suppressor blocks cell cycle progression and inhibits cyclin D1 accumulation. *Mol. Cell. Biol.* **2002**, *22*, 4309–4318. [CrossRef] [PubMed]
20. Song, S.J.; Song, M.S.; Kim, S.J.; Kim, S.Y.; Kwon, S.H.; Kim, J.G.; Calvisi, D.F.; Kang, D.; Lim, D.S. Aurora a regulates prometaphase progression by inhibiting the ability of RASSF1 A to suppress APC-Cdc20 activity. *Cancer Res.* **2009**, *69*, 2314–2323. [CrossRef]
21. Yin, Y.; Shen, W.H. PTEN: A new guardian of the genome. *Oncogene* **2008**, *27*, 5443–5453. [CrossRef]
22. Tsanou, E.; Peschos, D.; Batistatou, A.; Charalabopoulos, A.; Charalabopoulos, K. The E-cadherin adhesion molecule and colorectal cancer. A global literature approach. *Anticancer Res.* **2008**, *28*, 3815–3826. [PubMed]
23. Corso, G.; Intra, M.; Trentin, C.; Veronesi, P.; Galimberti, V. CDH1 Germline mutations and hereditary lobular breast cancer. *Fam. Cancer* **2016**, *15*, 215–219. [CrossRef] [PubMed]
24. Luo, W.; Fedda, F.; Lynch, P.; Tan, D. CDH1 gene and hereditary diffuse gastric cancer syndrome: Molecular and histological alterations and implications for diagnosis and treatment. *Front. Pharmacol.* **2018**, *9*, 1421. [CrossRef] [PubMed]
25. Wang, Q.; Wang, B.; Zhang, Y.M.; Wang, W. The association between CDH1 promoter methylation and patients with ovarian cancer: A systematic meta-analysis. *J. Ovarian Res.* **2016**, *9*, 23. [CrossRef] [PubMed]
26. Robson, E.J.D.; He, S.J.; Eccles, M.R. A PANorama of PAX genes in cancer and development. *Nat. Rev. Cancer* **2006**, *6*, 52–62. [CrossRef]
27. Li, C.G.; Eccles, M.R. PAX genes in cancer; friends or foes? *Front. Genet.* **2012**, *3*, 1–6. [CrossRef]
28. Su, P.H.; Lai, H.C.; Huang, R.L.; Chen, L.Y.; Wang, Y.C.; Wu, T.I.; Chan, M.W.Y.; Liao, C.C.; Chen, C.W.; Lin, W.Y.; et al. Paired box-1 (PAX1) activates multiple phosphatases and inhibits kinase cascades in cervical cancer. *Sci. Rep.* **2019**, *9*, 9195. [CrossRef]
29. Hentze, J.; Høgdall, C.; Høgdall, E. Methylation and ovarian cancer: Can DNA methylation be of diagnostic use? *Mol. Clin. Oncol.* **2019**, *10*, 323–330. [CrossRef]
30. Losi, L.; Fonda, S.; Saponaro, S.; Chelbi, S.T.; Lancellotti, C.; Gozzi, G.; Alberti, L.; Fabbiani, L.; Botticelli, L.; Benhattar, J. Distinct DNA methylation profiles in ovarian tumors: Opportunities for novel biomarkers. *Int. J. Mol. Sci.* **2018**, *19*, 1559. [CrossRef]
31. Lund, R.J.; Huhtinen, K.; Salmi, J.; Rantala, J.; Nguyen, E.V.; Moulder, R.; Goodlett, D.R.; Lahesmaa, R.; Carpén, O. DNA methylation and transcriptome changes associated with cisplatin resistance in ovarian cancer. *Sci. Rep.* **2017**, *7*, 1469. [CrossRef]
32. De Caceres, I.I.; Battagli, C.; Esteller, M.; Herman, J.G.; Dulaimi, E.; Edelson, M.I.; Bergman, C.; Ehya, H.; Eisenberg, B.L.; Cairns, P. Tumor cell-specific *BRCA1* and *RASSF1A* hypermethylation in serum, plasma, and peritoneal fluid from ovarian cancer patients. *Cancer Res.* **2004**, *64*, 6476–6481. [CrossRef] [PubMed]

33. Wu, Q.; Lothe, R.A.; Ahlquist, T.; Silins, I.; Tropé, C.G.; Micci, F.; Nesland, J.M.; Suo, Z.; Lind, G.E. DNA methylation profiling of ovarian carcinomas and their in vitro models identifies *HOXA9*, *HOXB5*, *SCGB3 A1*, and *CRABP1* as novel targets. *Mol. Cancer* **2007**, *6*, 45. [CrossRef] [PubMed]
34. Yang, H.J.; Liu, V.W.S.; Wang, Y.; Tsang, P.C.K.; Ngan, H.Y.S. Differential DNA methylation profiles in gynecological cancers and correlation with clinico-pathological data. *BMC Cancer* **2006**, *6*, 212. [CrossRef] [PubMed]
35. Zuberi, M.; Mir, R.; Dholariya, S.; Najar, I.; Yadav, P.; Javid, J.; Guru, S.; Mirza, M.; Gandhi, G.; Khurana, N.; et al. *RASSF1* and *PTEN* promoter hypermethylation influences the outcome in epithelial ovarian cancer. *Clin. Ovarian other Gynecol. Cancer* **2014**, *7*, 33–39. [CrossRef]
36. Qi, Q.; Ling, Y.; Zhu, M.; Zhou, L.; Wan, M.; Bao, Y.; Liu, Y. Promoter region methylation and loss of protein expression of PTEN and significance in cervical cancer. *Biomed. Rep.* **2014**, *2*, 653–658. [CrossRef] [PubMed]
37. Su, H.Y.; Lai, H.C.; Lin, Y.W.; Chou, Y.C.; Liu, C.Y.; Yu, M.H. An epigenetic marker panel for screening and prognostic prediction of ovarian cancer. *Int. J. Cancer* **2009**, *124*, 387–393. [CrossRef] [PubMed]
38. Hassan, Z.K.; Zekri, A.R.N.; Hafez, M.M.; Kamel, M.M. Human papillomavirus genotypes and methylation of *CADM1*, *PAX1*, *MAL* and *ADCYAP1* genes in epithelial ovarian cancer patients. *Asian Pac. J. Cancer Prev.* **2017**, *18*, 169–176.
39. Lin, H.W.; Fu, C.F.; Chang, M.C.; Lu, T.P.; Lin, H.P.; Chiang, Y.C.; Chen, C.A.; Cheng, W.F. *CDH1*, *DLEC1* and *SFRP5* methylation panel as a prognostic marker for advanced epithelial ovarian cancer. *Epigenomics* **2018**, *10*, 1397–1413. [CrossRef]
40. Koukoura, O.; Spandidos, D.A.; Daponte, A.; Sifakis, S. DNA methylation profiles in ovarian cancer: Implication in diagnosis and therapy. *Mol. Med. Rep.* **2014**, *10*, 3–9. [CrossRef]
41. Ma, L.; Zhang, J.; Liu, F.; Zhang, X. Expression of RASSF1 A and RASSF1C transcripts in human primary ovarian cancers. *Chinese J. Pathol.* **2005**, *34*, 150–153.
42. Chen, Y.; Zheng, H.; Yang, X.; Sun, L.; Xin, Y. Effects of mutation and expression of PTEN gene MRNA on tumorigenesis and progression of epithelial ovarian cancer. *Chinese Med. Sci. J.* **2004**, *19*, 25–30.
43. Chen, Y.; Zhao, Y.J.; Zheng, H.C.; Yang, X.F.; Wang, G.L.; Xin, Y. MRNA expression of PTEN and VEGF genes in epithelial ovarian cancer. *Chinese J. Cancer Res.* **2003**, *15*, 252–256. [CrossRef]
44. Wu, X.; Zhuang, Y.X.; Hong, C.Q.; Chen, J.Y.; You, Y.J.; Zhang, F.; Huang, P.; Wu, M.Y. Clinical importance and therapeutic implication of E-cadherin gene methylation in human ovarian cancer. *Med. Oncol.* **2014**, *31*, 100. [CrossRef] [PubMed]
45. R Core Team R: A Language and Environment for Statistical Computing. R Foundation for Statistical Computing, Vienna, Austria. Available online: https://www.R-project.org/ (accessed on 10 July 2019).
46. Chang, L.; Ni, J.; Zhu, Y.; Pang, B.; Graham, P.; Zhang, H.; Li, Y. Liquid biopsy in ovarian cancer: Recent advances in circulating extracellular vesicle detection for early diagnosis and monitoring progression. *Theranostics* **2019**, *9*, 4130–4140. [CrossRef] [PubMed]
47. Henderson, J.T.; Webber, E.M.; Sawaya, G.F. Screening for ovarian cancer: Updated evidence report and systematic review for the US preventive services task force. *JAMA* **2018**, *319*, 595–606. [CrossRef] [PubMed]
48. Dong, X.; Men, X.; Zhang, W.; Lei, P. Advances in Tumor Markers of Ovarian Cancer for Early Diagnosis. *Indian J. Cancer* **2014**, *51*, e72–e76. [PubMed]
49. Giannopoulou, L.; Kasimir-Bauer, S.; Lianidou, E.S. Liquid biopsy in ovarian cancer: Recent advances on circulating tumor cells and circulating tumor DNA. *Clin. Chem. Lab. Med.* **2018**, *56*, 186–197. [CrossRef] [PubMed]
50. Valenti, G.; Vitale, S.G.; Tropea, A.; Biondi, A.; Laganà, A.S. Tumor markers of uterine cervical cancer: A new scenario to guide surgical practice? *Updates Surg.* **2017**, *69*, 441–449. [CrossRef] [PubMed]
51. Muinelo-Romay, L.; Casas-Arozamena, C.; Abal, M. Liquid biopsy in endometrial cancer: New opportunities for personalized oncology. *Int. J. Mol. Sci.* **2018**, *19*, 2311. [CrossRef] [PubMed]
52. Cheung, T.H.; Yim, S.F.; Yu, M.Y.; Worley, M.J.; Fiascone, S.J.; Chiu, R.W.K.; Lo, K.W.K.; Siu, N.S.S.; Wong, M.C.S.; Yeung, A.C.M.; et al. Liquid biopsy of HPV DNA in cervical cancer. *J. Clin. Virol.* **2019**, *114*, 32–36. [CrossRef] [PubMed]
53. Widschwendter, M.; Zikan, M.; Wahl, B.; Lempiäinen, H.; Paprotka, T.; Evans, I.; Jones, A.; Ghazali, S.; Reisel, D.; Eichner, J.; et al. The potential of circulating tumor DNA methylation analysis for the early detection and management of ovarian cancer. *Genome Med.* **2017**, *9*, 116. [CrossRef] [PubMed]

54. Giannopoulou, L.; Mastoraki, S.; Buderath, P.; Strati, A.; Pavlakis, K.; Kasimir-Bauer, S.; Lianidou, E.S. *ESR1* methylation in primary tumors and paired circulating tumor DNA of patients with high-grade serous ovarian cancer. *Gynecol. Oncol.* **2018**, *150*, 355–360. [CrossRef] [PubMed]
55. Wang, C.; Horiuchi, A.; Imai, T.; Ohira, S.; Itoh, K.; Nikaido, T.; Katsuyama, Y.; Konishi, I. Expression of BRCA1 protein in benign, borderline, and malignant epithelial ovarian neoplasms and its relationship to methylation and allelic loss of the *BRCA1* gene. *J. Pathol.* **2004**, *202*, 215–223. [CrossRef] [PubMed]
56. Czekierdowski, A.; Czekierdowska, S.; Szymanski, M.; Wielgos, M.; Kaminski, P.; Kotarski, J. Opioid-binding protein/cell adhesion molecule-like (OPCML) gene and promoter methylation status in women with ovarian cancer. *Neuroendocrinol. Lett.* **2006**, *27*, 609–613. [PubMed]
57. Xiao, X.; Cai, F.; Niu, X.; Shi, H.; Zhong, Y. Association between P16 INK4 a promoter methylation and ovarian cancer: A meta-analysis of 12 published studies. *PLoS ONE* **2016**, *11*, e0163257. [CrossRef] [PubMed]
58. Agathanggelou, A.; Honorio, S.; Macartney, D.P.; Martinez, A.; Dallol, A.; Rader, J.; Fullwood, P.; Chauhan, A.; Walker, R.; Shaw, J.A.; et al. Methylation associated inactivation of *RASSF1A* from region 3p21.3 in lung, breast and ovarian tumours. *Oncogene* **2001**, *20*, 1509–1518. [CrossRef] [PubMed]
59. Liggett, T.E.; Melnikov, A.; Yi, Q.; Replogle, C.; Hu, W.; Rotmensch, J.; Kamat, A.; Sood, A.K.; Levenson, V. Distinctive DNA methylation patterns of cell-free plasma DNA in women with malignant ovarian tumors. *Gynecol. Oncol.* **2011**, *120*, 113–120. [CrossRef]
60. Zhang, Q.; Hu, G.; Yang, Q.; Dong, R.; Xie, X.; Ma, D.; Shen, K.; Kong, B. A multiplex methylation-specific PCR assay for the detection of early-stage ovarian cancer using cell-free serum DNA. *Gynecol. Oncol.* **2013**, *130*, 132–139. [CrossRef]
61. Rattanapan, Y.; Korkiatsakul, V.; Kongruang, A.; Chareonsirisuthigul, T.; Rerkamnuaychoke, B.; Wongkularb, A.; Wilailak, S. *EGFL7* and *RASSF1* promoter hypermethylation in epithelial ovarian cancer. *Cancer Genet.* **2018**, *224*, 37–40. [CrossRef]
62. Schondorf, T.; Ebert, M.P.; Hoffmann, J.; Becker, M.; Moser, N.; Pur, S.; Göhring, U.J.; Weisshaar, M.P. Hypermethylation of the *PTEN* gene in ovarian cancer cell lines. *Cancer Lett.* **2004**, *207*, 215–220. [CrossRef]
63. Pfeifer, G.P. Defining driver DNA methylation changes in human cancer. *Int. J. Mol Sci.* **2018**, *19*, 1166. [CrossRef] [PubMed]
64. Fang, C.; Wang, S.Y.; Liou, Y.L.; Chen, M.H.; Ouyang, W.; Duan, K.M. The promising role of PAX1 (Aliases: HUP48, OFC2) gene methylation in cancer screening. *Mol. Genet. Genomic Med.* **2019**, *7*, e506. [CrossRef] [PubMed]
65. Kan, Y.Y.; Liou, Y.L.; Wang, H.J.; Chen, C.Y.; Sung, L.C.; Chang, C.F.; Liao, C.I. ATL. *Int. J. Gynecol. Cancer* **2014**, *24*, 928–934. [CrossRef] [PubMed]
66. Xu, J.; Xu, L.; Yang, B.; Wang, L.; Lin, X.; Tu, H. Assessing methylation status of PAX1 in cervical scrapings, as a novel diagnostic and predictive biomarker, was closely related to screen cervical cancer. *Int. J. Clin. Exp. Pathol.* **2015**, *8*, 1674–1681. [PubMed]
67. Warton, K.; Mahon, K.L.; Samimi, G. Methylated circulating tumor DNA in blood: Power in cancer prognosis and response. *Endocr.-Relat. Cancer* **2016**, *23*, R157–R171. [CrossRef] [PubMed]
68. Xu, R.; Wei, W.; Krawczyk, M.; Wang, W.; Luo, H.; Flagg, K.; Yi, S.; Shi, W.; Quan, Q.; Li, K.; et al. Circulating tumour DNA methylation markers for diagnosis and prognosis of hepatocellular carcinoma. *Nat. Mater.* **2017**, *16*, 1155–1161. [CrossRef] [PubMed]
69. Gorgannezhad, L.; Umer, M.; Islam, M.N.; Nguyen, N.-T.; Shiddiky, M.J.A. Circulating Tumor DNA and Liquid Biopsy: Opportunities, Challenges, and Recent Advances in Detection Technologies. *Lab Chip* **2018**, *18*, 1174–1196. [CrossRef] [PubMed]
70. Wan, J.C.M.; Massie, C.; Garcia-Corbacho, J.; Mouliere, F.; Brenton, J.D.; Caldas, C.; Pacey, S.; Baird, R.; Rosenfeld, N. Liquid biopsies come of age: Towards implementation of circulating tumour DNA. *Nat. Rev. Cancer* **2017**, *17*, 223–238. [CrossRef] [PubMed]
71. Ng, S.B.; Chua, C.; Ng, M.; Gan, A.; Poon, P.S.; Teo, M.; Fu, C.; Leow, W.Q.; Lim, K.H.; Chung, A.; et al. Individualised multiplexed circulating tumour DNA assays for monitoring of tumour presence in patients after colorectal cancer surgery. *Sci. Rep.* **2017**, *7*, 40737. [CrossRef] [PubMed]
72. Yang, N.; Li, Y.; Liu, Z.; Qin, H.; Du, D.; Cao, X.; Cao, X.; Li, J.; Li, D.; Jiang, B.; et al. The characteristics of CtDNA reveal the high complexity in matching the corresponding tumor tissues. *BMC Cancer* **2018**, *18*, 319. [CrossRef]

73. Guo, Q.; Wang, J.; Xiao, J.; Wang, L.; Hu, X.; Yu, W.; Song, G.; Lou, J.; Chen, J. Heterogeneous mutation pattern in tumor tissue and circulating tumor DNA warrants parallel NGS panel testing. *Mol. Cancer* **2018**, *17*, 131. [CrossRef] [PubMed]
74. Kammesheidt, A.; Tonozzi, T.R.; Lim, S.W.; Braunstein, G.D. Mutation Detection using plasma circulating tumor DNA (CtDNA) in a cohort of asymptomatic adults at increased risk for cancer. *Int. J. Mol. Epidemiol. Genet.* **2018**, *9*, 1–12. [PubMed]
75. Toor, O.M.; Ahmed, Z.; Bahaj, W.; Boda, U.; Cummings, L.S.; McNally, M.E.; Kennedy, K.F.; Pluard, T.J.; Hussain, A.; Subramanian, J.; et al. Correlation of somatic genomic alterations between tissue genomics and CtDNA employing next-generation sequencing: Analysis of lung and gastrointestinal cancers. *Mol. Cancer Ther.* **2018**, *17*, 1123–1132. [CrossRef] [PubMed]
76. Qin, Z.; Ljubimov, V.A.; Zhou, C.; Tong, Y.; Liang, J. Cell-free circulating tumor DNA in cancer. *Chin. J. Cancer* **2016**, *35*, 36. [CrossRef] [PubMed]
77. Elazezy, M.; Joosse, S.A. Techniques of using circulating tumor DNA as a liquid biopsy component in cancer management. *Comput. Struct. Biotechnol. J.* **2018**, *16*, 370–378. [CrossRef]
78. Shao, X.; He, Y.; Ji, M.; Chen, X.; Qi, J.; Shi, W.; Hao, T.; Ju, S. Quantitative analysis of cell-free DNA in ovarian cancer. *Oncol. Lett.* **2015**, *10*, 3478–3482. [CrossRef] [PubMed]
79. Teschendorff, A.E.; Menon, U.; Gentry-Maharaj, A.; Ramus, S.J.; Gayther, S.A.; Apostolidou, S.; Jones, A.; Lechner, M.; Beck, S.; Jacobs, I.J.; et al. An epigenetic signature in peripheral blood predicts active ovarian cancer. *PLoS ONE* **2009**, *4*, e8274. [CrossRef]
80. Flanagan, J.M.; Wilhelm-Benartzi, C.S.; Metcalf, M.; Kaye, S.B.; Brown, R. Association of somatic DNA methylation variability with progression-free survival and toxicity in ovarian cancer patients. *Ann. Oncol. Off. J. Eur. Soc. Med. Oncol.* **2013**, *24*, 2813–2818. [CrossRef]
81. Flanagan, J.M.; Wilson, A.; Koo, C.; Masrour, N.; Gallon, J.; Loomis, E.; Flower, K.; Wilhelm-Benartzi, C.; Hergovich, A.; Cunnea, P.; et al. Platinum-based chemotherapy induces methylation changes in blood DNA associated with overall survival in patients with ovarian cancer. *Clin. Cancer Res.* **2017**, *23*, 2213–2222. [CrossRef]
82. Lu, Y.; Li, S.; Zhu, S.; Gong, Y.; Shi, J.; Xu, L.L. Methylated DNA/RNA in Body Fluids as Biomarkers for Lung Cancer. *Biol. Proced. Online.* **2017**, *19*, 2. [CrossRef]
83. Lissa, D.; Robles, A.I. Methylation analyses in liquid biopsy. *Transl. Lung Cancer Res.* **2016**, *5*, 492–504. [CrossRef] [PubMed]
84. Balgkouranidou, I.; Chimonidou, M.; Milaki, G.; Tsarouxa, E.G.; Kakolyris, S.; Welch, D.R.; Georgoulias, V.; Lianidou, E.S. Breast cancer metastasis suppressor-1 promoter methylation in cell-free DNA provides prognostic information in non-small cell lung cancer. *Br. J. Cancer* **2014**, *110*, 2054–2062. [CrossRef] [PubMed]
85. Kristensen, L.S.; Hansen, J.W.; Kristensen, S.S.; Tholstrup, D.; Harsløf, L.B.S.; Pedersen, O.B.; De Nully Brown, P.; Grønbæk, K. Aberrant methylation of cell-free circulating DNA in plasma predicts poor outcome in diffuse large B cell lymphoma. *Clin. Epigenetics* **2016**, *8*, 95. [CrossRef] [PubMed]
86. Majchrzak-Celińska, A.; Paluszczak, J.; Kleszcz, R.; Magiera, M.; Barciszewska, A.M.; Nowak, S.; Baer-Dubowska, W. Detection of *MGMT*, *RASSF1A*, *P15INK4B*, and *P14ARF* promoter methylation in circulating tumor-derived DNA of central nervous system cancer patients. *J. Appl. Genet.* **2013**, *54*, 335–344. [CrossRef] [PubMed]
87. Zhou, Q.; Li, W.; Leng, B.; Zheng, W.; He, Z.; Zuo, M.; Chen, A. Circulating cell free DNA as the diagnostic marker for ovarian cancer: A systematic review and meta-analysis. *PLoS ONE* **2016**, *11*, e0155495. [CrossRef] [PubMed]
88. Kamat, A.A.; Baldwin, M.; Urbauer, D.; Dang, D.; Han, L.Y.; Godwin, A.; Karlan, B.Y.; Simpson, J.L.; Gershenson, D.M.; Coleman, R.L.; et al. Plasma cell-free DNA in ovarian cancer: An independent prognostic biomarker. *Cancer* **2010**, *116*, 1918–1925. [CrossRef] [PubMed]
89. Teo, Y.V.; Capri, M.; Morsiani, C.; Pizza, G.; Faria, A.M.C.; Franceschi, C.; Neretti, N. Cell-free DNA as a biomarker of aging. *Aging Cell* **2019**, *18*, e12890. [CrossRef] [PubMed]
90. Moss, J.; Magenheim, J.; Neiman, D.; Zemmour, H.; Loyfer, N.; Korach, A.; Samet, Y.; Maoz, M.; Druid, H.; Arner, P.; et al. Comprehensive human cell-type methylation atlas reveals origins of circulating cell-free DNA in health and disease. *Nat. Commun.* **2018**, *9*, 5068. [CrossRef] [PubMed]

91. Lapin, M.; Oltedal, S.; Tjensvoll, K.; Buhl, T.; Smaaland, R.; Garresori, H.; Javle, M.; Glenjen, N.I.; Abelseth, B.K.; Gilje, B.; et al. Fragment size and level of cell-free DNA provide prognostic information in patients with advanced pancreatic cancer. *J. Transl. Med.* **2018**, *16*. [CrossRef] [PubMed]
92. Han, C.; Bellone, S.; Siegel, E.R.; Altwerger, G.; Menderes, G.; Bonazzoli, E.; Egawa-Takata, T.; Pettinella, F.; Bianchi, A.; Riccio, F.; et al. A novel multiple biomarker panel for the early detection of high-grade serous ovarian carcinoma. *Gynecol. Oncol.* **2018**, *149*, 585–591. [CrossRef] [PubMed]
93. Pisanic, T.R.; Cope, L.M.; Lin, S.F.; Yen, T.T.; Athamanolap, P.; Asaka, R.; Nakayama, K.; Fader, A.N.; Wang, T.H.; Shih, I.M.; et al. Methylomic analysis of ovarian cancers identifies tumor-specific alterations readily detectable in early precursor lesions. *Clin. Cancer Res.* **2018**, *24*, 6536–6547. [CrossRef] [PubMed]
94. Guo, W.; Zhu, L.; Yu, M.; Zhu, R.; Chen, Q.; Wang, Q. A Five-DNA methylation signature act as a novel prognostic biomarker in patients with ovarian serous cystadenocarcinoma. *Clin. Epigenetics* **2018**, *10*, 142. [CrossRef] [PubMed]
95. Vitale, S.G.; Capriglione, S.; Zito, G.; Lopez, S.; Gulino, F.A.; Di Guardo, F.; Vitagliano, A.; Noventa, M.; La Rosa, V.L.; Sapia, F.; et al. Management of endometrial, ovarian and cervical cancer in the elderly: current approach to a challenging condition. *Arch. Gynecol. Obstet.* **2019**, *299*, 299–315. [CrossRef] [PubMed]
96. Schuurman, M.S.; Kruitwagen, R.F.P.M.; Portielje, J.E.A.; Roes, E.M.; Lemmens, V.E.P.P.; Van der Aa, M.A. Treatment and outcome of elderly patients with advanced stage ovarian cancer: A nationwide analysis. *Gynecol. Oncol.* **2018**, *149*, 270–274. [CrossRef] [PubMed]

8

Combination of Fetal Fraction Estimators Based on Fragment Lengths and Fragment Counts in Non-Invasive Prenatal Testing

Juraj Gazdarica [1,2,3,*], Rastislav Hekel [1,2,3], Jaroslav Budis [1,3,4], Marcel Kucharik [1], Frantisek Duris [5], Jan Radvanszky [1,6], Jan Turna [2,4,6] and Tomas Szemes [1,2,4]

1 Geneton Ltd., Bratislava 84104, Slovakia
2 Department of Molecular Biology, Faculty of Natural Sciences, Comenius University, Bratislava 84104, Slovakia
3 Slovak Centre of Scientific and Technical Information, Bratislava 81104, Slovakia
4 Comenius University Science Park, Bratislava 84104, Slovakia
5 Department of Computer Science, Faculty of Mathematics, Physics and Informatics, Comenius University, Bratislava 84248, Slovakia
6 Institute of Clinical and Translational Research, Biomedical Research Center, Slovak Academy of Sciences, Bratislava 84505, Slovakia
* Correspondence: juraj.gazdarica@geneton.sk

Abstract: The reliability of non-invasive prenatal testing is highly dependent on accurate estimation of fetal fraction. Several methods have been proposed up to date, utilizing different attributes of analyzed genomic material, for example length and genomic location of sequenced DNA fragments. These two sources of information are relatively unrelated, but so far, there have been no published attempts to combine them to get an improved predictor. We collected 2454 single euploid male fetus samples from women undergoing NIPT testing. Fetal fractions were calculated using several proposed predictors and the state-of-the-art SeqFF method. Predictions were compared with the reference Y-based method. We demonstrate that prediction based on length of sequenced DNA fragments may achieve nearly the same precision as the state-of-the-art methods based on their genomic locations. We also show that combination of several sample attributes leads to a predictor that has superior prediction accuracy over any single approach. Finally, appropriate weighting of samples in the training process may achieve higher accuracy for samples with low fetal fraction and so allow more reliability for subsequent testing for genomic aberrations. We propose several improvements in fetal fraction estimation with a special focus on the samples most prone to wrong conclusion.

Keywords: NIPT; fetal fraction; statistical methods; DNA; maternal serum screening; fetal cells

---

1. Introduction

Prenatal testing for genomic defects of a fetus before birth is an integral component of current obstetric practice [1]. The primary aim of the testing is screening for the presence of abnormal number of chromosomes. Testing is focused mainly on identifying an additional copy of a chromosome, the condition called trisomy [2], represented as an excessive proportion of genomic material from the aberrant chromosome. Discovery of fetal DNA fragments in plasma extracted from maternal blood, cell-free fetal DNA (cffDNA) [3], opened up new options in the field of prenatal screening called non-invasive prenatal testing (NIPT) [1,4]. In contrast to established screening methods [5], sampling of genetic material from the mother's circulation does not pose any direct risk for the fetus [6]. On the other hand, fetal DNA fragments constitute only a minor part of the sampled, mostly maternal, genomic material and so pose a challenge for reliable detection of present aberration.

The proportion of fetal fragments in analyzed DNA mixture is called fetal fraction (FF). A sample with a trisomic fetus typically has an aberrant proportion of genomic material from the trisomic chromosome, compared to healthy samples. A trisomic sample with a very low FF can, however, be incorrectly evaluated as healthy, since the aberrant chromosome may cause only a weak deviation from normal values that can be presumed to represent a normal measurement error [7], which is particularly problematic for the detection of sub-chromosomal aberrations. Reliable estimation of the FF is therefore a crucial step of NIPT analysis to avoid false-negative results [8,9]. This can be based on relevant anthropometric and laboratory processing attributes with significant correlation with FF, namely the gestational age and the body mass index of the mother [10]. They were, however, reported to be too weak for use as stand-alone information for reliable conclusions, and so more sophisticated technical parameters exploiting different characteristics between maternal and fetal DNA fragments are typically used [11].

Count based methods of FF determination calculate disproportion of number of reads mapped to chromosomes that differ in mother and fetus genotypes. Although they are quite reliable, they can be used only on samples with male (XX vs. XY genotype) [12] or trisomic (2 copies vs. 3 copies of an aberrant chromosome) fetuses [13,14]. In pregnancy with a healthy female fetus, the FF have to be estimated by alternative methods.

Methods based on sequence variation in DNA fragments are, on the other hand, highly reliable [15,16], although their application typically requires another laboratory assay to recover genetic map of parents, and so it is considered too expensive and time-consuming for routine diagnostics. Alternatively, FF may be estimated from the genomic location patterns that slightly differ between maternal and fetal fragments due to differences of their nucleosome positioning (SANEFALCON method) [17] and euchromatic DNA structure (SeqFF method) [18]. Due to high accuracy and no additional laboratory costs, the SeqFF method is a preferred method for samples with female fetus [19].

The length of DNA fragment is another promising attribute for differentiation of DNA fragments, since fetal fragments tend to be shorter than maternal ones (Figure 1). Although the fragment lengths improve prediction of NIPT tests by detection of false positive predictions [20–22], their potential for prediction of FF has been understudied. Only a simple length-based metric has been proposed in [23], where the ratio of the number of shorter (100–150 bases) and longer (163–168 bases) fragments has correlated with the FF.

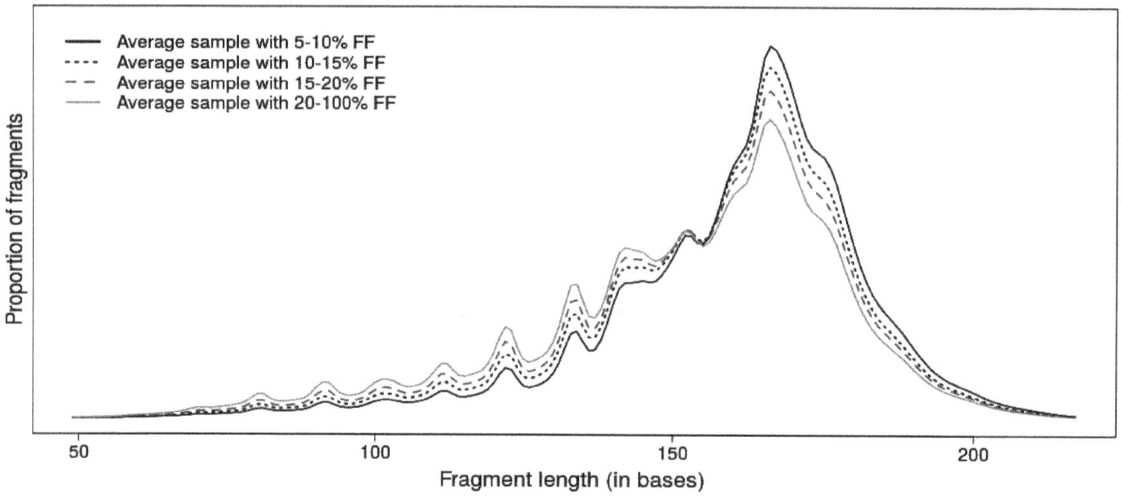

Figure 1. Average fragment-length profiles determined from samples with selected ranges of fetal fractions calculated across lengths. Samples with higher fetal fraction (FF) also have shorter fragments indicating that maternal fragments are longer than fetal ones.

We demonstrate in this study that fragment length profile may be utilized as a fairly accurate predictor of the FF. We investigate various methods and enhancements that lead to improvement of prediction accuracy and propose weighting training samples to improve accuracy in samples with low FF that are more prone to wrong testing conclusions. Although the proposed method is not as accurate as SeqFF, we show that their combination leads to more a reliable predictor than individual methods alone.

2. Results

2.1. Comparison of Length-Based Methods

We introduced a new method for estimating FF (NLRM) and also compared other statistical (LRM) and machine learning methods (NN, SVM) (see FF ESTIMATORS BASED ON FRAGMENT LENGTH in Materials and methods). The Y-based FF was used as a "ground truth" for each comparison. We used Pearson correlation and mean squared errors (MSE) of the predicted FF and the Y-based FF to compare the surveyed methods.

Correlations were higher when compared to ones obtained from a published method using the ratio of shorter and longer fragments (median correlation 0.568 on our data) [23]. After picking the best ratio of consecutive intervals of fragment lengths (FRAC ratio), we were able to significantly improve this correlation to 0.728. To get even better results, we applied nonlinear regression model (NLRM) to estimate parameters (or weights) for individual fragment lengths resulting from the FRAC ratio. Here we observed median correlation 0.812 for test set.

Although NN initially seemed to be a promising method for estimating FF using fragment length distribution, we observed only sub-par results (median correlation 0.778). LRM and SVM models achieved best correlation with median value 0.83 and 0.831, respectively. All results can be seen on Figure 2.

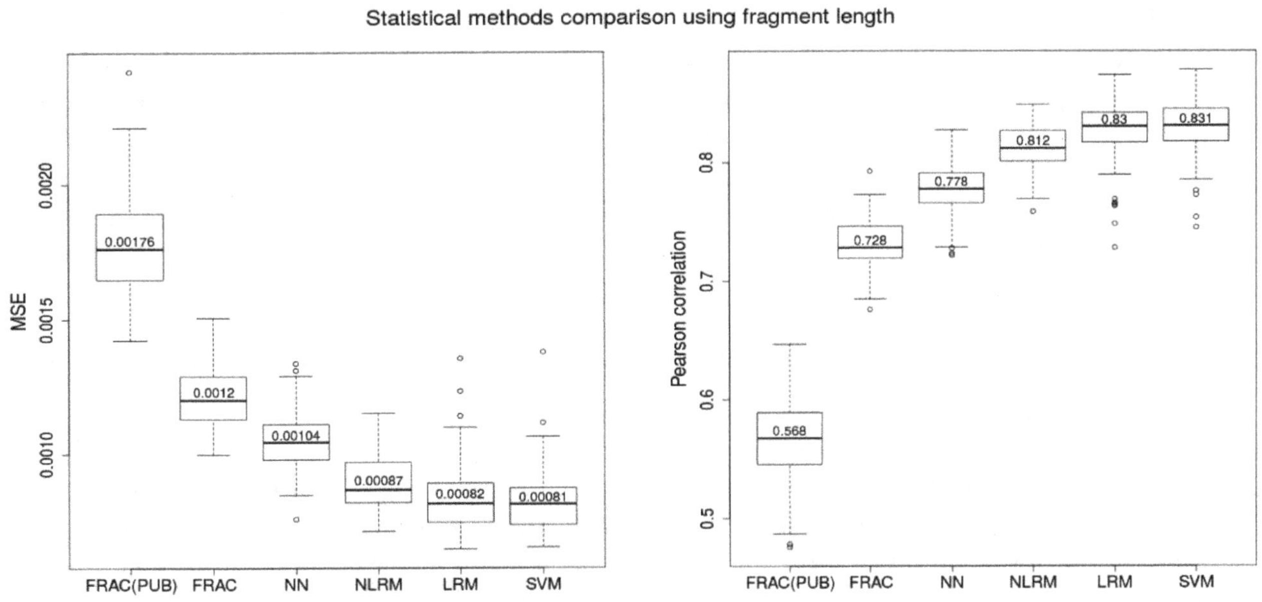

Figure 2. Boxplots of MSE and Pearson correlation of Y-based FF and estimated FF obtained from several different methods calculated for 100 testing sets. Training was performed on 100 complementary training sets. FRAC(PUB) represents the ratio of fragment length intervals designed by the study (Yu et al., 2014), FRAC means our best ratio of fragment length intervals derived by the best Pearson correlation with Y-based FF, NN—neural network, NLRM—non-linear regression model based on FRAC parameters, LRM—linear model using all fragment length parameters from 50–220 bp, SVM—support vector machine using all fragment length parameters from 50–220 bp.

2.2. Combination of Methods Improves Prediction Accuracy

We have chosen the SVM method (median correlation 0.831) as the representative method based on fragment-length profiles, according to the best prediction accuracy in the previous comparison. SeqFF performed slightly better on our dataset than the SVM method (0.831 vs. 0.877, see Figure 3). Since these two predictors use different attributes of fragments for prediction (lengths and positions), we combined these two approaches using a linear regression model.

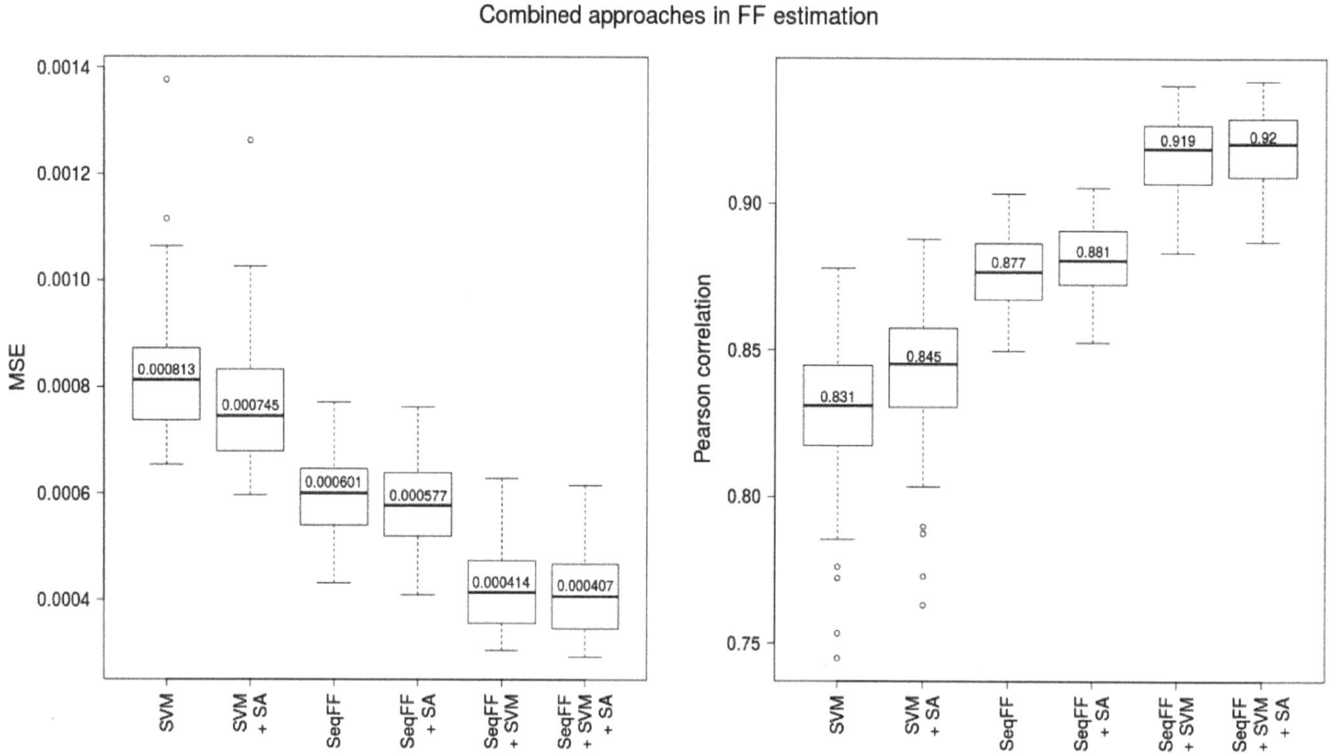

Figure 3. Boxplots of MSE and Pearson correlation of Y-based FF and estimated FF using SVM method calculated for 100 testing sets. Training was performed on 100 complementary training sets. Combined approach is denoted SeqFF + SVM. Improvement for every method was achieved by adding sample attributes (DNA library concentration, BMI, gestational age)—represented by "+ SA".

Furthermore, we examined the effect of additional predictive attributes from relevant anthropometric and laboratory processing attributes, namely the gestational age, the body mass index of the mother and the DNA library concentration to improve the models. In all examined conditions, the accuracy of the predictions was improved, although the improvement was only mild in some cases (Figure 3, with and without SA). Finally, using a linear regression model, we combined two approaches, SVM and SeqFF, whereby the best trained model almost reached 0.95 correlation when testing. Comparison of tested data from a randomly selected combined model with a Y-based method is presented in the supplement (Figure 4) showing no systematic difference between the two methods. Moreover, we compared FF prediction between male and female samples using the combined method. Similar distribution of male and female samples was observed (Supplementary Figures S4 and S5).

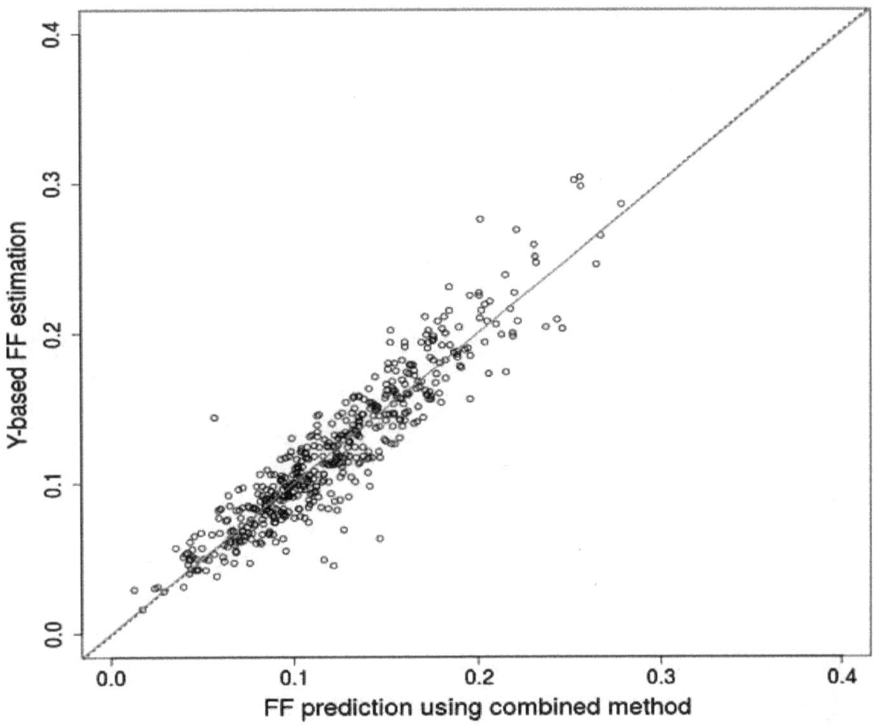

Figure 4. Linear regression of Y-based method with the combined method. Black circles denote individual testing samples. Dashed line represents overall trend of the prediction and the grey line is the 45° line.

Finally, we provide trained attribute (i.e., feature) weights of four examined linear models with different combinations of attributes in a separate table (Table 1).

Table 1. The table shows corresponding feature weights of four trained linear models. Each row represents a single feature of a linear model and each column represents a specific model. Each model has a different combination of features. If a feature is not part of the model, the value is empty (labeled by dash). Since one hundred models were trained for each combination, only the model with median correlation is displayed. SVM: support vector machine estimator prediction based on fragment length profile. SeqFF: prediction of the SeqFF model. BMI: body mass index of the mother. LC: DNA library concentration. GA: gestational age. SA: sample attributes (BMI + LC + GA).

Method	SeqFF + SA	SVM + SA	SVM + SeqFF	Seqff + SVM + SA
SVM	–	0.0418	0.0269	0.0243
SeqFF	1.1325	–	0.0237	0.0255
BMI	−0.0006	−0.0058	–	−0.0019
LC	−0.0020	−0.0005	–	−0.0015
GA	~0.0000	0.0037	–	0.0016
Intercept	0.0204	0.1222	0.1223	0.1227

2.3. Weighting of Samples with Low Fraction

From the perspective of NIPT, it is crucial to decrease the probability of false-negative results to a minimum. Thus, only results with predicted FF above a certain threshold are acceptable. In this sense, we examined adjustment of the model that would improve prediction accuracy on critical, low fraction samples (FF <10%) at the expense of prediction of high fraction samples (FF >10%). We achieved the improvement with inclusion of multiple copies of low fraction samples in training, increasing the significance of such samples on resulting parameters.

According to the expectation, we observed decline of prediction accuracy of samples with high fraction. On the other hand, the prediction accuracy of low fraction samples was gradually improved with increased significance of considered samples (Figure 5). This suggests that weighting samples may further improve the decision process with more precise prediction of critical samples.

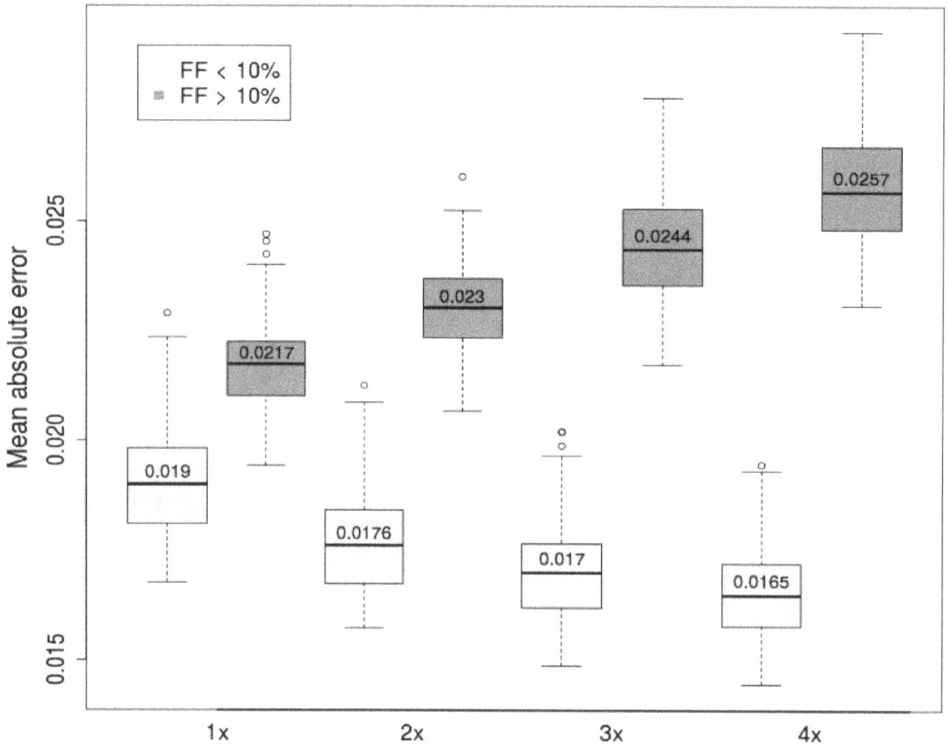

Figure 5. Mean absolute error Y-based FF and estimated FF using SVM method (lower is better). Only samples with FF < 10% used in training were sampled (weighted) with weights 2×, 3×, and 4×. Samples with FF > 10% were selected once. Scatterplots represented by linear regression with weighted samples are presented in the supplement (Supplementary Figures S1–S3).

3. Discussion

Accurate estimation of fetal fraction in analyzed mixtures of fetal and maternal DNA fragments is a critical part of non-invasive prenatal testing. The main challenge is to identify samples with too low a number of fetal fragments for reliable prediction of aneuploidy, since over-estimation of prediction may lead to false-negative results. Conversely, accuracy is not so important in typical settings for samples with a sufficiently high fetal fraction. Prediction models should be therefore trained and evaluated with a special focus on the importance of individual samples.

Several methods have been proposed for FF prediction, each with its own benefits and limitations. Early established methods were based on disproportion of number of fragments from sex chromosomes. They are simple to understand and implement and require a relatively small number of samples for training. Their calculation is therefore typically the first step in estimation, whether as the gold standard for other trained methods or initial determination of fetus sex. Testing may also utilize two separated prediction methods for male and female fetuses, possibly with different thresholds based on their precision. With the rising number of tested samples, the more complex models can be trained without the risk of overfitting. Since the number of samples should be much higher than the number of trained model parameters, laboratories should reach at least hundreds of samples for fragment length based models and tens of thousands for the position based models. This is a great advantage for simpler models, since the acquisition of such a number of samples may be limiting for small laboratories. This represents a problem for already established tests as well, since beneficial changes in laboratory processing may be halted due to a high cost of re-training.

Available methods are also highly diverse in the attributes they use for prediction, from the simplest models that utilize only patient attributes, to highly complex attributes aggregated from sequenced fragments. Combination of several independent, even weak predictors, can perform better than any of them alone, as was shown in many other domains [24]. Fetal fraction prediction benefited from such coupling as well, as presented in this study. Furthermore, the more accurate predictions including more methods may indicate or even correct the potential abnormal diversions of any single method. Present or future methods may be incrementally supplemented and so gradually improve the overall predictions. The effect of individual methods should be however weighted to prefer methods with higher accuracy. Assigned weights may also indicate that a gain from a less accurate predictor becomes negligible and so may be excluded without significant impact.

We propose several approaches that markedly improve accuracy of prediction of fetal fraction, making results of non-invasive prenatal testing more reliable. Firstly, the fragment length profiles may be utilized as a reliable predictor of FF and achieve similar precision to the favored methods based on positions of DNA fragments. Secondly, the combination of results from multiple predictors achieve far better predictions, at least when they are based on different attributes of input DNA fragments. The final combined method is free of systematic bias when compared to the Y-based method. We conclude that low quality predictors, like regressors based on relevant anthropometric and laboratory processing attributes, may contribute to the overall accuracy. Other methods utilizing their own set of distinctive attributes, both present and future, can be similarly included, and so improve prediction compared to their stand-alone usage. Based on these findings, we conclude that additional, possibly independent, information can significantly raise the prediction accuracy of FF prediction, and thus, should be used when possible.

Since the SeqFF publication does not provide a training option, we were forced to use the published parameters on our dataset. As a consequence, we were unable to reach correlation reported by the SeqFF study; however, our combined method reached similar scores. We presume that the combined method can significantly surpass the performance of the standalone SeqFF method if its parameters are properly trained on a similar dataset.

Finally, appropriate weighting of samples in the training process may achieve higher accuracy for samples with low FF, and so allow a more decision regarding which samples have enough fetal fragments for subsequent testing for genomic aberrations.

4. Materials and Methods

4.1. Sample Acquisition

We collected 2454 informative samples from women undergoing NIPT testing, with single euploid male fetus, for training and testing of FF models. These samples were concluded as informative based on their fetal fraction and manual examination. The hard coded FF threshold for an informative sample was 4% but in some cases a sample with lower FF was reclassified as informative by manual examination. Our work was part of two clinical studies approved by the Ethical Committee of the Bratislava Self-Governing Region (Sabinovska ul.16, 820 05 Bratislava): the first one called "NIPT study" (study ID 35900_2015 approved on 30 April of 2015 under the decision ID 03899_2015) and the second one called "SNiPT" (study ID 37136/2018 approved on 11 June of 2018 under the decision ID 07507/2018/HF). All patients that were included in the study signed written informed consents consistent with the Helsinki declaration which were approved by the above-mentioned ethics committee.

4.2. Sample Preparation and Sequencing

Peripheral blood from pregnant women was collected into EDTA tubes and kept at 4 °C until plasma separation. Blood plasma was separated within 36 h after collection and stored at −20 °C at a DNA isolation unit. DNA was isolated using a QIAgen DNA Blood Mini Kit (Hilden, Germany). Standard fragment libraries were prepared from isolated DNA using a modified protocol of the Illumina TruSeq Nano Kit (San Diego, CA, USA) as described previously [20]. Briefly, to decrease the laboratory costs we used reduced volumes of reagents, which was compensated for by completing 9 cycles of PCR instead of 8, as per protocol. Physical size selection of cfDNA fragments was performed using specific volumes of magnetic beads in order to enrich FF. Illumina NextSeq 500/550 High Output Kit v2 (San Diego, CA, USA) (75 cycles) was used for massively parallel sequencing of prepared libraries using pair-end sequencing with read length of 35 bp.

4.3. Mapping and GC Correction

The first stage of data processing was carried out as previously described [20]. NextSeq-produced fastq files (two per sample) were directly mapped using the Bowtie 2 [25] algorithm with –very-sensitive option to the human reference genome hg19 (GRCh37). Reads with mapping quality of 40 or higher were retained for further data processing. Length of a DNA fragment was determined as the difference of the leftmost and the rightmost mapped base of the corresponding read pair.

Next, we weighted mapped reads to eliminate the GC bias according to [26] (with the exclusion of intra-run normalization) to retrieve better estimates of underlying chromosomal distributions.

The corrected number of fragments per chromosome is determined by summing the corrected read counts over all bins of the specified chromosome. The exception is chromosome Y, which is presented only pregnancies with male fetuses. Even in such cases, low proportion of mappable regions does not contain enough reads for reliable correction. As a result, a vector of corrected number of fragments per autosomes, called autosomal counts, corrected number of fragments per chromosome X (chrX), and uncorrected number of fragments per chromosome Y (chrY) were passed to the calculation of the reference, Y-based FF (see below).

4.4. Data Preparation and Evaluation of Models

Fetal fractions were calculated using several proposed predictors and state-of-the-art method SeqFF. Predictions were compared with the reference Y-based method. The robustness of this comparison stems from using a multitude of training and testing datasets, thereby suppressing bias from a single random sampling. Each model presented in this paper was evaluated on 100 testing sets by Pearson correlation coefficient between Y-based values and predicted values of FF. For each such experiment, the whole dataset (2454 male samples) was divided into 80% for training and 20% for testing.

The proportion of fragments for each target length (50–220 bases) were calculated for each sample separately. Moreover, in the case of the support vector machine and neural network estimators, the resulting 171 feature vectors of training sets were normalized to have zero mean and unit variance.

Y-Based Estimator

The Y-based FF was calculated according to the equation

$$FY = \frac{\%chrY - female\%chrY}{male\%chrY - female\%chrY} \tag{1}$$

from [27], where male %chr Y and female %chr Y is the mean fraction of chromosome Y sequence reads of plasma samples obtained from 14 adult male individuals and pregnant women bearing euploid female fetuses, respectively. %chr Y is the fraction of chromosome Y sequence reads of the sample whose FF we want to calculate.

4.5. FF Estimators Based on Fragment Length

4.5.1. Linear Regression Model (LRM)

Fetal fractions were estimated using a standard linear regression model provided by R-software given by the equation

$$F = L\theta + \varepsilon, \tag{2}$$

where F is vector of fetal fractions, L is matrix of fragment lengths from 50 to 220 bp, θ is vector of parameters to be estimated, and ε is vector of errors. The training and testing sets were selected from the collected data. The parameters θ resulting from training were used to estimate FF for testing data.

4.5.2. Non-Linear Regression Model (NLRM)

We introduce a new approach for FF prediction—nonlinear function used in NLRM. Motivation behind this method originated from a published method using the ratio of shorter (100–150 bases) and longer (163–168 bases) fragments [23]. However, when we used a ratio of the same fragment lengths as reported in the publication, we obtained sub-par results for our data. Therefore, we think that the choice of fragment lengths is data specific and, in the first step, we were exhaustively searching for the best fragment lengths to use for our data.

Let I = $L_{50}, L_{51}, \ldots, L_{220}$ be set of proportions of fragments of length 50 bp to 220 bp. Let I_1, I_2 be subsets of consecutive elements of I. Ratio of fragments is defined as

$$FRAC = \frac{\sum_{l \in I_1} l}{\sum_{l \in I_2} l} \tag{3}$$

To effectively compute the best performing ratio, reduction of the number of parameters is needed when using optimization methods. To this end, we chose only consecutive intervals of fragment lengths I_1, I_2 for evaluation. The ratio with the best correlation with Y-based FF (FY) was then chosen for second step. We found out that $cor(FRAC, FY)$ was maximal for the subsets $I_1 = L_{131}, L_{132}, \ldots, L_{134}$ and $I2 = L_{97}, L_{98}, \ldots, L_{153}$.

In the second step, we applied a non-linear regression method for estimating model parameters (weights) corresponding to fragment lengths. In this case, NLRM is defined by the equation

$$F = \eta(\theta) + \varepsilon, \tag{4}$$

where $F = (F_1, \ldots, F_N)$ is the vector of fetal fractions, η is a nonlinear function of selected fragment lengths, $\theta = (M_{131}, \ldots, M_{134}, N_{97}, \ldots, N_{153})^T$ is vector of parameters to be estimated, and ε is vector of errors. Consider the non-linear regression model where for each sample (indexed by $i = 1, \ldots, N$), the FF estimation is given by

$$F_i = \frac{\sum_{r=131}^{134} M_r L_{r,i}}{\sum_{p=97}^{153} N_p L_{p,i}} + \varepsilon_i, i = 1, 2, \ldots, N. \tag{5}$$

Note that in the above, the nonlinear function η is given by the ratio of fragment lengths weighted by parameters $\theta = (M_{131}, \ldots, M_{134}, N_{97}, \ldots, N_{153})^T$.

Similarly to LRM, the training and testing sets were selected from the collected data. The *nlm* function from *R* software (https://www.r-project.org/) was used to estimate the parameters for NLRM. The parameters were then used to estimate FF for testing data. Other optimisation methods from *R* software like *nls* function and *optim* function using Broyden–Fletcher–Goldfarb–Shanno (BFGS) algorithm [28] were also applied to estimate parameters for NLRM, but resulting variance of the FF was much bigger compared to using *nlm*.

4.5.3. Neural Network (NN)

A multilayer perceptron (MLP) regression model was trained in several steps for which the datasets of normalized fragment length proportions were preprocessed. We selected *DNNRegressor* estimator from open source machine learning framework *TensorFlow* [8] as implementation of MLP regression. We tried two- and three-layer network architectures and several activation functions in the *DNNRegressor* method. As a result, we found the following configuration to have the best performance for the given input size: 55 units in input layer and 25 in two hidden layers each, *Adagrad* optimizer, hyperbolic tangent as activation function, and mean reduction of loss.

4.5.4. Support Vector Machine (SVM)

We selected a SVM estimator from open source machine learning library *sklearn* [29] as the implementation of SVM regression. Again, datasets of normalized fragment length proportions were used to predict fetal fraction. Parameters of the SVM were optimized through grid search using the method of sklearn (*GridSearchCV*) in conjunction with rigorous testing and empirical evidence. The following parameters were found to give the best results: linear kernel, epsilon 0.01, tolerance for stopping criterion (tol) 0.001, and penalty parameter of the error term (C) 1.0. Non-linear kernels were also tested but all had poorer performance than linear kernel.

4.6. FF Estimators Based on Read Counts

SeqFF

The number of fetal DNA fragments varies in different regions of the genome and significantly correlates with GC content and presence of coding regions. The position-based prediction method SeqFF [11] takes a vector of Loess-corrected fragment counts as input, partitioned into bins 50,000 bases long. The fetal fraction is then determined using standard multivariate regression models trained on a large number of samples (25,312 in original study) which exceeds this study close to tenfold. In addition, published software for SeqFF calculation does not provide a training option. For these reasons we used a pre-trained model and testing script from the original study to evaluate prediction accuracy on our dataset.

4.7. Correlation with Sample Attributes

Additional sample attributes from requisition were compared with the reference FF. Slight significant correlation was observed in three attributes, namely the gestational age, the body mass index of the mother, and the DNA library concentration, with Pearson correlation at the level 0.1, −0.33, and −0.22, respectively.

4.8. Combinaton of FF Estimators

A combined estimator for FF was built from SVM estimator and SeqFF estimator. Furthermore, we examined impact of additional predictive attributes from requisitions and laboratory processing of samples, namely the gestational age, the body mass index of the mother, and the DNA library concentration. Then, 5 variables, specifically FF estimates resulting from the SeqFF and SVM methods, the gestational age, the body mass index of the mother, and the DNA library concentration were inserted into linear regression with Y-based FF as a response variable. Finally, the parameters obtained from this linear regression were used to estimate FF for testing data.

4.9. Weighted Samples

The crucial step in the standard NIPT analysis is to identify samples with low fetal fractions that may not have enough fetal fragments for reliable identification of aneuploidy. Such samples are concluded as uninformative and may be subject to repeated sampling, typically few weeks later.

Overestimation of FF may spuriously inflate confidence in a ploidy call for a sample where the assay is not expected to be sensitive. On the other hand, for high-FF samples the accuracy of FF is less important because the aneuploidy-calling algorithm is sufficiently sensitive anyway.

With this fact in mind, we tried to improve predictions in the lower range of Y-based FF targets. All samples with fetal fraction below 10% (according to the Y-based estimator) within the dataset were duplicated, triplicated, and quadruplicated, giving three distinct datasets. To evaluate this approach, we trained prediction models on these datasets and compared them with models trained on the original dataset. Prediction ability of FF in lower and higher range was evaluated separately.

Author Contributions: Conceptualization, J.G., J.B., M.K., F.D., J.R., and T.S.; data curation, R.H.; formal analysis, J.G.; funding acquisition, T.S.; investigation, J.G., R.H., J.B., and M.K.; methodology, J.G., R.H., J.B., and M.K.; project administration, J.T. and T.S.; resources, J.T. and T.S.; software, J.G. and R.H.; supervision, J.B., J.T., and T.S.; validation, J.G. and R.H.; visualization, J.G.; writing—original draft, J.G. and R.H.; writing—review and editing, J.G., R.H., J.B., and J.R.

Abbreviations

NIPT	non-invasive prenatal testing
FF	fetal fraction
NN	neural network
SVM	support vector machine
cfDNA	cell free DNA
cffDNA	cell free fetal DNA
LRM	linear regression model
NLRM	non-linear regression model

References

1. Pös, O.; Biró, O.; Szemes, T.; Nagy, B. Circulating cell-free nucleic acids: Characteristics and applications. *Eur. J. Hum. Genet.* **2018**, *26*, 937–945. [CrossRef] [PubMed]
2. Allyse, M.; Minear, M.A.; Berson, E.; Sridhar, S.; Rote, M.; Hung, A.; Chandrasekharan, S. Non-invasive prenatal testing: A review of international implementation and challenges. *Int. J. Womens Health* **2015**, *7*, 113–126. [CrossRef] [PubMed]
3. Lo, Y.M.D.; Dennis Lo, Y.M.; Corbetta, N.; Chamberlain, P.F.; Rai, V.; Sargent, I.L.; Redman, C.W.G.; Wainscoat, J.S. Presence of fetal DNA in maternal plasma and serum. *Lancet* **1997**, *350*, 485–487. [CrossRef]
4. Budiš, J.; Kucharík, M.; Duriš, F.; Gazdarica, J.; Zrubcová, M.; Ficek, A.; Szemes, T.; Brejová, B.; Radvanszky, J. Dante: Genotyping of known complex and expanded short tandem repeats. *Bioinformatics* **2018**, *35*, 1310–1317. [CrossRef] [PubMed]
5. Alfirevic, Z.; Mujezinovic, F.; Sundberg, K. Amniocentesis and chorionic villus sampling for prenatal diagnosis. *Cochrane Database Syst. Rev.* **2003**. [CrossRef]
6. Bianchi, D.W.; Lamar Parker, R.; Wentworth, J.; Madankumar, R.; Saffer, C.; Das, A.F.; Craig, J.A.; Chudova, D.I.; Devers, P.L.; Jones, K.W.; et al. DNA Sequencing Versus Standard Prenatal Aneuploidy Screening. *Obstet. Gynecol. Surv.* **2014**, *69*, 319–321. [CrossRef]
7. Ashoor, G.; Syngelaki, A.; Poon, L.C.Y.; Rezende, J.C.; Nicolaides, K.H. Fetal fraction in maternal plasma cell-free DNA at 11–13 weeks' gestation: Relation to maternal and fetal characteristics. *Ultrasound Obstet. Gynecol.* **2013**, *41*, 26–32. [CrossRef] [PubMed]
8. Palomaki, G.E.; Kloza, E.M.; Lambert-Messerlian, G.M.; Haddow, J.E.; Neveux, L.M.; Ehrich, M.; van den Boom, D.; Bombard, A.T.; Deciu, C.; Grody, W.W.; et al. DNA sequencing of maternal plasma to detect Down syndrome: An international clinical validation study. *Genet. Med.* **2011**, *13*, 913–920. [CrossRef] [PubMed]
9. Canick, J.A.; Palomaki, G.E.; Kloza, E.M.; Lambert-Messerlian, G.M.; Haddow, J.E. The impact of maternal plasma DNA fetal fraction on next generation sequencing tests for common fetal aneuploidies. *Prenat. Diagn.* **2013**, *33*, 667–674. [CrossRef] [PubMed]
10. Rava, R.P.; Srinivasan, A.; Sehnert, A.J.; Bianchi, D.W. Circulating Fetal Cell-Free DNA Fractions Differ in Autosomal Aneuploidies and Monosomy X. *Clin. Chem.* **2013**, *60*, 243–250. [CrossRef] [PubMed]

11. Peng, X.L.; Jiang, P. Bioinformatics Approaches for Fetal DNA Fraction Estimation in Noninvasive Prenatal Testing. *Int. J. Mol. Sci.* **2017**, *18*, 453. [CrossRef] [PubMed]
12. Fan, H.C.; Blumenfeld, Y.J.; Chitkara, U.; Hudgins, L.; Quake, S.R. Noninvasive diagnosis of fetal aneuploidy by shotgun sequencing DNA from maternal blood. *Proc. Natl. Acad. Sci. USA* **2008**, *105*, 16266–16271. [CrossRef] [PubMed]
13. Chiu, R.W.K.; Akolekar, R.; Zheng, Y.W.L.; Leung, T.Y.; Sun, H.; Chan, K.C.A.; Lun, F.M.F.; Go, A.T.J.I.; Lau, E.T.; To, W.W.K.; et al. Non-invasive prenatal assessment of trisomy 21 by multiplexed maternal plasma DNA sequencing: Large scale validity study. *BMJ* **2011**, *342*, c7401. [CrossRef] [PubMed]
14. Duris, F.; Gazdarica, J.; Gazdaricova, I.; Strieskova, L.; Budis, J.; Turna, J.; Szemes, T. Mean and variance of ratios of proportions from categories of a multinomial distribution. *J. Stat. Distrib. Appl.* **2018**, *5*, 2. [CrossRef]
15. Jiang, P.; Chan, K.C.A.; Liao, G.J.W.; Zheng, Y.W.L.; Leung, T.Y.; Chiu, R.W.K.; Lo, Y.M.D.; Sun, H. FetalQuant: Deducing fractional fetal DNA concentration from massively parallel sequencing of DNA in maternal plasma. *Bioinformatics* **2012**, *28*, 2883–2890. [CrossRef]
16. Jiang, P.; Peng, X.; Su, X.; Sun, K.; Yu, S.C.Y.; Chu, W.I.; Leung, T.Y.; Sun, H.; Chiu, R.W.K.; Lo, Y.M.D.; et al. FetalQuantSD: Accurate quantification of fetal DNA fraction by shallow-depth sequencing of maternal plasma DNA. *NPJ Genom. Med.* **2016**, *1*, 16103. [CrossRef] [PubMed]
17. Straver, R.; Oudejans, C.B.M.; Sistermans, E.A.; Reinders, M.J.T. Calculating the fetal fraction for noninvasive prenatal testing based on genome-wide nucleosome profiles. *Prenat. Diagn.* **2016**, *36*, 614–621. [CrossRef]
18. Kim, S.K.; Hannum, G.; Geis, J.; Tynan, J.; Hogg, G.; Zhao, C.; Jensen, T.J.; Mazloom, A.R.; Oeth, P.; Ehrich, M.; et al. Determination of fetal DNA fraction from the plasma of pregnant women using sequence read counts. *Prenat. Diagn.* **2015**, *35*, 810–815. [CrossRef]
19. van Beek, D.M.; Straver, R.; Weiss, M.M.; Boon, E.M.J.; Huijsdens-van Amsterdam, K.; Oudejans, C.B.M.; Reinders, M.J.T.; Sistermans, E.A. Comparing methods for fetal fraction determination and quality control of NIPT samples. *Prenat. Diagn.* **2017**, *37*, 769–773. [CrossRef]
20. Minarik, G.; Repiska, G.; Hyblova, M.; Nagyova, E.; Soltys, K.; Budis, J.; Duris, F.; Sysak, R.; Gerykova Bujalkova, M.; Vlkova-Izrael, B.; et al. Utilization of Benchtop Next Generation Sequencing Platforms Ion Torrent PGM and MiSeq in Noninvasive Prenatal Testing for Chromosome 21 Trisomy and Testing of Impact of In Silico and Physical Size Selection on Its Analytical Performance. *PLoS ONE* **2015**, *10*, e0144811. [CrossRef]
21. Budis, J.; Gazdarica, J.; Radvanszky, J.; Szucs, G.; Kucharik, M.; Strieskova, L.; Gazdaricova, I.; Harsanyova, M.; Duris, F.; Minarik, G.; et al. Innovative method for reducing uninformative calls in non-invasive prenatal testing. *arXiv* **2018**, arXiv:1806.08552.
22. Shubina, J.; Trofimov, D.Y.; Barkov, I.Y.; Stupko, O.K.; Goltsov, A.Y.; Mukosey, I.S.; Tetruashvili, N.K.; Kim, L.V.; Bakharev, V.A.; Karetnikova, N.A.; et al. In silico size selection is effective in reducing false positive NIPS cases of monosomy X that are due to maternal mosaic monosomy X. *Prenat. Diagn.* **2017**, *37*, 1305–1310. [CrossRef] [PubMed]
23. Yu, S.C.Y.; Chan, K.C.A.; Zheng, Y.W.L.; Jiang, P.; Liao, G.J.W.; Sun, H.; Akolekar, R.; Leung, T.Y.; Go, A.T.J.I.; van Vugt, J.M.G.; et al. Size-based molecular diagnostics using plasma DNA for noninvasive prenatal testing. *Proc. Natl. Acad. Sci. USA* **2014**, *111*, 8583–8588. [CrossRef] [PubMed]
24. Ferreira, A.J.; Figueiredo, M.A.T. Boosting Algorithms: A Review of Methods, Theory, and Applications. In *Ensemble Machine Learning*; Springer: Boston, MA, USA, 2012; pp. 35–85.
25. Langmead, B.; Trapnell, C.; Pop, M.; Salzberg, S.L. Ultrafast and memory-efficient alignment of short DNA sequences to the human genome. *Genome Biol.* **2009**, *10*, R25. [CrossRef] [PubMed]
26. Liao, C.; Yin, A.-H.; Peng, C.-F.; Fu, F.; Yang, J.-X.; Li, R.; Chen, Y.-Y.; Luo, D.-H.; Zhang, Y.-L.; Ou, Y.-M.; et al. Noninvasive prenatal diagnosis of common aneuploidies by semiconductor sequencing. *Proc. Natl. Acad. Sci. USA* **2014**, *111*, 7415–7420. [CrossRef] [PubMed]
27. Hudecova, I.; Sahota, D.; Heung, M.M.S.; Jin, Y.; Lee, W.S.; Leung, T.Y.; Lo, Y.M.D.; Chiu, R.W.K. Maternal Plasma Fetal DNA Fractions in Pregnancies with Low and High Risks for Fetal Chromosomal Aneuploidies. *PLoS ONE* **2014**, *9*, e88484. [CrossRef] [PubMed]
28. Zhu, C.; Byrd, R.H.; Lu, P.; Nocedal, J. Algorithm 778: L-BFGS-B: Fortran subroutines for large-scale bound-constrained optimization. *ACM Trans. Math. Softw.* **1997**, *23*, 550–560. [CrossRef]
29. Garreta, R.; Moncecchi, G. *Learning Scikit-Learn: Machine Learning in Python*; Packt Publishing Ltd.: Birmingham, UK, 2013; ISBN 9781783281947.

9

Cell-Free Nucleic Acids and their Emerging Role in the Pathogenesis and Clinical Management of Inflammatory Bowel Disease

Zuzana Kubiritova [1,2], Jan Radvanszky [1,*] and Roman Gardlik [3]

1 Institute for Clinical and Translational Research, Biomedical Research Center, Slovak Academy of Sciences, Dubravska cesta 9, 84505 Bratislava, Slovakia
2 Department of Molecular Biology, Faculty of Natural Sciences, Comenius University, Ilkovicova 6, 84215 Bratislava, Slovakia
3 Institute of Molecular Biomedicine, Faculty of Medicine, Comenius University, 81372 Bratislava, Slovakia
* Correspondence: jradvanszky@gmail.com

Abstract: Cell-free nucleic acids (cfNAs) are defined as any nucleic acids that are present outside the cell. They represent valuable biomarkers in various diagnostic protocols such as prenatal diagnostics, the detection of cancer, and cardiovascular or autoimmune diseases. However, in the current literature, little is known about their implication in inflammatory bowel disease (IBD). IBD is a group of multifactorial, autoimmune, and debilitating diseases with increasing incidence worldwide. Despite extensive research, their etiology and exact pathogenesis is still unclear. Since cfNAs were observed in other autoimmune diseases and appear to be relevant in inflammatory processes, their role in the pathogenesis of IBD has also been suggested. This review provides a summary of knowledge from the available literature about cfDNA and cfRNA and the structures involving them such as exosomes and neutrophil extracellular traps and their association with IBD. Current studies showed the promise of cfNAs in the management of IBD not only as biomarkers distinguishing patients from healthy people and differentiating active from inactive disease state, but also as a potential therapeutic target. However, the detailed biological characteristics of cfNAs need to be fully elucidated in future experimental and clinical studies.

Keywords: cell-free nucleic acids; circulating nucleic acids; cell-free DNAs; cell-free RNAs; exosomes; inflammatory bowel disease; neutrophil extracellular traps; NETosis

---

1. Introduction

Inflammatory bowel disease (IBD), which involves Crohn's disease (CD) and ulcerative colitis (UC), is a group of multifactorial disorders characterized by chronic inflammation affecting the gastrointestinal tract in variable extent, typically leading to multiple symptoms such as weight loss, abdominal pain, recurrent diarrhea, and bleeding. Despite extensive research in this field, the exact pathogenesis of IBD is still unknown. According to current knowledge, IBD develops due to deregulated complex interactions between genetic, environmental, and immunological factors, including the complex interactions between the gastrointestinal microbiota and the host organism [1,2]. The clinical diagnosis of IBDs is based on numerous investigations, including ultrasonography, endoscopy, as well as laboratory analyses such as biochemical blood and stool markers, or histological tests of affected tissues. The diagnostic procedures can be time consuming and uncomfortable for patients, especially for the continuous monitoring of their health status, disease progression, or therapeutic effectiveness. Moreover, therapeutic possibilities are limited to the suppression of acute inflammation and maintaining remission, while a marked interindividual variability of effectiveness was reported for each line of actually available therapeutics [3,4]. Therefore, current research related to IBD is strongly focused on various aspects of the disease with the aim to elucidate the exact mechanisms of the pathogenesis, as well as to identify either novel biomarkers (for both disease status evaluations, prognostics, and therapeutic response predictions and monitoring) or new therapeutic targets and agents. One of the highly progressing fields that has an impact in each of these aspects is genomics and nucleic acids research in general. Besides the direct analyses of genomic material obtained from cells, the possibility of identification and analysis of cell-free nucleic acids (cfNAs) in the circulation, or in other body fluids, is now attracting particular attention in several biomedical fields. In a broader context of complex care, the analysis of nuclear genomic material offers the identification of yet asymptomatic individuals having high risk for certain types of IBD [5], while the identification of cfNAs seems to represent promising non-invasive biomarkers to distinguish patients having active disease from those in inactive phase or from healthy people, allowing to track disease onset, progression, and remission following therapy. However, still little is known about the specific cfNAs and their exact roles and associations in the pathogenesis of IBD. Therefore, the aim of this review is to collect and integrate available information about these important aspects.

2. Cell-Free Nucleic Acids and their General Recognition and Use

Our knowledge about the existence of cfNAs is not new. Since their first description in 1948 [6], their presence was observed in various biological fluids, such as blood, urine, saliva, but also in stool. Their biological significance and usability as biomarkers of health status is given by several factors. The two main advantages are their differential presence in healthy individuals versus patients having certain diseases, or specific physiological conditions, as well as their convenient accessibility from liquid biopsy, saliva, or stool sampling. Moreover, their differential patterns of methylation status or fragmentation may serve as information about the organism, organ, tissue, or mechanism of origin of particular fractions of cfNAs [7–10].

Although not the first, probably the most popular and most rapidly adopted use of cfNAs globally is the so-called non-invasive prenatal testing (NIPT) [11]. This method opened a completely new era of prenatal care, especially following its housing with advanced DNA analytical technologies, such as massively parallel genome-scale DNA sequencing. However, prenatal testing-based analyses of cfNAs started to show also other utilities, even going beyond conventional prenatal testing. The spread of cfNAs analyses already took place also into unrelated clinical fields, such as oncology, in a form of liquid biopsy-based non-invasive cancer detection (NICD), since cfDNA levels are typically elevated in cancer patients [12]. The possible relevance of cfNAs has been previously reported also for a range of other diseases and pathological conditions, such as trauma, myocardial infarction, stroke, transplantation, diabetes, sickle cell disease, sepsis, and aseptic inflammation [13]. Moreover, in addition to cancer and inflammation, other aging-related degenerative processes, such as cellular senescence, were also suggested to be associated with age-related increasing levels of cfNAs (specifically cfDNA), possibly through DNA damage response-induced higher genome instability [14].

3. Origin and Basic Types of Cell-Free Nucleic Acids

There are plenty of different types of cfNAs molecules that can both originate from various biological processes and participate on various physiological and pathophysiological processes in a complex yet poorly understood manner. According to their origin, cfNAs can be divided into various categories. At first, it is important to differentiate between endogenous and exogenous cfNAs. While endogenous cfNAs originate from tissues and the cells of the organism of interest itself, exogenous cfNAs may typically come from the host microbiome, from different infections and parasites, as well as from the ingested food of the host organism [15]. On the border between these are cfNAs originating from developing fetuses in maternal organisms [16] which, although having endogenous origin, are coming from a different individual of the same biological species. Moreover, in humans, the transfusion origin of cfNAs should also be taken into account [17].

When considering the type of the macromolecule of interest, cfNAs are generally divided into DNAs or RNAs. Endogenous cfDNA includes nuclear (genomic) DNA (cf-ncDNA) as well as mitochondrial DNA (cf-mtDNA), whereas exogenous DNA is usually represented by microbial and specifically bacterial DNA. It is hard to universally describe the length range and exact composition of cfDNAs, since their extremely dynamic nature and various extraction methods result in variable recovery rates.

However, they were described in the range from ultrashort, <100-bp fragments, up to several thousands of bps, both for nuclear as well as for mitochondrial DNA [12,18]. Nuclear cfDNA fragments have predominancy under physiological conditions, and are highly fragmented molecules to the size of approximately 166 bp, which is approximately the length of a segment of DNA wound around a protective histone octamer.

They mainly come from the apoptosis of hematopoietic cells [19]; however, they can also be released from necrotic cells—but only through phagocytosis [20]. Another process of the formation of circulating cfDNA is NETosis (name comes from neutrophil extracellular traps) [21], but the active release of newly synthesized DNA via vesicles and lipoproteonucleotide complexes was also reported [22].

In a comparison to cf-ncDNA, cf-mtDNA is much smaller. Its predominant size ranges approximately from 30 bp to 80 bp, because mtDNA is, unlike ncDNA, a small molecule that is not protected by histones. Similar to cf-ncDNA, cf-mtDNA can be transported by extracellular membrane vesicles (EMVs) [12].

Similar to cfDNAs, cfRNAs can also be divided into many other types. According to the functionality, we differentiate coding RNAs (mRNAs) and non-coding RNAs. Non-coding RNAs are further divided into many subtypes, including lncRNAs (long non-coding), miRNAs (micro), tRNAs (transfer), YRNAs, piRNAs (PIWI-interacting), circRNAs (circular), other small non-coding RNAs (ribosomal, small nuclear, small nucleolar); however, these are present in amounts less than 1% [12].

Among cfRNAs, the cf-mRNAs are fragmented and less abundant; therefore, many studies focus on the analysis of small non-coding RNAs, which are more stable and therefore also more abundant. As cfRNAs are relatively unstable molecules that are susceptible to degradation by ribonucleases, they can generally be found encapsulated within EMVs (apoptotic bodies, exosomes, microvesicles) or form ribonucleoprotein complexes (by binding to high-density lipoprotein or RNA-binding proteins) [23,24].

Based on some of the above-mentioned properties, both endogenous and exogenous cfNAs in circulation can be further divided into free cfNA fragments (naked sequences having no specific vesicles), vesicle-bound cfNA fragments (nucleic acids in EMVs divided into exosomes, microvesicles, and apoptotic bodies) and cfNA macromolecular complexes (e.g., nucleosomes, virtosomes, and neutrophil extracellular traps), all of which have specific origin and routes of transport (Figure 1) [12].

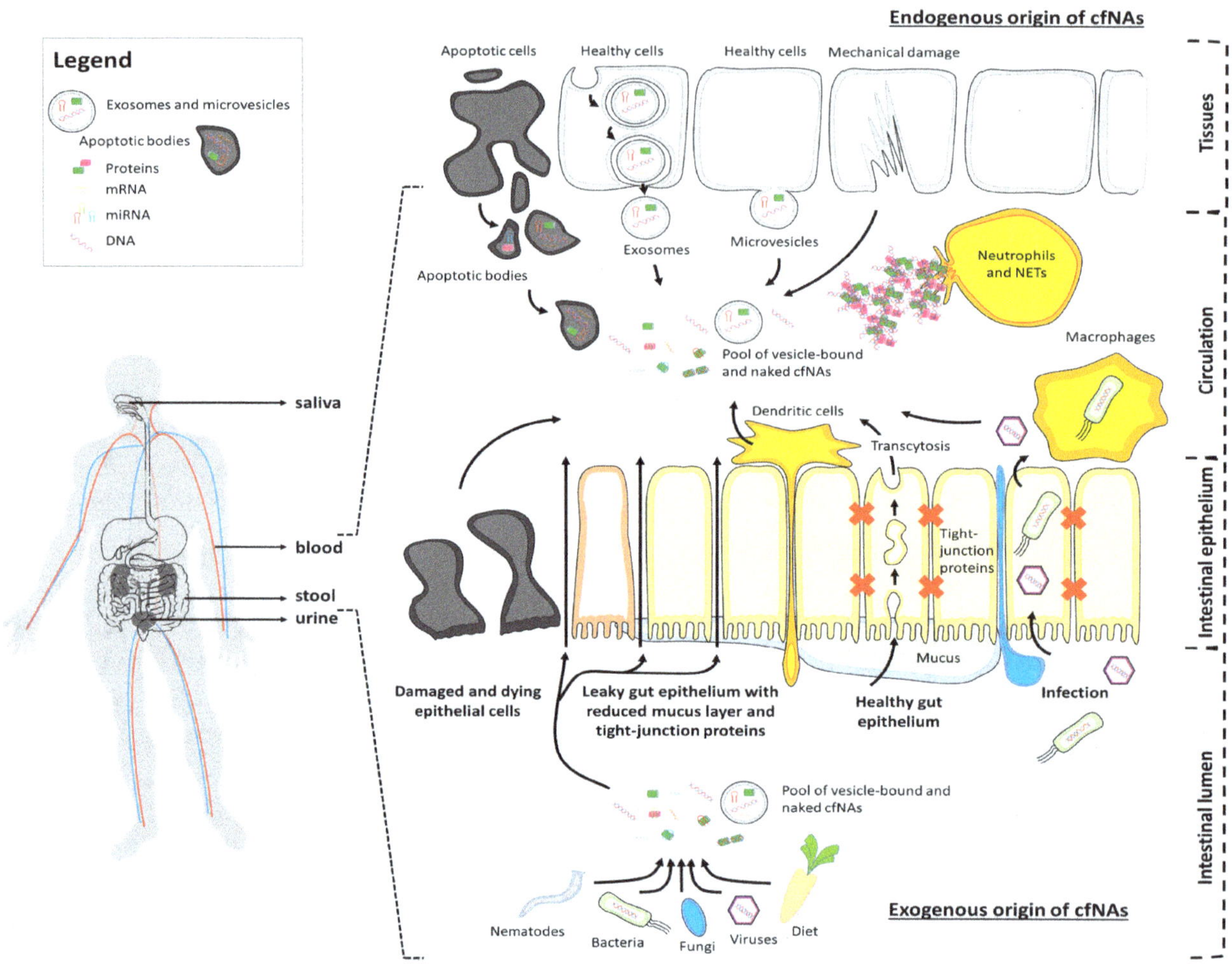

Figure 1. Origin, organization, routes of transfer, and availability for convenient sampling of cell-free nucleic acids (cfNAs). CfNAs can be readily isolated from various body fluids, including blood, saliva, urine, and stool, in which they are either naked or carried by various types of vesicles. Depicted are the most common sources of cfNAs in the circulation, which are relevant for inflammatory bowel disease (IBD), including endogenous sources (such as apoptotic bodies, exosomes, microvesicles, neutrophil extracellular traps (NETs), necrosis) as well as exogenous sources (such as diet and the intestinal microflora, including bacteria, fungi, nematodes, and viruses). Endogenous cfNAs most commonly originate from regulated or unregulated cell death (necrosis, mechanical damage vs. apoptosis, NETosis), or are secreted during processes of cell–cell communications (exosomes, microvesicles). Those of exogenous origin may reach the circulation by several processes such as hijacking the physiological transport machineries, transcytosis (vesicular uptake on one side of the epithelial barrier and release on the opposite side), continuous immunological sampling of antigens by dendritic cells, but also as a result of inefficient epithelial barrier functions during pathological processes. Common sources such as developing a fetus or tumors are not depicted, although the detection of colitis-to-cancer transition may have specific relevance for IBD patients. Apoptotic bodies may include also cellular organelles; however, these are not depicted. Since the figure focuses on the cfNAs content of the blood, content having endogenous origin in the intestinal lumen is not depicted, even though stool analysis may have specific relevance for the monitoring of health status and therapeutic effectiveness in IBD patients. Note that this depiction did not differentiate between single-layered and double-layered vesicles, and also that all the graphical components are illustrative and are not representative of real dimensions.

4. Exosomes in IBD

Exosomes are endosomal-derived nanovesicles that are secreted from many types of cells in both physiological and pathological conditions. They are not produced by "simple" budding from the plasma membrane, as microvesicles are; rather, their highly regulated biogenesis is connected to the endolysosomal pathway. This starts with an endocytosis-based generation of an early-endosome that matures to a late-endosome, a multivesicular body, through endosomal membrane budding. By exocytosis, such multivesicular bodies can fuse with the plasma membrane, releasing their individual vesicles (exosomes) from the cell. Alternatively, multivesicular bodies can fuse with lysosomes to become degraded, and the function of their cargos can be lost [25,26]. Since these exosomes can transport various molecules long distances inside the body, such as proteins, lipids, and nucleic acids (mainly miRNAs and mRNAs), they are generally involved in various biological activities, such as cell–cell communication and cell–environment interactions. These include the modulation of immune responses, intestinal barrier functions, and regulation of the intestinal microbiome. The specific functions of exosomes in these complex interactions, not only in IBD, depend primarily on their functional components and exosome structure [27]. While the transferred proteins can directly activate or inhibit biological pathways [28], transferred regulatory NAs can modulate pathways through the specific regulation of gene expression in target cells [29]. Exosomes have been previously isolated from plasma, colonic luminal fluid aspirates, intestinal epithelial cells, and saliva [30–33]. As they are released by different tissues and can be collected from several body fluids and biological materials, the information carried by their specific cargos may represent potential disease biomarkers, while they themselves may represent possible therapeutic targets or therapeutic agents [34].

Several studies focus on the analysis of protein content of exosomes from IBD patients and healthy controls or mouse models of colitis. When comparing their cargos, exosomes from the saliva of IBD patients, for example, revealed several proteins presenting exclusively in just one of the studied groups, i.e., only in UC patients, CD patients, or healthy controls. From these, for example, proteasome subunit alpha type 7 (PSMA7) was selected to be a promising biomarker, since this exosomal protein is related to proteasome activity and inflammatory response [33]. Tens of differentially expressed proteins in serum exosomes were also found in mice with acute colitis when compared to exosomes isolated from control mice. The majority of them were involved in the complement and coagulation cascade, which has been implicated in macrophage activation, pointing thus to specific roles in the pathogenesis of IBD [35]. Moreover, alterations of intestinal exosome proteomes were associated also with aberrant host–microbiota interactions in IBD. It was demonstrated that exosomes released from intestinal epithelial cells are involved in activating host innate immune responses and in subverting the intracellular replication control of adherent-invasive *Escherichia coli*. It is known that these strains are in a high prevalence in the intestinal mucosa of patients with CD, and enhance intestinal epithelial permeability by modulating and/or disorganizing cell junction proteins [36,37]. Along with altered protein profiles, distinct mRNA profiles were also found in exosomes shed from sites of inflammation in patients with IBD. These were shown to have pro-inflammatory effects on the colonic epithelium in vitro, which is mainly due to an increase of interleukin-8 (IL-8) [31].

The above-mentioned associations point toward an increased involvement of exosomes in pro-inflammatory cascade in IBD pathogenesis. However, exosomes from normal intestine were found to have immunosuppressive effects with a potential to prevent the development of IBD in a murine model of colitis, for example by inducing T-regulatory and dendritic cells. On the other hand, their inhibition can exacerbate murine IBD [30]. Besides exosomes from intestinal epithelial cells, exosomes secreted by mesenchymal stem cells derived from a human umbilical cord were shown to have a repairing effect on inflamed intestinal tissue. It was proven that they have profound effects on alleviating a dextran sulfate sodium (DSS)-induced colitis through the modulation of IL-7 expression in macrophages, or by regulating the ubiquitin modification level [38,39]. In the serum of patients with active IBD, elevated levels of secreted annexin A1-(ANXA1-) containing extracellular exosomes were detected, most likely as a result of systemic distribution in response to the inflammatory process.

Endogenous ANXA1 is released as a component of extracellular exosomes derived from intestinal epithelial cells, while these ANXA1-containing exosomes have the potential to activate wound repair circuits [32].

Taken together, these findings suggest that exosomes, through the specific constellation of their cargos, may exert both pro-inflammatory as well as anti-inflammatory effects on tissues, including intestinal tissues. Since exosomes with pro-inflammatory contents seems to be enriched during active IBD (or experimentally induced colitis), while those with anti-inflammatory cargos are dominant in healthy patients (or control mice), their precise balance in different physiological situations likely plays a crucial role in inducing, maintaining, or regulating the required functions of intestinal tissues. Therefore, exosomes or their specific cargos can be considered when looking for biomarkers of intestinal mucosal inflammation, but also when looking for potential therapeutic strategies in situations of chronic mucosal injury. Moreover, in the latter case, their involvement can be considered both as possible targets for inhibition (in case of pro-inflammatory contents), or as possible therapeutic agents (in case of anti-inflammatory contents), to induce and/or maintain intestinal homeostasis. The biological significance and clinical potential of exosomes as markers and therapeutic tools in IBD has been recently reviewed [26,27,40]. However, knowledge of the role of exosome-specific NAs has been generally limited to microRNAs as the major human plasma-derived exosomal RNA species. The role of miRNAs is discussed later in this text.

5. Neutrophil Exracellular Traps in IBD

Neutrophil extracellular traps (NETs) are complex structures released from neutrophils due to chromatin decondensation and spreading. These so-called "traps" are being released into the environment, consisting of a tangle of released nucleic acids, histones, and proteases [41]. Then, these traps are able to capture the bacteria and kill them. Despite being macromolecular complexes, NETs, or their degradation products, were found to contribute to the total pool of cfDNAs [21]. They were described by Takei et al. in 1996 as a pathway of cellular death that is different from apoptosis and necrosis [42], and in 2004, Brinkmann et al. described a process named NETosis, which is one of the mechanisms that neutrophils undertake for host defense [43]. There are several known models of NETosis, which are either dependent or independent of reactive oxygen species (ROS), and also they differ in releasing either nuclear or mitochondrial DNA. Neutrophils and NETs protect hosts from infectious diseases, while the aberrant formation and/or clearance of NETs may have pro-inflammatory characteristics and may be implicated in many infectious and non-infectious diseases, such as cancer, cardiologic, metabolic, inflammatory, and lung diseases [44]. The nuclear material released from NETs could be more immunogenic than the apoptotic one. Both native and oxidized endogenous DNA bound to NETs activate dendritic cells to synthetize interferon-α in a TLR (toll-like receptor)-dependent manner. NETs also increase the T-cell response to antigens and activate B cells to induce immunoglobulin (Ig) class switching and antibody production. Oxidized DNAs are also more resistant to degradation, contributing thus to sustain a dysregulated immune response, while the NET-mediated activation of the inflammasome further amplifies the inflammatory response through a feed-forward loop [41].

Despite the implication of NETs in various pathologies, very little is known about their association with IBD, although several studies have proposed their role in the pathogenesis of IBD of both adults [45,46] and pediatric patients [47]. Gut proteome analysis in patients with ulcerative colitis showed a high expression of proteins that are associated with NETs [45]. Whether NETs formation is causative, or is rather a result of the inflammation, is still not completely known. Although ROS production is enhanced in CD, and neutrophils may be more prone to NET formation [48], their formation in CD is still controversially described. For example, a recently published small-scale study proved the enhanced presence of NETs in the intestinal tissues of pediatric CD and UC patients [47]. Lehmann et al. also reported upregulated proteins belonging to the main components of NETs in both diseases (UC and CD) compared to controls [49]. However, in other studies NETs have been observed and correlated with inflammation in UC, but not in CD, pointing to the stimulation of the innate immune system in the etiology of UC [45]. There are also studies suggesting that the presence of NETs in IBD does not have to be necessarily considered as a detrimental factor [50]. On the other hand, there are studies pointing to a detrimental role of NETs in IBD. Anti-neutrophil cytoplasmic autoantibodies (ANCAs) are biomarkers for the diagnosis and prognosis of IBD that are considered to activate, complement, and cause endothelial damage. They are known to target neutrophil proteins, which all are released during NET formation, suggesting this process might be the general cause of ANCAs production in IBD [51]. For example, ANCAs against myeloperoxidase were detected in the serum of many IBD patients [52], and against leukocyte proteinase 3 were also detected in IBD patients, but more frequently in UC than in CD [53]. The results of Dinallo et al. showed that NET-associated proteins were over-expressed in the inflamed colon of UC patients as compared to CD patients, suggesting a role for NETs in sustaining mucosal inflammation in UC. The same authors described also NETs' production from the circulating neutrophils of UC patients in response to stimulation by TNF-α, with diminishing NETs formation following successful anti-TNF-α treatment [46,54]. Interesting findings came from He et al., who found that NETs facilitate pro-coagulant activity in patients with IBD as well as thromboembolic events that are known to exacerbate this disease [55]. In their extended study, they found that patients with active UC or CD had significantly increased levels of cfDNA, nucleosomes, and NETs formation compared to patients with inactive disease. They demonstrated that NETs represent a central component in the initiation and progression of colitis through mediating inflammation cell infiltration, driving cytokines release and thrombotic tendency.

In this context, the presence of NETs, their relative abundance, as well as their specific content seems to provide candidate biomarkers for the differential diagnosis of various types of IBD. Moreover, they may also represent the possible therapeutic targets of IBD and UC specifically [56], possibly through the specific inhibition of NET release, which was shown to be able to attenuate DSS-induced colitis in mice [46]. However, a detailed association of NET formation with different types and severities of IBD remains to be elucidated.

6. Cell-Free DNA in IBD

As was mentioned above, cfDNAs are present in various body fluids and biological materials. They are detectable under physiological conditions, but their presence during various diseases and physiological states is more interesting, at least for biomedical implications [57,58]. However, there are only a few publications regarding cfDNA in association with IBD.

It is known that cfDNA activates innate immunity through the activation of several DNA-sensing pathways, including toll-like receptor 9 (TLR9), stimulator of interferon protein (STING), and a protein called absent in melanoma 2 (AIM2) [59]. CfDNAs are able to bind to TLR9 and induce the cascade, leading to an inflammatory response, indicating that cfDNA could serve as a potential marker of inflammation. Experiments to study the role of these pathways in IBD have already been performed on knock-out mice that are deficient in these receptors. Some of them have paradoxically confirmed the protective role of these pathways in IBD [60,61]. This effect is probably mediated by intestinal microbiota, although detailed mechanisms are not yet known. Likewise, it was shown that the activation of TLR9 with the agonist administered prior to the induction of colitis induced anti-inflammatory effects [62]. On the other hand, a different study showed that the administration of oligodeoxynucleotides with CpG motifs (bacterial cfDNA) that activate TLR9 significantly exacerbates the course of DSS-induced colitis [63]. It is clear that bacterial ecDNA derived from intestinal microbiota is a heterogeneous group of various DNA molecules with potentially diverse roles in the pathogenesis of IBD.

Molnár et al. indicated that the intravenous administration of colitic cfDNA into healthy mice displays protective effects against DSS-induced colitis by altering the expression of several TLR9-related and inflammatory cytokine genes [64,65]. Thus, cfDNA found in the plasma of mice with DSS-induced colitis seems to have anti-inflammatory properties, but only in a preventive manner, as was shown by its transfer to healthy mice before the onset of the disease. It is known that the activation of TLRs in response to pathogen or damage-associated molecular patterns is associated with autophagy [66], and that autophagy contributes to NETosis [67]. This suggests a direct link between cfDNA and autophagy. In fact, TLR9-mediated autophagy might be the underlying phenomenon behind the protective effect of colitic cfDNA preconditioning [68]. However, the detailed mechanisms of action, as well as the origin of such cfDNA, is not fully understood.

In 2003, Rauh et al. showed for the first time that it is possible to detect cfDNA in the serum of UC patients, and that this cfDNA contained a microsatellite alteration previously identified in mucosa cells from UC patients [69]. A valuable study came from Koike et al., who detected a significantly higher amount of cfDNA in the circulation of mice with DSS-induced colitis compared to the control group. They also found a positive correlation between plasma cfDNA concentration and clinical severity of UC [70]. These findings are consistent with our recent study, which demonstrated a higher concentration of plasma cfDNA in mice with DSS-induced colitis group compared to the control. Moreover, the levels of plasma cfDNA negatively correlated with deoxyribonuclease (DNase) activity in the colon tissue [71]. These findings indicate that colon cells might represent one of the major sources of colitic plasma cfDNA, which further triggers downstream events that contribute to the pathogenesis. However, these findings are in discrepancy with a previous study in which no difference in cfDNA concentrations was found between DSS colitic mice and healthy controls [72]. On the other hand, recent results further demonstrated that the concentration of total cfDNA in the plasma of mice is increasing in parallel with the progression of the disease [73]. However, the increase of total plasma cfDNA levels did not correspond with an increase in plasma cf-ncDNA or cf-mtDNA. On the other hand, the total amount of cfDNA produced specifically by colon tissue was only increased in the early stages of the disease, and the increase corresponded to the increase of nuclear and mitochondrial cfDNA subtypes. These data suggest the crucial role of local colonic processes in triggering the inflammation. However, the later stage-associated increase in plasma cfDNA seems to be of a different than colonic origin.

Low amounts of human DNA released from the epithelial cells of the gastrointestinal wall can be detected in human fecal matter. However, in the state of inflammation or presence of infectious agents, when greater amounts of damaged and dead cells are exfoliated from the intestinal wall, the amount of DNA rises. Vincent et al. demonstrated that the excretion of large amounts of human DNA in feces is a general outcome of intestinal inflammation and is associated with the risk of *C. difficile* infection [74]. Casellas et al. showed that fecal and also gut lavage fluid DNA correlated with the clinical index and endoscopic score in patients with UC. Such fecal DNA excretion was significantly higher in patients with active disease, suggesting that it could be used as a non-invasive technique for the assessment of disease activity [75] and, because fecal DNA concentration increased in relapsed patients, also as an objective instrument to use in the follow-up of patients [76]. In addition, it was recently revealed that DNA methylation plays an important role in autoimmune-related chronic inflammatory diseases [77]. DNA methylation is an important epigenetic modification, which can silence genes, but also can lead to the increase of gene copy number and induce tumors. It is also known that IBD increases the risk of colorectal cancer, which is known as colitis-associated cancer [78]. The study of Lehmann-Werman et al. suggested that methylation patterns can be used to detect cfDNA derived from intestinal epithelial cells. Based on their results, intestinal DNA markers in healthy plasma and stool reflect the established route of clearance of intestinal DNA via the lumen of the gut. However, unlike Casellas et al., they detected only a minimal baseline intestinal cfDNA signal in IBD patients, which was indistinguishable from that of healthy individuals, compared to patients with advanced colorectal cancer, which had a strong intestinal signal [79]. On the other hand, Bai et al. detected a gradientally increased level of cfDNA in colitis and colon cancer mice, compared with control. They also studied the level of circulating DNA methylation, which decreased in colitis and colon cancer compared with control. Their results suggest that cfDNA and its methylation level can be considered new markers for colitis-to-cancer transformation [80].

7. Mitochondrial and Nuclear Cell-Free DNA in IBD

Mitochondria play an important role in inflammatory processes as they participate in metabolism and cell death signaling. Stress conditions can lead to damage to mitochondria and formation and the release of mitochondrial-derived vesicles (MDVs) containing mtDNA [81]. Such cf-mtDNA can act as a damage-associated molecular pattern (DAMP) that activates neutrophils through TLR9 because of the similarity with bacterial DNA [82,83]. Inflamed gut mucosa in IBD represents an enriched source of DAMPs; however, the role of cf-mtDNA in IBD is relatively unknown [84]. It has been proven that cf-mtDNA can activate various inflammatory responses via TLR9 receptors, including NLRP3 inflammasomes and neutrophils, as well as other downstream pro-inflammatory signaling proteins such as TNFα and NFκB [85,86]. Circulating cf-mtDNA was observed in a variety of inflammatory diseases and in patients with acute injury [83,87–91]. Boyapati et al. published the first report showing that cf-mtDNA is released during active IBD [92]. Increased levels of cf-mtDNA in UC and CD patients were detected, which significantly correlated with blood, clinical, and endoscopic markers and diseases activity. Inflammatory cells in lamina propria expressing TLR9 were also higher in active IBD patients. Apart from that, mitochondrial damage in inflamed UC mucosa and higher levels of fecal cf-mtDNA were observed. In parallel, these results were identified in a mouse model of DSS-induced colitis and recently, they confirmed these findings and demonstrated that deletion of the Tlr9 gene in mice results in the attenuation of acute DSS-induced colitis, suggesting cf-mtDNA-TLR9 signaling as an important and targetable pathway in IBD [92]. Another study showed that in mice with DSS-induced colitis, levels of plasma cf-ncDNA increased with the increased duration of colitis, and were directly proportional to the number of NETs [70].

Our group has recently tried to determine the dynamics of total cfDNA as well as cf-ncDNA and cf-mtDNA during an animal model of IBD. However, the concentration of circulating cf-mtDNA and cf-ncDNA in mice with colitis was not significantly different throughout the course of colitis compared to control mice [73]. On the other hand, cf-mtDNA released specifically from the colon tissue increased significantly during the early stages of the disease, indicating that colon-derived mt-DNA might be a significant factor that drives the local colonic inflammation, whereas other subtypes are primarily involved in the systemic inflammation.

8. Cell-Free DNA as a Therapeutic Target

DNases are enzymes cleaving DNA, which were proposed for the therapy of diseases with increased levels of cfDNA [93]. DNase I deficiency results in the difficulty of removing DNA from nuclear antigens, and consequently promotes susceptibility to autoimmune disorders [94]. DNase I deficiency was found in patients with systematic lupus erythematosus [95], and the administration of DNase I has been shown to be an efficient therapeutic agent in cystic fibrosis [96]. A reduced DNase I activity was also observed in patients with IBD [97]. Thus, in the view of the above-mentioned findings, the targeted digestion of cfDNA using DNase represents a potential novel therapeutic approach for IBD. DNase can directly cleave the circulating cfDNA, but can also break the structure of NETs, thereby reducing their pro-inflammatory properties [98,99]. The administration of DNase was shown to be effective in several immune-mediated experimental models, including sepsis [98] and hepatorenal injury [100]. In our recent study, intravenous DNase I injection was tested as potential therapy of DSS-induced colitis. Despite some improvement in the biochemical markers of colitis, the overall therapeutic effect was not proved, possibly due to the rapid half-life of the enzyme in circulation [72]. In the light of decreased DNase activity in the colon of mice with colitis [71], the topical colonic administration of DNase might be an interesting approach to further clarify the role of colon-derived cfDNA and test the rationale of cfDNA-targeted therapy.

NETs as higher structures may also represent possible therapeutic targets in IBD, and UC specifically [56], possibly through the specific inhibition of NET release, which was shown to be able to attenuate DSS-induced colitis in mice [46]. One of the principles to prevent the action of NETs is also disrupting their structure using DNase. Other approaches include preventing the oxidative burst of neutrophils that are necessary for the release of decondensed chromatin from neutrophils into the environment (e.g., by inhibitors of NADPH oxidase), or preventing the citrullination of histones (inhibitors of PAD4-peptidyl arginine deiminase type 4) [101].

At last, another means of possible cfDNA-based therapy relies on the administration of exosomes from healthy intestine, which were found to have immunosuppressive effects with a potential to prevent the development of IBD in a murine model of colitis. On the other hand, their inhibition can exacerbate murine IBD [30]. In light of recent findings proving the absence of cfDNA in exosomes, the possible therapeutic effect of exosomes might as well have been mediated by structures other than cfDNA [102]. In addition, exogenous exosome-like extracellular vesicles released by nematodes were also described to possess immunoregulatory molecules, proteins, and specific miRNAs, which were able to protect mice against chemically-induced colitis. Such specific proteins and miRNAs have great potential in the development of drugs to prevent chronic inflammatory diseases such as IBD [103].

9. Cell-Free miRNA in IBD

The studies focusing on the role of cfRNA mostly analyze the role of small non-coding RNAs, which are more stable than mRNA. Hence, the majority of publications regarding the role of cfRNA in IBD are related mainly to miRNA. MiRNAs are short (18 to 24 nucleotides in length), endogenous, non-coding single-stranded RNAs [104]. The sequences of miRNAs are evolutionary conserved across species, and at the post-transcriptional level, they regulate the expression of nearly one-third of the genes in the human genome and are involved in many biological processes (e.g., development, cell differentiation, proliferation, apoptosis), suggesting their important role in the pathogenesis of various diseases [105]. MiRNAs act primarily as post-transcriptional regulators via mRNA degradation and/or translational repression via the formation of miRNA-induced silencing complex (miRISC). MiRISC post-transcriptional control occurs through the inhibition of translation elongation, protein degradation, ribosome drop-off, or reducing the number of ribosomes on target mRNAs. However, miRNAs also have specific nuclear functions, including the miRNA-guided transcriptional control of gene expression. The mechanisms of miRNA-mediated transcriptional control of gene expression have not been completely elucidated [106].

The aberrant expression of miRNAs in IBD was, for the first time, reported in 2008 by Wu et al. Since then, many other studies were published that focused on the altered expression of miRNA in IBD, on the specific miRNAs and their association with target genes, or on single nucleotide polymorphism present in miRNAs implicated in the pathogenesis of IBD [107]. Among them, the most interesting studies (preliminary) are related to circulating miRNAs (here referred to as cf-miRNAs) identified in IBD patients, as they are considered promising biomarkers of disease severity, activity, or as a potential therapeutic target in IBD (for more information, see below). Cf-miRNAs are stable compared with mRNAs, and because they are packaged in exosomes or microvesicles, they are resistant to nuclease digestion. They were detected in serum and plasma samples, but also in urine or saliva [108,109].

In association with IBD, many studies have been published that detected either increased or decreased levels of specific cf-miRNA in patients with UC or CD compared with healthy controls. Some studies have focused on the relationships between altered miRNA expression in circulation and those at the diseased tissues, since sometimes, the altered miRNA expression in the diseased tissue is different from that observed in the peripheral areas where the cf-miRNA were quantified [110]. In 2011, Wu et al. showed for the first time that cf-miRNAs can be used to distinguish active CD and UC from healthy controls [111]. Since then, others have identified various dysregulated cf-miRNAs [112–116], and confirmed that there are several cf-miRNAs that are able to distinguish CD from UC or IBD from healthy controls (Table 1). Among them, some of these cf-miRNAs identified by Paraskevi et al. were consistent with those identified by Wu et al. and two cf-miRNAs that can differentiate CD from UC, identified by Netz et al., seem to have a biological basis for their differential expression based on the literature and acquired knowledge. An increase of specific cf-miRNAs was identified also in pediatric CD patients, which significantly decreased after six months of treatment, suggesting their role as non-invasive biomarkers of disease state [117]. However, Jensen et al. did not confirm findings that cf-miRNA can differentiate CD patients from healthy controls. They identified six downregulated and three upregulated cf-miRNAs in patients with CD compared with controls. However, in their validation cohort, only one from these, has-miR-16, was significantly downregulated in CD patients, with an inadequate discriminative power [118]. Interesting findings came from Polytarchou et al., who not only identified differentially expressed cf-miRNAs in patients with UC compared with controls, but also detected four cf-miRNAs correlating with disease activity, which were found to have higher sensitivity and specificity values than C-reactive protein (CRP) [114]. Similar results were found during analysis of patients with CD, where 10 cf-miRNAs were differentially expressed in patients compared with controls; two of these cf-miRNAs also correlated with CD disease activity and exhibited higher correlation values compared with CRP. Furthermore, distinct miRNA signatures between CD patients with ileal and colonic involvement were also revealed [115]. A distinct signature was also observed in mouse models and consequently validated using sera from UC patients. Based on obtained results, it seems that such a signature could be an ideal biomarker for IBD, since it can distinguish individuals at risk, predict the type of inflammation and disease status in patients, and evaluate the response to therapeutics [119]. These findings point to increasing evidence that cf-miRNAs play a key role in the pathogenesis of IBD, as many specific cf-miRNAs were identified to have different expression within UC compared with CD or within active phase in comparison to inactive disease, and some of these cf-miRNAs were shown to better reflect mucosal inflammation than CRP (Table 1) [120–124].

Table 1. Summary of identified nuclear (genomic) miRNAs (cf-miRNAs) deregulated in IBD. CD: Crohn's disease, UC: ulcerative colitis.

cf-miRNA	Observed Change	Reference
miRs-199a-5p, miRs-362-3p, miRs-532-3p, miRplus-E1271	↑ in active CD patients	Wu et al., 2011 [111]
miRplus-F1065	↓ in active CD patients	Wu et al., 2011 [111]
miR-340*	↑ in CD patients	Wu et al., 2011 [111]
miR-149*	↓ in CD patients	Wu et al., 2011 [111]
miRs-28-5p, miRs-151-5p, miRs-199a-5p, miRs-340*, miRplus-E1271, miRs-3180-3p, miRplus-E1035, miRplus-F1159	↑ in active UC patients	Wu et al., 2011 [111]
miRs-103-2*, miRs-362-3p, miRs-532-3p,	↑ in UC patients	Wu et al., 2011 [111]
miR-505*	↓ in UC patients	Wu et al., 2011 [111]
miR-16, miR-23a, miR-29a, miR-106a, miR-107, miR-126, miR-191, miR-199a-5p, miR-200c, miR-362-3p, miR-532-3p	↑ in CD patients	Paraskevi et al., 2012 [113]
miR-16, miR-21, miR-28-5p, miR-151-5p, miR-155, miR-199a-5p	↑ in UC patients	Paraskevi et al., 2012 [113]
miRs-195, miR-16, miR-93, miR-140, miR-30e, miR-20a, miR-106a, miR-192, miR-21, miR-484, miR-let-7b	↑ in active CD patients	Zahm et al., 2011 [117]
miR-miRs-188-5p, miR-422a, miR-378, miR-500, miR-501-5p, miR-769-5p, miR-874	↑ in UC patients	Duttagupta et al., 2012 [112]

Table 1. *Cont.*

cf-miRNA	Observed Change	Reference
hsa-miR-369-3p, hsa-miR-376a, hsa-miR-376, hsa-miR-411#, hsa-miR-411, mmu-miR-379	↓ in CD patients	Jensen et al., 2015 [118]
hsa-miR-200c, hsa-miR-181-2 #, hsa-miR-125a-5p	↑ in CD patients	Jensen et al., 2015 [118]
miR-223a-3p, miR-23a-3p, miR-302-3p, miR-191-5p, miR-22-3p, miR-17-5p, miR-30e-5p, miR-148b-3p, miR-320e	↑ in UC patients	Polytarchou et al., 2015 [114]
miR-1827, miR-612, miR-188-5p	↓ in UC patients	Polytarchou et al., 2015 [114]
hsa-miR-1183, hsa-miR-1827, hsa-miR-1286, hsa-miR-504, hsa-miR-188-5p, hsa-miR-574-5p, hsa-miR-192-5p, hsa-miR-149-5p, and hsa-miR-378e	↓ in CD patients	Oikonomopoulos et al., 2016 [115]
hsa-miR-30e-5p	↑ in CD patients	Oikonomopoulos et al., 2016 [115]
miR-598, miR-642	↑ in UC patients	Netz et al., 2017 [116]
miR-595, miR-1246	↑ in active IBD	Krissansen et al., 2015 [120]
miR-223	↑ in IBD	Wang et al., 2016 [121]
miR-16, miR-21, miR-223	↑ in IBD, strongly in CD patients	Schonauen et al., 2017 [122]
miR-106a, miR-362-3p	↑ in IBD	Omidbakhsh et al., 2018 [123]
miR-146b-5p	↑ in IBD	Chen et al., 2019 [124]

10. Cell-Free lncRNA in IBD

LncRNAs represent molecules longer than 200 nucleotides, and are localized mainly in the nucleus, but also in the cytoplasm and in extracellular fluids. Their role in gene regulation includes the control of the flux of genetic information, such as chromosome structure modulation, transcription, splicing, mRNA stability, mRNA availability, and post-translational modifications. In addition, they present interaction domains for DNA, mRNAs, miRNAs, and proteins. The mechanism of action of lncRNAs is given by their cellular and temporal specificity. LncRNA also contribute to the mRNA and protein content in the cell by regulating adjacent protein-coding genes expression [125].

LncRNAs can be categorized into sense, antisense, intronic, bidirectional, and intergenic lncRNAs [126]. Their expression is lower than those of protein-coding genes and depends on tissue and development-stage characteristics, pointing to their regulatory role [127,128]. Therefore, they have been studied in association with various diseases, such as cancer, cardiovascular, and neurological diseases. To date, hundreds of lncRNAs have been shown to be differentially expressed in IBD patients in comparison with healthy controls, and in active versus inactive disease state [129–131]. In many of these genes, various single nucleotide polymorphisms were identified and were found to be associated with transcription binding factors, expression quantitative trait loci, or DNase peaks [132]. There is a growing body of evidence that cf-lncRNAs play a key role in the pathogenesis of various diseases, and could be used as non-invasive biomarkers [133–135]. Recently, deregulated lncRNAs in the plasma samples of CD patients were observed [136], and Wang et al. identified significantly upregulated cf-lncRNAs (KIF9-AS1, LINC01272) and significantly downregulated cf-lncRNA DIO3OS in IBD patients compared with healthy controls (Table 2) [137]. Therefore, cf-lncRNAs could be potentially used as non-invasive biomarkers also in IBD; however, further investigation and studies are needed.

Table 2. Summary of identified nuclear (genomic) long non-coding (cf-lncRNAs) deregulated in IBD.

cf-lncRNAs	Observed Change	Reference
KIF9-AS1, LINC01272	↑ in IBD	Wang et al 2018 [137]
DIO3OS	↓ in IBD	Wang et al 2018 [137]
GUSBP2, RP5-968D22.1, RP11-68L1.2, RP11-428F8.2, GAS5-AS1, RP11-923I11.5, DDX11-AS1, XLOC_005955, XLOC_005807, AC009133.20	↑ in CD	Chen et al., 2016 [136]
AF113016, ALOX12P2, AGSK1, CTC-338M12.3, AC064871.3, RP11-510H23.3, LOC729678, XLOC 010037, LOC283761, XLOC 013142	↓ in CD	Chen et al., 2016 [136]

11. Clinical Relevance of cfNAs in IBD Care

Beyond cfDNAs as therapeutic targets, which was discussed in previous sections, another potential utility of cfNAs in clinical care can be found in differential diagnostics of clinically not typical or borderline cases of IBD, in monitoring the subclinical phases of disease onset in patients having a high risk of IBD, or those in remission, and also in monitoring the effectiveness of therapy used in IBD patients. All of these are basically connected to the continuous non-invasive monitoring of disease activity from blood, stool, or salivary samples of IBD patients, or those at risk. Although yet not extensively covered in the literature for IBD, cfNAs as biomarkers of therapy response were described in diseases such as metastatic colorectal cancer [138], non-small cell lung cancer [139], minimal residual disease [140], or in organ transplant monitoring [141]. However, a high potential for IBD activity tracking, and therefore also for therapy monitoring, can be anticipated from several of the above-described findings, which are mainly based on the total amount of cfNAs or on the specific dynamics between different types and origins of cfNAs identified. The main findings are as follows (for citations, see the relevant paragraphs of this review): (1) elevated levels of total cfNAs during active IBD; (2) specific balance between exosomes having pro-inflammatory and anti-inflammatory contents; (3) presence, relative abundance, and specific content of NETs discriminating various types of IBD; (4) presence and relative abundance of exogenous cfNAs as a reaction to the impaired barrier function of the intestinal epithelium; and (5) higher amounts of human DNA in feces as a general outcome of intestinal inflammation and epithelial cell damage when IBD is active. Another possibility of therapeutic response monitoring can be hypothesized also through the detection of anti-inflammatory content (immunoregulatory proteins and cfNAs) of hookworm-released exosomes [103] in the stool of IBD patients undergoing helminthotherapy.

Specific cfNAs-based application, having extremely high value in the clinical care of IBD patients, can be the non-invasive monitoring of colitis-to-cancer transformation. Such application was shown to be feasible by the monitoring of cfDNA and its methylation level [80].

12. Conclusions

As we reviewed here, different types of cfNAs can be released to different body fluids by several regulated or unregulated processes under specific physiological and pathophysiological conditions. When considering the endogenous cfNAs in blood, these include "passive" processes of cell death (necrosis, mechanical damage, etc.), "active" processes of cell death (apoptosis, NETosis), and "active" processes of cell–cell communications (exosomes, microvesicles). In the case of exogenous cfNAs in circulation (from microbiome, parasites, or ingested food), on the other hand, these processes can include "active" transport (hijacking the physiological transport machineries of the host organism, or through the immunological sampling of antigens by dendritic cells) and "passive" transport (because of inefficient epithelial barrier functions resulting from pathology or injury). Other specific processes can take place in body fluids that are different from blood, such as lymph, cerebrospinal fluid, saliva, urine, or stool. In each case, they represent signals from the processes taking place in the organism as well as signals to make reactions of the organism happen. Knowing the exact meaning of these signals can be exploited to actively step in the relevant processes or passively observe these processes. In IBD, for example, this takes place through inhibiting pro-inflammatory signal generation/transduction, strengthening anti-inflammatory signal generation/transduction, or monitoring disease onset and therapeutic efficiency. Therefore, it is evident that cfNAs play important roles in various diseases, including IBD. These specific molecules can be used as potential markers of disease state, and can even differentiate between the active and inactive phases of the disease. Their analysis is relatively easy and cheap, given their presence in body fluids. Therefore, CfNAs represent a significant tool of liquid medicine, which is reaching clinical care. It promises a non-invasive, low-risk diagnostic technique for patients. However, a number of various cfNAs types are known; yet regarding their role in IBD, only a limited number of publications are available. Nevertheless, these represent a significant basis for future research, as they demonstrated that cfNAs can be a valuable tool to stratify patients and distinguish them from healthy people. Further research is needed in order to identify the overall pool of cfNAs, but it is equally, if not more important to characterize these molecules in terms of their information value, cellular and subcellular origin, molecular pathways, and biological aspects in general. By achieving this, they could contribute to the elucidation of the pathogenesis of IBD and to the development of new therapeutic strategies.

Author Contributions: Conceptualization, Writing—Original Draft Preparation, Investigation, Z.K.; Conceptualization, Writing—Review & Editing, Funding Acquisition, Supervision, J.R and R.G.

References

1. Molodecky, N.A.; Soon, I.S.; Rabi, D.M.; Ghali, W.A.; Ferris, M.; Chernoff, G.; Benchimol, E.I.; Panaccione, R.; Ghosh, S.; Barkema, H.W.; et al. Increasing incidence and prevalence of the inflammatory bowel diseases with time, based on systematic review. *Gastroenterology* **2012**, *142*, 46–54. [CrossRef] [PubMed]
2. Kim, D.H.; Cheon, J.H. Pathogenesis of Inflammatory Bowel Disease and Recent Advances in Biologic Therapies. *Immune Netw.* **2017**, *17*, 25–40. [CrossRef] [PubMed]
3. Colombel, J.F.; Sandborn, W.J.; Reinisch, W.; Mantzaris, G.J.; Kornbluth, A.; Rachmilewitz, D.; Lichtiger, S.; D'Haens, G.; Diamond, R.H.; Broussard, D.L.; et al. Infliximab, azathioprine, or combination therapy for Crohn's disease. *N. Engl. J. Med.* **2010**, *362*, 1383–1395. [CrossRef] [PubMed]
4. Fakhoury, M.; Negrulj, R.; Mooranian, A.; Al-Salami, H. Inflammatory bowel disease: Clinical aspects and treatments. *J. Inflamm. Res.* **2014**, *7*, 113–120. [CrossRef] [PubMed]
5. Moustafa, A.; Li, W.; Anderson, E.L.; Wong, E.H.M.; Dulai, P.S.; Sandborn, W.J.; Biggs, W.; Yooseph, S.; Jones, M.B.; Venter, J.C.; et al. Genetic risk, dysbiosis, and treatment stratification using host genome and gut microbiome in inflammatory bowel disease. *Clin. Transl. Gastroenterol.* **2018**, *9*, e132. [CrossRef] [PubMed]
6. Mandel, P.; Metais, P. Les acides nucléiques du plasma sanguin chez l'homme. *C R Seances Soc. Biol. Fil.* **1948**, *142*, 241–243. [PubMed]
7. Mansour, H. Cell-free nucleic acids as noninvasive biomarkers for colorectal cancer detection. *Front. Genet.* **2014**, *5*, 182. [CrossRef] [PubMed]
8. Steinman, C.R. Free DNA in serum and plasma from normal adults. *J. Clin. Invest.* **1975**, *56*, 512–515. [CrossRef] [PubMed]
9. Tan, E.M.; Schur, P.H.; Carr, R.I.; Kunkel, H.G. Deoxybonucleic acid (DNA) and antibodies to DNA in the serum of patients with systemic lupus erythematosus. *J. Clin. Invest.* **1966**, *45*, 1732–1740. [CrossRef]
10. Leon, S.A.; Green, A.; Yaros, M.J.; Shapiro, B. Radioimmunoassay for nanogram quantities of DNA. *J. Immunol. Methods* **1975**, *9*, 157–164. [CrossRef]
11. Lo, Y.M.; Corbetta, N.; Chamberlain, P.F.; Rai, V.; Sargent, I.L.; Redman, C.W.; Wainscoat, J.S. Presence of fetal DNA in maternal plasma and serum. *Lancet* **1997**, *350*, 485–487. [CrossRef]
12. Pos, O.; Biro, O.; Szemes, T.; Nagy, B. Circulating cell-free nucleic acids: Characteristics and applications. *J. Immunol.* **2018**, *26*, 937–945.
13. Butt, A.N.; Swaminathan, R. Overview of circulating nucleic acids in plasma/serum. *Ann. N. Y. Acad. Sci.* **2008**, *1137*, 236–242. [CrossRef] [PubMed]

14. Gravina, S.; Sedivy, J.M.; Vijg, J. The dark side of circulating nucleic acids. *Aging Cell* **2016**, *15*, 398–399. [CrossRef]
15. Fritz, J.V.; Heintz-Buschart, A.; Ghosal, A.; Wampach, L.; Etheridge, A.; Galas, D.; Wilmes, P. Sources and Functions of Extracellular Small RNAs in Human Circulation. *Annu. Rev. Nutr.* **2016**, *36*, 301–336. [CrossRef]
16. Ventura, W.; Nazario-Redondo, C.; Sekizawa, A. Non-invasive prenatal diagnosis from the perspective of a low-resource country. *Int. J. Gynaecol Obstet.* **2013**, *122*, 270–273. [CrossRef]
17. Botezatu, I.; Serdyuk, O.; Potapova, G.; Shelepov, V.; Alechina, R.; Molyaka, Y.; Ananev, V.; Bazin, I.; Garin, A.; Narimanov, M.; et al. Genetic analysis of DNA excreted in urine: A new approach for detecting specific genomic DNA sequences from cells dying in an organism. *Clin. Chem.* **2000**, *46*, 1078–1084.
18. Burnham, P.; Kim, M.S.; Agbor-Enoh, S.; Luikart, H.; Valantine, H.A.; Khush, K.K.; De Vlaminck, I. Single-stranded DNA library preparation uncovers the origin and diversity of ultrashort cell-free DNA in plasma. *Sci. Rep.* **2016**, *6*, 27859. [CrossRef]
19. Lui, Y.Y.; Chik, K.W.; Chiu, R.W.; Ho, C.Y.; Lam, C.W.; Lo, Y.M. Predominant hematopoietic origin of cell-free DNA in plasma and serum after sex-mismatched bone marrow transplantation. *Clin. Chem.* **2002**, *48*, 421–427.
20. Jahr, S.; Hentze, H.; Englisch, S.; Hardt, D.; Fackelmayer, F.O.; Hesch, R.D.; Knippers, R. DNA fragments in the blood plasma of cancer patients: Quantitations and evidence for their origin from apoptotic and necrotic cells. *Cancer Res.* **2001**, *61*, 1659–1665.
21. Fuchs, T.A.; Kremer Hovinga, J.A.; Schatzberg, D.; Wagner, D.D.; Lammle, B. Circulating DNA and myeloperoxidase indicate disease activity in patients with thrombotic microangiopathies. *Blood* **2012**, *120*, 1157–1164. [CrossRef] [PubMed]
22. Mouliere, F.; Thierry, A.R. The importance of examining the proportion of circulating DNA originating from tumor, microenvironment and normal cells in colorectal cancer patients. *Expert Opin. Biol. Ther.* **2012**, *12* (Suppl. 1), S209–S215. [CrossRef]
23. Souza, M.F.; Kuasne, H.; Barros-Filho, M.C.; Ciliao, H.L.; Marchi, F.A.; Fuganti, P.E.; Paschoal, A.R.; Rogatto, S.R.; Colus, I.M.S. Circulating mRNAs and miRNAs as candidate markers for the diagnosis and prognosis of prostate cancer. *PLoS ONE* **2017**, *12*, e0184094. [CrossRef] [PubMed]
24. Tzimagiorgis, G.; Michailidou, E.Z.; Kritis, A.; Markopoulos, A.K.; Kouidou, S. Recovering circulating extracellular or cell-free RNA from bodily fluids. *Cancer Epidemiol.* **2011**, *35*, 580–589. [CrossRef] [PubMed]
25. El Andaloussi, S.; Mager, I.; Breakefield, X.O.; Wood, M.J. Extracellular vesicles: Biology and emerging therapeutic opportunities. *Nat. Rev. Drug Discov.* **2013**, *12*, 347–357. [CrossRef] [PubMed]
26. Baghaei, K.; Tokhanbigli, S.; Asadzadeh, H.; Nmaki, S.; Reza Zali, M.; Hashemi, S.M. Exosomes as a novel cell-free therapeutic approach in gastrointestinal diseases. *J. Cell Physiol.* **2019**, *234*, 9910–9926. [CrossRef] [PubMed]
27. Zhang, H.; Wang, L.; Li, C.; Yu, Y.; Yi, Y.; Wang, J.; Chen, D. Exosome-Induced Regulation in Inflammatory Bowel Disease. *Front. Immunol.* **2019**, *10*, 1464. [CrossRef] [PubMed]
28. Li, X.; Corbett, A.L. Challenges and opportunities in exosome research-Perspectives from biology, engineering, and cancer therapy. *APL Bioeng.* **2019**, *3*, 011503. [CrossRef] [PubMed]
29. Tran, T.H.; Mattheolabakis, G.; Aldawsari, H.; Amiji, M. Exosomes as nanocarriers for immunotherapy of cancer and inflammatory diseases. *Clin. Immunol.* **2015**, *160*, 46–58. [CrossRef] [PubMed]
30. Jiang, L.; Shen, Y.; Guo, D.; Yang, D.; Liu, J.; Fei, X.; Yang, Y.; Zhang, B.; Lin, Z.; Yang, F.; et al. EpCAM-dependent extracellular vesicles from intestinal epithelial cells maintain intestinal tract immune balance. *Nat. Commun.* **2016**, *7*, 13045. [CrossRef] [PubMed]
31. Mitsuhashi, S.; Feldbrugge, L.; Csizmadia, E.; Mitsuhashi, M.; Robson, S.C.; Moss, A.C. Luminal Extracellular Vesicles (EVs) in Inflammatory Bowel Disease (IBD) Exhibit Proinflammatory Effects on Epithelial Cells and Macrophages. *Inflamm. Bowel Dis.* **2016**, *22*, 1587–1595. [CrossRef] [PubMed]
32. Leoni, G.; Neumann, P.A.; Kamaly, N.; Quiros, M.; Nishio, H.; Jones, H.R.; Sumagin, R.; Hilgarth, R.S.; Alam, A.; Fredman, G.; et al. Annexin A1-containing extracellular vesicles and polymeric nanoparticles promote epithelial wound repair. *J. Clin. Invest.* **2015**, *125*, 1215–1227. [CrossRef] [PubMed]
33. Zheng, X.; Chen, F.; Zhang, Q.; Liu, Y.; You, P.; Sun, S.; Lin, J. Salivary exosomal PSMA7: A promising biomarker of inflammatory bowel disease. *Protein Cell* **2017**, *8*, 686–695. [CrossRef] [PubMed]
34. Console, L.; Scalise, M.; Indiveri, C. Exosomes in inflammation and role as biomarkers. *Clin. Chim. Acta* **2019**, *488*, 165–171. [CrossRef]

35. Wong, W.Y.; Lee, M.M.; Chan, B.D.; Kam, R.K.; Zhang, G.; Lu, A.P.; Tai, W.C. Proteomic profiling of dextran sulfate sodium induced acute ulcerative colitis mice serum exosomes and their immunomodulatory impact on macrophages. *Proteomics* **2016**, *16*, 1131–1145. [CrossRef] [PubMed]
36. Carriere, J.; Bretin, A.; Darfeuille-Michaud, A.; Barnich, N.; Nguyen, H.T. Exosomes Released from Cells Infected with Crohn's Disease-associated Adherent-Invasive Escherichia coli Activate Host Innate Immune Responses and Enhance Bacterial Intracellular Replication. *Inflamm. Bowel Dis.* **2016**, *22*, 516–528. [CrossRef] [PubMed]
37. Zhang, X.; Deeke, S.A.; Ning, Z.; Starr, A.E.; Butcher, J.; Li, J.; Mayne, J.; Cheng, K.; Liao, B.; Li, L.; et al. Metaproteomics reveals associations between microbiome and intestinal extracellular vesicle proteins in pediatric inflammatory bowel disease. *Nat. Commun.* **2018**, *9*, 2873. [CrossRef] [PubMed]
38. Mao, F.; Wu, Y.; Tang, X.; Kang, J.; Zhang, B.; Yan, Y.; Qian, H. Exosomes Derived from Human Umbilical Cord Mesenchymal Stem Cells Relieve Inflammatory Bowel Disease in Mice. *Biomed Res. Int.* **2017**, *2017*, 5356760. [CrossRef] [PubMed]
39. Wu, Y.; Qiu, W.; Xu, X.; Kang, J.; Wang, J.; Wen, Y.; Tang, X.; Yan, Y.; Qian, H.; Zhang, X.; et al. Exosomes derived from human umbilical cord mesenchymal stem cells alleviate inflammatory bowel disease in mice through ubiquitination. *Biomed. Res. Int.* **2018**, *10*, 2026–2036.
40. Zhang, Y.; Liu, Y.; Liu, H.; Tang, W.H. Exosomes: Biogenesis, biologic function and clinical potential. *Cell Biosci.* **2019**, *9*, 19. [CrossRef]
41. Bonaventura, A.; Liberale, L.; Carbone, F.; Vecchie, A.; Diaz-Canestro, C.; Camici, G.G.; Montecucco, F.; Dallegri, F. The Pathophysiological Role of Neutrophil Extracellular Traps in Inflammatory Diseases. *Thromb. Haemost.* **2018**, *118*, 6–27. [CrossRef] [PubMed]
42. Takei, H.; Araki, A.; Watanabe, H.; Ichinose, A.; Sendo, F. Rapid killing of human neutrophils by the potent activator phorbol 12-myristate 13-acetate (PMA) accompanied by changes different from typical apoptosis or necrosis. *J. Leukoc. Biol.* **1996**, *59*, 229–240. [CrossRef] [PubMed]
43. Brinkmann, V.; Reichard, U.; Goosmann, C.; Fauler, B.; Uhlemann, Y.; Weiss, D.S.; Weinrauch, Y.; Zychlinsky, A. Neutrophil extracellular traps kill bacteria. *Science* **2004**, *303*, 1532–1535. [CrossRef] [PubMed]
44. Delgado-Rizo, V.; Martinez-Guzman, M.A.; Iniguez-Gutierrez, L.; Garcia-Orozco, A.; Alvarado-Navarro, A.; Fafutis-Morris, M. Neutrophil Extracellular Traps and Its Implications in Inflammation: An Overview. *Front. Immunol.* **2017**, *8*, 81. [CrossRef] [PubMed]
45. Bennike, T.B.; Carlsen, T.G.; Ellingsen, T.; Bonderup, O.K.; Glerup, H.; Bogsted, M.; Christiansen, G.; Birkelund, S.; Stensballe, A.; Andersen, V. Neutrophil Extracellular Traps in Ulcerative Colitis: A Proteome Analysis of Intestinal Biopsies. *Inflamm. Bowel Dis.* **2015**, *21*, 2052–2067. [CrossRef]
46. Dinallo, V.; Marafini, I.; Di Fusco, D.; Laudisi, F.; Franze, E.; Di Grazia, A.; Figliuzzi, M.M.; Caprioli, F.; Stolfi, C.; Monteleone, I.; et al. Neutrophil Extracellular Traps Sustain Inflammatory Signals in Ulcerative Colitis. *J. Crohns. Colitis* **2019**, *13*, 772–784. [CrossRef] [PubMed]
47. Gottlieb, Y.; Elhasid, R.; Berger-Achituv, S.; Brazowski, E.; Yerushalmy-Feler, A.; Cohen, S. Neutrophil extracellular traps in pediatric inflammatory bowel disease. *Pathol. Int.* **2018**, *68*, 517–523. [CrossRef]
48. Vong, L.; Lorentz, R.J.; Assa, A.; Glogauer, M.; Sherman, P.M. Probiotic Lactobacillus rhamnosus inhibits the formation of neutrophil extracellular traps. *J. Immunol.* **2014**, *192*, 1870–1877. [CrossRef]
49. Lehmann, T.; Schallert, K.; Vilchez-Vargas, R.; Benndorf, D.; Puttker, S.; Sydor, S.; Schulz, C.; Bechmann, L.; Canbay, A.; Heidrich, B.; et al. Metaproteomics of fecal samples of Crohn's disease and Ulcerative Colitis. *J. Proteomics* **2019**, *201*, 93–103. [CrossRef]
50. Honda, M.; Kubes, P. Neutrophils and neutrophil extracellular traps in the liver and gastrointestinal system. *Nat. Rev. Gastroenterol. Hepatol.* **2018**, *15*, 206–221. [CrossRef]
51. Sangaletti, S.; Tripodo, C.; Chiodoni, C.; Guarnotta, C.; Cappetti, B.; Casalini, P.; Piconese, S.; Parenza, M.; Guiducci, C.; Vitali, C.; et al. Neutrophil extracellular traps mediate transfer of cytoplasmic neutrophil antigens to myeloid dendritic cells toward ANCA induction and associated autoimmunity. *Blood* **2012**, *120*, 3007–3018. [CrossRef] [PubMed]
52. Sugi, K.; Saitoh, O.; Matsuse, R.; Tabata, K.; Uchida, K.; Kojima, K.; Nakagawa, K.; Tanaka, S.; Teranishi, T.; Hirata, I.; et al. Antineutrophil cytoplasmic antibodies in Japanese patients with inflammatory bowel disease: Prevalence and recognition of putative antigens. *Am. J. Gastroenterol.* **1999**, *94*, 1304–1312. [CrossRef] [PubMed]

53. Mahler, M.; Bogdanos, D.P.; Pavlidis, P.; Fritzler, M.J.; Csernok, E.; Damoiseaux, J.; Bentow, C.; Shums, Z.; Forbes, A.; Norman, G.L. PR3-ANCA: A promising biomarker for ulcerative colitis with extensive disease. *Clin. Chim. Acta* **2013**, *424*, 267–273. [CrossRef] [PubMed]
54. Dinallo, V.; Di Fusco, D.; Laudisi, F.; Di Grazia, A.; Franze, E.; Marafini, I.; Troncone, E.; Monteleone, I.; Monteleone, G. pad4 and neutrophil extracellular traps are increased in the inflamed colon mucosa of patients with ulcerative colitis. *Digestive and Liver Disease* **2017**, *49*, E117. [CrossRef]
55. He, Z.; Si, Y.; Jiang, T.; Ma, R.; Zhang, Y.; Cao, M.; Li, T.; Yao, Z.; Zhao, L.; Fang, S.; et al. Phosphotidylserine exposure and neutrophil extracellular traps enhance procoagulant activity in patients with inflammatory bowel disease. *Thromb. Haemost.* **2016**, *115*, 738–751. [PubMed]
56. Angelidou, I.; Chrysanthopoulou, A.; Mitsios, A.; Arelaki, S.; Arampatzioglou, A.; Kambas, K.; Ritis, D. REDD1/Autophagy Pathway Is Associated with Neutrophil-Driven IL-1beta Inflammatory Response in Active Ulcerative Colitis. *J. Immunol.* **2018**, *200*, 3950–3961. [CrossRef]
57. Ahmed, A.I.; Soliman, R.A.; Samir, S. Cell Free DNA and Procalcitonin as Early Markers of Complications in ICU Patients with Multiple Trauma and Major Surgery. *Clin. Lab.* **2016**, *62*, 2395–2404. [CrossRef] [PubMed]
58. Clementi, A.; Virzi, G.M.; Brocca, A.; Pastori, S.; de Cal, M.; Marcante, S.; Granata, A.; Ronco, C. The Role of Cell-Free Plasma DNA in Critically Ill Patients with Sepsis. *Blood Purif.* **2016**, *41*, 34–40. [CrossRef]
59. Sharma, S.; Fitzgerald, K.A. Innate immune sensing of DNA. *PLoS Pathog.* **2011**, *7*, e1001310. [CrossRef]
60. Rose, W.A., 2nd; Sakamoto, K.; Leifer, C.A. TLR9 is important for protection against intestinal damage and for intestinal repair. *Sci Rep.* **2012**, *2*, 574.
61. Hu, S.; Peng, L.; Kwak, Y.T.; Tekippe, E.M.; Pasare, C.; Malter, J.S.; Hooper, L.V.; Zaki, M.H. The DNA Sensor AIM2 Maintains Intestinal Homeostasis via Regulation of Epithelial Antimicrobial Host Defense. *Cell Rep.* **2015**, *13*, 1922–1936. [CrossRef]
62. Rachmilewitz, D.; Katakura, K.; Karmeli, F.; Hayashi, T.; Reinus, C.; Rudensky, B.; Akira, S.; Takeda, K.; Lee, J.; Takabayashi, K.; et al. Toll-like receptor 9 signaling mediates the anti-inflammatory effects of probiotics in murine experimental colitis. *Gastroenterology* **2004**, *126*, 520–528. [CrossRef]
63. Obermeier, F.; Dunger, N.; Deml, L.; Herfarth, H.; Scholmerich, J.; Falk, W. CpG motifs of bacterial DNA exacerbate colitis of dextran sulfate sodium-treated mice. *Eur. J. Immunol.* **2002**, *32*, 2084–2092. [CrossRef]
64. Sipos, F.; Muzes, G.; Furi, I.; Spisak, S.; Wichmann, B.; Germann, T.M.; Constantinovits, M.; Krenacs, T.; Tulassay, Z.; Molnar, B. Intravenous administration of a single-dose free-circulating DNA of colitic origin improves severe murine DSS-colitis. *Pathol. Oncol. Res.* **2014**, *20*, 867–877. [CrossRef] [PubMed]
65. Muzes, G.; Sipos, F.; Furi, I.; Constantinovits, M.; Spisak, S.; Wichmann, B.; Valcz, G.; Tulassay, Z.; Molnar, B. Preconditioning with intravenous colitic cell-free DNA prevents DSS-colitis by altering TLR9-associated gene expression profile. *Dig. Dis. Sci.* **2014**, *59*, 2935–2946. [CrossRef] [PubMed]
66. Delgado, M.A.; Elmaoued, R.A.; Davis, A.S.; Kyei, G.; Deretic, V. Toll-like receptors control autophagy. *Embo. J.* **2008**, *27*, 1110–1121. [CrossRef] [PubMed]
67. Kaplan, M.J.; Radic, M. Neutrophil extracellular traps: Double-edged swords of innate immunity. *J. Immunol.* **2012**, *189*, 2689–2695. [CrossRef]
68. Muzes, G.; Kiss, A.L.; Tulassay, Z.; Sipos, F. Cell-free DNA-induced alteration of autophagy response and TLR9-signaling: Their relation to amelioration of DSS-colitis. *Comp. Immunol. Microbiol. Infect. Dis.* **2017**, *52*, 48–57. [CrossRef]
69. Rauh, P.; Rickes, S.; Fleischhacker, M. Microsatellite alterations in free-circulating serum DNA in patients with ulcerative colitis. *Dig. Dis.* **2003**, *21*, 363–366. [CrossRef]
70. Koike, Y.; Uchida, K.; Tanaka, K.; Ide, S.; Otake, K.; Okita, Y.; Inoue, M.; Araki, T.; Mizoguchi, A.; Kusunoki, M. Dynamic pathology for circulating free DNA in a dextran sodium sulfate colitis mouse model. *Pediatr. Surg. Int.* **2014**, *30*, 1199–1206. [CrossRef]
71. Maronek, M.; Gromova, B.; Liptak, R.; Klimova, D.; Cechova, B.; Gardlik, R. Extracellular DNA is Increased in Dextran Sulphate Sodium-Induced Colitis in Mice. *Folia. Biol. (Praha.)* **2018**, *64*, 167–172. [PubMed]
72. Babickova, J.; Conka, J.; Janovicova, L.; Boris, M.; Konecna, B.; Gardlik, R. Extracellular DNA as a Prognostic and Therapeutic Target in Mouse Colitis under DNase I Treatment. *Folia. Biol. (Praha.)* **2018**, *64*, 10–15. [PubMed]
73. Gardlik, R.; Maronek, M.; Liptak, R.; Gromova, B.; Cechova, B. THE ROLE OF EXTRACELLULAR DNA IN THE PATHOGENESIS OF DSS-INDUCED COLITIS IN MICE. *Gastroenterology* **2019**, *156*, S712. [CrossRef]
74. Vincent, C.; Mehrotra, S. Excretion of Host DNA in Feces Is Associated with Risk of Clostridium difficile Infection. *J. Immunol. Res.* **2015**, *2015*, 246203. [CrossRef] [PubMed]

75. Casellas, F.; Antolin, M.; Varela, E.; Garcia-Lafuente, A.; Guarner, F.; Borruel, N.; Armengol Miro, J.R.; Malagelada, J.R. Fecal excretion of human deoxyribonucleic acid as an index of inflammatory activity in ulcerative colitis. *Clin. Gastroenterol. Hepatol.* **2004**, *2*, 683–689. [CrossRef]
76. Casellas, F.; Borruel, N.; Antolin, M.; Varela, E.; Torrejon, A.; Armadans, L.; Guarner, F.; Malagelada, J.R. Fecal excretion of deoxyribonucleic acid in long-term follow-up of patients with inactive ulcerative colitis. *Inflamm. Bowel Dis.* **2007**, *13*, 386–390. [CrossRef] [PubMed]
77. Xia, M.; Liu, J.; Wu, X.; Liu, S.; Li, G.; Han, C.; Song, L.; Li, Z.; Wang, Q.; Wang, J.; et al. Histone methyltransferase Ash1l suppresses interleukin-6 production and inflammatory autoimmune diseases by inducing the ubiquitin-editing enzyme A20. *Immunity* **2013**, *39*, 470–481. [CrossRef] [PubMed]
78. Axelrad, J.E.; Lichtiger, S.; Yajnik, V. Inflammatory bowel disease and cancer: The role of inflammation, immunosuppression, and cancer treatment. *World J. Gastroenterol.* **2016**, *22*, 4794–4801. [CrossRef]
79. Lehmann-Werman, R.; Zick, A.; Paweletz, C.; Welch, M.; Hubert, A.; Maoz, M.; Davidy, T.; Magenheim, J.; Piyanzin, S.; Neiman, D.; et al. Specific detection of cell-free DNA derived from intestinal epithelial cells using methylation patterns. *BioRxiv.* **2018**, 409219.
80. Bai, X.; Zhu, Y.; Pu, W.; Xiao, L.; Li, K.; Xing, C.; Jin, Y. Circulating DNA and its methylation level in inflammatory bowel disease and related colon cancer. *Int J. Clin Exp. Pathol.* **2015**, *8*, 13764–13769.
81. Sugiura, A.; McLelland, G.L.; Fon, E.A.; McBride, H.M. A new pathway for mitochondrial quality control: Mitochondrial-derived vesicles. *Embo J.* **2014**, *33*, 2142–2156. [CrossRef] [PubMed]
82. Dyall, S.D.; Brown, M.T.; Johnson, P.J. Ancient invasions: From endosymbionts to organelles. *Science* **2004**, *304*, 253–257. [CrossRef] [PubMed]
83. Zhang, Q.; Raoof, M.; Chen, Y.; Sumi, Y.; Sursal, T.; Junger, W.; Brohi, K.; Itagaki, K.; Hauser, C.J. Circulating mitochondrial DAMPs cause inflammatory responses to injury. *Nature* **2010**, *464*, 104–107. [CrossRef] [PubMed]
84. Boyapati, R.K.; Rossi, A.G.; Satsangi, J.; Ho, G.T. Gut mucosal DAMPs in IBD: From mechanisms to therapeutic implications. *Mucosal Immunol.* **2016**, *9*, 567–582. [CrossRef] [PubMed]
85. Shimada, K.; Crother, T.R.; Karlin, J.; Dagvadorj, J.; Chiba, N.; Chen, S.; Ramanujan, V.K.; Wolf, A.J.; Vergnes, L.; Ojcius, D.M.; et al. Oxidized mitochondrial DNA activates the NLRP3 inflammasome during apoptosis. *Immunity* **2012**, *36*, 401–414. [CrossRef] [PubMed]
86. Boyapati, R.; Dorward, D.; Tamborska, A.; Kalla, R.; Ventham, N.; Rossi, A.; Satsangi, J.; Ho, G.T. Mitochondrial DNA is a damage-associated molecular pattern released during active IBD promoting TLR9-mediated inflammation. *J. Crohns Colitis* **2017**, *11*, S105–S106. [CrossRef]
87. Alvarado-Vasquez, N. Circulating cell-free mitochondrial DNA as the probable inducer of early endothelial dysfunction in the prediabetic patient. *Exp. Gerontol.* **2015**, *69*, 70–78. [CrossRef]
88. Li, L.; Hann, H.W.; Wan, S.; Hann, R.S.; Wang, C.; Lai, Y.; Ye, X.; Evans, A.; Myers, R.E.; Ye, Z.; et al. Cell-free circulating mitochondrial DNA content and risk of hepatocellular carcinoma in patients with chronic HBV infection. *Sci. Rep.* **2016**, *6*, 23992. [CrossRef]
89. Liu, J.; Zou, Y.; Tang, Y.; Xi, M.; Xie, L.; Zhang, Q.; Gong, J. Circulating cell-free mitochondrial deoxyribonucleic acid is increased in coronary heart disease patients with diabetes mellitus. *J. Diabetes Investig.* **2016**, *7*, 109–114. [CrossRef]
90. Caielli, S.; Athale, S.; Domic, B.; Murat, E.; Chandra, M.; Banchereau, R.; Baisch, J.; Phelps, K.; Clayton, S.; Gong, M.; et al. Oxidized mitochondrial nucleoids released by neutrophils drive type I interferon production in human lupus. *J. Exp. Med.* **2016**, *213*, 697–713. [CrossRef]
91. Boyapati, R.K.; Tamborska, A.; Dorward, D.A.; Ho, G.T. Advances in the understanding of mitochondrial DNA as a pathogenic factor in inflammatory diseases. *F1000Res.* **2017**, *6*, 169. [CrossRef] [PubMed]
92. Boyapati, R.K.; Dorward, D.A.; Tamborska, A.; Kalla, R.; Ventham, N.T.; Doherty, M.K.; Whitfield, P.D.; Gray, M.; Loane, J.; Rossi, A.G.; et al. Mitochondrial DNA Is a Pro-Inflammatory Damage-Associated Molecular Pattern Released During Active IBD. *J. Clin. Lab. Anal.* **2018**, *24*, 2113–2122. [CrossRef] [PubMed]
93. Kreuder, V.; Dieckhoff, J.; Sittig, M.; Mannherz, H.G. Isolation, characterisation and crystallization of deoxyribonuclease I from bovine and rat parotid gland and its interaction with rabbit skeletal muscle actin. *Eur. J. Biochem.* **1984**, *139*, 389–400. [CrossRef] [PubMed]
94. Napirei, M.; Wulf, S.; Mannherz, H.G. Chromatin breakdown during necrosis by serum Dnase1 and the plasminogen system. *Arthritis Rheum.* **2004**, *50*, 1873–1883. [CrossRef] [PubMed]
95. Martinez Valle, F.; Balada, E.; Ordi-Ros, J.; Vilardell-Tarres, M. DNase 1 and systemic lupus erythematosus. *Autoimmun Rev.* **2008**, *7*, 359–363. [CrossRef] [PubMed]

96. Yang, C.; Chilvers, M.; Montgomery, M.; Nolan, S.J. Dornase alfa for cystic fibrosis. *Cochrane Database Syst. Rev.* **2016**, *4*, Cd001127. [CrossRef] [PubMed]
97. Malickova, K.; Duricova, D.; Bortlik, M.; Hruskova, Z.; Svobodova, B.; Machkova, N.; Komarek, V.; Fucikova, T.; Janatkova, I.; Zima, T.; et al. Impaired deoxyribonuclease I activity in patients with inflammatory bowel diseases. *Autoimmune Dis.* **2011**, *2011*, 945861.
98. Laukova, L.; Konecna, B.; Babickova, J.; Wagnerova, A.; Meliskova, V.; Vlkova, B.; Celec, P. Exogenous deoxyribonuclease has a protective effect in a mouse model of sepsis. *Biomed. Pharmacother.* **2017**, *93*, 8–16. [CrossRef]
99. Jimenez-Alcazar, M.; Rangaswamy, C. Host DNases prevent vascular occlusion by neutrophil extracellular traps. *Science* **2017**, *358*, 1202–1206. [CrossRef]
100. Vokalova, L.; Laukova, L.; Conka, J.; Meliskova, V.; Borbelyova, V.; Babickova, J.; Tothova, L.; Hodosy, J.; Vlkova, B.; Celec, P. Deoxyribonuclease partially ameliorates thioacetamide-induced hepatorenal injury. *Am. J. Physiol Gastrointest Liver Physiol.* **2017**, *312*, G457–g463. [CrossRef]
101. Park, J.; Wysocki, R.W.; Amoozgar, Z.; Maiorino, L.; Fein, M.R.; Jorns, J.; Schott, A.F.; Kinugasa-Katayama, Y.; Lee, Y.; Won, N.H.; et al. Cancer cells induce metastasis-supporting neutrophil extracellular DNA traps. *Sci. Transl. Med.* **2016**, *8*, 361ra138. [CrossRef] [PubMed]
102. Jeppesen, D.K.; Fenix, A.M.; Franklin, J.L.; Higginbotham, J.N.; Zhang, Q.; Zimmerman, L.J.; Liebler, D.C.; Ping, J.; Liu, Q.; Evans, R.; et al. Reassessment of Exosome Composition. *Cell* **2019**, *177*, 428–445.e18. [CrossRef] [PubMed]
103. Eichenberger, R.M.; Ryan, S.; Jones, L.; Buitrago, G.; Polster, R.; Montes de Oca, M.; Zuvelek, J.; Giacomin, P.R.; Dent, L.A.; Engwerda, C.R.; et al. Hookworm Secreted Extracellular Vesicles Interact with Host Cells and Prevent Inducible Colitis in Mice. *J. Cell Physiol.* **2018**, *9*, 850. [CrossRef] [PubMed]
104. Bartel, D.P. MicroRNAs: Genomics, biogenesis, mechanism, and function. *Cell* **2004**, *116*, 281–297. [CrossRef]
105. Schickel, R.; Boyerinas, B.; Park, S.M.; Peter, M.E. MicroRNAs: Key players in the immune system, differentiation, tumorigenesis and cell death. *Oncogene* **2008**, *27*, 5959–5974. [CrossRef] [PubMed]
106. Catalanotto, C.; Cogoni, C.; Zardo, G. MicroRNA in Control of Gene Expression: An Overview of Nuclear Functions. *Int. J. Mol. Sci.* **2016**, *17*. [CrossRef]
107. Coskun, M.; Bjerrum, J.T.; Seidelin, J.B.; Nielsen, O.H. MicroRNAs in inflammatory bowel disease–pathogenesis, diagnostics and therapeutics. *World J. Gastroenterol.* **2012**, *18*, 4629–4634. [CrossRef]
108. Gallo, A.; Tandon, M.; Alevizos, I.; Illei, G.G. The majority of microRNAs detectable in serum and saliva is concentrated in exosomes. *PLoS ONE* **2012**, *7*, e30679. [CrossRef]
109. Weber, J.A.; Baxter, D.H.; Zhang, S.; Huang, D.Y.; Huang, K.H.; Lee, M.J.; Galas, D.J.; Wang, K. The microRNA spectrum in 12 body fluids. *Clin. Chem.* **2010**, *56*, 1733–1741. [CrossRef]
110. Mi, S.; Zhang, J.; Zhang, W.; Huang, R.S. Circulating microRNAs as biomarkers for inflammatory diseases. *Microrna* **2013**, *2*, 63–71. [CrossRef]
111. Wu, F.; Guo, N.J.; Tian, H.; Marohn, M.; Gearhart, S.; Bayless, T.M.; Brant, S.R.; Kwon, J.H. Peripheral blood microRNAs distinguish active ulcerative colitis and Crohn's disease. *Inflamm. Bowel Dis.* **2011**, *17*, 241–250. [CrossRef] [PubMed]
112. Duttagupta, R.; DiRienzo, S.; Jiang, R.; Bowers, J.; Gollub, J.; Kao, J.; Kearney, K.; Rudolph, D.; Dawany, N.B.; Showe, M.K.; et al. Genome-wide maps of circulating miRNA biomarkers for ulcerative colitis. *PLoS ONE* **2012**, *7*, e31241. [CrossRef] [PubMed]
113. Paraskevi, A.; Theodoropoulos, G.; Papaconstantinou, I.; Mantzaris, G.; Nikiteas, N.; Gazouli, M. Circulating MicroRNA in inflammatory bowel disease. *J. Crohns Colitis* **2012**, *6*, 900–904. [CrossRef] [PubMed]
114. Polytarchou, C.; Oikonomopoulos, A.; Mahurkar, S.; Touroutoglou, A.; Koukos, G.; Hommes, D.W.; Iliopoulos, D. Assessment of Circulating MicroRNAs for the Diagnosis and Disease Activity Evaluation in Patients with Ulcerative Colitis by Using the Nanostring Technology. *Inflamm. Bowel Dis.* **2015**, *21*, 2533–2539. [CrossRef]
115. Oikonomopoulos, A.; Polytarchou, C.; Joshi, S.; Hommes, D.W.; Iliopoulos, D. Identification of Circulating MicroRNA Signatures in Crohn's Disease Using the Nanostring nCounter Technology. *Inflamm. Bowel Dis.* **2016**, *22*, 2063–2069. [CrossRef]
116. Netz, U.; Carter, J.; Eichenberger, M.R.; Feagins, K.; Galbraith, N.J.; Dryden, G.W.; Pan, J.; Rai, S.N.; Galandiuk, S. Plasma microRNA Profile Differentiates Crohn's Colitis From Ulcerative Colitis. *Inflamm. Bowel Dis.* **2017**, *24*, 159–165. [CrossRef]

117. Zahm, A.M.; Thayu, M.; Hand, N.J.; Horner, A.; Leonard, M.B.; Friedman, J.R. Circulating microRNA is a biomarker of pediatric Crohn disease. *J. Pediatr. Gastroenterol. Nutr.* **2011**, *53*, 26–33. [CrossRef]
118. Jensen, M.D.; Andersen, R.F.; Christensen, H.; Nathan, T.; Kjeldsen, J.; Madsen, J.S. Circulating microRNAs as biomarkers of adult Crohn's disease. *Eur. J. Gastroenterol. Hepatol.* **2015**, *27*, 1038–1044. [CrossRef]
119. Viennois, E.; Zhao, Y.; Han, M.K.; Xiao, B.; Zhang, M.; Prasad, M.; Wang, L.; Merlin, D. Serum miRNA signature diagnoses and discriminates murine colitis subtypes and predicts ulcerative colitis in humans. *Sci. Rep.* **2017**, *7*, 2520. [CrossRef]
120. Krissansen, G.W.; Yang, Y.; McQueen, F.M.; Leung, E.; Peek, D.; Chan, Y.C.; Print, C.; Dalbeth, N.; Williams, M.; Fraser, A.G. Overexpression of miR-595 and miR-1246 in the sera of patients with active forms of inflammatory bowel disease. *Inflamm. Bowel Dis.* **2015**, *21*, 520–530. [CrossRef]
121. Wang, H.; Zhang, S.; Yu, Q.; Yang, G.; Guo, J.; Li, M.; Zeng, Z.; He, Y.; Chen, B.; Chen, M. Circulating MicroRNA223 is a New Biomarker for Inflammatory Bowel Disease. *Medicine (Baltimore)* **2016**, *95*, e2703. [CrossRef] [PubMed]
122. Schonauen, K.; Le, N.; von Arnim, U.; Schulz, C.; Malfertheiner, P.; Link, A. Circulating and Fecal microRNAs as Biomarkers for Inflammatory Bowel Diseases. *Aliment. Pharmacol. Ther.* **2018**, *24*, 1547–1557. [CrossRef] [PubMed]
123. Omidbakhsh, A.; Saeedi, M.; Khoshnia, M.; Marjani, A.; Hakimi, S. Micro-RNAs -106a and -362-3p in Peripheral Blood of Inflammatory Bowel Disease Patients. *Open Biochem. J.* **2018**, *12*, 78–86. [CrossRef] [PubMed]
124. Chen, P.; Li, Y.; Li, L.; Yu, Q.; Chao, K.; Zhou, G.; Qiu, Y.; Feng, R.; Huang, S.; He, Y.; et al. Circulating microRNA146b-5p is superior to C-reactive protein as a novel biomarker for monitoring inflammatory bowel disease. *Aliment Pharmacol Ther.* **2019**, *49*, 733–743. [CrossRef] [PubMed]
125. Fernandes, J.C.R.; Acuna, S.M.; Aoki, J.I.; Floeter-Winter, L.M.; Muxel, S.M. Long Non-Coding RNAs in the Regulation of Gene Expression: Physiology and Disease. *Noncoding RNA* **2019**, *5*. [CrossRef] [PubMed]
126. Ponting, C.P.; Oliver, P.L.; Reik, W. Evolution and functions of long noncoding RNAs. *Cell* **2009**, *136*, 629–641. [CrossRef] [PubMed]
127. Mercer, T.R.; Dinger, M.E.; Mattick, J.S. Long non-coding RNAs: Insights into functions. *Nat. Rev. Genet.* **2009**, *10*, 155–159. [CrossRef] [PubMed]
128. Derrien, T.; Johnson, R.; Bussotti, G.; Tanzer, A.; Djebali, S.; Tilgner, H.; Guernec, G.; Martin, D.; Merkel, A.; Knowles, D.G.; et al. The GENCODE v7 catalog of human long noncoding RNAs: Analysis of their gene structure, evolution, and expression. *Genome Res.* **2012**, *22*, 1775–1789. [CrossRef]
129. Padua, D.; Mahurkar-Joshi, S.; Law, I.K. A long noncoding RNA signature for ulcerative colitis identifies IFNG-AS1 as an enhancer of inflammation. *Am. J. Physiol. Gastrointest Liver Physiol.* **2016**, *311*, G446–G457. [CrossRef]
130. Liu, F.; Chen, N.; Gong, Y.; Xiao, R.; Wang, W.; Pan, Z. The long non-coding RNA NEAT1 enhances epithelial-to-mesenchymal transition and chemoresistance via the miR-34a/c-Met axis in renal cell carcinoma. *Oncotarget* **2017**, *8*, 62927–62938. [CrossRef]
131. Yarani, R.; Mirza, A.H.; Kaur, S.; Pociot, F. The emerging role of lncRNAs in inflammatory bowel disease. *Exp. Mol. Med.* **2018**, *50*, 161. [CrossRef] [PubMed]
132. Mirza, A.H.; Kaur, S.; Brorsson, C.A.; Pociot, F. Effects of GWAS-associated genetic variants on lncRNAs within IBD and T1D candidate loci. *PLoS ONE* **2014**, *9*, e105723. [CrossRef] [PubMed]
133. Xu, Y.; Shao, B. Circulating lncRNA IFNG-AS1 expression correlates with increased disease risk, higher disease severity and elevated inflammation in patients with coronary artery disease. *J. Clin. Lab. Anal.* **2018**, *32*, e22452. [CrossRef] [PubMed]
134. Zhu, Y.J.; Mao, D.; Gao, W.; Hu, H. Peripheral whole blood lncRNA expression analysis in patients with eosinophilic asthma. *Medicine (Baltimore)* **2018**, *97*, e9817. [CrossRef] [PubMed]
135. Zeng, Q.; Wu, J.; Yang, S. Circulating lncRNA ITSN1-2 is upregulated, and its high expression correlates with increased disease severity, elevated inflammation, and poor survival in sepsis patients. *J. Clin. Lab. Anal.* **2019**, *33*, e22836. [CrossRef] [PubMed]
136. Chen, D.; Liu, J.; Zhao, H.Y.; Chen, Y.P.; Xiang, Z.; Jin, X. Plasma long noncoding RNA expression profile identified by microarray in patients with Crohn's disease. *World J. Gastroenterol.* **2016**, *22*, 4716–4731. [CrossRef] [PubMed]
137. Wang, S.; Hou, Y.; Chen, W.; Wang, J.; Xie, W.; Zhang, X.; Zeng, L. KIF9AS1, LINC01272 and DIO3OS lncRNAs as novel biomarkers for inflammatory bowel disease. *Mol. Med. Rep.* **2018**, *17*, 2195–2202.

138. Barault, L.; Amatu, A.; Siravegna, G.; Ponzetti, A.; Moran, S.; Cassingena, A.; Mussolin, B.; Falcomata, C.; Binder, A.M.; Cristiano, C.; et al. Discovery of methylated circulating DNA biomarkers for comprehensive non-invasive monitoring of treatment response in metastatic colorectal cancer. *Gut* **2018**, *67*, 1995–2005. [CrossRef]
139. Usui, K.; Yokoyama, T.; Naka, G.; Ishida, H.; Kishi, K.; Uemura, K.; Ohashi, Y.; Kunitoh, H. Plasma ctDNA monitoring during epidermal growth factor receptor (EGFR)-tyrosine kinase inhibitor treatment in patients with EGFR-mutant non-small cell lung cancer (JP-CLEAR trial). *Jpn J. Clin. Oncol.* **2019**, *49*, 554–558. [CrossRef]
140. Pantel, K.; Alix-Panabieres, C. Liquid biopsy and minimal residual disease - latest advances and implications for cure. *Nat. Rev. Clin. Oncol.* **2019**, *16*, 409–424. [CrossRef]
141. Burnham, P.; Khush, K.; De Vlaminck, I. Myriad Applications of Circulating Cell-Free DNA in Precision Organ Transplant Monitoring. *Ann. Am. Thorac. Soc.* **2017**, *14*, S237–S241. [CrossRef] [PubMed]

10

Sex, Age and Bodyweight as Determinants of Extracellular DNA in the Plasma of Mice

Ľubica Janovičová [1], Barbora Konečná [1], Lenka Vokálová [2,3], Lucia Lauková [1,4], Barbora Vlková [1] and Peter Celec [1,5,6,*]

1 Institute of Molecular Biomedicine, Faculty of Medicine, Comenius University, Sasinkova 4, 811 08 Bratislava, Slovakia
2 Institute of Physiology, Faculty of Medicine, Comenius University, Sasinkova 2, 813 72 Bratislava, Slovakia
3 Department of Biomedicine, University of Basel, Hebelstrasse 20, 4031 Basel, Switzerland
4 Center for Biomedical Technology, Department for Health Sciences and Biomedicine, Danube University, 3500 Krems, Austria
5 Institute of Pathophysiology, Faculty of Medicine, Comenius University, Sasinkova 4, 81108 Bratislava, Slovakia
6 Department of Molecular Biology, Faculty of Natural Sciences, Comenius University, Mlynská dolina, Ilkovičova 6, 842 15 Bratislava, Slovakia
* Correspondence: petercelec@gmail.com

Abstract: Extracellular DNA (ecDNA) is studied as a possible biomarker, but also as a trigger of the immune responses important for the pathogenesis of several diseases. Extracellular deoxyribonuclease (DNase) activity cleaves ecDNA. The aim of our study was to describe the interindividual variability of ecDNA and DNase activity in the plasma of healthy mice, and to analyze the potential determinants of the variability, including sex, age, and bodyweight. In this experiment, 58 adult CD1 mice (41 females and 31 males) of a variable age (3 to 16 months old) and bodyweight (females 25.7 to 52.1 g, males 24.6 to 49.6 g) were used. The plasma ecDNA was measured using a fluorometric method. The nuclear ecDNA and mitochondrial ecDNA were quantified using real-time PCR. The deoxyribonuclease activity was assessed using the single radial enzyme diffusion method. The coefficient of variance for plasma ecDNA was 139%, and for DNase 48%. Sex differences were not found in the plasma ecDNA (52.7 ± 73.0 ηg/mL), but in the DNase activity (74.5 ± 33.5 K.u./mL for males, and 47.0 ± 15.4 K.u./mL for females). There were no associations between plasma ecDNA and bodyweight or the age of mice. Our study shows that the variability of plasma ecDNA and DNase in adult healthy mice is very high. Sex, age, and bodyweight seem not to be major determinants of ecDNA variability in healthy mice. As ecDNA gains importance in the research of several diseases, it is of importance to understand its production and cleavage. Further studies should, thus, test other potential determinants, taking into account cleavage mechanisms other than DNase.

Keywords: cell-free DNA; nuclease activity; aging; obesity; gender differences

---

1. Introduction

Extracellular DNA (ecDNA) is DNA outside of cells. The presence of ecDNA in blood was discovered by Mandel and Metais in 1948 [1]. The ecDNA is proinflammatory [2] and procoagulatory [3]. Apoptosis and necrosis are the major sources of ecDNA [4]. EcDNA is used as a biomarker in non-invasive prenatal diagnostics [5] and oncology [6]. In addition, donor DNA can be detected in the blood of the transplant recipient, suggesting that ecDNA could be a biomarker of transplant rejection [7,8]. EcDNA is also a therapeutic target. In several diseases, it has been shown that ecDNA leads to inflammation and a release of more ecDNA—this vicious circle can be stopped by the removal of ecDNA, as shown for sepsis [9], hepatorenal injury [10], and metabolic syndrome [11,12].

The knowledge about the normal ecDNA concentration values and the variability of ecDNA is limited. Currently, there are no standard physiological values for ecDNA in plasma. A recent study evaluating human plasma ecDNA showed that both nuclear and mitochondrial DNA vary within a range of over three orders of magnitude [13]. Regarding the technical variability, the amount of ecDNA found in body fluids depends on the processing of samples, and on the technique used for DNA quantification. It has been shown that the extraction efficiency from plasma has a coefficient of variance of approximately 29% [14]. The discordant methods of sample processing and quantification make a comparison of ecDNA concentrations between studies difficult [15]. Apart from technical variability, there is also a high biological variability of plasma ecDNA. This is best described for fetal DNA circulating in maternal blood [16]. Deoxyribonuclease-I (DNase I) does not affect the fragmentation of ecDNA in plasma [17]. The pattern of fragmentation of ecDNA was described in mice deficient in DNase I and DNase 1L3. Mice without DNase 1l3 have more longer fragments of ecDNA [18]. However, if mice are deficient for both DNase I together with DNase 1L3, and their neutrophils are stimulated, they die because of the inability to cleave ecDNA in blood [19].

Interindividual differences in ecDNA can be caused by differences in the release of ecDNA, its cleavage, or protection against the cleavage. Deoxyribonucleases (DNases) are able to cleave plasma ecDNA [20]. Obesity inducing the release of ecDNA from adipocytes makes body weight one of the candidate determinants of ecDNA [11]. Aging is associated with an increased production of neutrophil extracellular traps—a major source of ecDNA [21]. Women are more prone to autoimmune diseases associated with anti-DNA antibodies [22]. This suggests that the concentration of ecDNA in females could be higher, although the mechanisms are unclear. Understanding the variability of ecDNA could be very helpful to overcome the diagnostic limitations of ecDNA, but it could also shed light on the role of ecDNA in the pathogenesis of inflammatory diseases. The aim of this study is to describe the variability of ecDNA and DNase, and to analyze its potential determinants. Our hypothesis was that older females with a higher bodyweight have higher ecDNA and lower DNase in their plasma.

2. Results

An analysis of variability revealed high coefficients of interindividual variations for total ecDNA (Coefficient of Variance (CV) = 139%), nuclear DNA (ncDNA; CV = 50%), and mitochondrial (mtDNA; CV = 149%). The total ecDNA minimum is 100-fold lower than the maximum concentration. The maximum ncDNA and mtDNA concentrations were 10-fold and 1000-fold higher than the minimum concentrations, respectively. For the DNase activity, the maximum value was 7.5-fold higher than the minimum. Based on the 5% and 95% percentile, the normal values of the total ecDNA, ncDNA, mtDNA, and DNase activity were determined. These are solely specific for the type of sample processing described here. The normal values for the total ecDNA in the plasma were 5 to 266 ng/mL. For the ecDNA fractions, normal values were also determined. The normal values for plasma mtDNA were 6.5×10^3 to 6.4×10^5 GE/mL, and for the plasma ncDNA they were from 1.6×10^3 to 7.5×10^3 GE/mL of plasma. The mtDNA copy number is, thus, approximately 4 to 85 per cell. The normal values of DNase activity in the plasma were from 32 to 125 K.u./mL (Table 1).

Sex seems not to be a major determinant of plasma ecDNA variability in mice. There were no sex differences in the ecDNA concentrations in the total ecDNA (t-test; $p = 0.52$, $t = 0.64$; Figure 1A,B), ncDNA (t-test; $p = 0.26$, $t = 1.14$), or mtDNA (t-test; $p = 0.41$, $t = 0.83$; Figure 2). A statistically significant sex difference was found in the DNase activity—the male mice had a higher DNase activity by 59% (Figure 1D; $p < 0.001$, $t = 4.65$).

Table 1. Descriptive statistics of the measured variables, namely: bodyweight, age, plasma extracellular DNA (ecDNA), and deoxyribonuclease (DNase) activity in all of the mice combined.

	Bodyweight (g)	Age (Days)	Total ecDNA (ng/mL)	ncDNA (GE/mL)	mtDNA (GE/mL)	DNase Activity (K.u./mL)
Mean	38.06	226	52.7	3.3×10^3	1.4×10^5	59.0
Standard Error	0.72	14	8.8	2.0×10^2	2.6×10^4	3.4
Median	38.53	195	31.1	2.9×10^3	6.0×10^4	51.0
Standard Deviation	6.16	122	73.0	1.6×10^3	2.1×10^5	28.4
Minimum	24.62	99	2.9	7.2×10^2	3.1×10^3	25.1
Maximum	52.11	468	423.2	9.9×10^3	1.1×10^6	186.4
Coefficient of Variance	16%	54%	139%	50%	149%	48%

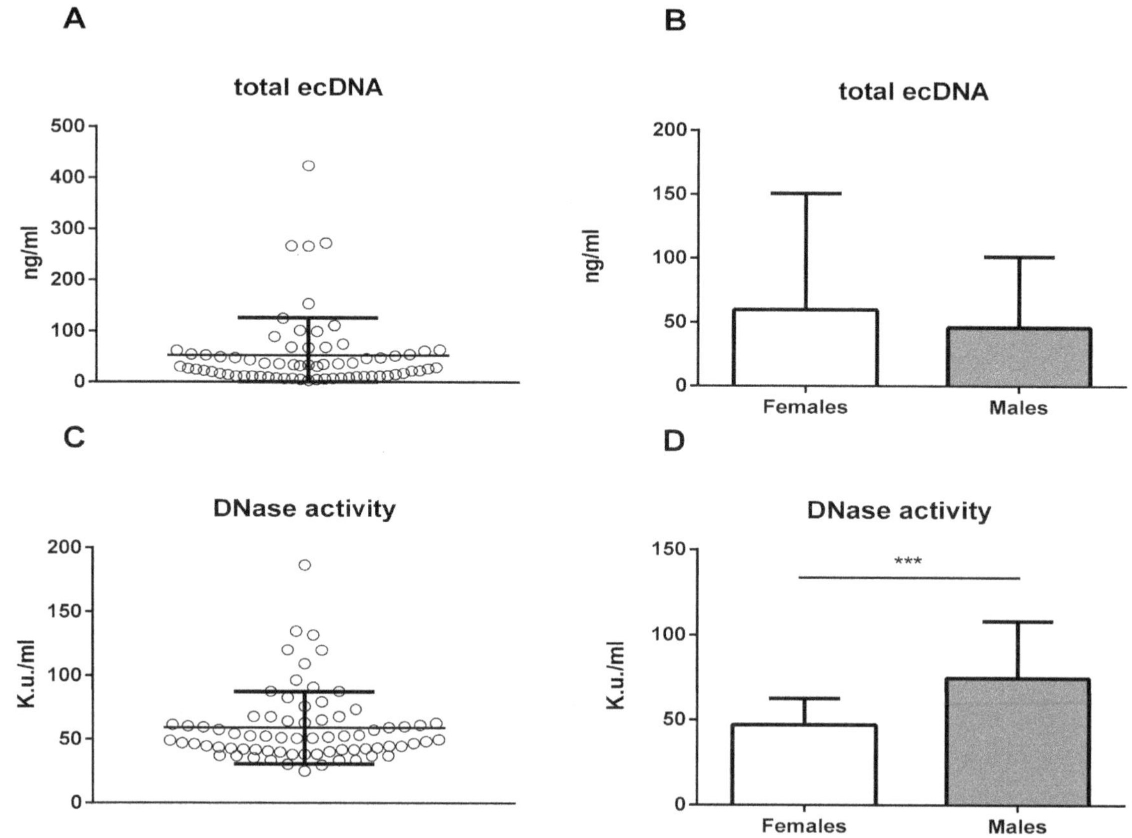

Figure 1. (**a**) Total extracellular DNA (ecDNA; $n = 57$) and (**b**) total ecDNA divided by sex (females $n = 34$, males $n = 23$) (t-test; $p = 0.52$, $t = 0.64$) and (**c**) deoxyribonuclease (DNase) activity ($n = 72$) and (**d**) DNase activity divided by sex (females $n = 41$, males $n = 31$) (t-test; $p < 0.001$, $t = 4.65$). DNase activity is expressed in Kunitz units (K.u.) per ml of plasma. The results are presented as mean + standard deviation. p-value is indicated as *** for $p < 0.001$.

The correlation analysis revealed that the total ecDNA was positively correlated with ncDNA ($p < 0.001$, Pearson's $r = 0.66$) and strongly with mtDNA ($p < 0.001$; Pearson's $r = 0.90$). NcDNA was also positively correlated with mtDNA ($p < 0.001$; Pearson's $r = 0.63$). The DNase activity did not correlate with the total ecDNA (n.s.; Pearson's $r = -0.11$). No association was found between the DNase activity and ncDNA (n.s.; Pearson's $r = -0.02$) or mtDNA (n.s.; Pearson's $r = -0.02$). The bodyweight of the mice did not correlate with the total ecDNA (n.s.; Pearson's $r = 0.01$), ncDNA (n.s.; Pearson's $r = 0.07$), or mtDNA (n.s.; Pearson's $r = -0.01$). Similarly, age did not correlate with the total ecDNA (n.s.; Pearson's $r = 0.03$), ncDNA (n.s.; Pearson's $r = 0.04$), and mtDNA (n.s.; Pearson's $r = 0.04$; Table 2).

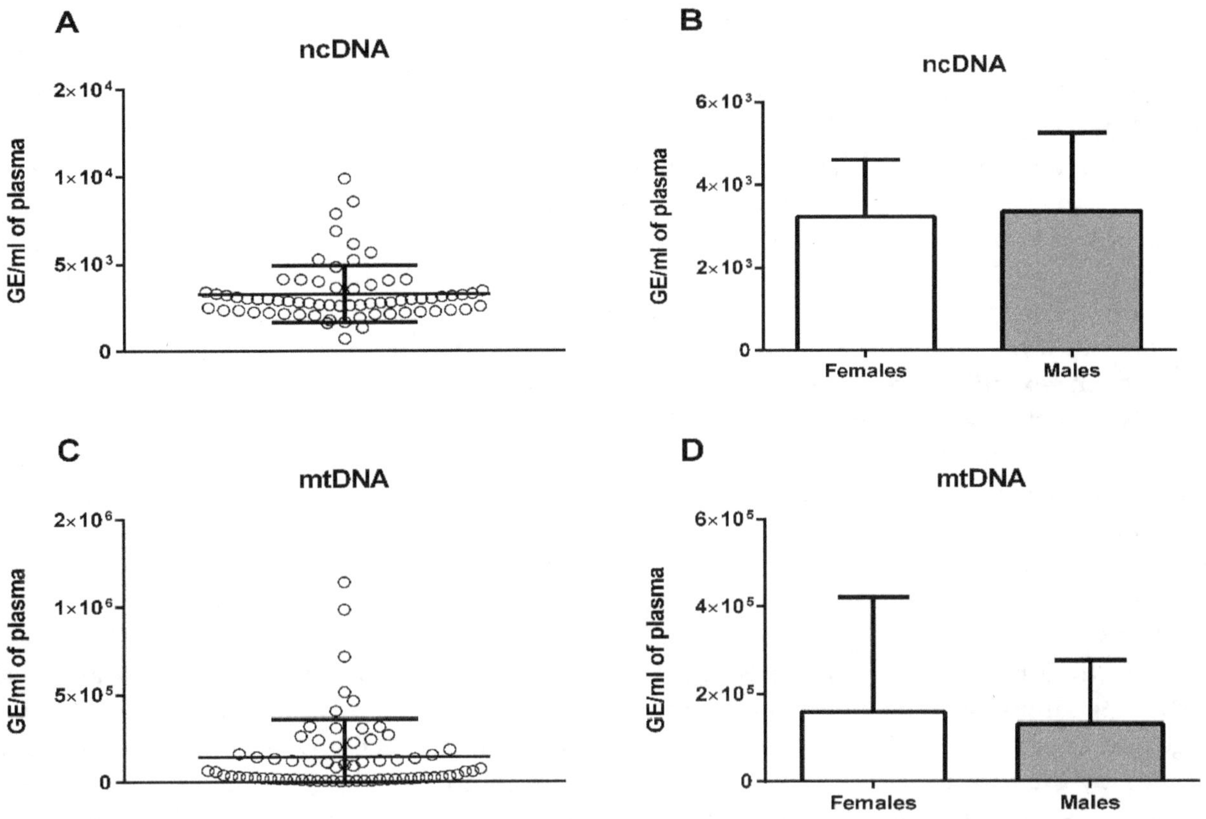

Figure 2. (**a**) Nuclear extracellular DNA (ncDNA; $n = 57$) and (**b**) nuclear extracellular DNA divided by sex (females $n = 34$, males $n = 23$; t-test; $p = 0.26$, $t = 1.14$) and (**c**) mitochondrial extracellular DNA (mtDNA; $n = 57$), and (**d**) mitochondrial extracellular DNA divided by sex (females $n = 34$, males $n = 23$; t-test; $p = 0.41$, $t = 0.83$). The extracellular DNA of both a nuclear and mitochondrial origin is expressed in genome equivalents (GE) per ml of plasma. The results are presented as mean + standard deviation.

Table 2. Correlation matrix describing the relationships between the analyzed variables, namely: bodyweight, age, total extracellular DNA (ecDNA), nuclear ecDNA (ncDNA), mitochondrial ecDNA (mtDNA), and deoxyribonuclease (DNase) activity. The data in the table are presented as Pearson's correlation coefficients with p-values indicated as $p < 0.001$ for ***.

	Bodyweight	Age	Total ecDNA	ncDNA	mtDNA	DNase Activity
Bodyweight		0.60 ***	0.01	0.07	−0.01	0.19
Age	0.60 ***		0.03	0.04	0.03	−0.11
Total ecDNA	0.01	0.03		0.66 ***	0.90 ***	−0.11
ncDNA	0.07	0.04	0.66 ***		0.63 ***	−0.02
mtDNA	−0.01	0.03	0.90 ***	0.63 ***		−0.02
DNase activity	0.19	−0.11	−0.11	−0.02	−0.02	

3. Discussion

In this cross-sectional study on healthy mice, we confirmed a very high variability in ecDNA and DNase activity. As ecDNA and its cleavage are likely involved in the pathogenesis of several diseases, it is of importance to analyze the determinants of this variability. However, using correlation analysis, in our study, we found no association of either ecDNA or DNase with the age or bodyweight of the mice. Similarly, no sex differences were found for ecDNA or its fractions—ncDNA and mtDNA. The only sex difference was found in the DNase activity, which was higher in male mice, confirming previous reports [23]. Thus, to explain the variability of ecDNA, cleavage mechanisms other than the action of DNases should be taken into account. These include the phagocytic activity of monocytes, cleavage by the liver, or excretion via kidney [24–26].

There are many age-related diseases that are associated with high concentrations of ecDNA, such as autoimmune diseases [27], myocardial infarction [28], obesity [11,29], and others. It has already been published that aging is associated with increased levels of ecDNA in humans [30]. Dying cells release ecDNA to the extracellular space, and this ecDNA can activate toll-like receptors and induce an immune system response [31]. Neutrophils can die while releasing their DNA. Neutrophil extracellular trap production is increased in aged mice [32]. Nevertheless, in our study in healthy mice, age did not affect ecDNA or DNase in the plasma.

Regarding ecDNA, there is one published study investigating sex differences in ecDNA concentration. Men were shown to have a higher nuclear ecDNA concentration in comparison to women, but there were no differences in ecDNA originating from the mitochondria [13]. In our study, no such sex differences in ecDNA were found. In comparison to humans, mice have a much higher DNase activity in the plasma. Whether this affects the interpretability of the results of the experiments on animal models of diseases related to ecDNA is unknown. From the available literature, the variability in ecDNA and DNase activity is high [23]. This was confirmed by our results in mice. However, except for the sex difference in the DNase, the determinants of this variability could not be identified. Other factors should be studied in the future. Magnesium and calcium ions, for example, are needed for DNases to be active, and can be one of the determinants of DNase activity that contribute to variability [33].

One of the proposed determinants of ecDNA is physical activity. DNase activity and ecDNA were shown to increase in response to high-intensity exercise [34]. Circadian biorhythms affect DNA damage and apoptosis [35]. Similarly, in females, the variability of ecDNA or DNase activity could be influenced by the estrous cycle. The hormonal changes might be relevant in autoimmune diseases related to ecDNA and their animal models [36]. Revealing the determinants of ecDNA variability both in humans and in mice, might help to understand the physiology and the pathophysiology related to ecDNA.

One important aspect not included in our study is technical variability. It has been shown that even the isolation of ecDNA contributes considerably to the overall variability in the outcome [14]. The optimization of protocols is crucial in achieving accurate and reproducible results. The centrifugation speed and time differ between studies, which makes it difficult to compare the results. The most commonly used protocol is two-step centrifugation, which removes apoptotic bodies and some microvesicles [37]. The choice of blood collection method is of importance. Serum is not suitable for ecDNA quantification, because of the release of genomic DNA from cells during coagulation [38]. EDTA plasma, on the other hand, cannot be used for the measurement of DNase activity, as it inhibits DNase I [39]. So, in our study, we have collected blood into EDTA and heparin to use plasma for DNA and DNase measurements, respectively. However, a further reduction of the technical variability, especially in the quantification of nuclear and mitochondrial DNA, is surely needed.

In conclusion, for future experiments, it is important to consider the high interindividual variability of ecDNA in healthy animals, which is independent of body weight, age, or sex. The observed sex difference in the DNase activity needs to be taken into consideration, especially if confirmed in other strains and species. In addition, the underlying endocrine or genetic mechanisms should be clarified. The major contributors to the biological variability of ecDNA are, however, yet to be identified.

4. Materials and Methods

4.1. Animal Procedures

The CD-1 mice were purchased from Anlab (Prague, Czech Republic). Both female ($n = 35$) and male ($n = 23$) healthy mice of varying age (7.5 ± 4 months) and of varying weights (females 34.1 ± 7.4 g and males 40.7 ± 6.2 g) had ad-libitum access to food and water. The animals were maintained in a temperature-controlled and light-controlled room with a 12-h light/dark cycle. Blood was collected from the retro-orbital plexus from mice in isoflurane anesthesia (AbbVie, Bratislava, Slovakia) into heparin and EDTA microvettes (Sarstedt, Nümbrecht, Germany), and centrifuged at 2000× g for 10 min at 4 °C. The mice were sacrificed by cervical dislocation. The plasma was stored at −20 °C until further analyses. All of the methods and procedures were conducted in accordance with Slovak legislation. The animals were housed and procedures were conducted in compliance with the Ethics committee of Institute of Molecular Biomedicine, Comenius University in Bratislava (Ro-536/18-221/3; date: 26 March 2018).

4.2. DNA Isolation and Quantification

DNA was isolated from the EDTA plasma using the QIAamp DNA Blood Mini kit (Qiagen, Hilden, Germany). The isolated DNA was quantified using a Qubit 3.0 fluorometer and Qubit dsDNA high sensitivity assay (Thermo Fisher Scientific, Waltham, MA, USA). The fractions of nuclear and mitochondrial DNA were estimated using real-time PCR on the Mastercycler realplex 4 (Eppendorf, Hamburg, Germany) with Sso Advanced Universal SYBR Green Supermix (Bio Rad Laboratories, Hercules, CA, USA). The following PCR program was used: one cycle of 98 °C for 3 min, 40 cycles of 98 °C for 15 s for denaturation, 60 °C for 30 s for annealing, and 72 °C for 30 s for extension. Primers were designed for the amplification of mouse mtDNA (F: 5′-CCCAGCTACTACCATCATTCAAGT-′, R: 5′-GATGGTTTGGGAGATTGGTTGATGT-3′) [40] and ncDNA (F: 5′-TGTCAGATATGTCCTTCAGCAAGG-3′, R: 5′-TGCTTAACTCTGCAGGCGTATG-3′) [41]. The quantified ecDNA is expressed in genome equivalents (GE).

4.3. DNase Activity

Plasma anticoagulated with heparin was used for the determination of DNase activity. The DNase activity was measured using the modified single radial enzyme-diffusion assay with the GoodView fluorescent dye [42]. Agarose gels (1%, 20 mM Tris-HCl, ph 7.5, 2 mM $MgCl_2$, 2 mM $CaCl_2$) containing DNA isolated from chicken livers (0.5 mg/mL of DNA) were visualized using iBOX (Vision works LP Analysis Software, UVP, Upland, CA, USA), and the radius of the circles were measured and compared to the calibration curve from the DNase standards. The DNase activity was recalculated and expressed in Kunitz units (K.u.).

4.4. Statistical Analysis

The results were analyzed using GraphPad Prism 6 (La Jolla, CA, USA). For the evaluation of the sex differences, the student t-test was used. Correlation analyses were conducted with the Pearson correlation test. The test results of $p < 0.05$ were considered statistically significant.

Author Contributions: L.J.—investigation, writing (original draft preparation), and visualization. B.K.—writing (original draft preparation, review, and editing). L.V.—investigation. L.L.—methodology and investigation. B.V.—methodology, investigation, supervision. P.C.—conceptualisation, validation, resources, writing (review and editing), supervision, and funding acquisition.

Abbreviations

DNase	Deoxyribonuclease
mtDNA	Mitochondrial DNA
ncDNA	Nuclear DNA
ecDNA	Extracellular DNA

References

1. Mandel, P.; Metais, P. Comptes rendus des seances de la Societe de biologie et de ses filiales. *Journal de la Société de Biologie* **1948**, *142*, 241–243.
2. Poli, C.; Augusto, J.F.; Dauve, J.; Adam, C.; Preisser, L.; Larochette, V.; Pignon, P.; Savina, A.; Blanchard, S.; Subra, J.F.; et al. IL-26 Confers Proinflammatory Properties to Extracellular DNA. *J. Immunol.* **2017**, *198*, 3650–3661. [CrossRef] [PubMed]

3. Li, B.; Liu, Y.; Hu, T.; Zhang, Y.; Zhang, C.; Li, T.; Wang, C.; Dong, Z.; Novakovic, V.A.; Hu, T.; et al. Neutrophil extracellular traps enhance procoagulant activity in patients with oral squamous cell carcinoma. *J. Cancer Res. Clin. Oncol.* **2019**, *145*, 1695–1707. [CrossRef] [PubMed]
4. Stroun, M.; Lyautey, J.; Lederrey, C.; Olson-Sand, A.; Anker, P. About the possible origin and mechanism of circulating DNA apoptosis and active DNA release. *Clin. Chim. Acta; Int. J. Clin. Chem.* **2001**, *313*, 139–142. [CrossRef]
5. Lo, Y.M.; Wainscoat, J.S.; Fleming, K.A. Non-invasive approach to prenatal diagnosis from maternal peripheral blood. *Prenat. Diagn.* **1992**, *12*, 547–548. [CrossRef] [PubMed]
6. Jahr, S.; Hentze, H.; Englisch, S.; Hardt, D.; Fackelmayer, F.O.; Hesch, R.D.; Knippers, R. DNA fragments in the blood plasma of cancer patients: Quantitations and evidence for their origin from apoptotic and necrotic cells. *Cancer Res.* **2001**, *61*, 1659–1665. [PubMed]
7. Burnham, P.; Kim, M.S.; Agbor-Enoh, S.; Luikart, H.; Valantine, H.A.; Khush, K.K.; De Vlaminck, I. Single-stranded DNA library preparation uncovers the origin and diversity of ultrashort cell-free DNA in plasma. *Sci. Rep.* **2016**, *6*, 27859. [CrossRef]
8. Lehmann-Werman, R.; Magenheim, J.; Moss, J.; Neiman, D.; Abraham, O.; Piyanzin, S.; Zemmour, H.; Fox, I.; Dor, T.; Grompe, M.; et al. Monitoring liver damage using hepatocyte-specific methylation markers in cell-free circulating DNA. *JCI Insight* **2018**, *3*. [CrossRef]
9. Hamaguchi, S.; Akeda, Y.; Yamamoto, N.; Seki, M.; Yamamoto, K.; Oishi, K.; Tomono, K. Origin of Circulating Free DNA in Sepsis: Analysis of the CLP Mouse Model. *Mediat. Inflamm.* **2015**, *2015*, 614518. [CrossRef]
10. Vokalova, L.; Laukova, L.; Conka, J.; Meliskova, V.; Borbelyova, V.; Babickova, J.; Tothova, L.; Hodosy, J.; Vlkova, B.; Celec, P. Deoxyribonuclease partially ameliorates thioacetamide-induced hepatorenal injury. *Am. J. Physiol. Gastrointest. Liver Physiol.* **2017**, *312*, G457–G463. [CrossRef]
11. Nishimoto, S.; Fukuda, D.; Higashikuni, Y.; Tanaka, K.; Hirata, Y.; Murata, C.; Kim-Kaneyama, J.R.; Sato, F.; Bando, M.; Yagi, S.; et al. Obesity-induced DNA released from adipocytes stimulates chronic adipose tissue inflammation and insulin resistance. *Sci. Adv.* **2016**, *2*, e1501332. [CrossRef] [PubMed]
12. Revelo, X.S.; Ghazarian, M.; Chng, M.H.; Luck, H.; Kim, J.H.; Zeng, K.; Shi, S.Y.; Tsai, S.; Lei, H.; Kenkel, J.; et al. Nucleic Acid-Targeting Pathways Promote Inflammation in Obesity-Related Insulin Resistance. *Cell Rep.* **2016**, *16*, 717–730. [CrossRef] [PubMed]
13. Meddeb, R.; Dache, Z.A.A.; Thezenas, S.; Otandault, A.; Tanos, R.; Pastor, B.; Sanchez, C.; Azzi, J.; Tousch, G.; Azan, S.; et al. Quantifying circulating cell-free DNA in humans. *Sci. Rep.* **2019**, *9*, 5220. [CrossRef] [PubMed]
14. O'Connell, G.C.; Chantler, P.D.; Barr, T.L. High Interspecimen Variability in Nucleic Acid Extraction Efficiency Necessitates the Use of Spike-In Control for Accurate qPCR-based Measurement of Plasma Cell-Free DNA Levels. *Lab. Med.* **2017**, *48*, 332–338. [CrossRef]
15. Vlkova, B.; Turna, J.; Celec, P. Fetal DNA in maternal plasma in preeclamptic pregnancies. *Hypertens. Pregnancy* **2015**, *34*, 36–49. [CrossRef]
16. Zhong, X.Y.; Burk, M.R.; Troeger, C.; Kang, A.; Holzgreve, W.; Hahn, S. Fluctuation of maternal and fetal free extracellular circulatory DNA in maternal plasma. *Obstet. Gynecol.* **2000**, *96*, 991–996.
17. Cheng, T.H.T.; Lui, K.O.; Peng, X.L.; Cheng, S.H.; Jiang, P.; Chan, K.C.A.; Chiu, R.W.K.; Lo, Y.M.D. DNase1 Does Not Appear to Play a Major Role in the Fragmentation of Plasma DNA in a Knockout Mouse Model. *Clin. Chem.* **2018**, *64*, 406–408. [CrossRef]
18. Serpas, L.; Chan, R.W.Y.; Jiang, P.; Ni, M.; Sun, K.; Rashidfarrokhi, A.; Soni, C.; Sisirak, V.; Lee, W.S.; Cheng, S.H.; et al. Dnase1l3 deletion causes aberrations in length and end-motif frequencies in plasma DNA. *Proc. Natl. Acad. Sci. USA* **2019**, *116*, 641–649. [CrossRef]
19. Jimenez-Alcazar, M.; Rangaswamy, C.; Panda, R.; Bitterling, J.; Simsek, Y.J.; Long, A.T.; Bilyy, R.; Krenn, V.; Renne, C.; Renne, T.; et al. Host DNases prevent vascular occlusion by neutrophil extracellular traps. *Science* **2017**, *358*, 1202–1206. [CrossRef]
20. Suck, D. DNA recognition by DNase I. *J. Mol. Recognit.* **1994**, *7*, 65–70. [CrossRef]
21. Ortmann, W.; Kolaczkowska, E. Age is the work of art? Impact of neutrophil and organism age on neutrophil extracellular trap formation. *Cell Tissue Res.* **2018**, *371*, 473–488. [CrossRef]
22. Giles, B.M.; Boackle, S.A. Linking complement and anti-dsDNA antibodies in the pathogenesis of systemic lupus erythematosus. *Immunol. Res.* **2013**, *55*, 10–21. [CrossRef]
23. Koizumi, T. Tissue distribution of deoxyribonuclease I (DNase I) activity level in mice and its sexual dimorphism. *Exp. Anim.* **1995** *44*, 181–185. [CrossRef]

24. Hochreiter-Hufford, A.; Ravichandran, K.S. Clearing the dead: Apoptotic cell sensing, recognition, engulfment, and digestion. *Cold Spring Harb. Perspect. Biol.* **2013**, *5*, a008748. [CrossRef]
25. Botezatu, I.; Serdyuk, O.; Potapova, G.; Shelepov, V.; Alechina, R.; Molyaka, Y.; Ananev, V.; Bazin, I.; Garin, A.; Narimanov, M.; et al. Genetic analysis of DNA excreted in urine: A new approach for detecting specific genomic DNA sequences from cells dying in an organism. *Clin. Chem.* **2000**, *46*, 1078–1084.
26. Hisazumi, J.; Kobayashi, N.; Nishikawa, M.; Takakura, Y. Significant role of liver sinusoidal endothelial cells in hepatic uptake and degradation of naked plasmid DNA after intravenous injection. *Pharm. Res.* **2004**, *21*, 1223–1228. [CrossRef]
27. Galeazzi, M.; Morozzi, G.; Piccini, M.; Chen, J.; Bellisai, F.; Fineschi, S.; Marcolongo, R. Dosage and characterization of circulating DNA: Present usage and possible applications in systemic autoimmune disorders. *Autoimmun Rev.* **2003**, *2*, 50–55. [CrossRef]
28. Lou, X.; Hou, Y.; Liang, D.; Peng, L.; Chen, H.; Ma, S.; Zhang, L. A novel Alu-based real-time PCR method for the quantitative detection of plasma circulating cell-free DNA: Sensitivity and specificity for the diagnosis of myocardial infarction. *Int. J. Mol. Med.* **2015**, *35*, 72–80. [CrossRef]
29. Ale, A.; Zhang, Y.; Han, C.; Cai, D. Obesity-associated extracellular mtDNA activates central TGFbeta pathway to cause blood pressure increase. *Am. J. Physiol. Endocrinol. Metab.* **2017**, *312*, E161–E174. [CrossRef]
30. Ermakov, A.V.; Konkova, M.S.; Kostyuk, S.V.; Izevskaya, V.L.; Baranova, A.; Veiko, N.N. Oxidized extracellular DNA as a stress signal in human cells. *Oxid. Med. Cell Longev.* **2013**, *2013*, 649747. [CrossRef]
31. Barton, G.M.; Kagan, J.C.; Medzhitov, R. Intracellular localization of Toll-like receptor 9 prevents recognition of self DNA but facilitates access to viral DNA. *Nat. Immunol.* **2006**, *7*, 49–56. [CrossRef]
32. Wang, Y.; Wang, W.; Wang, N.; Tall, A.R.; Tabas, I. Mitochondrial Oxidative Stress Promotes Atherosclerosis and Neutrophil Extracellular Traps in Aged Mice. *Arterioscler. Thromb. Vasc. Biol.* **2017**, *37*, e99–e107. [CrossRef]
33. Kishi, K.; Yasuda, T.; Takeshita, H. DNase I: Structure, function, and use in medicine and forensic science. *Leg. Med.* **2001**, *3*, 69–83. [CrossRef]
34. Velders, M.; Treff, G.; Machus, K.; Bosnyak, E.; Steinacker, J.; Schumann, U. Exercise is a potent stimulus for enhancing circulating DNase activity. *Clin. Biochem.* **2014**, *47*, 471–474. [CrossRef]
35. Sancar, A.; Lindsey-Boltz, L.A.; Kang, T.H.; Reardon, J.T.; Lee, J.H.; Ozturk, N. Circadian clock control of the cellular response to DNA damage. *FEBS Lett.* **2010**, *584*, 2618–2625. [CrossRef]
36. Jaini, R.; Altuntas, C.Z.; Loya, M.G.; Tuohy, V.K. Disruption of estrous cycle homeostasis in mice with experimental autoimmune encephalomyelitis. *J. Neuroimmunol.* **2015**, *279*, 71–74. [CrossRef]
37. Chim, S.S.; Tong, Y.K.; Chiu, R.W.; Lau, T.K.; Leung, T.N.; Chan, L.Y.; Oudejans, C.B.; Ding, C.; Lo, Y.M. Detection of the placental epigenetic signature of the maspin gene in maternal plasma. *Proc. Natl. Acad. Sci. USA* **2005**, *102*, 14753–14758. [CrossRef]
38. Zinkova, A.; Brynychova, I.; Svacina, A.; Jirkovska, M.; Korabecna, M. Cell-free DNA from human plasma and serum differs in content of telomeric sequences and its ability to promote immune response. *Sci. Rep.* **2017**, *7*, 2591. [CrossRef]
39. Kolarevic, A.; Yancheva, D.; Kocic, G.; Smelcerovic, A. Deoxyribonuclease inhibitors. *Eur. J. Med. Chem.* **2014**, *88*, 101–111. [CrossRef]
40. Rooney, J.P.; Ryde, I.T.; Sanders, L.H.; Howlett, E.H.; Colton, M.D.; Germ, K.E.; Mayer, G.D.; Greenamyre, J.T.; Meyer, J.N. PCR based determination of mitochondrial DNA copy number in multiple species. *Methods Mol. Biol.* **2015**, *1241*, 23–38. [CrossRef]
41. Wai, T.; Ao, A.; Zhang, X.; Cyr, D.; Dufort, D.; Shoubridge, E.A. The role of mitochondrial DNA copy number in mammalian fertility. *Biol. Reprod.* **2010**, *83*, 52–62. [CrossRef]
42. Nadano, D.; Yasuda, T.; Kishi, K. Measurement of deoxyribonuclease I activity in human tissues and body fluids by a single radial enzyme-diffusion method. *Clin. Chem.* **1993**, *39*, 448–452.

11

Circulating miRNA Profiling in Plasma Samples of Ovarian Cancer Patients

András Penyige [1,*], Éva Márton [2], Beáta Soltész [2], Melinda Szilágyi-Bónizs [2], Róbert Póka [3], János Lukács [3], Lajos Széles [4] and Bálint Nagy [2]

[1] Department of Human Genetics, Faculty of Medicine, Faculty of Pharmacy, University of Debrecen, Debrecen 4032, Hungary

[2] Department of Human Genetics, Faculty of Medicine, University of Debrecen, Debrecen 4032, Hungary; m.eva92@gmail.com (É.M.); soltesz.beata@med.unideb.hu (B.S.); szilagyi.melinda@med.unideb.hu (M.S.-B.); nagy.balint@med.unideb.hu (B.N.)

[3] Department of Obstetrics and Gynecology, Faculty of Medicine, University of Debrecen, Debrecen 4032, Hungary; pokar@med.unideb.hu (R.P.); lukacs.janos@med.unideb.hu (L.S.)

[4] Department of Biochemistry and Molecular Biology, Faculty of Medicine, University of Debrecen, Debrecen 4032, Hungary; szelesl@med.unideb.hu

* Correspondence: penyige@med.unideb.hu

Abstract: Ovarian cancer is one of the most common cancer types in women characterized by a high mortality rate due to lack of early diagnosis. Circulating miRNAs besides being important regulators of cancer development could be potential biomarkers to aid diagnosis. We performed the circulating miRNA expression analysis in plasma samples obtained from ovarian cancer patients stratified into FIGO I, FIGO III, and FIGO IV stages and from healthy females using the NanoString quantitative assay. Forty-five miRNAs were differentially expressed, out of these 17 miRNAs showed significantly different expression between controls and patients, 28 were expressed only in patients, among them 19 were expressed only in FIGO I patients. Differentially expressed miRNAs were ranked by the network-based analysis to assess their importance. Target genes of the differentially expressed miRNAs were identified then functional annotation of the target genes by the GO and KEGG-based enrichment analysis was carried out. A general and an ovary-specific protein–protein interaction network was constructed from target genes. Results of our network and the functional enrichment analysis suggest that besides HSP90AA1, MYC, SP1, BRCA1, RB1, CFTR, STAT3, E2F1, ERBB2, EZH2, and MET genes, additional genes which are enriched in cell cycle regulation, FOXO, TP53, PI-3AKT, AMPK, TGFβ, ERBB signaling pathways and in the regulation of gene expression, proliferation, cellular response to hypoxia, and negative regulation of the apoptotic process, the GO terms have central importance in ovarian cancer development. The aberrantly expressed miRNAs might be considered as potential biomarkers for the diagnosis of ovarian cancer after validation of these results in a larger cohort of ovarian cancer patients.

Keywords: ovarian cancer; circulating miRNA; blood plasma; NanoString; network analysis; biomarker

1. Introduction

Ovarian cancer (OC) is one of the most frequent gynecological malignancy among women. Despite recent progress that has been made in the treatment of patients, OC is still characterized by a high mortality rate. It is the fifth leading cause of cancer-induced death in females in the world; the overall five-year survival rate is still only 15%–30% according to recent reports [1–3]. The high mortality is at least partially due to the difficulties of early detection of the tumor. The routine diagnostic procedures (pelvic examination, transvaginal ultrasonography, CA125 antigen measurement) carried out in clinics are not suitable for early diagnosis [4–7]. During recent years several groups investigated the possibility of using circulating miRNAs as candidate biomarkers for diagnosis in several different human cancers, among them in OC. In addition to their diagnostic potential, the involvement of circulating miRNAs in tumor formation might be an equally important topic to study [8–10].

MiRNAs are endogenously expressed, short (~22 nucleotides) non-coding single-stranded regulatory RNA molecules, known to interfere with the translation of the mRNA coded by the target genes. MiRNAs are negative regulators of gene expression, upon their sequence-specific binding to the 3′-UTR region of mRNAs, they can repress the translation of target mRNA or facilitate its cleavage and elimination [11]. MiRNAs are promiscuous post-transcriptional regulators, due to their short "seed" sequence, a miRNA can interact with a large number of mRNAs, that way miRNAs are involved in the regulation of almost all important cellular, developmental, and pathological processes [12]. It is well established that miRNAs are present and can be reliably detected in blood plasma since these circulating miRNAs are very stable, which is the result of their packaging into vesicles or interaction with proteins that protect miRNAs from RNase digestion. These features could make them not only ideal candidates for non-invasive plasma-based biomarkers but also regulatory factors contributing to tumor progression [7,13].

Circulating miRNAs could be released from cells through an active, regulated secretion process packaged into exosomes or microvesicles [14]. Exosomes were shown to be involved in intercellular communication since they carry various bioactive molecules, among them miRNAs [7,15]. Exosomes can interact, fuse with the membrane of cells, and deliver their cargo to recipient target cells and thus modify their gene expression pattern [16–21]. Exosomes are considered as part of the tumor environment playing important roles in pre-metastatic environment formation, tumor progression, epithelial-to-mesenchymal transition, and the modulation of immune regulation [14]. MiRNAs could be secreted from cells via binding to protein containing complexes like AGO2 or HDL. Although the functional role of circulating miRNAs is still largely unknown, it is known that the dysregulation of miRNA expression and the presence of circulating miRNAs have been linked to the formation of cancer among them OC, as well [7,10,22]. It was shown that miRNAs could be oncomiRs or tumor suppressors, however this categorization is not straightforward, due to their extensive palette of target genes the same miRNA could play opposing roles in different processes [23].

Several differentially expressed miRNAs were found in the plasma of OC patients, like members of the miR-200 family, miR-141, miR-125b, miR-222-3p or Let-7 among others [14,24,25]. Despite the large number of reports no consensus regarding the circulating miRNA signature has been suggested so far, which would unambiguously distinguish OC patients from healthy individuals and an explanation of their potential biological significance in OC is still not clarified completely. The fact that ovarian cancer cells are rarely disseminated through the vasculature makes the interpretation of the pathophysiological role of circulating miRNAs difficult [1,15,26].

The aim of this study was to analyze the circulating miRNA expression profile in serous epithelial OC patients and compare it to that of healthy individuals in order to contribute to the growing body of data available to establish a useful miRNA set for OC diagnosis obtained by the non-invasive liquid biopsy. We attempted to identify the miRNA expression profile, which is specific for tumor samples. We have also aimed to investigate the possible involvement of circulating miRNAs in OC development by analyzing the biological importance of miRNA targets and by the functional enrichment profile of their target gene set.

2. Results

2.1. Expresion Levels of miRNAs in Plasma Samples of Ovarian Cancer Patients and Healthy Controls

In order to compare the plasma miRNA expression profile between healthy females and OC patients, 18 patients and six healthy control females were recruited in the study. The age range of the patients was from 43 to 75 years with a mean age of 58.3 year, the same values for controls were 59.2 ranging from 62 to 71. The clinical stage and tumor subtype were established for patients and based on their immunohistochemical analysis all patients belong to the serous epithelial ovarian cancer group. Patients were stratified into FIGO I, FIGO III, and FIGO IV stages according to the recommendations of the International Federation of Gynecology and Obstetrics. Six patients from each group were included in the study together with six healthy females and their plasma samples were collected for RNA extraction and miRNA analysis. The demographic and clinical data of the patients are shown in Supplementary Table S1. Plasma miRNA profiles were determined by using the nCounter Human v3 miRNA Panel of NanosString nCounter Analysis System (NanoString Technologies). Normalized counts differed considerably for miRNAs and individuals, however, most miRNA counts were low, especially in controls, with few exceptions. Hsa-miR-451a had the highest count with group mean values ranging from 1217 up to 3392, followed by hsa-miR-4454 and hsa-miR-499a-5p (group mean values: Between 36 and 89; 56 and 123, respectively). None of these highly expressed miRNAs showed significantly different expression among the cohorts. It might be important to note that the miRNA counts were the highest in samples of FIGO I stage patients.

Altogether 45 out of the 798 unique miRNAs showed significant differences in counts between tumor and normal plasma samples. Among them, 17 miRNAs showed significantly different expressions between controls and patients (Group 1 miRNAs). The rest, 28 miRNAs were expressed only in patients, 19 out of 28 were found only in FIGO I stage samples (Group 2 miRNAs) and nine were present in all FIGO stages (Group 3 miRNAs) (Table 1).

Table 1. List of microRNAs (miRNAs) showing significantly different expression patterns among controls and the three ovarian cancer (OC) patient groups.

Group 1 [1]	Group 2	Group 3
hsa-miRNA ID	hsa-miRNA ID	hsa-miRNA ID
hsa-miR-1185-2-3p	hsa-miR-1185-1-3p	hsa-miR-125a-3p
hsa-miR-553	hsa-miR-1197	hsa-miR-1281
hsa-miR-144-3p	hsa-miR-1266-5p	hsa-miR-128-2-5p
hsa-miR-146b-5p	hsa-miR-149-5p	hsa-miR-1305
hsa-miR-148b-3p	hsa-miR-23a-3p	hsa-miR-223-3p
hsa-miR-1976	hsa-miR-3161	hsa-miR-325
hsa-miR-19b-3p	hsa-miR-331-3p	hsa-miR-497-5p
hsa-miR-526a	hsa-miR-331-5p	hsa-miR-500a-5p
hsa-miR-219a-2-3p	hsa-miR-337-5p	hsa-miR-548h-3p
hsa-miR-25-3p	hsa-miR-3615	
hsa-miR-26b-5p	hsa-miR-409-3p	
hsa-miR-301a-3p	hsa-miR-4455	
hsa-miR-513a-3p	hsa-miR-498	
hsa-miR-552-3p	hsa-miR-520g-3p	
hsa-miR-584-5p	hsa-miR-584-5p	
hsa-miR-613	hsa-miR-590-5p	
hsa-miR-615-5p	hsa-miR-625-5p	
	hsa-miR-628-5p	
	hsa-miR-651-5p	

[1] Group 1: Significantly differentially expressed miRNAs. Group 2: MicroRNAs expressed only in FIGO I stage patients. Group 3: MicroRNAs expressed only in patients, but in all stages.

In Group 1 samples, the fold changes between the stage specific mean expression values were calculated for all miRNAs showing that 16 miRNAs were upregulated and hsa-miR-584-5p was downregulated in this group (Figure 1a). Typically, the largest fold change was found between the control and FIGO I stage expression values for most of the significantly differentially expressed miRNAs. Hierarchical clustering of the differentially expressed circulating miRNAs showed a distinct expression pattern between the three groups (Figure 1b).

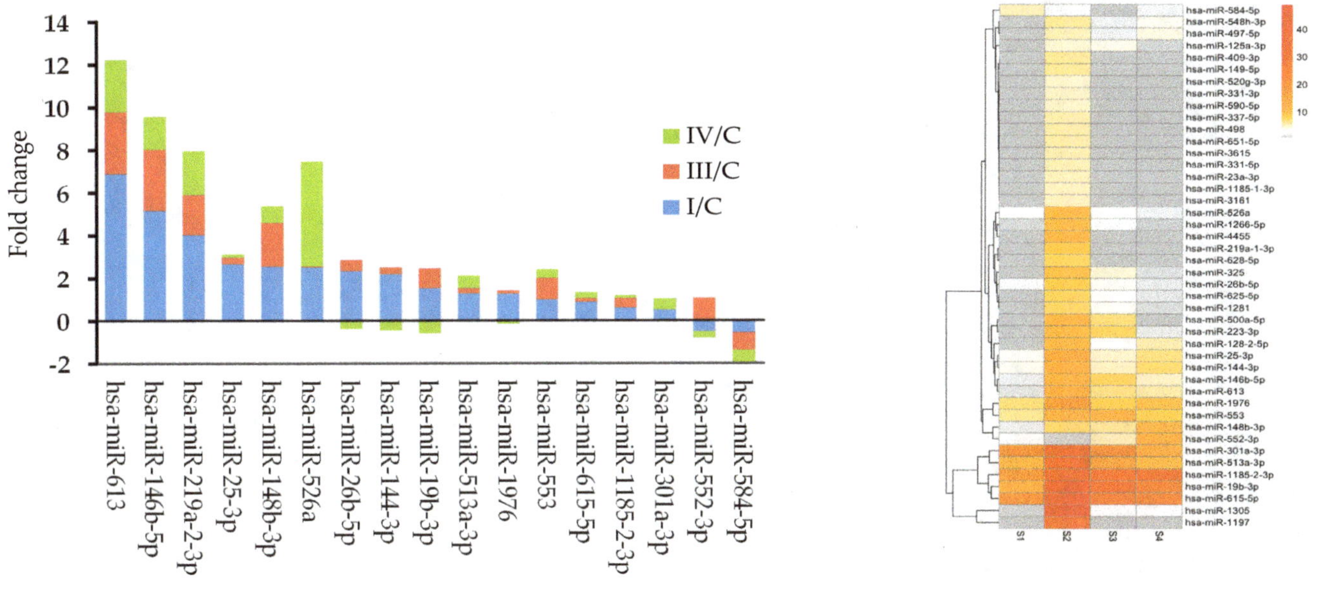

(a) (b)

Figure 1. (**a**) Stacked bar chart showing the fold changes between the stage specific expression of the significantly differentially expressed miRNAs. (**b**) Heat map of the differentially expressed circulating miRNAs. The expression cluster shows upregulated miRNAs in deeper color according to the scale on the right of the figure. S1 represent controls, S2, S3, and S4 represent the FIGO I, FIGO III, and FIGO IV samples, respectively.

2.2. Validation of Differentially Expressed Group 1 miRNAs by RT-qPCR

To validate our findings, six Group 1 miRNAs (hsa-miR-19b-3p, hsa-miR-25-3p, hsa-miR-26b-5p, hsa-miR-144-3p, hsa-miR-148b-3p, and hsa-miR-301a-3p) were randomly chosen together with hsa-miR-197, which did not show differential expression between the groups. The relative expression of these miRNAs was determined by RT-qPCR measurements using hsa-miR-103-3p as the reference miRNA. Expression of all six Group 1 miRNAs was significantly upregulated compared with those in the control samples, while the expression of hsa-miR-197 did not show significant difference between the groups (the Kruskal–Wallis p-values were the following: hsa-miR-19b-3p: 0.03572; hsa-miR-25-3p: 0.0071; hsa-miR-26b-5p: 0.0096; hsa-miR-144-3p: 0.01828; hsa-miR-148b-3p: 0.00014; hsa-miR-301a-3p: 0.01078; hsa-miR-197: 0,741) (Figure 2). Considering the perfect positive agreement between the results obtained with the nCounter Human v3 miRNA Panel and PCR measurements, the RT-qPCR method validated our results.

2.3. MIRNA Ranking, Target Gene Prediction and Analysis

To study the possible functional role of circulating miRNAs, the differentially expressed miRNAs and their targets were analyzed. The fact that a single miRNA has several target mRNAs and translation of a mRNA can be regulated by several miRNAs warrants a network-based analysis of the miRNA function. First, the miRNet tool was used to predict miRNA targets and interactions between miRNAs and targets in order to determine the importance of miRNAs in the network. The interaction networks were constructed separately for the three groups of miRNAs. Table 2 shows the lists of miRNAs ranked by their degree centrality value (betweenness centrality provided the same ranking) that represent the importance of given miRNAs in the interacting network. (The complete networks are shown in Supplementary Figure S1a–c)

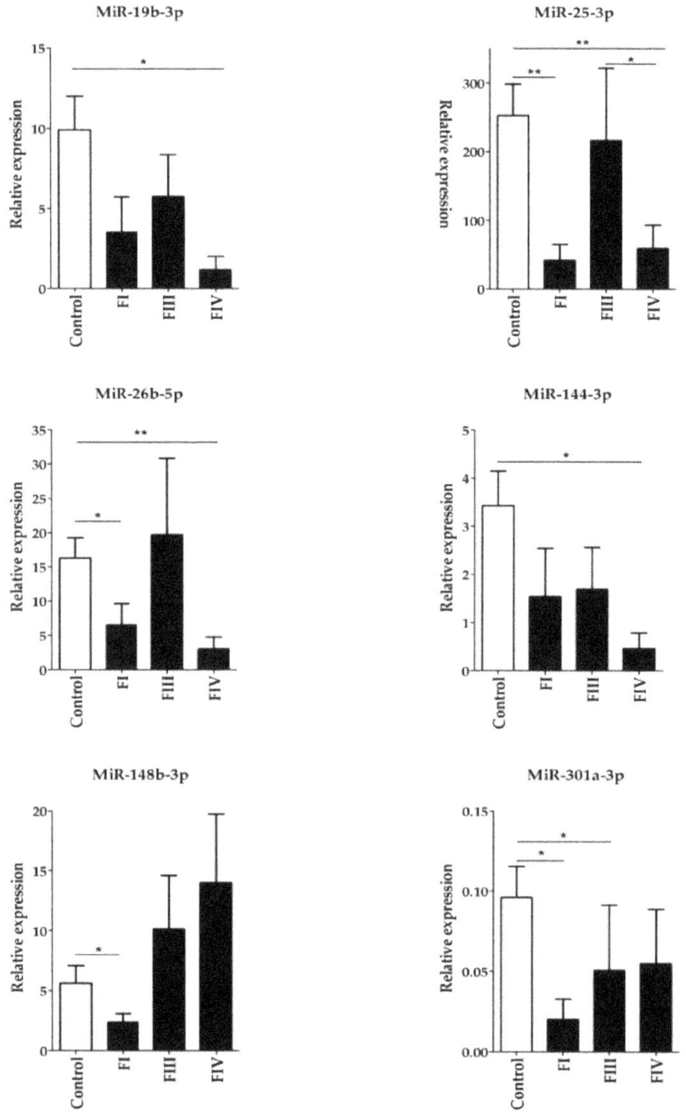

Figure 2. Validation of six randomly chosen significantly differentially expressed Group 1 miRNAs using the RT-qPCR measurement. Expression levels of plasma miRNAs were compared between ovarian cancer patients and control females. Total miRNA was isolated from plasma samples and the amounts of mature hsa-miRNAs was determined by the miScript PCR System. The expression of PCR products was normalized to hsa-miR-103-3p and relative miRNA expression levels were determined by the $2^{-\Delta Ct}$ method. All measurements were done in triplicate. Data distribution was analyzed by the Kruskal–Wallis one-way ANOVA test with Dunn's *post-hoc* analysis, p-values shown in the figure are as follows: *: $P < 0.05$; **: $P < 0.01$.

To demonstrate and visualize the most important miRNA–target interactions, we have constructed core miRNA–target networks to show the strongest interactions among the differentially expressed miRNAs and target genes by the mirTargeLink tool (Figure 3). Just like in the miRNet network, hsa-miR-19b-3p, hsa-miR-26b-5p, hsa-miR-25-3p, and hsa-miR-301a-3p have central importance in the core network of Group 1 miRNAs. Their targets, PTEN, EZH2, KAT2B, BCL2L11, TP53, SMAD4, and ERBB2 are known to be involved in tumorogenesis (Figure 3A).

Among Group 2 miRNAs, hsa-miR-23a-3p, hsa-miR-498, hsa-miR-331-3p, and hsa-miR-625-5p have central importance. Hsa-miR-23a-3p targets HMGN2, FOXO3, PPP2R5E, LDHA, and ATAT1. Hsa-miR-498, hsa-miR-331-3p, and hsa-miR-625-5p have two common targets, the tumor suppressor FHIT gene and the proto-oncogene ERBB2 (Figure 3B). In the third miRNA group

(Group 3) hsa-miR-223-3p and hsa-miR-497-5p have a central role. Their common target IGF1R is a proto-oncogene, which is highly expressed in several tumor cells, CDK4 and CDC25A are proto-oncogenes (Figure 3C) [27–30].

Table 2. Ranking of the differentially expressed miRNAs based on their degree centrality values in the miRNet network.

Group 1		Group 2		Group 3	
miRNA ID	**Degree**	**miRNA ID**	**Degree**	**miRNA ID**	**Degree**
hsa-mir-26b-5p	1874	hsa-mir-331-3p	406	hsa-mir-497-5p	461
hsa-mir-19b-3p	714	hsa-mir-520g-3p	404	hsa-mir-125a-3p	310
hsa-mir-25-3p	518	hsa-mir-149-5p	397	hsa-mir-548h-3p	292
hsa-mir-1976	501	hsa-mir-498	320	hsa-mir-1305	195
hsa-mir-148b-3p	403	hsa-mir-23a-3p	249	hsa-mir-1281	180
hsa-mir-301a-3p	395	hsa-mir-625-5p	227	hsa-mir-500a-5p	145
hsa-mir-144-3p	211	hsa-mir-4455	165	hsa-mir-223-3p	98
hsa-mir-513a-3p	187	hsa-mir-1185-1-3p	117	hsa-mir-325	32
hsa-mir-552-3p	167	hsa-mir-3161	115		6
hsa-mir-146b-5p	121	hsa-mir-409-3p	111		
hsa-mir-615-5p	70	hsa-mir-1197	74		
hsa-mir-584-5p	67	hsa-mir-584-5p	67		
hsa-mir-219a-2-3p	63	hsa-mir-590-5p	66		
hsa-mir-526a	61	hsa-mir-651-5p	65		
		hsa-mir-331-5p	63		
		hsa-mir-628-5p	51		
		hsa-mir-3615	39		
		hsa-mir-337-5p	7		

In addition to using miRNet, the TargetScan and miRTarBase databases were also used to predict target genes of the differentially expressed miRNAs, only experimentally validated targets were chosen from all three databases (Supplementary Table S2). In order to assess miRNA expression patterns in OC, we further analyzed our data to recognize common patterns of targets among the different samples to focus our analysis on their similarities. Intersections of common targets of the three differently expressed miRNA groups revealed 54 miRNA targets that were simultaneously differentially targeted by the different miRNA groups. Of these targets, we identified five genes that were targeted by all three miRNA groups. Figure 4 and Table 3 show the intersections of common targets of the three differently expressed miRNA groups.

A.

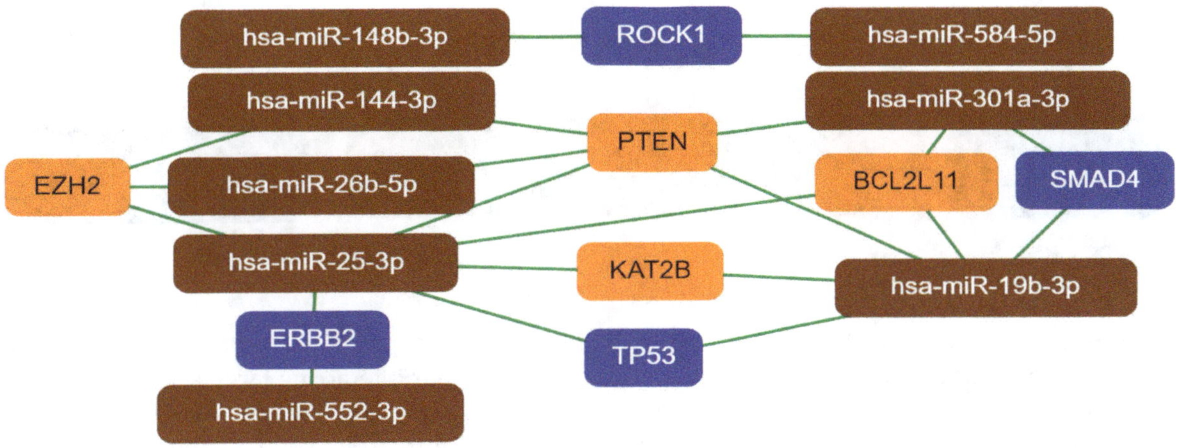

Figure 3. *Cont.*

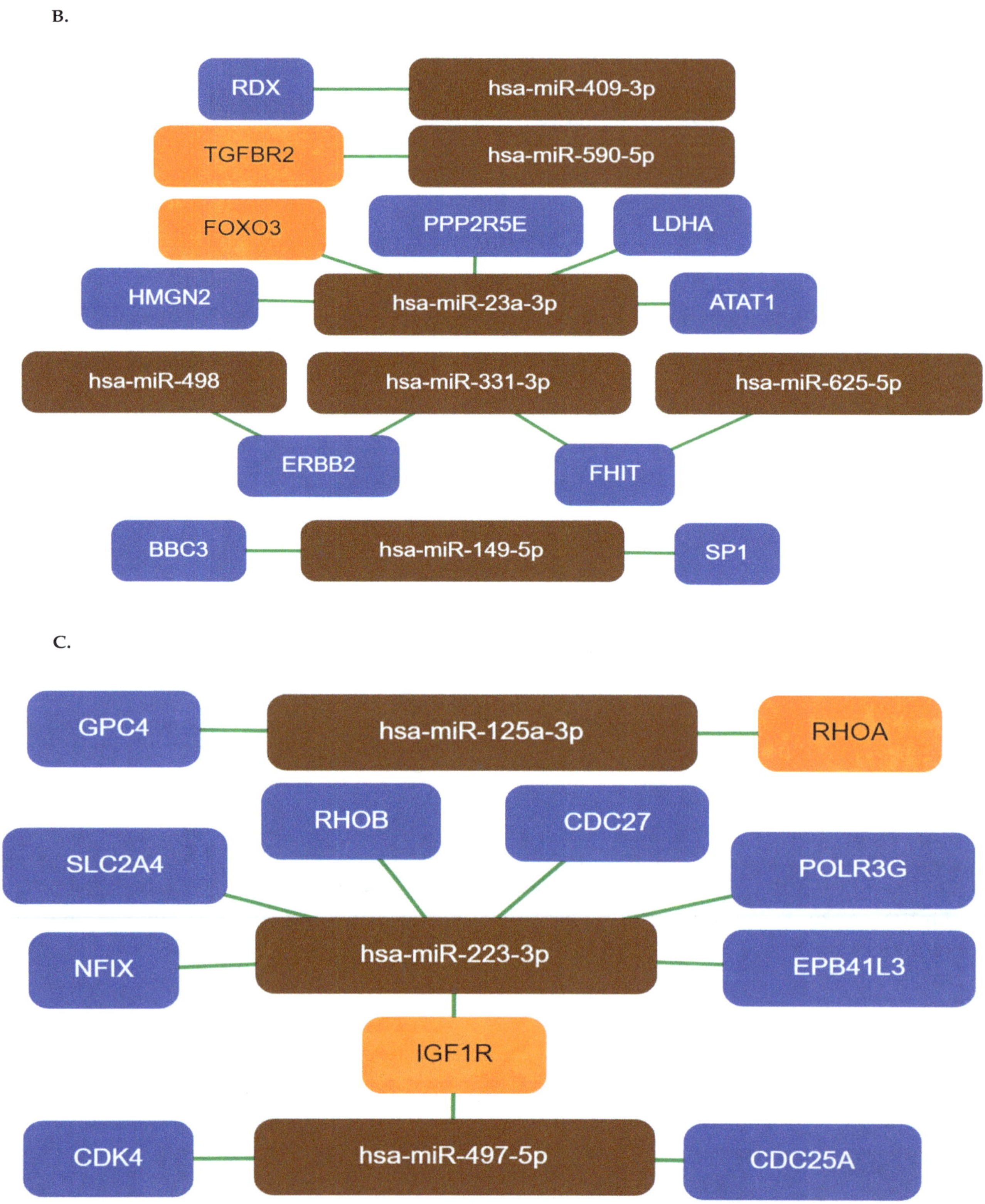

Figure 3. The core networks of differentially expressed miRNAs and their experimentally validated target genes. Group 1, Group 2, and Group 3 miRNAs and their interacting targets are represented in part **A**, **B**, and **C**, respectively. The networks were generated by the mirTargeLink tool using the strong interaction option. Isolated networks are also shown in the figure. Color code: Orange, more than two interactions; blue, two interactions in the full network.

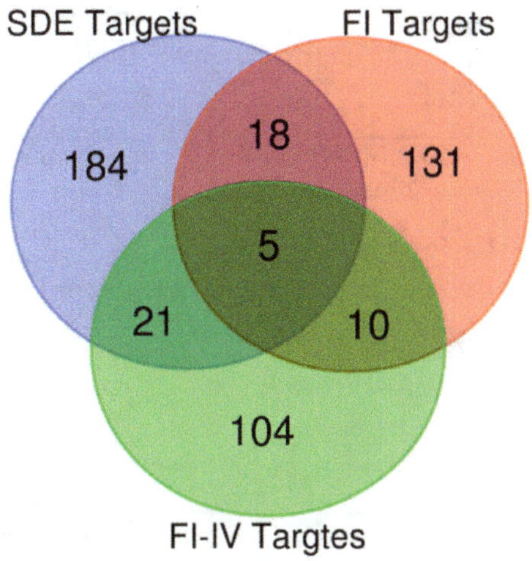

Figure 4. Venn diagram showing the common targets of the three differentially expressed miRNA groups. The SDE, FI, and FI-IV labels represent Group 1, Group 2, and Group 3 miRNAs, respectively. SDE: significantly differentially expressed miRNAs; FI: miRNAs expressed in FIGO I stage patients; FI-FIV: miRNAs expressed in all patients.

Table 3. Shared targets of the three differentially expressed miRNA groups.

Compared Groups	Common Targets	Genes
Gr1/Gr2/Gr3	5	MET SMAD7 EZH2 TERT IL6
Gr2/Gr1	18	TLR4 MTOR IGF1 ZEB1 SOX4 PTBP2 BIRC5 CCNE1 RHOB MMP16 IGF1R MYC PXK SOCS1 PBX3 PRKAA1 CFTR FBXW7
Gr1/Gr3	21	ROCK1 KCNJ6 PTEN TGFBR2 HOTAIR PLEKHA1 ERBB2 FGA CDH1 PPP2R5E FGG IRS1 RECK DDX3X FGB HLA-G WWP1 RB1 TCEAL1 HSP90AA1 ZFX
Gr2/Gr3	10	GIT1 NPM3 BRCA1 CHUK STAT3 SP1 FOXO3 VEGFA E2F1 MEF2C

In Table 3, the HOTAIR lncRNA is also present as the target of miRNAs. This is not the only identified lncRNA target, significantly differentially expressed miRNAs interact with XIST, HOTAIR, MALAT1, NEFL, KCNQ1OT1, and CTA-204B4.6, as major lncRNA hubs in the network (data not shown).

2.4. Pathway and Gene Ontology Enrichment Analysis of miRNA Targets

To obtain a more precise understanding of the potential pathophysiological role of the differentially expressed miRNAs in the OC development, a functional annotation and enrichment analysis of their target genes in the gene ontology biological process (GO-BP) terms and in canonical pathways (in the KEGG database) was performed using the database for annotation, visualization, and integrated discovery tool (DAVID) [31]. Our analysis resulted in a large number of enriched functional categories (pathways and GO terms, too), some of them very general including a large number of target genes such as hsa05200:Pathways in cancer, hsa05206:MicroRNAs in cancer, hsa04110:Cell cycle; or for the GO-BP terms: GO:0000165~MAPK cascade; GO:0045944~positive regulation of transcription from RNA polymerase II promoter; GO:0043066~negative regulation of apoptotic process or GO:0008283~cell proliferation. We have ranked the categories according to the significance of target enrichment and the top 25 KEGG pathways and GO-BP terms are shown in Supplementary Table S3. Several cancer types (glioma, prostate cancer, chronic myeloid leukemia, bladder cancer, non-small cell lung cancer, breast cancer, colorectal cancer, renal cell carcinoma, among others) and viral infectious pathways (Hepatitis C, HTLV-I infection, Epstein-Barr virus infection) were also identified in our analysis, however these categories are not included in the Supplementary table. Some of the enriched KEGG pathways and GO terms are shared by targets of more than one miRNA groups and these are shown in Tables 4 and 5, respectively.

Table 4. Functional annotation of target genes based on their enrichment in specific Kyoto Encyclopedia of Genes and Genomes (KEGG) pathways. The top 25 most significant pathways are shown, which are targeted by at least two miRNA groups.

Targets of All miRNA groups	Targets of Group 1 and Group 2 miRNAs	Targets of Group 2 and Group 3 miRNAs	Targets of Group 1 and Group 3 miRNAs
hsa04068:FoxO signaling pathway	hsa05213:Endometrial cancer	hsa04066:HIF-1 signaling pathway	hsa04110:Cell cycle
hsa04115:p53 signaling pathway	hsa04390:Hippo signaling pathway	hsa05206:MicroRNAs in cancer	hsa04012:ErbB signaling pathway
hsa04151:PI3K-Akt signaling pathway	hsa04722:Neurotrophin signaling pathway	hsa04621:NOD-like receptor signaling pathway	hsa04150:mTOR signaling pathway
hsa04152:AMPK signaling pathway		hsa04550:Signaling pathways regulating pluripotency of stem cells	hsa04917:Prolactin signaling pathway
hsa04350:TGF-beta signaling pathway		hsa05202:Transcriptional misregulation in cancer	
hsa04510:Focal adhesion			
hsa04931:Insulin resistance			
hsa05200:Pathways in cancer			
hsa05205:Proteoglycans in cancer			
hsa05230:Central carbon metabolism in cancer			

Table 5. Functional annotation of target genes based on their enrichment in gene ontology (GO_ biological processes. Those members of the top 25 GO terms are shown, which are enriched by all three miRNA group targets.

GO Biological Process
GO:0010628~positive regulation of gene expression
GO:0008284~positive regulation of cell proliferation
GO:0008285~negative regulation of cell proliferation
GO:0071456~cellular response to hypoxia
GO:0045892~negative regulation of transcription, DNA-templated
GO:0042517~positive regulation of tyrosine phosphorylation of Stat3 protein
GO:0043066~negative regulation of apoptotic process

At the same time targets of the three differentially expressed miRNA groups are also enriched in pathways unique to a given miRNA group. These unique pathways are shown in Table 6.

Table 6. The unique functional annotation of target genes of a given miRNA group based on their enrichment in specific KEGG pathways.

Targets of Group 1 miRNAs	Targets of Group 2 miRNAs	Targets of Group 3 miRNAs
hsa04919:Thyroid hormone signaling pathway	hsa04920:Adipocytokine signaling pathway	hsa04630:Jak-STAT signaling pathway
hsa05203:Viral carcinogenesis	hsa04520:Adherens junction	hsa04910:Insulin signaling pathway
hsa04620:Toll-like receptor signaling pathway	hsa04210:Apoptosis	hsa04660:T cell receptor signaling pathway
hsa04014:Ras signaling pathway	hsa04922:Glucagon signaling pathway	hsa04062:Chemokine signaling pathway
hsa04015:Rap1 signaling pathway	hsa04064:NF-kappa B signaling pathway	hsa04914:Progesterone-mediated oocyte maturation
hsa04071:Sphingolipid signaling pathway	hsa04915:Estrogen signaling pathway	
	hsa04010:MAPK signaling pathway	

2.5. Protein–Protein Interaction Network Analysis of miRNA Targets

Subsequently targets of the three different miRNA groups were collapsed into a single non-redundant target list, which was used to construct protein–protein interaction (PPI) networks. First, a general PPI network, then an ovary-specific PPI network was constructed from the target lists by using the NetworkAnalyst tool. Both networks proved to be a large fuzzy network, in the general

PPI there are 3168 nodes and 5379 edges, the ovary-specific network contains 2353 edges and 3361 nodes. In Figure 5a,b, the general and ovary-specific minimum networks are shown with major hubs labeled, respectively.

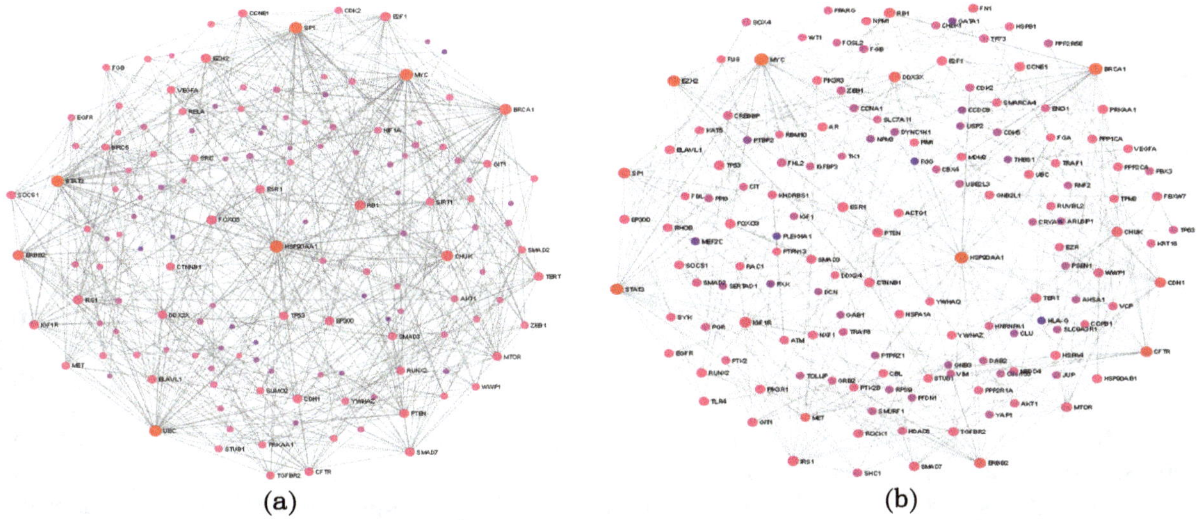

Figure 5. Topology of the general and ovary-specific protein–protein interaction (PPI) networks constructed from the common targets of differentially expressed miRNAs using the NetworkAnalyst tool. Part (**a**) and (**b**): The general and ovary-specific minimum PPI networks, respectively. Nodes represent proteins, only the major hub nodes are labeled in the networks.

The size of the nodes in Figure 5 corresponds to their degree centrality (and betweenness centrality) values and the nodes in the network with large degree centrality are considered to be key nodes or hubs with biological importance. Most of the major hubs overlap in the two networks and represent proteins, which are already known to be involved in tumorogenesis, such as HSP90AA1, MYC, SP1, BRCA1, RB1, CFTR, STAT3, E2F1, ERBB2, EZH2, and MET among others. The ranking of nodes based on degree centrality, however, is not identical in the two PPI networks: MYC, BRCA1, CFTR, EZH2, and STAT3 have higher ranks in the ovary-specific network. These interacting proteins occupy a central position in the network, so they could be considered as key biological factors in OC development. An advantage of the network-based approach is that it provides a mean to discover novel proteins, which interact physically and functionally with the seed proteins and may represent new cancer genes or cancer biomarkers.

The NetworkAnalyst tool could carry out a network-based functional enrichment and pathway analysis based on the gene ontology and KEGG databases, too. That way it was possible to compare results of the general analysis with results of an ovary-specific enrichment analysis. The results are shown in Tables 7 and 8.

These functional enrichment results suggest that circulating plasma miRNAs are not randomly released from cells since many of their predicted target genes are enriched in critically important pathways and biological processes contributing to tumorogenesis, such as TGFβ signaling pathway, NF-kappaB signaling pathway, VEGF signaling pathway, Rap1 signaling pathway, Ras signaling pathway, ErbB signaling pathway, Focal adhesion, MAPK signaling pathway FoxO signaling pathway, Proteoglycans in cancer, PI3K-Akt signaling pathway, Focal adhesion, AGE-RAGE signaling pathway, JNK cascade, Peptidyl-Tyr-phosphorylation, Phosphatidylinositol 3-kinase signaling, and SMAD protein signal transduction. In the case of OC, the estrogen signaling pathway could have specific importance [32]. The functional enrichment analysis also revealed several cancer types; however, those are not listed in Tables 7 and 8.

Table 7. KEGG pathways-based general and ovary-specific functional enrichment analysis of all target genes of differentially expressed miRNAs.

General Analysis [1] KEGG Pathway	P Value	Ovary-Specific Analysis KEGG Pathway	P Value
Pathways in cancer	2.3564×10^{-29}	Pathways in cancer	2.26×10^{-38}
Central carbon metabolism in cancer	3.8587×10^{-11}	Epstein-Barr virus infection	1.54×10^{-29}
Endometrial cancer	1.1410×10^{-8}	Cell cycle	5.79×10^{-25}
Insulin resistance	5.5166×10^{-8}	Cellular senescence	1.3×10^{-24}
TGF-beta signaling pathway	2.0410×10^{-7}	ErbB signaling pathway	3.01×10^{-24}
Toll-like receptor signaling path.	2.1829×10^{-7}	MAPK signaling pathway	5.21×10^{-24}
NF-kappa B signaling pathway	3.3876×10^{-7}	FoxO signaling pathway	3.67×10^{-23}
AMPK signaling pathway	4.6513×10^{-7}	Proteoglycans in cancer	8.67×10^{-23}
Prolactin signaling pathway	7.7049×10^{-7}	Ubiquitin mediated proteolysis	1.7×10^{-21}
Ras signaling pathway	1.4145×10^{-6}	PI3K-Akt signaling pathway	2.24×10^{-20}
ErbB signaling pathway	1.8772×10^{-6}	Prolactin signaling pathway	3.53×10^{-20}
Focal adhesion	2.9277×10^{-6}	Focal adhesion	7.12×10^{-20}
mTOR signaling pathway	2.8262×10^{-6}	AGE-RAGE signaling pathway	4.55×10^{-18}
Insulin signaling pathway	1.1171×10^{-5}	T cell receptor signaling pathway	1.02×10^{-16}
T cell receptor signaling pathway	1.1184×10^{-5}	Adherens junction	1.24×10^{-16}
Chemokine signaling pathway	6.4640×10^{-5}	TNF signaling pathway	1.47×10^{-16}
VEGF signaling pathway	9.9975×10^{-5}	Estrogen signaling pathway	2.35×10^{-16}
cAMP signaling pathway	1.2228×10^{-4}	NF-kappa B signaling pathway	3.44×10^{-16}
Rap1 signaling pathway	1.7607×10^{-4}	Transcriptional regulation in cancer	3.8×10^{-16}
Estrogen signaling pathway	0.0022	TGF-beta signaling pathway	1.84×10^{-14}

[1] In general analysis tissue specific expression is not considered.

Table 8. Gene ontology-based general and ovary-specific functional enrichment of all target genes of differentially expressed miRNAs.

General Analysis [1] GO Biological Process	P Value	Ovary-Specific Analysis GO Biological Process	P Value
Phosphatidylinositol-mediated signaling	4.5190×10^{-11}	Phosphorylation	4.62×10^{-61}
TGFβ receptor signaling pathway	1.9554×10^{-8}	Regulation of protein modification process	1.14×10^{-51}
MAPK cascade	4.6884×10^{-7}	Regulation of transferase activity	2.69×10^{-46}
Peptidyl-Tyr-phosphorylation	6.050×10^{-7}	Regulation of kinase activity	2.5×10^{-45}
Phosphatidylinositol 3-kinase signaling	7.1633×10^{-7}	Enzyme linked receptor protein signaling	1.45×10^{-43}
Positive regulation of Tyr- phosphorylation of Stat3 protein.	9.8981×10^{-7}	Regulation of cell cycle	1.46×10^{-41}
I-κB kinase/NF-kappaB signaling	2.9099×10^{-6}	Regulation of protein kinase activity	6.19×10^{-41}
IL6-mediated signaling pathway	5.8361×10^{-6}	Cell proliferation	2.65×10^{-40}
Positive regulation of pri-miRNA transcription	1.2812×10^{-7}	Cellular response to stress	1.15×10^{-37}
Positive regulation of EMT	2.7940×10^{-5}	Intracellular protein kinase cascade	2.57×10^{-37}
JNK cascade	5.6629×10^{-5}	Cell cycle	3.26×10^{-37}
Response to calcium ion	1.4710×10^{-4}	Positive regulation of RNA metabolic process	1.38×10^{-36}
SMAD protein signal transduction	2.0946×10^{-4}	Regulation of programmed cell death	3.09×10^{-35}
Insulin signaling pathway	2.0956×10^{-4}	Positive regulation of signal transduction	4.45×10^{-35}
T cell receptor signaling pathway	2.9929×10^{-4}	Regulation of cell proliferation	1.18×10^{-34}
Positive regulation of GTPase activity	8.5273×10^{-4}	Negative regulation of apoptotic process	6.4×10^{-34}
Toll-like receptor signaling pathway	9.5585×10^{-4}	Intracellular signal transduction	8.73×10^{-34}
Cell-matrix adhesion	0.0010	Reproduction	8.6×10^{-33}
Heterotypic cell-cell adhesion	0.0012	Negative regulation of transcription	1.78×10^{-32}
SMAD protein complex assembly	0.0017	Negative regulation of nucleobase_containing compound metabolic process	3.1×10^{-29}

[1] In general analysis tissue specific expression is not considered.

3. Discussion

In recent years several groups examined the biological importance of cell free miRNAs present in body fluids. It was suggested that circulating miRNAs have the potential to become non-invasive biomarkers for the early diagnosis of cancer [9,10,24]. It is equally interesting however, to study the pathophysiological role of these miRNAs, since circulating miRNAs released from cells are known to be involved in intercellular communication and dysregulation of miRNAs in tissues is known to be associated with several cancers [7,9,10,14].

We have compared the expression profiles of circulating miRNA in blood plasma samples of six healthy females and 18 OC patients. Patients were divided into FIGO I, FIGO III, and FIGO IV stages, having six patients in each group. The nCounter Human v3 miRNA Panel of the NanoString System was used to measure the miRNA levels. MiRNA counts were low for most of the miRNAs, especially in the control samples. This might be due to the detection method, the NanoString method does not require an amplification step, so it is clearly different from those methods that use PCR for the miRNA measurement. It might suggest that the NanoString method is less sensitive for low abundance miRNAs.

Comparing miRNA expression profiles in the control and OC patient samples we have identified 45 miRNAs showing different expression between controls and patients. Our data showed that 17 miRNAs out of 45 were present both in the control and patient plasma (Group 1 miRNAs), however their expression levels differed significantly between the four groups. With the exception of the tumor suppressor has-miR-584-5p all Group 1 miRNAs were upregulated. 19 miRNAs were found only in samples of FIGO I patients (Group 2 miRNA) and nine miRNAs were detected in all three patient groups but were absent in control samples (Group 3 miRNAs). The finding of miRNAs, which are present only in patient samples might be important from a diagnostic point of view, as it shows that circulating miRNAs have the potential to be used as non-invasive biomarkers. Our sample number, however, is too low to draw any firm conclusions.

The large majority of our differentially expressed miRNAs have been previously reported to play a role in different cancer types. A few of our differentially expressed miRNAs, however, were found to be associated with OC in previous reports (i.e., hsa-miR-144-3p, hsa-miR-337-5p, hsa-miR-500a-5p, hsa-miR-26b-5p, hsa-miR-125a-3p, hsa-miR-19b-3p) [7,14,24,26]. Using the miRNet tool and a network-based approach we have constructed a miRNA–target interaction network for the miRNA groups. MiRNAs were ranked based on their degree-centrality value in the network, which reflects their biological importance. Hsa-mir-26b-5p, hsa-mir-19b-3p, and hsa-mir-25-3p were the top three ranked miRNAs for Group 1, hsa-mir-331-3p, hsa-mir-520g-3p, and hsa-mir-149-5p for Group 2 and hsa-mir-497-5p, hsa-mir-125a-3p, and hsa-mir-223-3p were the top three miRNAs for Group 3. The key miRNA–target interactions were visualized by the miTargetLink tool. For the Group 1 miRNAs, PTEN, EZH2, KAT2B, BCL2L11, TP53, SMAD4, and ERBB2 are the main targets, all known to be involved in tumorogenesis. PTEN and TP53 are known tumor suppressor proteins—impairing of KAT2B activity may contribute to genome instability; both oncogenic and tumor suppressive effects of EZH2 have been demonstrated in different cancer types and its expression is known to be regulated by miRNAs [33,34]. BCL2L11 is a tumor suppressor, it is an important regulator of apoptosis; loss of the SMAD4 activity may disrupt DNA damage response and repair mechanisms and enhance genomic instability [35,36]. The downregulation of these genes is in agreement with tumor formation, however the receptor tyrosine kinase ERBB2 is a proto-oncogene, the role of its downregulation by has-miR-25-3p and hsa-miR-552-3p is not known [37].

HMGN2, FOXO3, PPP2R5E, LDHA, ATAT1, FHIT, and the ERBB2 genes are the main targets for top Group 2 miRNAs, while the IGF1R, CDK4, CDC25A, SLC2A4/GLUT4, RHOB, CDC27, and POLR3G are the major targets for miRNAs of the third group. HMGN2 is an anti-tumor effector molecule of $CD8^{+}T$ cells, FOXO3 is a core tumor suppressor in breast cancer; downregulation of PPP2R5E is a common event in acute myeloid leukemia [38–40]. However, the ATAT1 activity is required for microtubule organization, it is specifically upregulated in colon cancer tissue and LDHA has an aberrantly high expression in multiple cancers [41,42].

Overexpression of CDC25A is known to be associated with malignancy and poor prognosis in cancer patients [30]. Hsa-miR-223-3p targets the EPB41L3, a potential tumor suppressor gene, the NFIX gene that regulates both cell proliferation and migration, the SLC2A4/GLUT4 is a glucose transporter and a biomarker for many types of malignant tumors [43–45]. A dualistic role of RHOB was reported, it could be a proto-oncogene or a tumor suppressor depending on the context of cancer development and progression. CDC27 is a tumor suppressor, its downregulation inhibits the proliferation of cancer cells [46], POLR3G is required for proliferation, its depletion triggers proliferative arrest and differentiation of prostate cancer cells [47].

Considering the negative regulatory role of miRNAs, it is noteworthy to find that not only tumor suppressor genes but proto-oncogenes are also present among the interacting targets. However, it is known, that miRNAs could have dualistic effect, a given miRNA can be an oncomiR or tumor suppressor depending on the cellular context [23]. The miRTarBase and TargetLink databases were also used to predict experimentally validated targets of the differentially expressed miRNAs, several common targets were identified for three miRNA groups, the MET, SMAD7, EZH2, TERT, and IL6 genes for example are targeted by at least one member of each different miRNA group. The tyrosine–protein kinase MET is a proto-oncogene, its role in cell migration and in epithelial–mesenchymal transition (EMT) is well known [48]. SMAD7 has a tumor suppressing role through blocking the TGF-β-stimulated cancer progression by increasing angiogenesis and inducing EMT [49]. The telomerase reverse transcriptase TERT gene plays a crucial step in tumorogenesis, it is required to maintain the telomere length and telomerase activity to gain immortality [50]. IL6 is an inflammatory cytokine which promotes metastasis in OC [51].

The miRNA group specific target lists were used in a functional annotation analysis based on the enrichment of miRNA targets in the KEGG pathways and gene ontology biological processes terms. This analysis revealed that miRNA targets are enriched in known cancer pathways, signaling pathway which are crucial pathways in tumorogenesis. To name a few, FoxO signaling pathway, p53 signaling pathway, PI3-AKT pathway AMPK pathway, TGFβ signaling pathway, focal adhesion, proteoglycans in cancer, Hippo signaling pathway, ERBB signaling pathway, JAK-STAT signaling pathway, Estrogen signaling pathway, and MAPK signaling pathway were among the most significant ones. In GO-BP terms the positive regulation of gene expression, positive regulation of cell proliferation, negative regulation of cell proliferation, cellular response to hypoxia, negative regulation of transcription, positive regulation of tyrosine phosphorylation of Stat3 protein, and negative regulation of apoptotic process were the most significant ones based on target enrichment and over-representation. All of these processes and terms are known players of tumorogenesis. The results of the enrichment analysis show that most miRNA targets are involved in signaling pathways and biological processes, which are critical for tumor formation, suggesting that circulating miRNAs could be potential regulatory factors in tumorogenesis. At the same time these data also show that the identified enriched pathways and GO terms are not specific for a given tumor type, our identified miRNA targets are associated with regulatory and signaling processes which are important in several different tumor types.

The target lists generated by the three prediction tools were merged into a single list and possible interactions between the target proteins and their functionally important interacting protein partners were analyzed by constructing general and ovary-specific PPI networks. The major hub proteins—HSP90AA1, MYC, SP1, BRCA1, RB1, CFTR, STAT3, E2F1, ERBB2, EZH2, and MET—were basically the same in the two networks, suggesting that our differentially expressed miRNAs regulate target genes which are involved in basic processes of tumor formation. This notion is strengthened by our network-based functional enrichment analysis, which provided a very similar enrichment pattern (both in KEGG and GO-BP terms) to those which were recognized by the DAVID tool for the different miRNA target groups.

In conclusion, our pilot study identified significantly differentially expressed circulating miRNAs in plasma samples of OC patients. Our functional annotation analysis showed that the experimentally validated targets of the differentially expressed miRNAs are key regulators of tumor formation, suggesting that circulating miRNAs might play an important pathophysiological role in the formation of different tumor types. On the other hand, these results also show that the differentially expressed miRNAs identified in our study have limited usefulness in the diagnosis of OC. A clear limitation of our study is the low sample size, however, we feel that our results would warrant validation in a large cohort of OC patients.

4. Materials and Methods

4.1. Patients and Samples

Twenty-four blood samples (six disease-free healthy controls, 18 serous ovarian cancer patients) were collected from the Department of Obstetrics and Gynecology, Faculty of Medicine, University of Debrecen. All patients that underwent surgery and tissue samples were histologically diagnosed. Pathological characterization of tumor stages was assessed according to the International Federation of Gynecology and Obstetrics (FIGO) criteria. None of the patients received chemotherapy or radiotherapy treatment prior to participation in the study. Each subject provided written informed consent. The study was approved by the Scientific and Research Ethics Committee of the Medical Research Council of the Ministry of Health, Budapest, Hungary (ETT TUKEB), (Project Identification Code: 30231-2/2016/EKU, date: 06 June 2016) and was conducted in accordance with the Declaration of Helsinki. Controls were followed up, none of them received gynecological treatment during the study period.

Peripheral blood (9 mL) was drawn into EDTA anticoagulated tubes (BD Vacutainer) from each patient and from healthy volunteers and kept at 4 °C until further processing (within two hours of collection). Plasma samples were subjected to a two-step centrifugation protocol (2500× *g* and 16,000× *g*; 10–10 min, 4 °C) to obtain plasma. After separation, the cell-free plasma samples were homogenized, aliquoted, and stored at −80 °C until further processing.

4.2. RNA Isolation and Purification for the NanoString Device

Prior to RNA isolation blood samples were thawed on ice, then circulating RNA was isolated from 500 μL plasma samples using the miRNeasy Serum/Plasma RNA isolation kit (Qiagen, Hilden, Germany) according to the manufacturer's protocol. The quality of the RNA was analyzed using the Nanodrop device (Thermo Scientific, Waltham, MA, USA).

4.3. RNA Expression Analysis

The miRNA content of all samples was analyzed using the nCounter Human v3 miRNA Panel of NanoString nCounter Analysis System (NanoString Technologies, Seattle, WA, USA), which contains 798 unique hsa-miRNA barcodes. 100 ng RNA/sample was used as input for the measurements, hybridization was carried out for 18 h, and miRNA counts were collected by scanning on the HIGH mode. The background correction of data was performed by subtracting the mean ± 2 standard deviation of the negative control set. Lane-by-lane technical variation was corrected by using the geometric median value of the positive code-set. The complete data set was normalized by calculating the geometric mean of 10 "housekeeping" miRNA counts for each sample to generate the normalization factor.

4.4. Prediction and Analysis of Experimentally Validated Target Genes

First, a miRNA–target gene network was constructed using the web based miRNet tool [http://www.mirnet.ca]. Top miRNAs in the network were ranked by degree and betweenness centrality values. The prediction of experimentally validated target genes of miRNAs was carried out by using the web based miRNet, miRTarBase, and TargetScan software programs (http://miRTarBase.mbc.nctu.edu; www.targetscan.org). Target intersections were further validated by the miRWalk2 database (www.http://zmf.umm.uni-heidelberg.deg). The general and ovary-specific protein–protein interaction (PPI) network of target genes was constructed using the NetworkAnalyst 3.0 tool [www.networkanalyst.ca].

4.5. Functional Annotation and Pathway Enrichment Analysis

The lists of miRNA targets was used as input and the online Database for Annotation, Visualization, and Integrated Discovery (DAVID; https://david.ncifcerf.gov) software tool was used to perform gene ontology (GO) and Kyoto Encyclopedia of Genes and Genomes (KEGG) based functional pathway enrichment analysis for the predicted target genes of prioritized differentially expressed hsa-miRNAs. The NetworkAnalyst tool was used to carry out ovary-specific enrichment analysis. A p-value of < 0.05 was considered statistically significant.

4.6. Validation of hsa-miRNA Expression by Quantitative Real-Time Polymerase Chain Reaction (RT-qPCR) on Selected hsa-miRNAs

Circulating RNA was extracted from 200 μL plasma samples of 16 healthy control females and 18 OC patients by using the miRNeasy Serum/Plasma Kit (Qiagen, Hilden, Germany) including 3.5 μL miRNeasy Serum/Plasma Spike-In Control RNA, according to the manufacturer's instructions. A miRNA-specific fluorometric assay on a Qubit® 2.0 Fluorimeter (Thermo Fischer Scientific, USA) was used to determine the concentration of RNA. To detect and measure the amounts of mature miRNAs the miScript PCR System (Qiagen, Hilden, Germany) was used. The miScript II RT Kit (Qiagen) was used for reverse transcription of miRNAs. The quantitative real-time PCR reaction was used (LightCycler®96; Roche Molecular Systems Inc., Pleasanton, CA, USA) to determine the level of hsa-miR-25-3p, hsa-miR-26b-5p, hsa-miR-144-3p, hsa-miR-19b-3p, hsa-miR-301a-3p, hsa-miR-148b-3p, hsa-miR-553, and hsa-miR-197 by using the miScript SYBR Green PCR Kit (Qiagen). The PCR reaction mixture contained 500 pg reverse transcription products. The reaction mixtures were first denatured at 95 °C for 15 min, followed by 50 amplification cycles of 94 °C for 15 s, 55 °C for 30 s and 70 °C for 30 s. Finally, a melting curve was generated by taking fluorescent measurements every 0.2 °C for 25 s from 50 °C until 95 °C to detect a single PCR product. Cycle threshold (Ct) values above the determinable range (up to 45) were assigned a Ct of 45. All measurements were performed in triplicate and the amounts of PCR products were normalized to an internal control (hsa-miR103-3p). Relative expression levels were calculated by the $2^{-\Delta Ct}$ method.

4.7. Statistical Analysis

All data were analyzed using the GraphPad Prism statistical package (GraphPad Prism7, San Diego, CA, USA). Descriptive column statistics of each data set were performed and the distribution of data was analyzed by the Kolmogorov–Smirnov test. To assess the statistical significance of differences in miRNA counts between the control and patient groups the nonparametric one-way ANOVA Kruskal–Wallis test in combination with the post hoc Dunn's test to adjust for multiple comparisons was applied. In all tests the difference was considered significant at $p < 0.05$ value. Where applicable, the Dunn's p-values were indicated as: $p < 0.05$(*); $p < 0.01$(**). The fold change in the expression of a miRNA between the control data and a given FIGO stage data was calculated as: (FIGO stage mean count – Control mean count)/Control mean count.

Author Contributions: A.P., B.N., B.S., and M.S.-B. were involved in designing all experiments; A.P. and L.S. analyzed and interpreted the data. B.S. and É.M. performed the experiments; A.P. prepared the manuscript. R.P. and J.L. assisted in conducting the experiment. B.N., B.S., and M.S.-B. assisted in editing the manuscript. All authors read and approved the final manuscript.

Acknowledgments: We thank Kálmán Szenthe and the RT-Europe Nonprofit Kft for their helpful contribution.

Abbreviations

ANOVA	Analysis of variance
EMT	Epithelial–mesenchymal transition
FIGO	Multidisciplinary Digital Publishing Institute
GO	Gene ontology

KEGG Kyoto Encyclopedia of Genes and Genomes
OC Ovary cancer
PPI Protein–protein interaction

References

1. Webb, P.M.; Jordan, S.J. Epidemiology of epithelial ovarian cancer. *Best Pract. Res. Clin. Obstet. Gynaecol.* **2017**, *41*, 3–14. [CrossRef] [PubMed]
2. Duffy, M.J.; Bonfrer, J.M.; Kulpa, J.; Rustin, G.J.; Soletormos, G.; Torre, G.C.; Tuxen, M.K.; Zwirner, M. CA125 in ovarian cancer: European Group on Tumor Markers guidelines for clinical use. *Int. J. Gynecol. Cancer* **2005**, *15*, 679–691. [CrossRef] [PubMed]
3. Meany, D.L.; Sokoll, L.J.; Chan, D.W. Early detection of cancer: Immunoassays for plasma tumor markers. *Expert Opin. Med. Diagn.* **2009**, *3*, 597–605. [CrossRef] [PubMed]
4. Kosaka, N.; Iguchi, H.; Ochiya, T. Circulating microRNA in body fluid: A new potential biomarker for cancer diagnosis and prognosis. *Cancer Sci.* **2010**, *101*, 2087–2092. [CrossRef] [PubMed]
5. Li, W.; Wang, Y.; Zhang, Q.; Tang, L.; Liu, X.; Dai, Y.; Xiao, L.; Huang, S.; Chen, L.; Guo, Z.; et al. MicroRNA-486 as a biomarker for early diagnosis and recurrence of non-small cell lung cancer. *PLoS ONE* **2015**, *10*, e0134220. [CrossRef] [PubMed]
6. Cappelletti, V.; Appierto, V.; Tiberio, P.; Fina, E.; Callari, M.; Daidone, M.G. Circulating biomarkers for prediction of treatment response. *J. Natl. Cancer Inst. Monogr.* **2015**, *6*, 60–63. [CrossRef]
7. Shen, J.; Zhu, X.; Fei, J.; Shi, P.; Yu, S.; Zhou, J. Advances of exosome in the development of ovarian cancer and its diagnostic and therapeutic prospect. *Onco Targets Ther.* **2018**, *11*, 2831–2841. [CrossRef] [PubMed]
8. Kosaka, N.; Yoshioka, Y.; Fujita, Y.; Ochiya, T. Versatile roles of extracellular vesicles in cancer. *J. Clin. Investig.* **2016**, *126*, 1163–1172. [CrossRef]
9. Hayes, J.; Peruzzi, P.P.; Lawler, S. MicroRNAs in cancer: Biomarkers, functions and therapy. *Trends Mol. Med.* **2014**, *20*, 460–469. [CrossRef]
10. Leva, G.; Di Garofalo, M.; Croce, C.M. microRNAs in cancer. *Annu. Rev. Pathol.* **2014**, *9*, 287–314. [CrossRef]
11. Lages, E.; Guttin, A.; Nesr, H.; Lages, E.; Guttin, A.; Nesr, H. MicroRNAs - molecular features and role in cancer. *Front. Biosci.* **2012**, *17*, 2508–2540. [CrossRef]
12. Paul, P.; Chakraborty, A.; Sarkar, D.; Langthasa, M.; Rahman, M.; Bari, M.; Singha, R.S.; Malakar, A.K.; Chakraborty, S. Interplay between miRNAs and human diseases. *J. Cell Physiol.* **2018**, *233*, 2007–2018. [CrossRef] [PubMed]
13. Turchinovich, A.; Weiz, L.; Langheinz, A.; Burwinkel, B. Characterization of extracellular circulating microRNA. *Nucleic Acids Res.* **2011**, *39*, 7223–7233. [CrossRef] [PubMed]
14. Lin, C.; Wu, S.; Zhang, K.; Qing, Y.; Xu, A.T. Comprehensive overview of exosomes in ovarian cancer: Emerging biomarkers and therapeutic strategies. *J. Ovarian Res.* **2017**, *10*, 73. [CrossRef]
15. Minciacchi, V.R.R.; Freeman, M.R.; Di Vizio, D. Extracellular Vesicles in Cancer: Exosomes, Microvesicles and the Emerging Role of Large Oncosomes. *Semin. Cell Dev. Biol.* **2015**, *40*, 41–51. [CrossRef] [PubMed]
16. Valadi, H.; Ekström, K.; Bossios, A.; Sjöstrand, M.; Lee, J.J.; Lötvall, J.O. Exosome-mediated transfer of mRNAs and microRNAs is a novel mechanism of genetic exchange between cells. *Nat. Cell Biol.* **2007**, *9*, 654–659. [CrossRef] [PubMed]
17. Kosaka, N.; Iguchi, H.; Yoshioka, Y.; Takeshita, F.; Matsuki, Y.; Ochiya, T. Secretory mechanisms and intercellular transfer of microRNAs in living cells. *J. Biol. Chem.* **2010**, *285*, 17442–17452. [CrossRef] [PubMed]
18. Zhang, Y.; Liu, D.; Chen, X.; Li, J.; Li, L.; Bian, Z.; Sun, F.; Yin, Y.; Cai, X.; Xu, T.; et al. Secreted monocytic miR-150 enhances targeted endothelial cell migration. *Mol. Cell.* **2010**, *39*, 133–144. [CrossRef] [PubMed]
19. Grange, C.; Tapparo, M.; Collino, F.; Vitillo, L.; Damasco, C.; Deregibus, M.C.; Tetta, C.; Bussolati, B.; Camussi, G. Microvesicles released from human renal cancer stem cells stimulate angiogenesis and formation of lung premetastatic niche. *Cancer Res.* **2011**, *71*, 5346–5356. [CrossRef]
20. Challagundla, K.B.; Wise, P.M.; Neviani, P.; Chava, H.; Murtadha, M.; Xu, T.; Kennedy, R.; Ivan, C.; Zhang, X.; Vannini, I.; et al. Exosome-mediated transfer of microRNAs within the tumor microenvironment and neuroblastoma resistance to chemotherapy. *J. Natl. Cancer Inst.* **2015**, 107. [CrossRef]
21. Fong, M.Y.; Zhou, W.; Liu, L.; Chandra, M.; Chow, A.; Li, S.; Chin, A.R.; Tremblay, J.R.; Tsuyada, A.; Kong, M.; et al. Breast-cancer-secreted miR-122 reprograms glucose metabolism in premetastatic niche to promote metastasis. *Nat. Cell Biol.* **2015**, *17*, 183–194. [CrossRef] [PubMed]

22. Iorio, M.V.; Croce, C.M. MicroRNA dysregulation in cancer: Diagnostics, monitoring and therapeutics. A comprehensive review. *EMBO Mol. Med.* **2012**, *4*, 143–159. [CrossRef] [PubMed]
23. Svoronos, A.A.; Engelman, D.M.; Slack, F.J. OncomiR or Tumor Suppressor? The Duplicity of MicroRNAs in Cancer. *Cancer Res.* **2016**, *76*, 3666–3670. [CrossRef] [PubMed]
24. Akira, Y.; Matsuzaki, J.; Yamamoto, Y.; Yoneoka, Y.; Takahashi, K.; Shimizu, H.; Uehara, T.; Ishikawa, M.; Ikeda, S.-I.; Sonoda, T.; et al. Integrated extracellular microRNA profiling for ovarian cancer screening. *Nat. Commun.* **2018**, *9*, 4319.
25. Berindan-Neagoe, I.; Monroig, P.; Pasculli, B.; Calin, G.A. MicroRNAome genome: A treasure for cancer diagnosis and therapy. *CA Cancer J. Clin.* **2014**, *64*, 311–336. [CrossRef] [PubMed]
26. Lengyel, E. Ovarian cancer development and metastasis. *Am. J. Pathol.* **2010**, *177*, 1053–1064. [CrossRef] [PubMed]
27. Waters, C.E.; Saldivar, J.C.; Hosseini, S.A.; Huebner, K. The FHIT gene product: Tumor suppressor and genome 'caretaker'. *Cell Mol. Life Sci.* **2014**, *71*, 4577–4587. [CrossRef]
28. Kim, J.G.; Kang, M.J.; Yoon, Y.-K.; Kim, H.-P.; Park, J.; Song, H.-P.; Kang, G.H.; Kang, K.W.; Oh, D.Y.; Yi, E.C.; et al. Heterodimerization of Glycosylated Insulin-Like Growth Factor-1 Receptors and Insulin Receptors in Cancer Cells Sensitive to Anti-IGF1R Antibody. *PLoS ONE* **2012**, *7*, e33322. [CrossRef]
29. Dai, M.; Zhang, C.; Ali, A.; Hong, X.; Tian, J.; Lo, C.; Fils, N.-A.; Burgos, S.A.; Ali, S.; Lebrun, J.J. CDK4 regulates cancer stemness and is a novel therapeutic target for triple-negative breast cancer. *Sci. Rep.* **2016**, *6*, 35383. [CrossRef]
30. Shen, T.; Huang, S. The role of Cdc25A in the regulation of cell proliferation and apoptosis. *Anticancer Agents Med. Chem.* **2012**, *12*, 631–639.
31. Da Wei, H.; Sherman, B.T.; Lempicki, R.A. Systematic and integrative analysis of large gene lists using DAVID bioinformatics resources. *Nat. Protoc.* **2009**, *4*, 44–57.
32. Andersen, C.L.; Sikora, M.J.; Boisen, M.M.; Ma, T.; Christie, A.; Tseng, G.; Park, Y.; Luthra, S.; Chandran, U.; Haluska, P.; et al. Active Estrogen Receptor-alpha Signaling in Ovarian Cancer Models and Clinical Specimens. *Clin. Cancer Res.* **2017**, *23*, 3802–3812. [CrossRef] [PubMed]
33. Fournier, M.; Orpinell, M.; Grauffel, C.; Scheer, E.; Garnier, J.M.; Ye, T.; Chavant, V.; Joint, M.; Esashi, F.; Dejaegere, A.; et al. KAT2A/KAT2B-targeted acetylome reveals a role for PLK4 acetylation in preventing centrosome amplification. *Nat. Commun.* **2016**, *7*, 13227. [CrossRef] [PubMed]
34. Yan, K.S.; Lin, C.-Y.; Liao, T.-W.; Peng, C.-M.; Lee, S.-C.; Liu, Y.-J.; Chan, V.P.; Chou, R.H. EZH2 in Cancer Progression and Potential Application in Cancer Therapy: A Friend or Foe? *Int. J. Mol. Sci.* **2017**, *18*, 1172. [CrossRef] [PubMed]
35. Zhang, H.; Duan, J.; Qu, Y.; Deng, T.; Liu, R.; Zhang, L.; Bai, M.; Li, J.; Ning, T.; Ge, S.; et al. Onco-miR-24 regulates cell growth and apoptosis by targeting BCL2L11 in gastric cancer. *Protein Cell* **2016**, *7*, 141–151. [CrossRef] [PubMed]
36. Zhao, M.; Mishra, L.; Deng, C.X. The role of TGF-β/SMAD4 signaling in cancer. *Int. J. Biol. Sci.* **2018**, *14*, 111–123. [CrossRef] [PubMed]
37. Worzfeld, T.; Swiercz, J.M.; Looso, M.; Straub, B.K.; Sivaraj, K.K.; Offermanns, S. ErbB-2 signals through Plexin-B1 to promote breast cancer metastasis. *J. Clin. Investig.* **2012**, *122*, 1296–1305. [CrossRef]
38. Su, L.; Hu, A.; Luo, Y.; Zhou, W.; Zhang, P.; Feng, Y. HMGN2, a new anti-tumor effector molecule of CD8+ T cells. *Mol. Cancer* **2014**, *13*, 178. [CrossRef]
39. Gong, C.; Yao, S.; Gomes, A.R.; Man, E.P.S.; Lee, H.J.; Gong, G.; Chang, S.; Kim, S.B.; Fujino, K.; Kim, S.W.; et al. BRCA1 positively regulates FOXO3 expression by restricting FOXO3 gene methylation and epigenetic silencing through targeting EZH2 in breast cancer. *Oncogenesis* **2016**, *5*, e214. [CrossRef]
40. Cristóbal, I.; Cirauqui, C.; Castello-Cros, R.; Garcia-Orti, L.; Calasanz, M.J.; Odero, M.D. Downregulation of PPP2R5E is a common event in acute myeloid leukemia that affects the oncogenic potential of leukemic cells. *Haematologica* **2013**, *98*, e103. [CrossRef]
41. Oh, S.; You, E.; Ko, P.; Jeong, J.; Keum, S.; Rhee, S. Genetic disruption of tubulin acetyltransferase, αTAT1, inhibits proliferation and invasion of colon cancer cells through decreases in Wnt1/β-catenin signaling. *Biochem. Biophys. Res. Commun.* **2017**, *482*, 8–14. [CrossRef] [PubMed]
42. Feng, Y.; Xiong, Y.; Qiao, T.; Li, X.; Jia, L.; Han, Y. Lactate dehydrogenase A: A key player in carcinogenesis and potential target in cancer therapy. *Cancer Med.* **2018**, *7*, 6124–6136. [CrossRef] [PubMed]

43. Zeng, R.; Liu, Y.; Jiang, Z.-J.; Huang, J.-P.; Wang, Y.; Li, X.-F.; Xiong, W.-B.; Wu, X.-C.; Zhang, J.-R.; Wang, Q.-E.; et al. EPB41L3 is a potential tumor suppressor gene and prognostic indicator in esophageal squamous cell carcinoma. *Int. J. Oncol.* **2018**, *52*, 1443–1454. [CrossRef] [PubMed]
44. Heng, Y.H.E.; Zhou, B.; Harris, L.; Harvey, T.; Smith, A.; Horne, E.; Martynog, B.; Andersen, J.; Achimastou, A.; Cato, K.; et al. NFIX Regulates Proliferation and Migration Within the Murine SVZ Neurogenic Niche. *Cereb. Cortex* **2015**, *25*, 3758–3778. [CrossRef] [PubMed]
45. Chen, C.; Shen, H.; Zhang, L.G.; Liu, J.; Cao, X.G.; Yao, A.L.; Kang, S.S.; Gao, W.X.; Han, H.; Cao, F.H.; et al. Construction and analysis of protein-protein interaction networks based on proteomics data of prostate cancer. *Int. J. Mol. Med.* **2016**, *37*, 1576–1586. [CrossRef] [PubMed]
46. Qiu, L.; Wu1, J.; Pan, C.; Tan, X.; Lin, J.; Liu, R.; Chen, S.; Geng, R.; Huang, W. Downregulation of CDC27 inhibits the proliferation of colorectal cancer cells via the accumulation of p21 Cip1/Waf1. *Cell Death Dis.* **2016**, *7*, e2074. [CrossRef]
47. Petrie, J.L.; Swan, C.; Ingram, R.M.; Frame, F.M.; Collins, A.T.; Dumay-Odelot, H.; Teichmann, M.; Maitland, N.J.; White, R.J. Effects on prostate cancer cells of targeting RNA polymerase III. *Nucleic Acids Res.* **2019**, *47*, 3937–3956. [CrossRef]
48. Tang, C.; Jardim, D.L.; Hong, D. MET in ovarian cancer Metastasis and resistance? *Cell Cycle* **2014**, *13*, 1220–1221. [CrossRef]
49. Li, Y.; Gong, W.; Ma, X.; Sun, X.; Jiang, H.; Chen, T. Smad7 maintains epithelial phenotype of ovarian cancer stem-like cells and supports tumor colonization by mesenchymal-epithelial transition. *Mol. Med. Rep.* **2015**, *11*, 309–316. [CrossRef]
50. Pilsworth, J.A.; Cochrane, D.R.; Xia, Z.; Aubert, G.; Färkkilä, A.E.M.; Horlings, H.M.; Yanagida, S.; Yang, W.; Lim, J.L.P.; Wang, Y.K.; et al. TERT promoter mutation in adult granulosa cell tumor of the ovary. *Mod. Pathol.* **2018**, *7*, 1107–1115. [CrossRef]
51. Browning, L.; Patel, M.R.; Horvath, E.B.; Tawara, K.; Jorcyk, C.L. IL-6 and ovarian cancer: Inflammatory cytokines in promotion of metastasis. *Cancer Manag. Res.* **2018**, *10*, 6685–6693. [CrossRef] [PubMed]

12

Exosomes-Associated DNA—New Marker in Pregnancy Complications?

Barbora Konečná [1,*], Ľubomíra Tóthová [1] and Gabriela Repiská [2]

1 Institute of Molecular Biomedicine, Faculty of Medicine, Comenius University in Bratislava, Bratislava 81108, Slovakia; tothova.lubomira@gmail.com
2 Institute of Physiology, Faculty of Medicine, Comenius University in Bratislava, Bratislava 81372, Slovakia; gabika.repiska@gmail.com
* Correspondence: basa.konecna@gmail.com

Abstract: Despite a large number of studies, the etiology of pregnancy complications remains unknown. The involvement of cell-free DNA or fetal cell-free DNA in the pathogenesis of pregnancy complications is currently being hypothesized. Cell-free DNA occurs in different forms—free; part of neutrophil extracellular traps; or as recently discovered, carried by extracellular vesicles. Cell-free DNA is believed to activate an inflammatory pathway, which could possibly cause pregnancy complications. It could be hypothesized that DNA in its free form could be easily degraded by nucleases to prevent the inflammatory activation. However, recently, there has been a growing interest in the role of exosomes, potential protectors of cell-free DNA, in pregnancy complications. Most of the interest from recent years is directed towards the micro RNA carried by exosomes. However, exosome-associated DNA in relation to pregnancy complications has not been truly studied yet. DNA, as an important cargo of exosomes, has been so far studied mostly in cancer research. This review collects all the known information on the topic of not only exosome-associated DNA but also some information on vesicles-associated DNA and the studies regarding the role of exosomes in pregnancy complications from recent years. It also suggests possible analysis of exosome-associated DNA in pregnancy from plasma and emphasizes the importance of such analysis for future investigations of pregnancy complications. A major obstacle to the advancement in this field is the proper uniformed technique for exosomes isolation. Similarly, the sensitivity of methods analyzing a small fraction of DNA, potentially fetal DNA, carried by exosomes is variable.

Keywords: cell-free DNA; exosomes; extracellular vesicles; fetal DNA; preeclampsia; growth retardation; gestational diabetes mellitus

---

1. Introduction

Pregnancy complications are often associated with spontaneous preterm birth and might result in mortality or morbidity of the mother and/or the child. Despite significant improvements in monitoring and prevention in other areas of health care, the etiology of complications associated with pregnancy, such as preeclampsia (PE), intra-uterine growth retardation (IUGR), spontaneous abortion, or gestational diabetes mellitus (GDM), still remains unknown. However, a discovery of cell-free fetal DNA in the circulation of the mother represents the most important finding in this research field [1]. Since then, the noninvasive prenatal testing for chromosomal abnormalities can be performed using only a blood sample collected from the mother, which brings about new possibilities for prenatal diagnostics. Such a new technique is beneficial also when the increased concentrations of cell-free fetal DNA are detected in the circulation of mothers with various pregnancy complications [2]. Before this discovery, prenatal tests were performed by invasive methods such as amniocentesis or chorionic villus sampling. Unfortunately, these invasive procedures are associated with 1–2% fetal loss [3].

The noninvasive prenatal DNA diagnosis based on the analysis of cfDNA isolated from maternal plasma is rapidly evolving the research area; and using the latest methods for DNA analysis, various noninvasive prenatal tests are implemented into clinical practice. Currently, the determination of fetal gender, fetal Rhesus D (RhD) genotyping, aneuploidies, micro-deletions, and the detection of paternally-inherited monogenic disorders are available for pregnant women worldwide [4]. However, the clinical utility of cfDNA for noninvasive prenatal testing is even higher. In recent years, advanced investigation approaches have allowed to reconstruct the whole fetal genome [5,6] and methylome [7]. Based on the biological characteristic of cfDNA (its origin, fragment size and methylation), it presents also a potential new approach for pregnancy complication prediction and diagnosis [8].

Cell-free DNA (cfDNA) can be found in the circulation of every healthy individual. It has either nuclear or mitochondrial origin. It is released after cell death, such as necrosis or apoptosis [9], and is released as free fragments. Additionally, during inflammatory states, neutrophils produce net-like structures, which also cause the death of a neutrophil. These so-called neutrophil extracellular traps (NETs) are formed in order to kill pathogens. CfDNA, including histone proteins attached to the DNA, is incorporated into these structures [10]. CfDNA can also be released from the living cells through the process of exocytosis. Exocytosis is the process of releasing small vesicles, termed exosomes [11]. Exosomes are a subtype of extracellular vesicles (EVs). Recently, it was found that among other structures, EVs also contain DNA [12]. Since the DNA inside of vesicles is protected by the vesicular double membrane, EVs have been hypothesized to function in horizontal gene transfer.

Fetal cfDNA is produced mainly by the apoptosis of placental cells of the trophoblast [13]. It is produced during normal pregnancies; however, the process of apoptosis is likely increased during complicated pregnancies, due to increased oxidative stress and inflammatory response [14]. Phillipe et al. [15] were the first to hypothesize that fetal cfDNA could have a pro-inflammatory effect. Similar to bacterial DNA, fetal cfDNA is hypomethylated [2,16]. Toll-like receptor 9, expressed mainly on immune cells, is sensitive to hypomethylated DNA. The interaction of toll-like receptor 9 with fetal cfDNA triggers the innate immune response. This was experimentally proven on a mice model, in which free fetal DNA was injected into pregnant mice [16].

Living cells of the trophoblast also release EVs. It was published by Gupta et al. [17] that such vesicles contain both DNA and RNA. The aim of this review is to examine the role of fetal cfDNA from EVs in the pathogenesis of pregnancy complications. This review will present a summary of potential effects of fetal cfDNA in pregnancy complications; evidence of cfDNA being contained in EVs, specifically in exosomes; and a role of these exosomes in pregnancy complications. Since most recent cell-free nucleic acids papers focus on RNA, these studies are also mentioned; however, RNA is not a main focus of this review. The end this review concludes the possible techniques for the isolation of placental EVs and potential options for analysis of EVs that are beneficial for the field of pregnancy complications.

2. CfDNA and Pregnancy Complications

Preeclampsia (PE) is the most common of pregnancy complications. It is associated with the highest rate of morbidity and mortality among pregnant women worldwide. Clinically, it is characterized by hypertension, proteinuria and edema [18]. Spiral artery remodeling, altered angiogenesis and endothelial damage are believed to be involved in pathogenesis [19]. Recent meta-analysis [20] pointed out that cell-free fetal DNA is a predictive marker of both early and late onset PE. However, the most reliable detection time is at the beginning of the second trimester since the fetal fraction is the highest. It means that already in the first trimester, not much fetal DNA circulates in maternal blood compared to maternal DNA; and, on the other hand, in the third trimester, maternal DNA concentration increases with weight gain and concentration of fetal DNA becomes extinct [20]. Fetal RNA has also been studied as a potential biomarker of PE, especially at the beginning of the second trimester. A panel of specific mRNAs can serve as a predictor of PE by monitoring endothelial growth factor and endoglin, which are highly sensitive biomarkers [21,22]. Recently, the deportation of trophoblast was studied, and it was found that fetal trophoblast cells as well as immune cells were present in the lungs of the mother, which may have an immunomodulatory effect [23]. Interestingly, placental NETs have also been hypothesized to trigger the autoimmune reaction in PE [24]. Unfortunately, it is still not clear what causes PE and whether a high concentration of fetal cfDNA is a consequence or a cause.

Intrauterine growth restriction (IUGR) is characterized when the fetus does not achieve the normal growth due to placental causes, most often by placental insufficiency, which is associated either with hypoxia or with blood vessel morphology [25]. Other common causes of IUGR include maternal malnutrition as in vegan or vegetarian mothers [26] or exposure to toxic substances such as drugs or alcohol during pregnancy [27]. The causes cannot be exhaustively completed in this review, since except for the above-mentioned, more than 150 genetic disorders might be linked to IUGR [28]. Endocrine dysregulations along with other epigenetic factors might also contribute to IUGR. Nevertheless, the IUGR condition is often linked with PE. Maternal causes such as too young or too old maternal age; various maternal diseases and fetal causes such as chromosomal abnormalities; or multiple gestations and genetic factors are known key factors in IUGR. IUGR is another pregnancy complication as well as PE that might be associated with increased number of fetal cells and subsequently, an increased concentration of fetal nucleic acids in the circulation of a mother. Hahn et al. [29] described how fetal cfDNA might be used to study alterations in placentation, which is believed to be a cause of IUGR. However, the primary cause of altered placentation is probably due to a modification of maternal spiral arteries. Again, it is not clear whether high concentrations of cell-free fetal DNA contribute to the complication or if they result from it. A recent study discussed that even though cfDNA is higher during IUGR, it is important to consider the activity of deoxyribonuclease I, which is important for cfDNA elimination. This study showed that the activity of this enzyme is increased and therefore it is difficult to monitor cfDNA concentrations properly [30]. Considering the possible severe consequences of IUGR such as vascular, pulmonary, cardiac or neurological problems it is necessary to search for a predictor of IUGR and improvement of antenatal treatments [31].

Pregnancies are often complicated by spontaneous abortions, which are associated with increased stress or inflammation [32,33]. Although the terminology is not clear and standardized, spontaneous abortion may be defined as noninduced loss of an embryo or fetus or passage of conception products before the 20 weeks of gestation [34]. The major cause is chromosomal abnormalities [35]. Recently, it was shown that low progesterone metabolite concentrations contribute to spontaneous abortions due to triggering of inflammation [33,36]. Lim et al. [37] analyzed blood from 268 women and observed the association of spontaneous abortions with high fetal cfDNA concentrations. Similarly, Jakobsen et al. [38] observed that high concentrations of fetal cfDNA at the 25th week of pregnancy could predict the preterm delivery. The latest study of NETosis associated with spontaneous abortion showed that chorioamniotic NETs were found to be increased in women with chorioamnionitis and preterm delivery [39]. Similarly, it was recently found that cfDNA of both mitochondrial and nuclear origins were elevated in amniotic fluids in pregnancies complicated by preterm prelabor rupture of membranes [40]. Certain inflammatory cytokines and chemokines contribute to the pathogenesis of spontaneous abortions, which is associated with inflammatory mechanisms, and are possibly associated with cfDNA [36].

Gestational diabetes mellitus (GDM) is a spontaneous glucose intolerance developed during gestation. Chronic insulin resistance leads to β cells dysfunction, leading to glucotoxicity and later to macrosomia and lasting type II diabetes after delivery [41,42]. Nine miRNAs involved in placental and fetal development were discovered to be dysregulated during GDM. These miRNAs were of placental tissue origin [43]. However, other serum miRNAs were also found to be involved in the development of GDM and even possibly used as predictors of insulin resistance [44] or predictors of GDM in pregnancies associated with obesity [45,46]. CfDNA in the form of NETs has been associated with GDM [47]. A study by Stoikou et al. [48] showed that tumor necrosis factor alpha as a proinflammatory factor elevates NET formation in women with GDM compared to healthy controls. Also, Thurik et al. [49] pointed out the association between cfDNA and GDM in the first trimester of pregnancy. To the best of our knowledge, there are no other studies investigating cfDNA in GDM; however, such an association is often mentioned as a co-factor of other pregnancy complications.

Even though prenatal testing has been practiced for years already, using cfDNA as a marker from maternal blood is still in its initial studies. All the above-mentioned studies are only from recent years. However, this raises many questions and hypotheses that are yet to be answered or tested.

3. CfDNA Associated with Exosomes

EVs can be classified into three broad classes based on their size, endocytic origin, biogenesis and sedimentation properties—exosomes, microvesicles and apoptotic bodies. The exosome subclass is commonly defined as bilipid membrane-bound nanovesicles (size 40–120 nm in diameter) that are derived from multivesicular bodies. Under physiological and pathophysiological conditions, they are actively released from almost all types of cells into the extracellular space and body fluids [50,51]. Current research on circulating nucleic acids has shown that the circulating mRNA and miRNA molecules detected in plasma, serum and other biofluids are packaged in EVs [52,53], reflecting cells of origin and the disease state of the tissue. For example, exosomes from the plasma of pregnant preeclamptic women were analyzed, and it was demonstrated that their proteins [54] and microRNAs [55,56] could be used as biomarkers to predict PE. It was also found that preeclamptic placentas produce more extracellular vesicles in comparison to normal pregnancy [57,58], which could also be used as an indicator of the disease. For further comprehensive detailed information, there are several review articles available that deal with the biogenesis of all kinds of EVs [59–62].

Several studies have shown that exosomes also contain DNA molecules [63–66]. Thakur et al. revealed for the first time that tumor-derived exosomes contain double-stranded DNA, indicating that exosomal DNA reflects the mutational status of paternal tumor cells [66]. In other studies, double-stranded DNA was identified in exosomes isolated from the serum or plasma of pancreatic and prostate cancer patients [64,65]. This finding illustrates the translational potential of DNA isolated from exosomes as a circulating biomarker for the early detection of cancers and the monitoring of treatment response. Another newer study compared the concentrations of total DNA isolated from plasma and plasma exosomes and provided evidence that a large proportion (more than 90%) of plasma cfDNA is localized in exosomes rather than being truly free and circulating in plasma [67]. Apart from nuclear DNA, studies have also shown that cells release exosomes containing mitochondrial DNA (mtDNA) [68,69]. Guescini et al. reported that exosomes, constitutively released by glioblastoma cells and astrocytes, carry mtDNA, which can be further transferred between the cells [69]. The full mitochondrial genome was identified in plasma exosomes from patients with hormonal therapy-resistant metastatic breast cancer. It was further demonstrated that the horizontal transfer of mtDNA from vesicles acts as an oncogenic signal promoting an exit from the dormancy of therapy-induced cancer stem-like cells [70].

Despite the evidence of cfDNA association with exosomes, the discussion about its localization is still ongoing. Until now, most studies have been focused on intra-exosomal DNA. The exosomes have been shown to carry intra-vesicular DNA protected by a phospholipid bilayer membrane and the mutation status of this DNA was comparable to that of the cell origin [64–66,71]. In contrast, Shelke et al. [72] showed that DNase I-sensitive DNA could be associated with the outside of exosomes isolated from a human mast cell line. They have suggested that DNA can cause aggregation of these vesicles, possibly influencing their effects in recipient cells. Also, Nemeth et al. showed that ciprofloxacin-induced release of mitochondrial and chromosomal DNA is associated with the surface of exosomes [73]. Nevertheless, the above-mentioned studies showed that exosomes-associated DNA might be important in terms of physiological or pathophysiological states. CfDNA associated with EVs, especially exosomes, could be utilized as a biomarker in cancer, signal molecule or a molecule triggering the process of aggregation. However, the latest publication questions the real presence of DNA in exosomes and co-isolation of histone-bound DNA together with the isolation of exosomes [74].

It seems that the smallest vesicles, i.e., exosomes, do not carry DNA. On the contrary, DNA is released through autophagy, and endosomal mechanisms are independent from exosomal release [75].

4. Exosomes in Pregnancy Complications

Secretion of EVs, including exosomes, has also been reported in placental cells as a response to changes in the extracellular environment. The release of exosomes during pregnancy is modulated by particular features of the cellular microenvironment such as low oxygen tension (i.e., hypoxia) or high glucose concentration [76]. It was shown that placenta-derived exosomes regulate the migration and invasion of target cells; play an important role in intracellular communication; and potentially contribute to the placentation and development of maternal-fetal vascular exchange [77,78]. Placental exosomes are distinguished according to the placental alkaline phosphatase (PLAP) marker and are supposed to be released under placental dysfunctions [79]. Even animal models demonstrate fetal-maternal trafficking and signaling via exosomes [80]. It was shown that exosomes deliver paracrine cargo that signals labor and delivery of fetuses in mice [81]. Additionally, another study showed by quantitative proteomics that human maternal plasma exosomes physiology could possibly reflect the homeostasis in pregnancy and indicate preterm birth [82].

Placental-specific exosomes have been identified in maternal blood of healthy pregnant women. Beside PLAP, placental exosomes can be distinguished from maternally-derived exosomes by the presence of specific miRNAs [83–85]. Their concentration significantly increases during the first trimester of pregnancy and as early as the sixth gestational week [86], but the contribution of placental exosomes to total plasma exosomes and exosomes' bioactivity decreases in late pregnancy [87]. Recent studies suggest a role of exosomes in pregnancy complications development. It has been shown that in pregnancy complications, such as PE and GDM, the number of secreted vesicles change [88–90]. A higher number of exosomes in the blood of pregnant women with early and late onset-PE compared with normal pregnancies was detected, which also suggests a possible pathophysiological role of placenta-derived exosomes in PE. The study showed a higher relative number of placental-derived exosomes in early onset-PE, but a lower relative number of placental-derived exosomes in late onset-PE in relation to total exosomes in maternal circulation [89]. Similarly, a 10-fold higher concentration of exosomes is present in the circulation of obese patients leading to proinflammatory pathways and insulin resistance when compared with normal weight patients [91].

As well as the placental origin, the origin of vesicles can be determined in more detail. The vesicles could be divided into embryo, oviduct, endometrial epithelium, and stroma-derived vesicles, all of which are involved in the communication with cells of trophoblast [92]. Extravillous trophoblast releases extracellular vesicles containing specifically human leukocyte antigen-G (HLA-G) that is only expressed in these cells. It was demonstrated that these vesicles are both exosomes and microvesicles [50,93]. HLA-G is known to protect the fetus from invasion by maternal immune system [94]; therefore, it is supposed that its function in exosomes is to signal and interact with responsible receptors [93]. It is believed that at the beginning of pregnancy EVs are mainly produced by cells of the endometrium, embryo and trophoblasts but later they are produced by syncytiotrophoblast cells, all having an immune-suppressing effect on the maternal immune system [95].

Many studies have focused on the differential expression of miRNAs in extracellular vesicles in placental tissue and blood from normal and preeclamptic pregnancies. They are summarized in a recently published review article of Chairello et al. [96]. A recent study showed various miRNAs cargo in maternal plasma exosomes obtained in the first trimester of pregnancy, which regulates different signaling pathways in pregnancy [97]. Also, Salomon et al. in their newest study of miRNA profiles of preeclamptic pregnant women suggested that the quantification of placenta-derived exosomes present in maternal blood and the measurement of specific miRNAs expression may improve our ability to identify asymptomatic women at risk for development of PE [98]. In addition to miRNAs, other potential molecular markers, such as placental proteins, angiogenic factors and fetal cfDNA, were investigated as potential predictive markers for PE development. The overview of all studies investigating small vesicles, exosomes, during pregnancy complications is summarized in Table 1.

Table 1. Chronological overview of studies regarding exosomes and pregnancy complications from years 2016–2018.

Study Year (Reference)	Pregnancy Complication	Material	Aim
Chang 2018 [99]	PE	Plasma	Antiangiogenic factors
Jayabalan 2018 [100,101]	GDM	Plasma, adipose tissue	Proteomics
Wang 2018 [102]	PE	Serum	miR-548c-5p
Menon 2018 [103]	Preterm birth	Plasma	Micro RNA
Luo 2018 [104]	IUGR	Umbilical cord blood	Micro RNA
Motawi 2018 [105]	PE	Umbilical cord blood and cell media	Micro RNA
Nair 2018 [106]	GDM	Placenta, plasma, skeletal muscle	Micro RNA
Hu 2018 [107]	PE	Urine	Expression of renal sodium transporters
Zhao 2018 [108]	IUGR, abortion	Plasma	Micro RNA
Miranda 2018 [109]	IUGR	Plasma	Basic characterization
Beretti 2018 [110]	Immune response	Amniotic fluid stem cell exosomes	Basic characterization
Shen 2018 [111]	PE	Serum	miR-155
Saez 2018 [112]	GDM	Plasma	Cargo
Biro 2017 [113]	Hypertension, PE	Plasma	Micro RNA
Ermini 2017 [114]	PE	Plasma	Cargo
Rodosthenous 2017 [115]	IUGR	Plasma	Micro RNA
Motta-Mejia 2017 [116]	PE	Plasma	Endothelial factors
Salomon 2017 [98]	PE	Plasma	Micro RNA
Truong 2017 [97]	PE, preterm birth	Plasma	Micro RNA
Elfeky 2017 [117]	Obesity	Plasma	Basic characterization
Shi 2017 [118]	GDM	Plasma	Micro RNA
Pillay 2016 [89]	PE	Plasma	Concentration
Sheller 2016 [119]	Preterm birth	Amniotic membrane	Cargo
Gysler 2016 [120]	Autoimmune disorders	Plasma	Micro RNA
Ospina-Prieto 2016 [121]	PE	Trophoblast cells	miRNA-144
Sandrim 2016 [56]	PE	Plasma	miR-885-5p
Salomon 2016 [90]	GDM	Plasma	Concentration
Panfoli 2016 [122]	Preterm birth	Umbilical cord cells	Characterization

PE—preeclampsia, IUGR—intra-uterine growth retardation, GDM—gestational diabetes mellitus.

To the best of our knowledge, there are no studies showing the presence of fetal cfDNA associated with exosomes circulating in maternal blood. Only a recent review described in detail the crosstalk between mother and the fetus, however, they discussed extracellular vesicles in general, not exosomes specifically [92]. A pilot study from our group showed that exosomes isolated from the blood of pregnant women contain not only DNA of maternal but also of fetal origin [123]. These results are raising new questions and provide the basis for further investigation of the localization of the vesicles and their function in healthy and complicated pregnancies. Nevertheless, the overview of studies regarding DNA in any relation to EVs or exosomes and pregnancy are summarized in Table 2.

Table 2. Chronological overview of studies regarding DNA and pregnancy complications.

Study Year (Reference)	Pregnancy Complication	Material	Aim
Fernando 2018 [124]	Pregnant vs. Non-pregnant	Plasma	Fragment size pattern of cfDNA
Sheller-Miller 2017 [125]	Term labor	Amnion ephithelial cells	Cargo
Tong 2017 [126]	PE	Placental explants	Concentration of EVs
Sheller 2016 [119]	Term delivery	Amnion ephithelial cells	Cargo
Orozco 2009 [127]	PE	Plasma	EVs containing DNA concentrations
Orozco 2008 [128]	PE	Extravillous trophoblast and plasma	EVs containing DNA basic characterization
Gupta 2005 [129]	PE	Placental explants and neutrophils	NETs formation

5. Methods to Analyze cfDNA from Exosomes and Implications for Pregnancy Complications

High variability in the isolation as well as exosome characterization techniques used contributes to the diversity of published data [130]. Thakur et al. found that nucleic acid content in exosomes reflects the whole genomic DNA of parental cell lines [66]. They also compared different cancer cell lines suggesting that exosomal DNA could be used as a potential indicator of the disease reflecting its origin depending on the type of the cancer. It was already known that there is fetal DNA inside of exosomes isolated from placenta [119]. We hypothesize that exosome-associated fetal DNA in the circulation could also be used as a potential diagnostic marker for pregnancy complications. However, since no well-established method exists there have been problems with reproducibly isolating plasma exosomes. Also, the high abundance of plasma albumin interferes with several analyses of exosomes [131]. The most common isolation method is ultracentrifugation; however, size-exclusion chromatography [132] or different commercial kits [133–135] are also used.

Until now, the cancer research field has used cfDNA as a potential liquid biopsy biomarker to reflect the status of the organism [136]. There already exist protocols to obtain the highest yield of cfDNA since the concentration has to be high enough to make the analyses [137,138]. It was already published that 90% of cfDNA in plasma is hidden in exosomes [67]. So far, the common techniques of plasma exosome isolation followed by DNA analysis have been ExoQuick exosome precipitation solution [139], ExoLution Plus extraction technology [140] and differential centrifugation [141]. However, there have not been many studies analyzing fetal cfDNA in exosomes isolated from plasma. Other groups also hypothesize that fetal DNA could be carried in exosomes, which could therefore serve as suitable material for Y chromosome detection and fetal DNA quantification [127,142]. Our preliminary data showed that fetal DNA is packaged inside of exosomes and using the ultracentrifugation protocol for exosomes isolation we were able to quantify it [123]. Our study was the first one that attempted experimentally to quantify fetal DNA in exosomes. The results showed that the concentration of fetal DNA in plasma was 10 times higher than in plasma exosomes after the usage of DNase, that degraded any surface-bound DNA. The dynamics of fetal cfDNA in pregnancy is already known. Fetal DNA continuously rises throughout trimesters [143]. It is debatable whether cfDNA from the fetus is localized in exosomes or other vesicles. Perhaps more fetal DNA could be present in exosomes in the second trimester in comparison to other trimesters since it was found that the concentration of exosomes is the highest during this part of gestation [144]. However, potentially more fetal DNA could be localized inside of exosomes rather than being free, although, better optimization of isolation techniques is needed as well as more sensitive quantification methods. Extracellular RNA Communication Consortium has already started projects dealing with the isolation methods focused on RNA and of

EVs in general [145]. Pregnancy complications are believed to be associated with an increased number of exosomes in maternal blood; therefore, it would be interesting to analyze plasma samples from women with PE, GDM or IUGR by a standardized method and obtain comprehensive information.

6. Conclusions

Every cell releases cfDNA. Fetal DNA is released from the cells of the placenta and circulates in the blood of the mother. Currently, it is known that fetal cfDNA could be used as a potential biomarker of several pregnancy complications. PE, IUGR, spontaneous abortions, or GDM are usually associated with increased concentrations of cell-free fetal DNA. Researchers in the field of exosomes have recently discovered that these small EVs also carry DNA. Such DNA is probably protected by their double membrane against the degradation by nucleases. As it is probably more stable, it could be used as a biomarker, and, therefore, is a source of information. However, the proper isolation method of placental exosomes followed by a proper isolation of vesicular fetal DNA needs to be optimized. As of now, the fetal cfDNA of exosomal origin is believed to have the potential to become a biomarker of pregnancy complications, along with the potential to explain the pathogenesis of various pregnancy complications in the future.

Abbreviations

CfDNA	cell-free DNA
NETs	neutrophil extracellular traps
EVs	extracellular vesicles

References

1. Lo, Y.M.; Corbetta, N.; Chamberlain, P.F.; Rai, V.; Sargent, I.L.; Redman, C.W.; Wainscoat, J.S. Presence of fetal DNA in maternal plasma and serum. *Lancet* **1997**, *350*, 485–487. [CrossRef]
2. Van Boeckel, S.R.; Davidson, D.J.; Norman, J.E.; Stock, S.J. Cell-free fetal DNA and spontaneous preterm birth. *Reproduction* **2018**, *155*, R137–R145. [CrossRef] [PubMed]
3. Norwitz, E.R.; Levy, B. Noninvasive prenatal testing: The future is now. *Rev. Obstet. Gynecol.* **2013**, *6*, 48–62. [PubMed]
4. Breveglieri, G.; D'Aversa, E.; Finotti, A.; Borgatti, M. Non-invasive prenatal testing using fetal DNA. *Mol. Diagn. Ther.* **2019**, *23*, 291–299. [CrossRef] [PubMed]
5. Chan, K.C.; Jiang, P.; Sun, K.; Cheng, Y.K.; Tong, Y.K.; Cheng, S.H.; Wong, A.I.; Hudecova, I.; Leung, T.Y.; Chiu, R.W.; et al. Second generation noninvasive fetal genome analysis reveals de novo mutations, single-base parental inheritance, and preferred DNA ends. *Proc. Natl. Acad. Sci. USA* **2016**, *113*, E8159–E8168. [CrossRef] [PubMed]
6. Lo, Y.M.; Chan, K.C.; Sun, H.; Chen, E.Z.; Jiang, P.; Lun, F.M.; Zheng, Y.W.; Leung, T.Y.; Lau, T.K.; Cantor, C.R.; et al. Maternal plasma DNA sequencing reveals the genome-wide genetic and mutational profile of the fetus. *Sci. Transl. Med.* **2010**, *2*, 61ra91. [CrossRef] [PubMed]
7. Sun, K.; Lun, F.M.F.; Leung, T.Y.; Chiu, R.W.K.; Lo, Y.M.D.; Sun, H. Noninvasive reconstruction of placental methylome from maternal plasma DNA: Potential for prenatal testing and monitoring. *Prenat. Diagn.* **2018**, *38*, 196–203. [CrossRef] [PubMed]
8. Sun, K.; Jiang, P.; Wong, A.I.C.; Cheng, Y.K.Y.; Cheng, S.H.; Zhang, H.; Chan, K.C.A.; Leung, T.Y.; Chiu, R.W.K.; Lo, Y.M.D. Size-tagged preferred ends in maternal plasma DNA shed light on the production mechanism and show utility in noninvasive prenatal testing. *Proc. Natl. Acad. Sci. USA* **2018**, *115*, E5106–E5114. [CrossRef]
9. Jung, K.; Fleischhacker, M.; Rabien, A. Cell-free DNA in the blood as a solid tumor biomarker—A critical appraisal of the literature. *Clin. Chim. Acta* **2010**, *411*, 1611–1624. [CrossRef]
10. Fuchs, T.A.; Brill, A.; Duerschmied, D.; Schatzberg, D.; Monestier, M.; Myers, D.D., Jr.; Wrobleski, S.K.; Wakefield, T.W.; Hartwig, J.H.; Wagner, D.D. Extracellular DNA traps promote thrombosis. *Proc. Natl. Acad. Sci. USA* **2010**, *107*, 15880–15885. [CrossRef]
11. Gould, S.J.; Raposo, G. As we wait: Coping with an imperfect nomenclature for extracellular vesicles. *J. Extracell. Vesicles* **2013**, *2*, 20389. [CrossRef] [PubMed]

12. Cai, J.; Wu, G.; Jose, P.A.; Zeng, C. Functional transferred DNA within extracellular vesicles. *Exp. Cell Res.* **2016**, *349*, 179–183. [CrossRef] [PubMed]
13. Huppertz, B.; Kingdom, J.C. Apoptosis in the trophoblast—Role of apoptosis in placental morphogenesis. *J. Soc. Gynecol. Investig.* **2004**, *11*, 353–362. [CrossRef] [PubMed]
14. Tjoa, M.L.; Cindrova-Davies, T.; Spasic-Boskovic, O.; Bianchi, D.W.; Burton, G.J. Trophoblastic oxidative stress and the release of cell-free feto-placental DNA. *Am. J. Pathol.* **2006**, *169*, 400–404. [CrossRef] [PubMed]
15. Phillippe, M. Cell-free fetal DNA, telomeres, and the spontaneous onset of parturition. *Reprod. Sci.* **2015**, *22*, 1186–1201. [CrossRef]
16. Scharfe-Nugent, A.; Corr, S.C.; Carpenter, S.B.; Keogh, L.; Doyle, B.; Martin, C.; Fitzgerald, K.A.; Daly, S.; O'Leary, J.J.; O'Neill, L.A. Tlr9 provokes inflammation in response to fetal DNA: Mechanism for fetal loss in preterm birth and preeclampsia. *J. Immunol.* **2012**, *188*, 5706–5712. [CrossRef] [PubMed]
17. Gupta, A.K.; Holzgreve, W.; Huppertz, B.; Malek, A.; Schneider, H.; Hahn, S. Detection of fetal DNA and rna in placenta-derived syncytiotrophoblast microparticles generated in vitro. *Clin. Chem.* **2004**, *50*, 2187–2190. [CrossRef]
18. Steegers, E.A.; von Dadelszen, P.; Duvekot, J.J.; Pijnenborg, R. Pre-eclampsia. *Lancet* **2010**, *376*, 631–644. [CrossRef]
19. Phipps, E.A.; Thadhani, R.; Benzing, T.; Karumanchi, S.A. Pre-eclampsia: Pathogenesis, novel diagnostics and therapies. *Nat. Rev. Nephrol.* **2019**, *15*, 275–289. [CrossRef]
20. Contro, E.; Bernabini, D.; Farina, A. Cell-free fetal DNA for the prediction of pre-eclampsia at the first and second trimesters: A systematic review and meta-analysis. *Mol. Diagn. Ther.* **2017**, *21*, 125–135. [CrossRef]
21. Purwosunu, Y.; Sekizawa, A.; Okazaki, S.; Farina, A.; Wibowo, N.; Nakamura, M.; Rizzo, N.; Saito, H.; Okai, T. Prediction of preeclampsia by analysis of cell-free messenger rna in maternal plasma. *Am. J. Obstet. Gynecol.* **2009**, *200*, e381–e387. [CrossRef] [PubMed]
22. Sekizawa, A.; Purwosunu, Y.; Farina, A.; Shimizu, H.; Nakamura, M.; Wibowo, N.; Rizzo, N.; Okai, T. Prediction of pre-eclampsia by an analysis of placenta-derived cellular mrna in the blood of pregnant women at 15–20 weeks of gestation. *BJOG* **2010**, *117*, 557–564. [CrossRef] [PubMed]
23. Pritchard, S.; Wick, H.C.; Slonim, D.K.; Johnson, K.L.; Bianchi, D.W. Comprehensive analysis of genes expressed by rare microchimeric fetal cells in the maternal mouse lung. *Biol. Reprod.* **2012**, *87*, 42. [CrossRef] [PubMed]
24. Hahn, S.; Giaglis, S.; Hoesli, I.; Hasler, P. Neutrophil nets in reproduction: From infertility to preeclampsia and the possibility of fetal loss. *Front. Immunol.* **2012**, *3*, 362. [CrossRef] [PubMed]
25. Galan, H.L.; Rigano, S.; Radaelli, T.; Cetin, I.; Bozzo, M.; Chyu, J.; Hobbins, J.C.; Ferrazzi, E. Reduction of subcutaneous mass, but not lean mass, in normal fetuses in denver, colorado. *Am. J. Obstet. Gynecol.* **2001**, *185*, 839–844. [CrossRef]
26. Sebastiani, G.; Herranz Barbero, A.; Borras-Novell, C.; Alsina Casanova, M.; Aldecoa-Bilbao, V.; Andreu-Fernandez, V.; Pascual Tutusaus, M.; Ferrero Martinez, S.; Gomez Roig, M.D.; Garcia-Algar, O. The effects of vegetarian and vegan diet during pregnancy on the health of mothers and offspring. *Nutrients* **2019**, *11*, 557. [CrossRef]
27. Nardozza, L.M.; Caetano, A.C.; Zamarian, A.C.; Mazzola, J.B.; Silva, C.P.; Marcal, V.M.; Lobo, T.F.; Peixoto, A.B.; Araujo Junior, E. Fetal growth restriction: Current knowledge. *Arch. Gynecol. Obstet.* **2017**, *295*, 1061–1077. [CrossRef]
28. Giabicani, E.; Pham, A.; Brioude, F.; Mitanchez, D.; Netchine, I. Diagnosis and management of postnatal fetal growth restriction. *Best Pract. Res. Clin. Endocrinol. Metab.* **2018**, *32*, 523–534. [CrossRef]
29. Hahn, S.; Huppertz, B.; Holzgreve, W. Fetal cells and cell free fetal nucleic acids in maternal blood: New tools to study abnormal placentation? *Placenta* **2005**, *26*, 515–526. [CrossRef]
30. Ershova, E.; Sergeeva, V.; Klimenko, M.; Avetisova, K.; Klimenko, P.; Kostyuk, E.; Veiko, N.; Veiko, R.; Izevskaya, V.; Kutsev, S.; et al. Circulating cell-free DNA concentration and dnase i activity of peripheral blood plasma change in case of pregnancy with intrauterine growth restriction compared to normal pregnancy. *Biomed. Rep.* **2017**, *7*, 319–324. [CrossRef]
31. Malhotra, A.; Allison, B.J.; Castillo-Melendez, M.; Jenkin, G.; Polglase, G.R.; Miller, S.L. Neonatal morbidities of fetal growth restriction: Pathophysiology and impact. *Front. Endocrinol.* **2019**, *10*, 55. [CrossRef] [PubMed]
32. Goldenberg, R.L.; Culhane, J.F.; Iams, J.D.; Romero, R. Epidemiology and causes of preterm birth. *Lancet* **2008**, *371*, 75–84. [CrossRef]

33. Ku, C.W.; Tan, Z.W.; Lim, M.K.; Tam, Z.Y.; Lin, C.H.; Ng, S.P.; Allen, J.C.; Lek, S.M.; Tan, T.C.; Tan, N.S. Spontaneous miscarriage in first trimester pregnancy is associated with altered urinary metabolite profile. *BBA Clin.* **2017**, *8*, 48–55. [CrossRef] [PubMed]
34. Kolte, A.M.; Bernardi, L.A.; Christiansen, O.B.; Quenby, S.; Farquharson, R.G.; Goddijn, M.; Stephenson, M.D. Terminology for pregnancy loss prior to viability: A consensus statement from the eshre early pregnancy special interest group. *Hum. Reprod.* **2015**, *30*, 495–498. [CrossRef] [PubMed]
35. Nagaishi, M.; Yamamoto, T.; Iinuma, K.; Shimomura, K.; Berend, S.A.; Knops, J. Chromosome abnormalities identified in 347 spontaneous abortions collected in japan. *J. Obstet. Gynaecol. Res.* **2004**, *30*, 237–241. [CrossRef]
36. Di Renzo, G.C.; Tosto, V.; Giardina, I. The biological basis and prevention of preterm birth. *Best Pract. Res. Clin. Obstet. Gynaecol.* **2018**, *52*, 13–22. [CrossRef] [PubMed]
37. Lim, J.H.; Kim, M.H.; Han, Y.J.; Lee, D.E.; Park, S.Y.; Han, J.Y.; Kim, M.Y.; Ryu, H.M. Cell-free fetal DNA and cell-free total DNA levels in spontaneous abortion with fetal chromosomal aneuploidy. *PLoS ONE* **2013**, *8*, e56787. [CrossRef]
38. Jakobsen, T.R.; Clausen, F.B.; Rode, L.; Dziegiel, M.H.; Tabor, A. High levels of fetal DNA are associated with increased risk of spontaneous preterm delivery. *Prenat. Diagn.* **2012**, *32*, 840–845. [CrossRef]
39. Gomez-Lopez, N.; Romero, R.; Leng, Y.; Garcia-Flores, V.; Xu, Y.; Miller, D.; Hassan, S.S. Neutrophil extracellular traps in acute chorioamnionitis: A mechanism of host defense. *Am. J. Reprod. Immunol.* **2017**, *77*, e12617. [CrossRef]
40. Kacerovsky, M.; Vlkova, B.; Musilova, I.; Andrys, C.; Pliskova, L.; Zemlickova, H.; Stranik, J.; Halada, P.; Jacobsson, B.; Celec, P. Amniotic fluid cell-free DNA in preterm prelabor rupture of membranes. *Prenat. Diagn.* **2018**, *38*, 1086–1095. [CrossRef]
41. Gilmartin, A.B.; Ural, S.H.; Repke, J.T. Gestational diabetes mellitus. *Rev. Obstet. Gynecol.* **2008**, *1*, 129–134. [PubMed]
42. Plows, J.F.; Stanley, J.L.; Baker, P.N.; Reynolds, C.M.; Vickers, M.H. The pathophysiology of gestational diabetes mellitus. *Int. J. Mol. Sci.* **2018**, *19*, 3342. [CrossRef] [PubMed]
43. Li, J.; Song, L.; Zhou, L.; Wu, J.; Sheng, C.; Chen, H.; Liu, Y.; Gao, S.; Huang, W. A microrna signature in gestational diabetes mellitus associated with risk of macrosomia. *Cell. Physiol. BioChem.* **2015**, *37*, 243–252. [CrossRef] [PubMed]
44. Zhao, C.; Dong, J.; Jiang, T.; Shi, Z.; Yu, B.; Zhu, Y.; Chen, D.; Xu, J.; Huo, R.; Dai, J.; et al. Early second-trimester serum mirna profiling predicts gestational diabetes mellitus. *PLoS ONE* **2011**, *6*, e23925. [CrossRef] [PubMed]
45. Ibarra, A.; Vega-Guedes, B.; Brito-Casillas, Y.; Wagner, A.M. Diabetes in pregnancy and micrornas: Promises and limitations in their clinical application. *Noncoding RNA* **2018**, *4*, 32. [CrossRef] [PubMed]
46. Wander, P.L.; Boyko, E.J.; Hevner, K.; Parikh, V.J.; Tadesse, M.G.; Sorensen, T.K.; Williams, M.A.; Enquobahrie, D.A. Circulating early- and mid-pregnancy micrornas and risk of gestational diabetes. *Diabetes Res. Clin. Pract.* **2017**, *132*, 1–9. [CrossRef] [PubMed]
47. Vokalova, L.; van Breda, S.V.; Ye, X.L.; Huhn, E.A.; Than, N.G.; Hasler, P.; Lapaire, O.; Hoesli, I.; Rossi, S.W.; Hahn, S. Excessive neutrophil activity in gestational diabetes mellitus: Could it contribute to the development of preeclampsia? *Front. Endocrinol.* **2018**, *9*, 542. [CrossRef] [PubMed]
48. Stoikou, M.; Grimolizzi, F.; Giaglis, S.; Schafer, G.; van Breda, S.V.; Hoesli, I.M.; Lapaire, O.; Huhn, E.A.; Hasler, P.; Rossi, S.W.; et al. Gestational diabetes mellitus is associated with altered neutrophil activity. *Front. Immunol.* **2017**, *8*, 702. [CrossRef] [PubMed]
49. Thurik, F.F.; Lamain-de Ruiter, M.; Javadi, A.; Kwee, A.; Woortmeijer, H.; Page-Christiaens, G.C.; Franx, A.; van der Schoot, C.E.; Koster, M.P. Absolute first trimester cell-free DNA levels and their associations with adverse pregnancy outcomes. *Prenat. Diagn.* **2016**, *36*, 1104–1111. [CrossRef]
50. Atay, S.; Gercel-Taylor, C.; Kesimer, M.; Taylor, D.D. Morphologic and proteomic characterization of exosomes released by cultured extravillous trophoblast cells. *Exp. Cell Res.* **2011**, *317*, 1192–1202. [CrossRef]
51. Thery, C.; Amigorena, S.; Raposo, G.; Clayton, A. Isolation and characterization of exosomes from cell culture supernatants and biological fluids. *Curr. Protoc. Cell Biol.* **2006**, *30*, 3–22. [CrossRef] [PubMed]
52. Chim, S.S.; Shing, T.K.; Hung, E.C.; Leung, T.Y.; Lau, T.K.; Chiu, R.W.; Lo, Y.M. Detection and characterization of placental micrornas in maternal plasma. *Clin. Chem.* **2008**, *54*, 482–490. [CrossRef] [PubMed]
53. Gallo, A.; Tandon, M.; Alevizos, I.; Illei, G.G. The majority of micrornas detectable in serum and saliva is concentrated in exosomes. *PLoS ONE* **2012**, *7*, e30679. [CrossRef] [PubMed]

54. Tan, K.H.; Tan, S.S.; Ng, M.J.; Tey, W.S.; Sim, W.K.; Allen, J.C.; Lim, S.K. Extracellular vesicles yield predictive pre-eclampsia biomarkers. *J. Extracell. Vesicles* **2017**, *6*, 1408390. [CrossRef] [PubMed]
55. Escudero, C.A.; Herlitz, K.; Troncoso, F.; Acurio, J.; Aguayo, C.; Roberts, J.M.; Truong, G.; Duncombe, G.; Rice, G.; Salomon, C. Role of extracellular vesicles and micrornas on dysfunctional angiogenesis during preeclamptic pregnancies. *Front. Physiol.* **2016**, *7*, 98. [CrossRef] [PubMed]
56. Sandrim, V.C.; Luizon, M.R.; Palei, A.C.; Tanus-Santos, J.E.; Cavalli, R.C. Circulating microrna expression profiles in pre-eclampsia: Evidence of increased mir-885-5p levels. *BJOG* **2016**, *123*, 2120–2128. [CrossRef] [PubMed]
57. Goswami, D.; Tannetta, D.S.; Magee, L.A.; Fuchisawa, A.; Redman, C.W.; Sargent, I.L.; von Dadelszen, P. Excess syncytiotrophoblast microparticle shedding is a feature of early-onset pre-eclampsia, but not normotensive intrauterine growth restriction. *Placenta* **2006**, *27*, 56–61. [CrossRef]
58. Marques, F.K.; Campos, F.M.; Filho, O.A.; Carvalho, A.T.; Dusse, L.M.; Gomes, K.B. Circulating microparticles in severe preeclampsia. *Clin. Chim. Acta* **2012**, *414*, 253–258. [CrossRef]
59. Colombo, M.; Raposo, G.; Thery, C. Biogenesis, secretion, and intercellular interactions of exosomes and other extracellular vesicles. *Annu. Rev. Cell Dev. Biol.* **2014**, *30*, 255–289. [CrossRef]
60. Mathieu, M.; Martin-Jaular, L.; Lavieu, G.; Thery, C. Specificities of secretion and uptake of exosomes and other extracellular vesicles for cell-to-cell communication. *Nat. Cell Biol.* **2019**, *21*, 9–17. [CrossRef]
61. Van Niel, G.; D'Angelo, G.; Raposo, G. Shedding light on the cell biology of extracellular vesicles. *Nat. Rev. Mol. Cell Biol.* **2018**, *19*, 213–228. [CrossRef] [PubMed]
62. Yanez-Mo, M.; Siljander, P.R.; Andreu, Z.; Zavec, A.B.; Borras, F.E.; Buzas, E.I.; Buzas, K.; Casal, E.; Cappello, F.; Carvalho, J.; et al. Biological properties of extracellular vesicles and their physiological functions. *J. Extracell. Vesicles* **2015**, *4*, 27066. [CrossRef] [PubMed]
63. Cai, J.; Han, Y.; Ren, H.; Chen, C.; He, D.; Zhou, L.; Eisner, G.M.; Asico, L.D.; Jose, P.A.; Zeng, C. Extracellular vesicle-mediated transfer of donor genomic DNA to recipient cells is a novel mechanism for genetic influence between cells. *J. Mol. Cell Biol.* **2013**, *5*, 227–238. [CrossRef] [PubMed]
64. Kahlert, C.; Melo, S.A.; Protopopov, A.; Tang, J.; Seth, S.; Koch, M.; Zhang, J.; Weitz, J.; Chin, L.; Futreal, A.; et al. Identification of double-stranded genomic DNA spanning all chromosomes with mutated kras and p53 DNA in the serum exosomes of patients with pancreatic cancer. *J. Biol. Chem.* **2014**, *289*, 3869–3875. [CrossRef] [PubMed]
65. Lazaro-Ibanez, E.; Sanz-Garcia, A.; Visakorpi, T.; Escobedo-Lucea, C.; Siljander, P.; Ayuso-Sacido, A.; Yliperttula, M. Different gdna content in the subpopulations of prostate cancer extracellular vesicles: Apoptotic bodies, microvesicles, and exosomes. *Prostate* **2014**, *74*, 1379–1390. [CrossRef] [PubMed]
66. Thakur, B.K.; Zhang, H.; Becker, A.; Matei, I.; Huang, Y.; Costa-Silva, B.; Zheng, Y.; Hoshino, A.; Brazier, H.; Xiang, J.; et al. Double-stranded DNA in exosomes: A novel biomarker in cancer detection. *Cell Res.* **2014**, *24*, 766–769. [CrossRef] [PubMed]
67. Fernando, M.R.; Jiang, C.; Krzyzanowski, G.D.; Ryan, W.L. New evidence that a large proportion of human blood plasma cell-free DNA is localized in exosomes. *PLoS ONE* **2017**, *12*, e0183915. [CrossRef] [PubMed]
68. Boudreau, L.H.; Duchez, A.C.; Cloutier, N.; Soulet, D.; Martin, N.; Bollinger, J.; Pare, A.; Rousseau, M.; Naika, G.S.; Levesque, T.; et al. Platelets release mitochondria serving as substrate for bactericidal group iia-secreted phospholipase a2 to promote inflammation. *Blood* **2014**, *124*, 2173–2183. [CrossRef] [PubMed]
69. Guescini, M.; Genedani, S.; Stocchi, V.; Agnati, L.F. Astrocytes and glioblastoma cells release exosomes carrying mtdna. *J. Neural Transm.* **2010**, *117*, 1–4. [CrossRef] [PubMed]
70. Sansone, P.; Savini, C.; Kurelac, I.; Chang, Q.; Amato, L.B.; Strillacci, A.; Stepanova, A.; Iommarini, L.; Mastroleo, C.; Daly, L.; et al. Packaging and transfer of mitochondrial DNA via exosomes regulate escape from dormancy in hormonal therapy-resistant breast cancer. *Proc. Natl. Acad. Sci. USA* **2017**, *114*, E9066–E9075. [CrossRef]
71. Miranda, K.C.; Bond, D.T.; McKee, M.; Skog, J.; Paunescu, T.G.; Da Silva, N.; Brown, D.; Russo, L.M. Nucleic acids within urinary exosomes/microvesicles are potential biomarkers for renal disease. *Kidney Int.* **2010**, *78*, 191–199. [CrossRef] [PubMed]
72. Shelke, G.; Jang, S.C.; Yin, Y.; Lässer, C.; Lötvall, J. Human mast cells release extracellular vesicle-associated DNA. *Matters* **2016**. [CrossRef]
73. Nemeth, A.; Orgovan, N.; Sodar, B.W.; Osteikoetxea, X.; Paloczi, K.; Szabo-Taylor, K.E.; Vukman, K.V.; Kittel, A.; Turiak, L.; Wiener, Z.; et al. Antibiotic-induced release of small extracellular vesicles (exosomes) with surface-associated DNA. *Sci. Rep.* **2017**, *7*, 8202. [CrossRef]

74. Pluchino, S.; Smith, J.A. Explicating exosomes: Reclassifying the rising stars of intercellular communication. *Cell* **2019**, *177*, 225–227. [CrossRef] [PubMed]
75. Jeppesen, D.K.; Fenix, A.M.; Franklin, J.L.; Higginbotham, J.N.; Zhang, Q.; Zimmerman, L.J.; Liebler, D.C.; Ping, J.; Liu, Q.; Evans, R.; et al. Reassessment of exosome composition. *Cell* **2019**, *177*, 428–445. [CrossRef] [PubMed]
76. Rice, G.E.; Scholz-Romero, K.; Sweeney, E.; Peiris, H.; Kobayashi, M.; Duncombe, G.; Mitchell, M.D.; Salomon, C. The effect of glucose on the release and bioactivity of exosomes from first trimester trophoblast cells. *J. Clin. Endocrinol. Metab.* **2015**, *100*, E1280–E1288. [CrossRef] [PubMed]
77. Salomon, C.; Kobayashi, M.; Ashman, K.; Sobrevia, L.; Mitchell, M.D.; Rice, G.E. Hypoxia-induced changes in the bioactivity of cytotrophoblast-derived exosomes. *PLoS ONE* **2013**, *8*, e79636. [CrossRef]
78. Salomon, C.; Ryan, J.; Sobrevia, L.; Kobayashi, M.; Ashman, K.; Mitchell, M.; Rice, G.E. Exosomal signaling during hypoxia mediates microvascular endothelial cell migration and vasculogenesis. *PLoS ONE* **2013**, *8*, e68451. [CrossRef] [PubMed]
79. Jin, J.; Menon, R. Placental exosomes: A proxy to understand pregnancy complications. *Am. J. Reprod. Immunol.* **2018**, *79*, e12788. [CrossRef] [PubMed]
80. Sheller-Miller, S.; Lei, J.; Saade, G.; Salomon, C.; Burd, I.; Menon, R. Feto-maternal trafficking of exosomes in murine pregnancy models. *Front. Pharm.* **2016**, *7*, 432. [CrossRef] [PubMed]
81. Sheller-Miller, S.; Trivedi, J.; Yellon, S.M.; Menon, R. Exosomes cause preterm birth in mice: Evidence for paracrine signaling in pregnancy. *Sci. Rep.* **2019**, *9*, 608. [CrossRef] [PubMed]
82. Menon, R.; Dixon, C.L.; Sheller-Miller, S.; Fortunato, S.J.; Saade, G.R.; Palma, C.; Lai, A.; Guanzon, D.; Salomon, C. Quantitative proteomics by swath-ms of maternal plasma exosomes determine pathways associated with term and preterm birth. *Endocrinology* **2019**, *160*, 639–650. [CrossRef] [PubMed]
83. Bullerdiek, J.; Flor, I. Exosome-delivered micrornas of "chromosome 19 microrna cluster" as immunomodulators in pregnancy and tumorigenesis. *Mol. Cytogenet.* **2012**, *5*, 27. [CrossRef] [PubMed]
84. Donker, R.B.; Mouillet, J.F.; Chu, T.; Hubel, C.A.; Stolz, D.B.; Morelli, A.E.; Sadovsky, Y. The expression profile of c19mc micrornas in primary human trophoblast cells and exosomes. *Mol. Hum. Reprod.* **2012**, *18*, 417–424. [CrossRef] [PubMed]
85. Luo, S.S.; Ishibashi, O.; Ishikawa, G.; Ishikawa, T.; Katayama, A.; Mishima, T.; Takizawa, T.; Shigihara, T.; Goto, T.; Izumi, A.; et al. Human villous trophoblasts express and secrete placenta-specific micrornas into maternal circulation via exosomes. *Biol. Reprod.* **2009**, *81*, 717–729. [CrossRef] [PubMed]
86. Salomon, C.; Rice, G.E. Role of exosomes in placental homeostasis and pregnancy disorders. *Prog. Mol. Biol. Transl. Sci.* **2017**, *145*, 163–179.
87. Salomon, C.; Torres, M.J.; Kobayashi, M.; Scholz-Romero, K.; Sobrevia, L.; Dobierzewska, A.; Illanes, S.E.; Mitchell, M.D.; Rice, G.E. A gestational profile of placental exosomes in maternal plasma and their effects on endothelial cell migration. *PLoS ONE* **2014**, *9*, e98667. [CrossRef]
88. Mitchell, M.D.; Peiris, H.N.; Kobayashi, M.; Koh, Y.Q.; Duncombe, G.; Illanes, S.E.; Rice, G.E.; Salomon, C. Placental exosomes in normal and complicated pregnancy. *Am. J. Obstet. Gynecol.* **2015**, *213*, S173–S181. [CrossRef]
89. Pillay, P.; Maharaj, N.; Moodley, J.; Mackraj, I. Placental exosomes and pre-eclampsia: Maternal circulating levels in normal pregnancies and, early and late onset pre-eclamptic pregnancies. *Placenta* **2016**, *46*, 18–25. [CrossRef] [PubMed]
90. Salomon, C.; Scholz-Romero, K.; Sarker, S.; Sweeney, E.; Kobayashi, M.; Correa, P.; Longo, S.; Duncombe, G.; Mitchell, M.D.; Rice, G.E.; et al. Gestational diabetes mellitus is associated with changes in the concentration and bioactivity of placenta-derived exosomes in maternal circulation across gestation. *Diabetes* **2016**, *65*, 598–609. [CrossRef]
91. Pardo, F.; Villalobos-Labra, R.; Sobrevia, B.; Toledo, F.; Sobrevia, L. Extracellular vesicles in obesity and diabetes mellitus. *Mol. Asp. Med.* **2018**, *60*, 81–91. [CrossRef] [PubMed]
92. Kurian, N.K.; Modi, D. Extracellular vesicle mediated embryo-endometrial cross talk during implantation and in pregnancy. *J. Assist. Reprod. Genet.* **2019**, *36*, 189–198. [CrossRef] [PubMed]
93. Adam, S.; Elfeky, O.; Kinhal, V.; Dutta, S.; Lai, A.; Jayabalan, N.; Nuzhat, Z.; Palma, C.; Rice, G.E.; Salomon, C. Review: Fetal-maternal communication via extracellular vesicles—Implications for complications of pregnancies. *Placenta* **2017**, *54*, 83–88. [CrossRef] [PubMed]

94. LeMaoult, J.; Rouas-Freiss, N.; Carosella, E.D. Hla-g5 expression by trophoblast cells: The facts. *Mol. Hum. Reprod.* **2005**, *11*, 719–722. [CrossRef] [PubMed]
95. Nair, S.; Salomon, C. Extracellular vesicles and their immunomodulatory functions in pregnancy. *Semin. Immunopathol.* **2018**, *40*, 425–437. [CrossRef]
96. Chiarello, D.I.; Salsoso, R.; Toledo, F.; Mate, A.; Vazquez, C.M.; Sobrevia, L. Foetoplacental communication via extracellular vesicles in normal pregnancy and preeclampsia. *Mol. Asp. Med.* **2017**, *60*, 69–80. [CrossRef] [PubMed]
97. Truong, G.; Guanzon, D.; Kinhal, V.; Elfeky, O.; Lai, A.; Longo, S.; Nuzhat, Z.; Palma, C.; Scholz-Romero, K.; Menon, R.; et al. Oxygen tension regulates the mirna profile and bioactivity of exosomes released from extravillous trophoblast cells—Liquid biopsies for monitoring complications of pregnancy. *PLoS ONE* **2017**, *12*, e0174514. [CrossRef]
98. Salomon, C.; Guanzon, D.; Scholz-Romero, K.; Longo, S.; Correa, P.; Illanes, S.E.; Rice, G.E. Placental exosomes as early biomarker of preeclampsia: Potential role of exosomal micrornas across gestation. *J. Clin. Endocrinol. Metab.* **2017**, *102*, 3182–3194. [CrossRef]
99. Chang, X.; Yao, J.; He, Q.; Liu, M.; Duan, T.; Wang, K. Exosomes from women with preeclampsia induced vascular dysfunction by delivering sflt (soluble fms-like tyrosine kinase)-1 and seng (soluble endoglin) to endothelial cells. *Hypertension* **2018**, *72*, 1381–1390. [CrossRef]
100. Jayabalan, N.; Lai, A.; Nair, S.; Guanzon, D.; Scholz-Romero, K.; Palma, C.; McIntyre, H.D.; Lappas, M.; Salomon, C. Quantitative proteomics by swath-ms suggest an association between circulating exosomes and maternal metabolic changes in gestational diabetes mellitus. *Proteomics* **2018**, e1800164. [CrossRef]
101. Jayabalan, N.; Lai, A.; Ormazabal, V.; Adam, S.; Guanzon, D.; Palma, C.; Scholz-Romero, K.; Lim, R.; Jansson, T.; McIntyre, H.D.; et al. Adipose tissue exosomal proteomic profile reveals a role on placenta glucose metabolism in gestational diabetes mellitus. *J. Clin. Endocrinol. Metab.* **2018**, *104*, 1735–1752. [CrossRef] [PubMed]
102. Wang, Z.; Wang, P.; Wang, Z.; Qin, Z.; Xiu, X.; Xu, D.; Zhang, X.; Wang, Y. Mirna-548c-5p downregulates inflammatory response in preeclampsia via targeting ptpro. *J. Cell. Physiol.* **2018**, *234*, 11149–11155. [CrossRef] [PubMed]
103. Menon, R.; Debnath, C.; Lai, A.; Guanzon, D.; Bhatnagar, S.; Pallavi, S.K.; Sheller-Miller, S.; Garbhini Study, t.; Salomon, C. Circulating exosomal mirna profile during term and preterm birth pregnancies—A longitudinal study. *Endocrinology* **2018**, *160*, 249–275. [CrossRef] [PubMed]
104. Luo, J.; Fan, Y.; Shen, L.; Niu, L.; Zhao, Y.; Jiang, D.; Zhu, L.; Jiang, A.; Tang, Q.; Ma, J.; et al. The pro-angiogenesis of exosomes derived from umbilical cord blood of intrauterine growth restriction pigs was repressed associated with mirnas. *Int. J. Biol. Sci.* **2018**, *14*, 1426–1436. [CrossRef] [PubMed]
105. Motawi, T.M.K.; Sabry, D.; Maurice, N.W.; Rizk, S.M. Role of mesenchymal stem cells exosomes derived micrornas; mir-136, mir-494 and mir-495 in pre-eclampsia diagnosis and evaluation. *Arch. BioChem. Biophys.* **2018**, *659*, 13–21. [CrossRef] [PubMed]
106. Nair, S.; Jayabalan, N.; Guanzon, D.; Palma, C.; Scholz-Romero, K.; Elfeky, O.; Zuniga, F.; Ormazabal, V.; Diaz, E.; Rice, G.E.; et al. Human placental exosomes in gestational diabetes mellitus carry a specific set of mirnas associated with skeletal muscle insulin sensitivity. *Clin. Sci.* **2018**, *132*, 2451–2467. [CrossRef] [PubMed]
107. Hu, C.C.; Katerelos, M.; Choy, S.W.; Crossthwaite, A.; Walker, S.P.; Pell, G.; Lee, M.; Cook, N.; Mount, P.F.; Paizis, K.; et al. Pre-eclampsia is associated with altered expression of the renal sodium transporters nkcc2, ncc and enac in urinary extracellular vesicles. *PLoS ONE* **2018**, *13*, e0204514. [CrossRef]
108. Zhao, G.; Yang, C.; Yang, J.; Liu, P.; Jiang, K.; Shaukat, A.; Wu, H.; Deng, G. Placental exosome-mediated bta-mir-499-lin28b/let-7 axis regulates inflammatory bias during early pregnancy. *Cell Death Dis.* **2018**, *9*, 704. [CrossRef]
109. Miranda, J.; Paules, C.; Nair, S.; Lai, A.; Palma, C.; Scholz-Romero, K.; Rice, G.E.; Gratacos, E.; Crispi, F.; Salomon, C. Placental exosomes profile in maternal and fetal circulation in intrauterine growth restriction—Liquid biopsies to monitoring fetal growth. *Placenta* **2018**, *64*, 34–43. [CrossRef]
110. Beretti, F.; Zavatti, M.; Casciaro, F.; Comitini, G.; Franchi, F.; Barbieri, V.; La Sala, G.B.; Maraldi, T. Amniotic fluid stem cell exosomes: Therapeutic perspective. *Biofactors* **2018**, *44*, 158–167. [CrossRef]
111. Shen, L.; Li, Y.; Li, R.; Diao, Z.; Yany, M.; Wu, M.; Sun, H.; Yan, G.; Hu, Y. Placentaassociated serum exosomal mir155 derived from patients with preeclampsia inhibits enos expression in human umbilical vein endothelial cells. *Int. J. Mol. Med.* **2018**, *41*, 1731–1739.
112. Saez, T.; Salsoso, R.; Leiva, A.; Toledo, F.; de Vos, P.; Faas, M.; Sobrevia, L. Human umbilical vein

endothelium-derived exosomes play a role in foetoplacental endothelial dysfunction in gestational diabetes mellitus. *BioChim. Biophys. Acta Mol. Basis Dis.* **2018**, *1864*, 499–508. [CrossRef]

113. Biro, O.; Alasztics, B.; Molvarec, A.; Joo, J.; Nagy, B.; Rigo, J., Jr. Various levels of circulating exosomal total-mirna and mir-210 hypoxamir in different forms of pregnancy hypertension. *Pregnancy Hypertens.* **2017**, *10*, 207–212. [CrossRef] [PubMed]
114. Ermini, L.; Ausman, J.; Melland-Smith, M.; Yeganeh, B.; Rolfo, A.; Litvack, M.L.; Todros, T.; Letarte, M.; Post, M.; Caniggia, I. A single sphingomyelin species promotes exosomal release of endoglin into the maternal circulation in preeclampsia. *Sci. Rep.* **2017**, *7*, 12172. [CrossRef] [PubMed]
115. Rodosthenous, R.S.; Burris, H.H.; Sanders, A.P.; Just, A.C.; Dereix, A.E.; Svensson, K.; Solano, M.; Tellez-Rojo, M.M.; Wright, R.O.; Baccarelli, A.A. Second trimester extracellular micrornas in maternal blood and fetal growth: An exploratory study. *Epigenetics* **2017**, *12*, 804–810. [CrossRef] [PubMed]
116. Motta-Mejia, C.; Kandzija, N.; Zhang, W.; Mhlomi, V.; Cerdeira, A.S.; Burdujan, A.; Tannetta, D.; Dragovic, R.; Sargent, I.L.; Redman, C.W.; et al. Placental vesicles carry active endothelial nitric oxide synthase and their activity is reduced in preeclampsia. *Hypertension* **2017**, *70*, 372–381. [CrossRef] [PubMed]
117. Elfeky, O.; Longo, S.; Lai, A.; Rice, G.E.; Salomon, C. Influence of maternal bmi on the exosomal profile during gestation and their role on maternal systemic inflammation. *Placenta* **2017**, *50*, 60–69. [CrossRef] [PubMed]
118. Shi, R.; Zhao, L.; Cai, W.; Wei, M.; Zhou, X.; Yang, G.; Yuan, L. Maternal exosomes in diabetes contribute to the cardiac development deficiency. *BioChem. Biophys. Res. Commun.* **2017**, *483*, 602–608. [CrossRef] [PubMed]
119. Sheller, S.; Papaconstantinou, J.; Urrabaz-Garza, R.; Richardson, L.; Saade, G.; Salomon, C.; Menon, R. Amnion-epithelial-cell-derived exosomes demonstrate physiologic state of cell under oxidative stress. *PLoS ONE* **2016**, *11*, e0157614. [CrossRef]
120. Gysler, S.M.; Mulla, M.J.; Guerra, M.; Brosens, J.J.; Salmon, J.E.; Chamley, L.W.; Abrahams, V.M. Antiphospholipid antibody-induced mir-146a-3p drives trophoblast interleukin-8 secretion through activation of toll-like receptor 8. *Mol. Hum. Reprod.* **2016**, *22*, 465–474. [CrossRef]
121. Ospina-Prieto, S.; Chaiwangyen, W.; Herrmann, J.; Groten, T.; Schleussner, E.; Markert, U.R.; Morales-Prieto, D.M. Microrna-141 is upregulated in preeclamptic placentae and regulates trophoblast invasion and intercellular communication. *Transl. Res.* **2016**, *172*, 61–72. [CrossRef] [PubMed]
122. Panfoli, I.; Ravera, S.; Podesta, M.; Cossu, C.; Santucci, L.; Bartolucci, M.; Bruschi, M.; Calzia, D.; Sabatini, F.; Bruschettini, M.; et al. Exosomes from human mesenchymal stem cells conduct aerobic metabolism in term and preterm newborn infants. *FASEB J.* **2016**, *30*, 1416–1424. [CrossRef] [PubMed]
123. Repiska, G.; Konecna, B.; Shelke, G.V.; Lasser, C.; Vlkova, B.I.; Minarik, G. Is the DNA of placental origin packaged in exosomes isolated from plasma and serum of pregnant women? *Clin. Chem. Lab. Med.* **2018**, *56*, e150–e153. [CrossRef] [PubMed]
124. Fernando, M.R.; Jiang, C.; Krzyzanowski, G.D.; Ryan, W.L. Analysis of human blood plasma cell-free DNA fragment size distribution using evagreen chemistry based droplet digital pcr assays. *Clin. Chim. Acta Int. J. Clin. Chem.* **2018**, *483*, 39–47. [CrossRef] [PubMed]
125. Sheller-Miller, S.; Urrabaz-Garza, R.; Saade, G.; Menon, R. Damage-associated molecular pattern markers hmgb1 and cell-free fetal telomere fragments in oxidative-stressed amnion epithelial cell-derived exosomes. *J. Reprod. Immunol.* **2017**, *123*, 3–11. [CrossRef] [PubMed]
126. Tong, M.; Johansson, C.; Xiao, F.; Stone, P.R.; James, J.L.; Chen, Q.; Cree, L.M.; Chamley, L.W. Antiphospholipid antibodies increase the levels of mitochondrial DNA in placental extracellular vesicles: Alarmin-g for preeclampsia. *Sci. Rep.* **2017**, *7*, 16556. [CrossRef] [PubMed]
127. Orozco, A.F.; Jorgez, C.J.; Ramos-Perez, W.D.; Popek, E.J.; Yu, X.; Kozinetz, C.A.; Bischoff, F.Z.; Lewis, D.E. Placental release of distinct DNA-associated micro-particles into maternal circulation: Reflective of gestation time and preeclampsia. *Placenta* **2009**, *30*, 891–897. [CrossRef]
128. Orozco, A.F.; Jorgez, C.J.; Horne, C.; Marquez-Do, D.A.; Chapman, M.R.; Rodgers, J.R.; Bischoff, F.Z.; Lewis, D.E. Membrane protected apoptotic trophoblast microparticles contain nucleic acids: Relevance to preeclampsia. *Am. J. Pathol.* **2008**, *173*, 1595–1608. [CrossRef]
129. Gupta, A.K.; Hasler, P.; Holzgreve, W.; Gebhardt, S.; Hahn, S. Induction of neutrophil extracellular DNA lattices by placental microparticles and il-8 and their presence in preeclampsia. *Hum. Immunol.* **2005**, *66*, 1146–1154. [CrossRef]
130. Lasser, C. Mapping extracellular rna sheds lights on distinct carriers. *Cell* **2019**, *177*, 228–230. [CrossRef]

131. Baranyai, T.; Herczeg, K.; Onodi, Z.; Voszka, I.; Modos, K.; Marton, N.; Nagy, G.; Mager, I.; Wood, M.J.; El Andaloussi, S.; et al. Isolation of exosomes from blood plasma: Qualitative and quantitative comparison of ultracentrifugation and size exclusion chromatography methods. *PLoS ONE* **2015**, *10*, e0145686. [CrossRef] [PubMed]
132. Taylor, D.D.; Lyons, K.S.; Gercel-Taylor, C. Shed membrane fragment-associated markers for endometrial and ovarian cancers. *Gynecol. Oncol.* **2002**, *84*, 443–448. [CrossRef] [PubMed]
133. Enderle, D.; Spiel, A.; Coticchia, C.M.; Berghoff, E.; Mueller, R.; Schlumpberger, M.; Sprenger-Haussels, M.; Shaffer, J.M.; Lader, E.; Skog, J.; et al. Characterization of rna from exosomes and other extracellular vesicles isolated by a novel spin column-based method. *PLoS ONE* **2015**, *10*, e0136133. [CrossRef] [PubMed]
134. Hong, C.S.; Funk, S.; Muller, L.; Boyiadzis, M.; Whiteside, T.L. Isolation of biologically active and morphologically intact exosomes from plasma of patients with cancer. *J. Extracell. Vesicles* **2016**, *5*, 29289. [CrossRef] [PubMed]
135. Lobb, R.J.; Becker, M.; Wen, S.W.; Wong, C.S.; Wiegmans, A.P.; Leimgruber, A.; Moller, A. Optimized exosome isolation protocol for cell culture supernatant and human plasma. *J. Extracell. Vesicles* **2015**, *4*, 27031. [CrossRef] [PubMed]
136. Kadam, S.K.; Farmen, M.; Brandt, J.T. Quantitative measurement of cell-free plasma DNA and applications for detecting tumor genetic variation and promoter methylation in a clinical setting. *J. Mol. Diagn.* **2012**, *14*, 346–356. [CrossRef] [PubMed]
137. Meddeb, R.; Pisareva, E.; Thierry, A.R. Guidelines for the preanalytical conditions for analyzing circulating cell-free DNA. *Clin. Chem.* **2019**, *65*, 623–633. [CrossRef] [PubMed]
138. Trigg, R.M.; Martinson, L.J.; Parpart-Li, S.; Shaw, J.A. Factors that influence quality and yield of circulating-free DNA: A systematic review of the methodology literature. *Heliyon* **2018**, *4*, e00699. [CrossRef] [PubMed]
139. Ye, W.; Tang, X.; Yang, Z.; Liu, C.; Zhang, X.; Jin, J.; Lyu, J. Plasma-derived exosomes contribute to inflammation via the tlr9-nf-kappab pathway in chronic heart failure patients. *Mol. Immunol.* **2017**, *87*, 114–121. [CrossRef]
140. Krug, A.K.; Enderle, D.; Karlovich, C.; Priewasser, T.; Bentink, S.; Spiel, A.; Brinkmann, K.; Emenegger, J.; Grimm, D.G.; Castellanos-Rizaldos, E.; et al. Improved egfr mutation detection using combined exosomal rna and circulating tumor DNA in nsclc patient plasma. *Ann. Oncol.* **2017**, *29*, 700–706. [CrossRef]
141. Helmig, S.; Fruhbeis, C.; Kramer-Albers, E.M.; Simon, P.; Tug, S. Release of bulk cell free DNA during physical exercise occurs independent of extracellular vesicles. *Eur. J. Appl. Physiol.* **2015**, *115*, 2271–2280. [CrossRef] [PubMed]
142. Saadeldin, I.M.; Oh, H.J.; Lee, B.C. Embryonic-maternal cross-talk via exosomes: Potential implications. *Stem Cells Cloning* **2015**, *8*, 103–107. [PubMed]
143. Karapetyan, A.O.; Baev, O.R.; Krasnyi, A.M.; Sadekova, A.A.; Mullabaeva, S.M. Extracellular DNA in the dynamics of uncomplicated pregnancy. *Bull. Exp. Biol. Med.* **2018**, *166*, 92–95. [CrossRef] [PubMed]
144. Sarker, S.; Scholz-Romero, K.; Perez, A.; Illanes, S.E.; Mitchell, M.D.; Rice, G.E.; Salomon, C. Placenta-derived exosomes continuously increase in maternal circulation over the first trimester of pregnancy. *J. Transl. Med.* **2014**, *12*, 204. [CrossRef] [PubMed]
145. Das, S.; Extracellular, R.N.A.C.C.; Ansel, K.M.; Bitzer, M.; Breakefield, X.O.; Charest, A.; Galas, D.J.; Gerstein, M.B.; Gupta, M.; Milosavljevic, A.; et al. The extracellular rna communication consortium: Establishing foundational knowledge and technologies for extracellular rna research. *Cell* **2019**, *177*, 231–242. [CrossRef] [PubMed]

13

Liquid Biopsy as a Tool for Differentiation of Leiomyomas and Sarcomas of Corpus Uteri

Dana Dvorská [1], Henrieta Škovierová [1], Dušan Braný [1,*], Erika Halašová [1] and Zuzana Danková [2]

1 Division of Molecular Medicine, Biomedical Center Martin, Jessenius Faculty of Medicine in Martin, Comenius University in Bratislava, 036 01 Martin, Slovakia
2 Division of Oncology, Biomedical Center Martin, Jessenius Faculty of Medicine in Martin, Comenius University in Bratislava, 036 01 Martin, Slovakia
* Correspondence: dusan.brany@uniba.sk

Abstract: Utilization of liquid biopsy in the management of cancerous diseases is becoming more attractive. This method can overcome typical limitations of tissue biopsies, especially invasiveness, no repeatability, and the inability to monitor responses to medication during treatment as well as condition during follow-up. Liquid biopsy also provides greater possibility of early prediction of cancer presence. Corpus uteri mesenchymal tumors are comprised of benign variants, which are mostly leiomyomas, but also a heterogenous group of malignant sarcomas. Pre-surgical differentiation between these tumors is very difficult and the final description of tumor characteristics usually requires excision and histological examination. The leiomyomas and malignant leiomyosarcomas are especially difficult to distinguish and can, therefore, be easily misdiagnosed. Because of the very aggressive character of sarcomas, liquid biopsy based on early diagnosis and differentiation of these tumors would be extremely helpful. Moreover, after excision of the tumor, liquid biopsy can contribute to an increased knowledge of sarcoma behavior at the molecular level, especially on the formation of metastases which is still not well understood. In this review, we summarize the most important knowledge of mesenchymal uterine tumors, the possibilities and benefits of liquid biopsy utilization, the types of molecules and cells that can be analyzed with this approach, and the possibility of their isolation and capture. Finally, we review the typical abnormalities of leiomyomas and sarcomas that can be searched and analyzed in liquid biopsy samples with the final aim to pre-surgically differentiate between benign and malignant mesenchymal tumors.

Keywords: liquid biopsy; cell-free nucleic acids; circulating tumor cells; leiomyomas; sarcomas; leiomyosarcomas; exosomes

---

1. Introduction

Uterine mesenchymal tumors (UMT) are a very heterogeneous group comprised of benign variants, which are mostly leiomyomas (ULM) and also malignant sarcomas. ULM are very common, affecting over 50% of Caucasian women and up to 80% of African origin women and they are one of the main causes of hysterectomy [1,2]. The incidence may be even higher, particularly in developing countries, because ULM can be asymptomatic in more than 50% of women who are, therefore, unaware of their presence [1,2]. Although they do not primarily endanger patients´ lives, ULM can markedly decrease the quality of life if they become symptomatic [3]. In contrast, sarcomas are very rare, with an estimated incidence of three to seven per 100,000 women [4], comprising only up to 8% of all corpus uteri malignancies [5]. Sarcomas are also very aggressive. They spread rapidly and have a low five-year survival rate and high rates of recurrence with low therapeutic efficacy [5–7].

UMT heterogeneity causes complicated classification. The most common form of UMT is "conventional" or "typical" ULM. This covers 90–95% of all lesions and these are considered benign without signs of malignancy [1,3,8]. This group is also heterogeneous, so that "conventional" ULM can differ in the following characteristics: (1) in histological degenerative changes such as myxoid, hyaline, fatty, hemorrhagic, and cystic changes [8]; (2) their localization in the uterus [3]; and (3) the type of genetic driver alternations [9].

In addition to strictly benign ULM forms, the remaining 5–10% can mimic malignancy and, therefore, display some of typical malignant traits such as higher mitotic index, nuclear atypia or the presence of coagulant necrosis [8,10]. There are also smooth muscle proliferations with unusual growth patterns, including benign metastatic ULM or intravenous leiomyomatosis [11]. The most common variants of non-conventional ULM are atypical, bizarre-cellular and mitotically active [8,10]. The most unpredictable, but still considered non-malignant UMT, are the smooth muscle tumors of uncertain potential, abbreviated STUMP [12]. In addition, inflammatory myofibroblastic tumors are quite rare in the uterus, but can closely mimic both benign and malignant variants of UMT [13]. Debate also continues concerning whether uterine leiomyosarcomas (LMS) arise from malignant transformation of ULM [14–16], but, herein, we consider ULM and LMS two separate entities, as generally assumed [5].

The malignant sarcoma group is also very heterogeneous; comprising histologically different tumors of various origin. These are divided into nonepithelial and mixed epithelial-nonepithelial categories dependent on the type of cancerous cell and its presumed originating tissue [7]. However, a better explanation of the variability of these tumors is characterized by sarcoma division into the four previously traditional categories: (1) homologous sarcomas composed only of uterine tissue; (2) heterogeneous types with mixed uterine and non-uterine tissues, typically striated muscle, bone, and cartilage; (3) pure types which comprise only mesodermal structures; and (4) mixed types with mesodermal and non-mesodermal structures; mainly epidermal [17]. These categories could have been combined, for example, LMS were categorized as homologous pure tumors. Nonetheless, this traditional classification incorrectly included several mixed mesenchymal tumors such as carcinosarcomas which are currently listed as tumors of endometrial origin [18]. Moreover, the variability of sarcomas means that future reclassification cannot be excluded. According to present-day classification, the most common type of sarcomas are LMS, comprising up to 60%, followed by up to 20% endometrial stromal sarcomas (ESS), up to 10% adenosarcomas, and the remaining 10% comprises other types of sarcomas [19].

UMT heterogeneity makes it very difficult to pre-surgically distinguish particular types without histopathologic confirmation after excision; including diagnosis of mitotic index, coagulant necrosis, and cytological atypia [5,6]. It is still difficult to distinguish mesenchymal lesions, even with the help of standard imaging devices—variable types of ultrasonography, magnetic resonance or computed tomography (1) malignant sarcomas from benign ULM [1,3] and (2) one type of malignant tumor from other types [5], and even when helpful clinical indications, for example, usual sarcomas occurrence and ULM nonoccurrence after menopause are available [20]. Moreover, non-imaging clinical diagnostic approaches also cannot identify UMT type with certainty; these methods are more suited to identifying endometrial lesions [5,21].

Importantly, ULM are especially difficult to differentiate from the LMS [3,5], because imaging and collateral symptoms are very similar; notably those of abnormal uterine bleeding, abdominal and pelvic pain, dysmenorrhea, and urinary problems [1,7]. Furthermore, ULM and LMS have some common features connected with their formation, especially higher ER receptor expression and activity [22], higher incidence in African origin and obese women [2,7], and both tumor types are more frequent in women with diabetes mellitus [23]. The many similarities between ULM and LMS include also their same origin by transformation of the same cells [24]. Therefore, it would be very beneficial to utilize molecular approaches to distinguish between particular UMT types and this could overcome insufficiencies in pre-surgical clinical diagnosis, especially in ULM and LMS [25]. These factors have inspired the main focus of this article on the use of a noninvasive approach to differentiate between these two tumor groups and we also present brief information on other sarcoma types. This mainly includes ESS, which can be differentiated from other UMT types, and, especially, LMS through molecular approaches [26].

2. Liquid Biopsy

The presence of tumorous cells released into the bloodstream was first described in the nineteenth century by Thomas Ashworth [27]. This can be considered the discovery of the phenomenon now known as liquid biopsy (LB). In general terms, LB is described as a noninvasive technique which enables analysis of molecular biomarkers sampled from body fluids. While this primarily concerns blood [28], it also includes cerebrospinal fluid, pleural effusion, broncho-alveolar lavage fluid, saliva, urine, and sputum [29]. The main concept of LB is to enable more simple, efficient, faster and less-expensive monitoring of disease status, response to treatment, metastasis progression, and early differentiation of particular tumor types. All this can be achieved by analyzing the body fluid targets which are described in greater detail in the following section. Finally, LB is a revolutionary technique that uses modern and precise medical principles, which can help clinicians in therapeutic decisions and make the diagnostic process less stressful for patients. LB has mainly arisen because of invasive method limitations which include the following: (1) it can have unexpected complications and cannot be performed when clinical conditions have worsened or when a tumor is inaccessible [30], (2) it is often unrepeatable and provides only information on the tumor at one point in time and precludes monitoring responses to medication during both treatment and follow up, (3) it cannot be used for screening or early prediction of cancerous disease, and (4) it is generally performed after incidental early-stage tumor identification or following clinical examination resulting from exacerbated patient symptomatology [28,31]. In addition, many patients fear invasive procedures, prefer noninvasive diagnosis, and medicated non-surgical treatment. Furthermore, over 50% women with UMT want to preserve their uterus and fertility [32].

In contrast, the test results from LB are more suitable for early diagnosis because these are usually available much earlier, they have potential to estimate the risk of metastasis progression and relapse more precisely, and they are usually less expensive [28,31]. Simply taking the blood or other body fluids and analyzing target molecules and cells is, therefore, beneficial in many ways, but despite the above mentioned benefits it is important to realize that LB is often used only for assessment of tumor characteristics, but when the LB diagnoses the tumor as malignant it still requires surgical excision [33,34]. In addition, the detection ability of LB still remains challenging and further studies are required to test its accuracy and ability to correctly identify various tumor types. It also remains uncertain if LB constantly provides a representative sampling of the whole tumor mass or only its parts. However, intratumoral and spatial heterogeneity can also often limit precise interpretation of the molecular profile in tissue biopsies [33]. Finally, there is, in some aspects a lack of established consensus to guide its utilization [33,34] and despite the LB approach being beneficial and becoming more and more popular, in some aspects, confirmation of its precise, advanced, and reliable use in clinical praxis is required.

3. Cells and Molecules That Can Be Analyzed from LB Samples

The following cells and molecules can be analyzed using the LB approach: circulating tumor DNA (ctDNA); circulating tumor cells (CTC); exosome vesicles; circulating RNAs, i.e., messenger RNA (mRNA); microRNA (miRNA); long non-coding RNA (lncRNA); and also proteins, peptides, and metabolites [29,31].

3.1. ctDNA

The ctDNA is tumor derived and fragmented DNA which should preserve the abnormal molecular pattern present in the primary tumor [35–39]. Nonetheless, some studies also claim that ctDNA's pattern of abnormalities generally differs from those in primary tumors [40–42]. The ctDNA is primarily released from tumors by apoptosis and necrosis (programmed and not programmed cell death), but also by tumor secretion [43]. Various factors affect the total composition of ctDNA circulating in the bloodstream; it is especially dependent on tumor status, burden, and histopathology [44].

3.2. CTC

CTC can be valuable disease indicators because they are exfoliated from both primary tumors and metastatic lesions. Their numbers and concentration can be used as a predictive marker, and as with exosomes, they contain variable nucleic acids, proteins, and metabolites which can be utilized in molecular analyses at the mutational, transcriptional, and epigenetical levels [45,46].

3.3. Exosomes

Exosomes are membrane-bound phospholipid vesicles, actively secreted by a variety of cells, including cancer cells, and they contain proteins as well as the following nucleic acids: miRNAs, lncRNAs, and mRNAs from tumors. Although the half-life of exosomes in bloodstream is still analyzed and debated, Boukoris et Mathivanan [47] considered exosomes highly stable in stored conditions, and therefore compared to freely circulating nucleic acids (CNA), those molecules encapsulated in exosomes should also be more stable in stored conditions. However, miRNAs, which can also be contained in exosomes, are generally considered very stable and their activity should remain the same, regardless if they are encapsulated in exosomes or incorporated in ribonuclear complexes [48].

3.4. mRNAs, miRNAs and lncRNAs

mRNAs, miRNAs and lncRNAs are present in the bloodstream through both active secretion in exosomes and apoptotic/necrotic cell lysis. However, apart from miRNAs, the half-life of other RNAs in the bloodstream is presumed to be very short and their utilization for LB, therefore, requires further analysis [49,50].

All the above elements, however, are very rare. For example, ctDNA comprises only 0.1% to 10% of all circulating free DNA (cfDNA) released into the bloodstream from all human cells. Especially leucocytes [51] and cfDNA concentration is itself quite low at 10 to 100 ng/mL [52]. The ctDNA concentration, therefore, ranges from 0.01 ng/mL to 10 ng/mL and while the half-life of unencapsulated particles ranges usually from 15 minutes to 2.5 h [31,53], ctDNA half-time is considered relatively long, i.e., at least two hours [51].

In addition, CTC have a frequency as low as 1 CTC per 10^6–10^7 leucocytes which is usually less than 1 CTC per milliliter of blood [46]. Therefore, it is necessary to sample higher blood volumes, usually ranging from 5 to 9 mL to detect at least one CTC. Their occurrence also depends on the tumor stage, where lower numbers are more common in early-stage disease and this increases only slightly in patients with advanced metastases [54,55].

Rajagopal et al. [56] considered that increased exosome number is a potent biomarker for abnormal physiological states and that their total numbers increase in cancer patients as compared to healthy subjects. However, precise exosome quantification is difficult to determine because, in addition to total number, the variations in size, and protein and nucleic acid content, per vesicle can vary and "quantification" of these latter variations should be considered in further molecular analysis [57].

4. Capture, Enrichment, and Isolation of Freely Circulating Vesicles, Molecules and Cells

Traditional exosome capture and enrichment includes ultrafiltration and size exclusion chromatography, precipitation with polymers, and immunoaffinity purification by magnetic beads [58]. These methods can be utilized because exosomes have specific biological patterns, and their cellular origin provides typical surface markers. These include the CD9, CD63, and CD81 members of the tetraspanin family, heat-shock proteins such as HSP70, and the Rab protein family [59]. However, the exosome surface protein profile can vary with both the character and origin of the cell that released them, and therefore establishing more specific membrane biomarkers and the genetic profile of exosomes would be beneficial for using exosomes as agents in novel targeted therapies [60].

Microfluid-based systems are now becoming more popular than traditional exosome capturing systems, because these can directly analyze blood samples and are, therefore, more practical for clinical praxis [45]. However, regardless of the capture technique used, it is very important to the utilization of exosomes in clinical diagnostics, and potentially also in tumor differentiation, that the nucleic acids and metabolites encapsulated in these vesicles can be analyzed following the electron microscope validation and isolation [60,61]. While most current analyzes have targeted the expression of RNA transcripts, which are abundantly present in these exosomes [61,62], there is scant knowledge of the genomic and epigenetic profile of DNAs encapsulated in these vesicles.

ctDNA isolation is relatively easier than capturing exosomes. The main ctDNA isolation limiting factor is its separation from other cfDNA molecules, however, both of the following targeted and untargeted approaches are now available for this separation [36].

(1) Targeted approaches are based on identifying the following mutations/abnormalities in the ctDNA elements: (a) previously accepted abnormal patterns in ctDNAs (e.g., in different cancer type or based on *in silico* prediction) and (b) the same pattern in the primary tumor and the released ctDNA. Here, the "standard" methods of molecular biology such as variable PCR modifications and pyrosequencing can be used [63–66], but precise and more sensitive methods including BEAMing, droplet digital PCR, and next generation sequencing (NGS) are more suitable because of the very low ctDNA levels as compared to cfDNAs [67–69].

(2) Nontargeted approaches require no prior knowledge of molecular alternatives and while these are generally based on NGS techniques, the digital karyotyping and PARE methods can also be utilized [36]. Whereas, both targeted and nontargeted approaches are used mainly for detecting point mutations, insertions and deletions, amplifications, translocations, and copy number alterations [36,67]; identifying abnormal epigenetic patterns and especially changes in promoter region methylation levels can be a suitable alternative to genomic alterations. It is supposed that ctDNA epigenetic patterns should also usually remain the same as in the tumors from which they were released and also be specific for both cancer type and progression [35,67,70–75].

The capture of CTC remains extremely challenging because of its rarity and short half-life. There are still affinity-based methods used which take advantage of antigens that are differentially expressed by CTC [46]. These include: (1) epCAM which is normally present on the surface of cells released from epithelial carcinomas [45,46]; (2) the MUC1 present in breast and ovary carcinomas [76,77]; and (3) the EphB4 which is typically over-expressed in advanced head and neck cancers [46]. However, this approach remains specific for limited cancer types because these markers are suitable for positive selection, and therefore knowledge of further CTC surface antigens typical for different cancer types is essential. A further limitation is imposed by epithelial cells which are undergoing epithelial-mesenchymal transition (EMT) and cannot be captured by positive selection [45,46]. Negative selection is used to avoid selection bias and is based on capturing and subsequent removal of non-CTC cells in the blood, most importantly, leucocytes. These are commonly depleted through immunomagnetic targeting and removal against CD45 and other leucocyte antigens [45,78]. To the best of our knowledge, no studies have focused on identifying CTC released from any type of uterine sarcoma or ULM. However, there are some options for the identification of CTC from different non-uterine sarcoma types, and this provides opportunities to determine CTC from uterine sarcomas. For example, Benini et al. [79] were able to identify CTC released from Ewing sarcomas by immunoseparation with CD99 antibody and magnetic microbeads. The origin of the cells was verified by the presence of fusions typical for this tumor. The expression of cell-surface vimentin provided an even better chance to identify CTC of various sarcomatous origins [80]. Most recently, Hayashi et al. [81] recorded CTC identification from various sarcoma types based on positive vimentin staining, negative CD45 staining, and nuclear morphology distinct from normal white blood cells.

The standard approaches for identifying CTC include using magnetic beads armed with antibodies for positive or negative separation and employing columns or cartridges or microchips coated with antibodies [82]. However, the CTC physical properties, such as electric charges, can also be used to enrich and isolate these cells with more practical microfluid cell sorters [83,84] or Microhall platforms [85,86]. Finally, microfiltration systems, such as CellSieve^{TM}, also provide great possibilities for the identification of variable CTC with only minimal cell processing [81]. Interestingly, CTC were previously thought to be released only from metastases, but it is now accepted that two CTC "types" exist. This is contrary to previous claims, because the first type acts as a metastatic factor and the other is only passively released into the bloodstream, and although it should not promote metastases formation, its role in tumor spread remains undetermined [87].

5. Known Abnormalities in UMT

LB methodology enables analysis of various molecular targets over longer periods and this makes it possible to monitor disease progression and response to treatment. Furthermore, it also provides the possibility of differentiating particular tumor types. This can be very beneficial for early diagnosis of cancer. For differentiation of UMT with the LB approach, there are two options: (1) assessment of the mutational/transcriptional/epigenetic pattern in solid tumors and subsequent identification of the very same abnormalities in LB samples and (2) determination of specific patterns of change in the candidate markers directly in the LB samples and their connection with particular cancer diseases.

Unfortunately, specific LB biomarkers with the potential to non-invasively distinguish particular UMT types have not been assessed. Further analysis is, therefore, required to find a successful marker, but this is quite promising because the presence of ctDNA released from ULM and malignant sarcomas has been noted in non-invasive prenatal testing (NIPT) [88–90]. Furthermore, Hemming et al. [91,92] described the presence of ctDNA from LMS in LB samples and these correlated with tumor progression. Moreover, many of the biomarkers of solid UMT described below have previously been reported in LB from patients with different cancer diseases and these provide the possibility of early diagnosis, prognosis, and therapeutic outcome. These include the identical microRNAs [93–97], methylation changes [95,98,99], mutational changes [100–102], and even similar chromosomal rearrangements [103–106]. We, therefore, suggest a way to identify LB biomarkers suitable for UMT differentiation by searching for abnormal patterns already established in solid UMT in LB samples. This possibility is very promising also because reports identify that solid tumor abnormalities should remain unchanged in CNA [35–39,67,70–75] and eventually also in vesicles and CTC. Therefore, all abnormal patterns listed below are noted in solid tumors and their summary can form the basis for designing LB-based analysis in patients with UMT.

6. miRNAs

The miRNAs are short non-coding RNAs which can bind to target mRNAs and induce gene silencing and repression of transcription [107]. The 2000 miRNA genes already identified in the human genome is expected to increase in number and their abnormal activity has been connected with the development of various cancerous diseases because they regulate the important cellular functions of development, differentiation, growth and metabolism [107]. Analysis of miRNA abnormal activity is currently attractive, however, in ULM and LMS tissues, it is still not sufficiently known. The most notable abnormally expressed microRNAs in ULM and LMS tissues are listed in Table 1. There are some interesting characteristics of miRNAs in these tumors including that treatment with 17β-oestradiol and medroxyprogesterone acetate should result in the regulation of variable miRNAs' activity, at least in the ULM and LMS cell lines [108]. Liu et al. [109] noted that various miRNAs are differentially active in African and Caucasian women. Wang et al. [110] considered that the abnormalities in microRNAs activity should be associated with tumor size as well as ethnicity. Moreover, ULM typically have an abnormal extracellular matrix, and miR-15b is more active in ULM tissue samples than in healthy myometrium (MM), and also more active in ULM cell lines than in MM cell lines [111]. This miRNA is supposed to regulate RECK protein expression in normal physiological status to ensure negative regulation of matrix metalloproteinases [112,113]. In addition, the let-7 family miRNA family has the

HMGA2 gene as a target and *HMGA2* abnormal expression, especially from translocations, is considered a typical ULM mark [114].

The activity of most analyzed miRNAs has been assessed in ULM. MicroRNA activity is less understood in LMS and studies that could compare the activity of identical microRNAs in both tumor types are widely lacking. Exceptions, however, include studies by de Almeida et al. [115] and Schiavon et al. [116], which focused on assessing miRNA activity in both tumor types. In the former study, the authors identified 24 oncomirs with de-regulated expression profile in both ULM and LMS. The latter work recorded similar results in establishing widely different ULM and LMS expression patterns. Danielsson et al. [117] also suggested that different miRNA activity should contribute to separate ULM and LMS development, and they created a clustering system differentiating ULM and LMS based on miRNA activity. Unfortunately, however, they accomplished this only in cell lines. The most notably different ULM and LMS miRNAs recorded in these studies are listed in Table 1, under the "observed in LMS/ULM tissue" section, and these can be considered the best candidates for distinguishing between ULM and LMS on miRNA activity using LB. However, many of the analyzed ULM miRNAs were not analyzed in malignant LMS, even in the two separate studies. Therefore, the miRNAs summarized in Table 1 and reported as abnormal in ULM (listed under the "observed in ULM/MM tissues"), could prove beneficial if analyzed in LMS. There are also studies which assessed different miRNAs expression profiles in ESS. For example, Kowalewska et al. [118] reported four miRNAs which were differentially active in ESS and the control group, but it is quite interesting that they noted no significant differences in miRNA expression in LMS and control healthy tissues. The additional study by dos Anjos et al. [119] compared LMS, carcinosarcoma, and ESS miRNA profiles and identified both variable miRNA activity in these three tumor types and a variety of miRNAs connected with lower cancer-specific survival rates. This may, however, be due to tumor differences, whereas LMS are primarily smooth-muscle tumors. ESS, even of mesenchymal origin, are mixed tumors with a large proportion of endometrial elements and carcinosarcomas are classified as being of endometrial origin [18].

Table 1. Abnormally active microRNAs in ULM and LMS.

ULM			LMS		
miRNA	**Observed in**	**Expression in Tumorous Tissue**	**miRNA**	**Observed in**	**Expression in Tumorous Tissue**
Let-7 family * [110]	ULM/MM tissue	Up	miR-15a * [120]	Primary/metastatic LMS tissue	Up in metastases
miR-27a * [110]	ULM/MM tissue	Up	miR-92a * [120]	Primary/metastatic LMS tissue	Up in metastases
miR-21 * [110]	ULM/MM tissue	Up	miR-31 * [120]	Primary/metastatic LMS tissue	Up in primary
miR-23b * [110]	ULM/MM tissue	Up	miR-122-5p [116]	LMS/ULM tissue	Up
miR-200a * [121]	UtLM-hTert	Up	miR-206 * [116]	LMS/ULM tissue	Up
miR-542-3b [122]	ULM/MM tissue	Up	miR-373-3p * [116]	LMS/ULM tissue	Up
miR-377 [122]	ULM/MM tissue	Up	miR-144-3p [116]	LMS/ULM tissue	Up
miR-363 [123]	ULM/MM tissue	Up	miR-372-3p * [116]	LMS/ULM tissue	Up
miR-490 * [123]	ULM/MM tissue	Up	miR-34a-5p [116]	LMS/ULM tissue	Down
miR-137 [123]	ULM/MM tissue	Up	miR-27b-3p [116]	LMS/ULM tissue	Down
miR-15b * [111]	ULM/MM tissue; ULM/MM cell lines	Up	miR-135b-5p [116]	LMS/ULM tissue	Down
miR-30a [121]	ULM/MM tissue; UtlM-hTERT cell lines	Up	miR-9-5p [116]	LMS/ULM tissue	Down
miR-32 [110]	ULM/MM tissue	Down	miR-10b-5p * [115]	LMS/ULM tissue	Up
miR-29b * [110,124]	ULM/MM tissue	Down	miR-125b-1-3p [115]	LMS/ULM tissue	Up
miR-542-5p [122]	ULM/MM tissue	Down	miR-140-5p * [115]	LMS/ULM tissue	Up

Table 1. *Cont.*

ULM			LMS		
miRNA	**Observed in**	**Expression in Tumorous Tissue**	**miRNA**	**Observed in**	**Expression in Tumorous Tissue**
miR-642 [122]	ULM/MM tissue	Down	miR-145-5p [115]	LMS/ULM tissue	Up
miR-93/106 [125]	ULM/MM tissue	Down	miR-130b-3p * [92]	LMS/ULM tissue	Down
miR-486-5p * [123]	ULM/MM tissue	Down	miR-148-3p [115]	LMS/ULM tissue	Down
miR-217 [123]	ULM/MM tissue	Down	miR-204-5p [92]	LMS/ULM tissue	Down
miR-4792 [123]	ULM/MM tissue	Down	miR-203a-3p [92]	LMS/ULM tissue	Down
miR-200a * [109]	ULM/MM tissue; UtlM-hTERT cell lines	Down	miR-152 * [126]	LMS/ULM tissue /SKLMS1cell lines	Down
mir-143 * [108]	ULM/MM tissue	Down			
miR-200c * [127]	ULM/MM tissue	Down			
miR-197 * [128]	ULM/MM tissue	Down			
miR-212 [121]	ULM/MM tissue; UtlM-hTERT cell lines	Down			

* These were reported and analysed in LB samples in patients with different cancer diseases [93–97]; ULM—uterine leiomyomas, MM—healthy myometrium, LMS—uterine leiomyosarcomas.

7. DNA Mutations and Chromosomal Aberrations in ULM and Sarcomas

ULM often result from gene mutations, but this occurs only on a small scale. The most mutated is the *MED12* gene observed in almost 70% of ULM, and the second most frequent is *FH* gene mutations which comprise only approximately 1% [129,130]. In addition, non-sporadic germline mutations of this gene occurred more often in women with both ULM and LMS [131,132]. Interestingly, malignant uterine LMS also have *MED12* gene mutation in almost 70% of tumors [133], and this can be, therefore, considered a typical mark for these tumors. While this common mutation potentially distinguishes the two tumor groups from healthy tissue, unfortunately it cannot be used as a distinguishing factor of these tumors and is, therefore, unsuitable for LB diagnostics. In contrast to ULM, LMS have quite often mutated *TP53* gene [6,134]. Although mutations in this gene are typical for many cancerous diseases, LB determination of *TP53* mutation in women with previous image-diagnosed UMT should always provoke rapid tissue biopsy, subsequent histological examination, and close observation of the patient. Moreover, this gene should also be considered a potential candidate for gene therapy in women with LMS [135].

The *ATRX* gene is also relatively often mutated in LMS tissues [134], and also in STUMP [136]. Furthermore, the *RB1* gene is typically mutated in LMS [26] and in ULM with bizarre nuclei [137]. However, the percentage of mutations of both genes in LMS is less than 50% and the extent of *ATRX* mutation in leiomyomatous tissues remains undetermined. In addition, ULM with bizarre nuclei are quite rare and the *RB1* gene is often mutated in a wide range of cancers [138]. In summary, the mutation profiles of ULM and sarcomas are not well understood, and currently demonstrated abnormalities provide only limited possibilities for LB differentiation of these tumors. However, mutation changes in the *TP53*, *ATRX*, and *RB1* genes have been detected and analyzed in LB samples in patients with different cancer diseases [100–102]. In contrast to mutational changes, the ULM and LMS chromosomal aberrations are quite different and better understood, so current knowledge suggests that LB differentiation of these tumors based on chromosomal aberrations is more suitable than just relying on mutational changes. Moreover, this identification from LB samples is feasible, for example, with ultra-low passage whole genome sequencing [91,139,140]. The most important recognized chromosomal aberrations in ULM and LMS are listed in Table 2.

Table 2. Most typical chromosomal rearrangements in ULM, LMS, and ESS.

Chromosomal Aberrations in ULM			Chromosomal Aberrations in LMS			Chromosomal Aberrations in ESS		
Chromosome/Locus	**Type**	**Affected Genes**	**Chromosome/Locus**	**Type**	**Affected Genes**	**Chromosome/Locus**	**Type**	**Affected Genes**
12q15 [141]	Translocation	*HMGA2*	12q13-15 [142]	Amplification	*RB1*	(7;17)(p15;q21) [143]	Translocation	*JAZF1,SUZ12*
14q24 [144]	Translocation	*RAD51B*	10q21.3 * [145]	Loss	*PTEN*	(6;7)(p21;p15) [146]	Translocation	*JAZF1-PHF1*
7(q22q32) [147]	Deletion	*CUX1*	13q14.2-q14.3 * [145]	Loss	*LEU*	(6;10)(p21;p11) [146]	Translocation	*EPC1-PHF1*
6p21 [148]	Translocation	*HMGA1*	7q36.3 [142]	Gain	*PTPRN2*	(1;6)(p34;p21) [149]	Translocation	*MEAF6-PHF1*
10q22 [150]	Translocation	*KAT6B*	7q33-q35 [142]	Gain	*HAVCR1*	(X;17)(p11;q21) [151]	Translocation	*CXorf67-MBTD1*
1(q31q43) [130]	Deletion	*FH*	1p21.1 * [142]	Loss	*AMY2A*	(X;22)(p11;q13) [152]	Translocation	*ZC3H7B-BCOR*
Xq22 [153]	Deletion	*COL4A5/COl4A6*	9p.21	Gain	*CDKN2*	(10;17)(q22;p13) [154]	Translocation	*YWHAE-NUTM*
1p36 [155]	Translocation/deletion	*AJAPI, NPHP*	12q.15 [156]	Gain	*MDM2*	der(22)t(X;22)(p11;q13) [152]	Deletion	/
8q12 [157]	Insertion/translocation	*PLAG1*	1q21 * [158]	Amplification	*FLF, PRUNE*	del(16)(q22) [149]	Deletion	/
19q * [159]	Deletion	/	5p14-pter [158,160]	Amplification	/			
12 [161]	Trisomy	/	13q31 [158,160]	Amplification	/			
10 [147]	Monosomy	/	19p13 [158,160]	Amplification	/			
22q * [159]	Deletion/monosomy	/	20q13 [158,160]	Amplification	/			

* Similar chromosomal rearrangements were reported in LB samples from patients with different cancer diseases [103–106]; ULM—uterine leiomyomas, LMS—uterine leiomyosarcomas, ESS—endometrial stromal sarcomas.

8. Methylation Changes in ULM and LMS

The methylation profiles of ULM and sarcoma promoter regions provide further possibilities for tumor differentiation based on LB, because changes in DNA methylation patterns can be identified in ctDNA and have been proposed as potential biomarkers for tumor staging, prognosis, and monitoring of the treatment response [31,98,162]. It has also been suggested that abnormal methylation patterns should remain identical in both the primary tumors and the ctDNAs released by the tumor [35,67,70–75]. Maekawa et al. [163] reported great variability in ULM methylation patterns. These authors reported 120 genes with different methylation pattern in ULM and MM, and 22 of these were estrogen receptor alpha (ERα) target genes, thus, indicating that their abnormal methylation could contribute to abnormal estrogen response. One of these genes is *DAPK1* which is often abnormally methylated in various cancer types [164] and importantly, it has also been reported abnormally methylated in the serum of women with ULM [165].

Following bisulfide-sequencing of the ERα receptor gene promoter region, Asada et al. [166] reported that seven CpGs in distal sites are often variably methylated in healthy and tumorous tissues. It is quite likely that methylation variations in distal ERα regions are dependent on menstrual cycle phases [167], and since this has already been assessed in solid tissues, it would be very interesting and beneficial to analyze the status in LB samples. ULM also have differentially methylated sex chromosomes [168] where the most notably hypomethylated was the *TSPYL2* gene. In addition, women of African origin have hypermethylated *DLEC1*, *KRT1,9* and *KLF11* genes in ULM as compared to MM [169].

Importantly, methylation markers have already been used by Sato et al. [170] to distinguish between ULM and LMS, especially by establishing a hierarchical clustering system able to discriminate between these tumors with 70% accuracy. The system is based on different methylation levels in ten genes. This study is also unique because it assesses the methylation status in both ULM and LMS tumor types. In contrast, the remaining methylation markers listed in Table 3 were assessed either for benign or malignant lesions, but not for both, and further analyses comparing the status of these genes in both tumor types would be beneficial to determine how well they coud distinguish between these tumor types. For LMS are typical aberrations altering the cell cycle [6], whereby promoter region hypermethylation is supposed to induces loss of function of the important cell cycle regulator *CDKN2* [171] and hypermethylation also most likely causes the *BRCA1* gene´s protein reduction in almost 30% of uterine LMS [172].

Malignant LMS have abnormal estrogen receptor activity and sensitivity, but these are lower than in benign ULM [22]. The LMS also have often hypomethylated *ESR1* target genes [26] and the polycomb group target genes [173]. Importantly, authors Miyata et al. [173] claimed that ULM and LMS can be distinguished by their global DNA methylation levels, and also presumed that the methylation pattern is different on a genome-wide level in ULM subtypes with different genetic driver aberration, i.e., *MED12* mutation, *HMGA2* activation, and *FH* mutation [9,174]. These results make tumor differentiation based on their ctDNA methylation levels very interesting, however, the bloodstream ctDNAs are very fragmented and poorly represented, and therefore ctDNA methylation analyses on either a genome-wide scale or a locus-specific level should preferably use very sensitive novel sequencing approaches [67,98]. The hyper- and hypomethylated genes in ULM and LMS tissues are listed in Table 3.

Table 3. Abnormally methylated genes in ULM and LMS.

ULM		LMS	
Gene	**Methylation Change**	**Gene**	**Methylation Change**
IRS1 [163]	Hypermethylation	*MGMT* * [175]	Hypermethylation
COL4A1 [163]	Hypermethylation	*BRCA1* * [172]	Hypermethylation
GSTM5 [163]	Hypermethylation	*CDKN2* * [137]	Hypermethylation
DAPK1 * [165]	Hypermethylation	*PTEN* [6]	Hypermethylation
KLF11 [169]	Hypermethylation	*RASSF1A* *a [176]	Hypermethylation
DLEC1 * [169]	Hypermethylation	*DAPK1* *a [177]	Hypermethylation
KRT19 [169]	Hypermethylation		
ALX1 ‡ [170]	Hypermethylation		
CBLN1 ‡ [170]	Hypermethylation		
CORIN ‡ [170]	Hypermethylation		
DUSP6 [170]	Hypermethylation		
FOXP1 ‡ [170]	Hypermethylation		
GATA2 ‡* [170]	Hypermethylation		
IGLON5 ‡ [170]	Hypermethylation		
NPTX2 ‡* [170]	Hypermethylation		
NTRK2 ‡ [170]	Hypermethylation		
STEAP4 ‡* [170]	Hypermethylation		
PRL ‡ [170]	Hypomethylation		
PART1 [170]	Hypomethylation		
TSPYL2 [168]	Hypomethylation		
OCRL [168]	Hypomethylation		

* These were reported and analyzed in LB samples in patients with different cancer diseases [95,98,99]; ‡ Methylation status of these genes was compared between ULM and LMS in a study by Sato et al. [170]; [a] Methylation changes were noted in variable types of leiomyosarcomas, including non-uterine. *DAPK1* gene was reported as hypermethylated in ULM and sarcomas as compared with healthy tissues, but it was not compared between ULM and LMS; ULM—uterine leiomyomas, LMS—uterine leiomyosarcomas.

9. lncRNAs and mRNAs

ULM expression profiles are relatively well assessed in studies analyzing gene activity on the mRNA levels in solid tumors [25,178–180]. The LMS mRNA expression profile is also relatively sufficiently known [6,25,181,182] and variably expressed genes are recognized also in primary and metastatic LMS [183] and in LMS as comparing to ESS [184]. Although mRNA molecules directly released into the bloodstream after apoptosis or necrosis can technically reflect intracellular processes in tumor cells and their levels can have some predictive value of tumor character, their very short half-life, instability, low abundance, and regular contamination with other intracellular mRNAs severely limits them as useful biomarkers [185]. In contrast, assessment of mRNA expression in CTC and exosomes appears more suitable because of the newly developed approaches [186–189].

Genome-wide analyses of ULM lncRNA expression have demonstrated that hundreds of these molecules are aberrantly active in ULM as compared to MM [190]. The number of abnormally active lncRNAs correlates with tumor size, and lncRNA H19 was considered as the most important abnormally active lncRNA in ULM [190]. lncRNA H19 contributes to *MED12* and *HMGA2* gene regulation. It is assumed that lncRNA H19 should also affects the activity of various extracellular matrix remodeling genes [191]. Therefore, further study of this lncRNA should shed more light on the regulation of important ULM marker expression and this lncRNA could potentially be used as a reliable distinguishing marker of these tumors. Finally, further research of lncRNAs' abnormal activity in rare malignant sarcomas should also prove advantageous, but this is still currently lacking.

10. ULM and Sarcoma Metabolites

ULM are also the subject of metabolic studies because of their high occurrence, the search for efficacious medication, and their great current cost. The very interesting study performed by Heinonen et al. [192] assessed abnormal metabolomes and metabolic pathways in ULM, and they associated these with particular molecular subtypes, which harbor different driver genetic abnormalities. These were divided into the following four groups: those with the mutated *MED12* gene, those with *HMGA2* gene translocations, the group with *FH* gene biallelic inactivation, and *COL4A5-COL4A6* deletions [9]. In summary, Heinonen et al. [192] recorded 70 dysregulated metabolites in all ULM types, and the characteristic dysregulations in particular ULM subtypes included: (1) the FH subtype which has characteristic metabolic alterations in the tricarboxylic acid cycle and pentose phosphate pathways and increased levels of multiple lipids and amino acids and (2) the MED12 subtype which has markedly reduced levels of vitamin A and dysregulation of ascorbate metabolism. Interestingly, the retinoid acid receptor pathways have previously been observed to be very dysregulated in ULM, and this caused lower levels of biologically active retinoid acid and increased levels of its abnormal metabolites [9,193,194]. In addition, it has been suggested that premenopausal parous women with ULM share a wide range of metabolic syndrome features such as higher serum triglycerides levels (TG $\geq$ 150 mg/dL), low serum-high density lipoproteins (<50 mg/dL), and hyperglycemia (FPG $\geq$ 100 mg/dL) [195]. The obesity and high blood pressure observed in this syndrome have also been previously and repetitively associated with ULM occurrence [1–3]. However, while it is generally accepted [3] that vitamin D deficiency is strongly associated with ULM occurrence, its abnormal metabolites were not confirmed in these tumors [192].

The uterine sarcoma metabolite profile is poorly understood due to its rare occurrence. Paradoxically, some authors regard metabolomic markers more reliable than molecular markers in sarcoma diagnostics, especially in the high-grade type which are more difficult to distinguish histologically. Moreover, individual tumors often contain areas of different histologic grade, necrotic regions, and variable inflammatory cell infiltrate, and this heterogeneity hinders the search for molecular markers [196,197]. However, the preference for metabolomic markers over the molecular remains disputed and subjective.

Lastly, the same pattern of precursor ion differences has been found in LMS, myxofibrosarcomas, and undifferentiated pleiomorphic sarcomas. The two metabolite signals correlate with overall survival, *m/z* 180.9436 and *m/z* 241.0118; and one with metastasis-free survival, *m/z* 160.8417. In addition, FTICR-MSI identified *m/z* 241.0118 as inositol cyclic phosphate and *m/z* 160.8417 as carnitine [197].

11. Discussion and Conclusions

In this review, we based the differentiation of UMT on the LB approach because of the diverse action of non-coding RNAs, and mutational and methylation changes and differences in metabolic activity in ULM, other UMT, and MM. We also suggested searching for these previously known solid tumor abnormalities in LB samples. However, the differentiation of two or more tumor types by LB necessitates, in the final step, assessment of their altered patterns in the body fluid samples. For example, hypermethylation or mutational changes can be considered proof of ULM rather than LMS when a LB sample is abnormally hypermethylated or has particular mutations, and this abnormal pattern was previously observed with high predictive value in ULM but not in LMS. This explanation, however, is simplistic and follows only basic principles because the entire procedure for establishing a particular marker to its utilization in clinical testing is a long and multistep process. While it is debatable if there is sufficient knowledge of the differences in ULM, it is certainly lacking in sarcomas. Therefore, to enable LB to determine tumor character and to monitor the response to treatment and tumor progression in malignant variants, it is essential to increase current knowledge of changes in ULM and also to perform these analyses in uterine sarcomas, especially LMS.

Moreover, we based our suggestions on the assumption that changes in CNA should remain the same as they are in the primary tumors which release them [35–39,67,70–75]. However, it should be mentioned, some studies suggest that the changes can vary [40–42]. In addition, it is necessary to realize that metastatic lesions can differ from primary tumors in their molecular patterns [183]. It would, therefore, be beneficial to perform more comparative analyses of primary and metastatic sarcomas to understand the changes in their progression. In addition, it would also be helpful to use more gene assays with high predictive value in LB screening, instead of sole ones. Finally, there is still no successful tool for differentiating UMT, so even achieving lower predictive value would be advantageous. An example here is that assays correctly assessing tumor type in eight out of 10 women would be better than none.

It is quite intriguing that there is no interest in developing techniques that could differentiate such frequent tumors as ULM and such aggressive tumors as sarcomas. The main reason appears to be the very low incidence of sarcomas and most UMT lesions are ULM. These latter are not life threatening and often asymptomatic. In addition, the statistical threat to life caused by uterine sarcomas is low, and almost negligible in comparison to the "deadliest" cancer types such as breast, lung and colon cancer. However, if sarcomatous lesions are primarily misdiagnosed as ULM, this can have fatal consequences, especially when women refuse surgical excision and prefer medicated treatment. Moreover, although uterine sarcomas are rare, these lesions still affect thousands of women globally and can be easily misdiagnosed. Therefore, early noninvasive diagnosis could provide certainty that lesions are really benign; and it could significantly prolong full-quality life even when the lesion is malign. Moreover, appropriate LB approaches can monitor both the metastatic progression of malignant lesions and response to treatment, and thus contribute to their better understanding.

Unfortunately, minimal analyses have focused directly on capturing and identifying circulating elements from UMT and most findings have been incidental and collateral. It is known that NIPT results can be affected by the presence of tumorous circulating elements [88]. The first study of these discrepancies we are aware of was conducted by Osoborne et al. [198]. Woman with metastatic neuroendocrine carcinoma gave birth to a healthy infant, and although both invasive testing during pregnancy and postnatal placental histology revealed no abnormalities, NIPT testing indicated trisomy in chromosomes 13 and 18. These trisomes, therefore, arose from release of tumor DNA into the bloodstream and not due to infant chromosomal aberrations. A summary of a three-year period of NIPT testing enabled identification of 55 samples with altered genomic profiles and thus not reportable NIPT results [89]. All these discrepancies are presumed to arise due to the presence of various neoplasm types, but importantly almost half of the identified discrepancies arose as a result of the presence of benign ULM and only one because of uterine sarcomas presence.

One study focused directly on identifying circulating ctDNAs from uterine LMS and confirmed ctDNA presence in five of 10 cases [91]. This number is considered quite low, but sarcomatous ctDNA was identified in five of six patients with progressive disease. Therefore, ctDNA identification correlated with tumor burden, and it was present in patients with both primary and metastatic lesions but not identified in disease-free or stable subjects. This was a pilot study, so repetition and finding new approaches and targets should increase the numbers of positively identified LMS ctDNAs.

In addition, Eastley et al. [199] confirmed the possibility of detecting and subsequently analyzing ctDNAs released from soft tissue tumors, including two leiomyosarcoma samples. The authors found *TP53* and *HRAS* mutations in both primary tumors and plasma samples. However, it is necessary to remember that the percentage of these mutations differs in tissue and plasma samples and the authors do not mention if these leiomyosarcomas arise from uterine tissues or not.

The understanding of CTC and exosome release from ULM and sarcomas is still poor, and to the best of our knowledge no miRNA profiles have been recovered from these tumors in LB samples despite their abundance in primary tumors. More studies should, therefore, focus on assessing non-coding

RNAs, especially miRNAs in LB samples and on the character of CTC and exosomes released from UMT. From a positive perspective, however, combined approaches have already been developed for the purposes of LB-based tumor diagnosis and prognosis, and these detect desired ctDNA changes in mesenchymal tumors, including uterine LMS [200].

In conclusion, there are no currently known LB biomarkers that can be used for early and noninvasive distinction between particular UMT types. However, (1) there are recorded nucleic acid molecules released from both benign and malignant variants into the bloodstream [89–92]; (2) many molecular biomarkers reported in solid UMT have been noted in LB samples in different cancer types and these can be used for early diagnosis, prognosis, and therapeutic outcomes [93–102]; and (3) it is presumed that abnormal patterns in solid tumors should remain the same in CNA [35–39,67,73–75]. On the basis of these facts, we can state that it is theoretically possible, and highly likely in practice, to distinguish clinically between UMT types using the LB approach when appropriate molecular markers are employed. Our review should, therefore, inspire further research and analysis of known abnormalities in solid tissues in LB samples. The implication of the markers in these processes could also help monitoring of both the progression of malignant variants and the response to treatment. It is, therefore, essential in the future to establish as many molecular markers as possible for each type of UMT and to evaluate them using LB approaches. This will be most beneficial when sufficiently large patient cohort screening can be performed with a variety of advanced molecular methods. Performing NGS-targeted approaches for identification of the markers highlighted in this review is suggested as a most suitable option. Moreover, genome-wide studies would also be welcome to identify useful markers, including epigenetic markers. Finally, any possible improvements in LB approaches for women with UMT will benefit both patients and clinicians, because clinicians would then be able to more precisely measure disease burden and better coordinate clinical treatment, providing greater certainty to affected women that their lesion remains benign.

Author Contributions: D.D., D.B., and H.S. wrote the original draft; Z.D. and E.H. supervised the project, revised, and approved the manuscript.

Abbreviations

cfDNA	circulating free DNA
CNA	freely circulating nucleic acids
CTC	circulating tumor cells
ctDNA	circulating tumor DNA
EMT	epithelial-mesenchymal transition
ERα	estrogen receptor alpha
ESS	endometrial stromal sarcomas
LB	liquid biopsy
LMS	uterine leiomyosarcomas
lncRNA	long non-coding RNA
mRNA	messenger RNA
miRNA	microRNA
MM	healthy myometrium
NGS	next generation sequencing
NIPT	non-invasive prenatal testing
STUMP	smooth muscle tumors of uncertain malignant potential
ULM	uterine leiomyomas
UMT	uterine mesenchymal tumors

References

1. Flake, G.P.; Andersen, J.; Dixon, D. Etiology and Pathogenesis of Uterine Leiomyomas: A Review. *Environ. Health Perspect.* **2003**, *111*, 1037–1054. [CrossRef] [PubMed]
2. Baird, D.D.; Dunson, D.B.; Hill, M.C.; Cousins, D.; Schectman, J.M. High Cumulative Incidence of Uterine Leiomyoma in Black and White Women: Ultrasound Evidence. *Am. J. Obstet. Gynecol.* **2003**, *188*, 100–107. [CrossRef] [PubMed]
3. Islam, M.S.; Protic, O.; Giannubilo, S.R.; Toti, P.; Tranquilli, A.L.; Petraglia, F.; Castellucci, M.; Ciarmela, P. Uterine Leiomyoma: Available Medical Treatments and New Possible Therapeutic Options. *J. Clin. Endocrinol. Metab.* **2013**, *98*, 921–934. [CrossRef] [PubMed]
4. Brooks, S.E.; Zhan, M.; Cote, T.; Baquet, C.R. Surveillance, Epidemiology, and End Results Analysis of 2677 Cases of Uterine Sarcoma 1989–1999. *Gynecol. Oncol.* **2004**, *93*, 204–208. [CrossRef] [PubMed]
5. Santos, P.; Cunha, T.M. Uterine Sarcomas: Clinical Presentation and MRI Features. *Diagnostic Interv. Radiol.* **2015**, *21*, 4–9. [CrossRef] [PubMed]
6. Kobayashi, H.; Uekuri, C.; Akasaka, J.; Ito, F.; Shigemitsu, A.; Koike, N.; Shigetomi, H. The Biology of Uterine Sarcomas: A Review and Update. *Mol. Clin. Oncol.* **2013**, *1*, 599–609. [CrossRef] [PubMed]
7. Benson, C.; Miah, A.B. Uterine Sarcoma – Current Perspectives. *Int. J. Womens. Health* **2017**, *9*, 597–606. [CrossRef] [PubMed]
8. Arleo, E.K.; Schwartz, P.E.; Hui, P.; McCarthy, S. Review of Leiomyoma Variants. *Am. J. Roentgenol.* **2015**, *205*, 912–921. [CrossRef]
9. Mehine, M.; Kaasinen, E.; Heinonen, H.-R.; Mäkinen, N.; Kämpjärvi, K.; Sarvilinna, N.; Aavikko, M.; Vähärautio, A.; Pasanen, A.; Bützow, R.; et al. Integrated Data Analysis Reveals Uterine Leiomyoma Subtypes with Distinct Driver Pathways and Biomarkers. *Proc. Natl. Acad. Sci.* **2016**, *113*, 1315–1320. [CrossRef] [PubMed]
10. Zhang, Q.; Ubago, J.; Li, L.; Guo, H.; Liu, Y.; Qiang, W.; Kim, J.J.; Kong, B.; Wei, J.-J. Molecular Analyses of 6 Different Types of Uterine Smooth Muscle Tumors: Emphasis in Atypical Leiomyoma. *Cancer* **2014**, *120*, 3165–3177. [CrossRef]
11. Prat, J.; 'Nomonde, M. Uterine Sarcomas. *Int. J. Gynecol. Obstet.* **2015**, *131*, S105–S110. [CrossRef] [PubMed]
12. Kalogiannidis, I.; Stavrakis, T.; Dagklis, T.; Petousis, S.; Nikolaidou, C.; Venizelos, I.; Rousso, D. A Clinicopathological Study of Atypical Leiomyomas: Benign Variant Leiomyoma or Smooth-Muscle Tumor of Uncertain Malignant Potential. *Oncol. Lett.* **2016**, *11*, 1425. [CrossRef] [PubMed]
13. Pickett, J.L.; Chou, A.; Andrici, J.A.; Clarkson, A.; Sioson, L.; Sheen, A.; Reagh, J.; Najdawi, F.; Kim, Y.; Riley, D.; et al. Inflammatory Myofibroblastic Tumors of the Female Genital Tract Are Under-Recognized: A Low Threshold for ALK Immunohistochemistry Is Required. *Am. J. Surg. Pathol.* **2017**, *41*, 1433–1442. [CrossRef] [PubMed]
14. Indraccolo, U.; Luchetti, G.; Indraccolo, S.R. Malignant Transformation of Uterine Leiomyomata. *Eur. J. Gynaecol. Oncol.* **2008**, *29*, 543–544.
15. Patacchiola, F.; Palermo, P.; Di Luigi, G. Leiomyosarcoma: A Rare Malignant Transformation of a Uterine Leiomyoma. *Eur. J. Gynaecol. Oncol.* **2015**, *36*, 84–87.
16. Holzmann, C.; Saager, C.; Mechtersheimer, G.; Koczan, D.; Helmke, B.M.; Bullerdiek, J. Malignant Transformation of Uterine Leiomyoma to Myxoid Leiomyosarcoma after Morcellation Associated with ALK rearrangement and Loss of 14q. *Oncotarget* **2018**, *9*, 27595–27604. [CrossRef] [PubMed]
17. Blythe, J.G.; Bari, W.A. Uterine Sarcoma: Histology, Classification and Prognosis. In *Gynecology and Obstetrics*; Harper Row: Hagerstown, MD, USA, 2008; Volume 4, pp. 1–19. [CrossRef]
18. Artioli, G.; Wabersich, J.; Ludwig, K.; Gardiman, M.P.; Borgato, L.; Garbin, F. Rare Uterine Cancer: Carcinosarcomas. Review from Histology to Treatment. *Crit. Rev. Oncol. Hematol.* **2015**, *94*, 98–104. [CrossRef]
19. D'Angelo, E.; Prat, J. Uterine Sarcomas: A Review. *Gynecol. Oncol.* **2010**, *116*, 131–139. [CrossRef]
20. Chen, I.; Firth, B.; Hopkins, L.; Bougie, O.; Xie, R.-H.; Singh, S. Clinical Characteristics Differentiating Uterine Sarcoma and Fibroids. *JSLS J. Soc. Laparoendosc. Surg.* **2018**, *22*, e2017.00066. [CrossRef]
21. Falcone, T.; Parker, W.H. Surgical Management of Leiomyomas for Fertility or Uterine Preservation. *Obstet. Gynecol.* **2013**, *121*, 856–868. [CrossRef] [PubMed]

22. Bodner, K.; Bodner-Adler, B.; Kimberger, O.; Czerwenka, K.; Mayerhofer, K. Estrogen and Progesterone Receptor Expression in Patients with Uterine Smooth Muscle Tumors. *Fertil. Steril.* **2004**, *81*, 1062–1066. [CrossRef] [PubMed]
23. Felix, A.S.; Cook, L.S.; Gaudet, M.M.; Rohan, T.E.; Schouten, L.J.; Setiawan, V.W.; Wise, L.A.; Anderson, K.E.; Bernstein, L.; De Vivo, I.; et al. The Etiology of Uterine Sarcomas: A Pooled Analysis of the Epidemiology of Endometrial Cancer Consortium. *Br. J. Cancer* **2013**, *108*, 727–734. [CrossRef] [PubMed]
24. Singh, Z. Leiomyosarcoma: A Rare Soft Tissue Cancer Arising from Multiple Organs. *J. Cancer Res. Pract.* **2018**, *5*, 1–8. [CrossRef]
25. Mas, A.; Simón, C. Molecular Differential Diagnosis of Uterine Leiomyomas and Leiomyosarcomas. *Biol. Reprod.* **2018**, *0*, 1–9. [CrossRef] [PubMed]
26. Tsuyoshi, H.; Yoshida, Y. Molecular Biomarkers for Uterine Leiomyosarcoma and Endometrial Stromal Sarcoma. *Cancer Sci.* **2018**, *109*, 1743–1752. [CrossRef] [PubMed]
27. Ashworth, T. A Case of Cancer in Which Cells Similar to Those in the Tumours Were Seen in the Blood after Death. *Aust. Med. J.* **1869**, *14*, 146–149. [CrossRef]
28. Crowley, E.; Di Nicolantonio, F.; Loupakis, F.; Bardelli, A. Liquid Biopsy: Monitoring Cancer-Genetics in the Blood. *Nat. Rev. Clin. Oncol.* **2013**, *10*, 472–484. [CrossRef] [PubMed]
29. Lodewijk, I.; Dueñas, M.; Rubio, C.; Munera-Maravilla, E.; Segovia, C.; Bernardini, A.; Teijeira, A.; Paramio, J.M.; Suárez-Cabrera, C. Liquid Biopsy Biomarkers in Bladder Cancer: A Current Need for Patient Diagnosis and Monitoring. *Int. J. Mol. Sci.* **2018**, *19*, 2514. [CrossRef]
30. Perakis, S.; Speicher, M.R. Emerging Concepts in Liquid Biopsies. *BMC Med.* **2017**, *15*, 75. [CrossRef]
31. Palmirotta, R.; Lovero, D.; Cafforio, P.; Felici, C.; Mannavola, F.; Pellè, E.; Quaresmini, D.; Tucci, M.; Silvestris, F. Liquid Biopsy of Cancer: A Multimodal Diagnostic Tool in Clinical Oncology. *Ther. Adv. Med. Oncol.* **2018**, *10*, 175883591879463. [CrossRef]
32. Borah, B.J.; Nicholson, W.K.; Bradley, L.; Stewart, E.A. The Impact of Uterine Leiomyomas: A National Survey of Affected Women. *Am. J. Obstet. Gynecol.* **2013**, *209*, 319.e1–319.e20. [CrossRef]
33. Ilié, M.; Hofman, P. Pros: Can Tissue Biopsy Be Replaced by Liquid Biopsy? *Transl. lung cancer Res.* **2016**, *5*, 420–423. [CrossRef]
34. Arneth, B. Update on the Types and Usage of Liquid Biopsies in the Clinical Setting: A Systematic Review. *BMC Cancer* **2018**, *18*, 527. [CrossRef]
35. Warton, K.; Mahon, K.L.; Samimi, G. Methylated Circulating Tumor DNA in Blood: Power in Cancer Prognosis and Response. *Endocr. Relat. Cancer* **2016**, *23*, R157–R171. [CrossRef]
36. Gorgannezhad, L.; Umer, M.; Islam, M.N.; Nguyen, N.-T.; Shiddiky, M.J.A. Circulating Tumor DNA and Liquid Biopsy: Opportunities, Challenges, and Recent Advances in Detection Technologies. *Lab Chip* **2018**, *18*, 1174–1196. [CrossRef]
37. Wan, J.C.M.; Massie, C.; Garcia-Corbacho, J.; Mouliere, F.; Brenton, J.D.; Caldas, C.; Pacey, S.; Baird, R.; Rosenfeld, N. Liquid Biopsies Come of Age: Towards Implementation of Circulating Tumour DNA. *Nat. Rev. Cancer* **2017**, *17*, 223–238. [CrossRef]
38. Ng, S.B.; Chua, C.; Ng, M.; Gan, A.; Poon, P.S.; Teo, M.; Fu, C.; Leow, W.Q.; Lim, K.H.; Chung, A.; et al. Individualised Multiplexed Circulating Tumour DNA Assays for Monitoring of Tumour Presence in Patients after Colorectal Cancer Surgery. *Sci. Rep.* **2017**, *7*, 40737. [CrossRef]
39. Yang, N.; Li, Y.; Liu, Z.; Qin, H.; Du, D.; Cao, X.; Cao, X.; Li, J.; Li, D.; Jiang, B.; et al. The Characteristics of CtDNA Reveal the High Complexity in Matching the Corresponding Tumor Tissues. *BMC Cancer* **2018**, *18*, 319. [CrossRef]
40. Guo, Q.; Wang, J.; Xiao, J.; Wang, L.; Hu, X.; Yu, W.; Song, G.; Lou, J.; Chen, J. Heterogeneous Mutation Pattern in Tumor Tissue and Circulating Tumor DNA Warrants Parallel NGS Panel Testing. *Mol. Cancer* **2018**, *17*, 131. [CrossRef]
41. Kammesheidt, A.; Tonozzi, T.R.; Lim, S.W.; Braunstein, G.D. Mutation Detection Using Plasma Circulating Tumor DNA (CtDNA) in a Cohort of Asymptomatic Adults at Increased Risk for Cancer. *Int. J. Mol. Epidemiol. Genet.* **2018**, *9*, 1–12.
42. Toor, O.M.; Ahmed, Z.; Bahaj, W.; Boda, U.; Cummings, L.S.; McNally, M.E.; Kennedy, K.F.; Pluard, T.J.; Hussain, A.; Subramanian, J.; et al. Correlation of Somatic Genomic Alterations Between Tissue Genomics and CtDNA Employing Next-Generation Sequencing: Analysis of Lung and Gastrointestinal Cancers. *Mol. Cancer Ther.* **2018**, *17*, 1123–1132. [CrossRef]

43. Schwarzenbach, H.; Hoon, D.S.B.; Pantel, K. Cell-Free Nucleic Acids as Biomarkers in Cancer Patients. *Nat. Rev. Cancer* **2011**, *11*, 426–437. [CrossRef]
44. Leung, F.; Kulasingam, V.; Diamandis, E.P.; Hoon, D.S.B.; Kinzler, K.; Pantel, K.; Alix-Panabieres, C. Circulating Tumor DNA as a Cancer Biomarker: Fact or Fiction? *Clin. Chem.* **2016**, *62*, 1054–1060. [CrossRef]
45. Shao, H.; Chung, J.; Issadore, D. Diagnostic Technologies for Circulating Tumour Cells and Exosomes. *Biosci. Rep.* **2016**, *36*, e00292. [CrossRef]
46. Agarwal, A.; Balic, M.; El-Ashry, D.; Cote, R.J. Circulating Tumor Cells. *Cancer J.* **2018**, *24*, 70–77. [CrossRef]
47. Boukouris, S.; Mathivanan, S. Exosomes in Bodily Fluids Are a Highly Stable Resource of Disease Biomarkers. *Proteom.-Clin. Appl.* **2015**, *9*, 358–367. [CrossRef]
48. Tian, F.; Shen, Y.; Chen, Z.; Li, R.; Ge, Q. No Significant Difference between Plasma MiRNAs and Plasma-Derived Exosomal MiRNAs from Healthy People. *Biomed. Res. Int.* **2017**, *2017*, 1304816. [CrossRef]
49. Kawaguchi, T.; Komatsu, S.; Ichikawa, D.; Tsujiura, M.; Takeshita, H.; Hirajima, S.; Miyamae, M.; Okajima, W.; Ohashi, T.; Imamura, T.; et al. Circulating MicroRNAs: A Next-Generation Clinical Biomarker for Digestive System Cancers. *Int. J. Mol. Sci.* **2016**, *17*, 1459. [CrossRef]
50. De Souza, M.F.; Kuasne, H.; de Camargo Barros-Filho, M.; Cilião, H.L.; Marchi, F.A.; Fuganti, P.E.; Paschoal, A.R.; Rogatto, S.R.; de Syllos Cólus, I.M. Circulating MRNAs and MiRNAs as Candidate Markers for the Diagnosis and Prognosis of Prostate Cancer. *PLoS ONE* **2017**, *12*, e0184094. [CrossRef]
51. Diehl, F.; Schmidt, K.; Choti, M.A.; Romans, K.; Goodman, S.; Li, M.; Thornton, K.; Agrawal, N.; Sokoll, L.; Szabo, S.A.; et al. Circulating Mutant DNA to Assess Tumor Dynamics. *Nat. Med.* **2008**, *14*, 985–990. [CrossRef]
52. Fleischhacker, M.; Schmidt, B. Circulating Nucleic Acids (CNAs) and Cancer—A Survey. *Biochim. Biophys. Acta-Rev. Cancer* **2007**, *1775*, 181–232. [CrossRef]
53. Yao, W.; Mei, C.; Nan, X.; Hui, L. Evaluation and Comparison of in Vitro Degradation Kinetics of DNA in Serum, Urine and Saliva: A Qualitative Study. *Gene* **2016**, *590*, 142–148. [CrossRef]
54. Young, R.; Pailler, E.; Billiot, F.; Drusch, F.; Barthelemy, A.; Oulhen, M.; Besse, B.; Soria, J.-C.; Farace, F.; Vielh, P. Circulating Tumor Cells in Lung Cancer. *Acta Cytol.* **2012**, *56*, 655–660. [CrossRef]
55. Su, Z.; Zhao, J.; Ke, S.; Zhang, J.; Liu, X.; Wang, Y.; Sun, Q.; Pan, Q. Clinical Significance of Circulating Tumor Cells via Combined Whole Exome Sequencing in Early Stage Cancer Screening: A Case Report. *Exp. Ther. Med.* **2018**, *16*, 2527–2533. [CrossRef]
56. Rajagopal, C.; Harikumar, K.B. The Origin and Functions of Exosomes in Cancer. *Front. Oncol.* **2018**, *8*, 66. [CrossRef]
57. Koritzinsky, E.H.; Street, J.M.; Star, R.A.; Yuen, P.S.T. Quantification of Exosomes. *J. Cell. Physiol.* **2017**, *232*, 1587–1590. [CrossRef]
58. Li, P.; Kaslan, M.; Lee, S.H.; Yao, J.; Gao, Z. Progress in Exosome Isolation Techniques. *Theranostics* **2017**, *7*, 789–804. [CrossRef]
59. Taylor, D.D.; Gercel-Taylor, C. Exosomes/Microvesicles: Mediators of Cancer-Associated Immunosuppressive Microenvironments. *Semin. Immunopathol.* **2011**, *33*, 441–454. [CrossRef]
60. Yang, Y.; Hong, Y.; Cho, E.; Kim, G.B.; Kim, I.-S. Extracellular Vesicles as a Platform for Membrane-Associated Therapeutic Protein Delivery. *J. Extracell. Vesicles* **2018**, *7*, 1440131. [CrossRef]
61. Wu, Y.; Deng, W.; Klinke II, D.J. Exosomes: Improved Methods to Characterize Their Morphology, RNA Content, and Surface Protein Biomarkers. *Analyst* **2015**, *140*, 6631–6642. [CrossRef]
62. Sheridan, C. Exosome Cancer Diagnostic Reaches Market. *Nat. Biotechnol.* **2016**, *34*, 359–360. [CrossRef]
63. Lissa, D.; Robles, A.I. Methylation Analyses in Liquid Biopsy. *Transl. Lung Cancer Res.* **2016**, *5*, 492. [CrossRef]
64. Balgkouranidou, I.; Chimonidou, M.; Milaki, G.; Tsarouxa, E.G.; Kakolyris, S.; Welch, D.R.; Georgoulias, V.; Lianidou, E.S. Breast Cancer Metastasis Suppressor-1 Promoter Methylation in Cell-Free DNA Provides Prognostic Information in Non-Small Cell Lung Cancer. *Br. J. Cancer* **2014**, *110*, 2054–2062. [CrossRef]
65. Majchrzak-Celińska, A.; Paluszczak, J.; Kleszcz, R.; Magiera, M.; Barciszewska, A.-M.; Nowak, S.; Baer-Dubowska, W. Detection of MGMT, RASSF1A, P15INK4B, and P14ARF Promoter Methylation in Circulating Tumor-Derived DNA of Central Nervous System Cancer Patients. *J. Appl. Genet.* **2013**, *54*, 335–344. [CrossRef]
66. Kristensen, L.S.; Hansen, J.W.; Kristensen, S.S.; Tholstrup, D.; Harsløf, L.B.S.; Pedersen, O.B.; De Nully Brown, P.; Grønbæk, K. Aberrant Methylation of Cell-Free Circulating DNA in Plasma Predicts Poor Outcome in Diffuse Large B Cell Lymphoma. *Clin. Epigenetics* **2016**, *8*, 95. [CrossRef]

67. Elazezy, M.; Joosse, S.A. Techniques of Using Circulating Tumor DNA as a Liquid Biopsy Component in Cancer Management. *Comput. Struct. Biotechnol. J.* **2018**, *16*, 370–378. [CrossRef]
68. Zhang, R.; Chen, B.; Tong, X.; Wang, Y.; Wang, C.; Jin, J.; Tian, P.; Li, W. Diagnostic Accuracy of Droplet Digital PCR for Detection of EGFR T790M Mutation in Circulating Tumor DNA. *Cancer Manag. Res.* **2018**, *10*, 1209–1218. [CrossRef]
69. Calapre, L.; Warburton, L.; Millward, M.; Ziman, M.; Gray, E.S. Circulating Tumour DNA (CtDNA) as a Liquid Biopsy for Melanoma. *Cancer Lett.* **2017**, *404*, 62–69. [CrossRef]
70. Widschwendter, M.; Zikan, M.; Wahl, B.; Lempiäinen, H.; Paprotka, T.; Evans, I.; Jones, A.; Ghazali, S.; Reisel, D.; Eichner, J.; et al. The Potential of Circulating Tumor DNA Methylation Analysis for the Early Detection and Management of Ovarian Cancer. *Genome Med.* **2017**, *9*, 116. [CrossRef]
71. Le, A.; Szaumkessel, M.; Tan, T.; Thiery, J.-P.; Thompson, E.; Dobrovic, A. DNA Methylation Profiling of Breast Cancer Cell Lines along the Epithelial Mesenchymal Spectrum—Implications for the Choice of Circulating Tumour DNA Methylation Markers. *Int. J. Mol. Sci.* **2018**, *19*, 2553. [CrossRef]
72. Xu, R.; Wei, W.; Krawczyk, M.; Wang, W.; Luo, H.; Flagg, K.; Yi, S.; Shi, W.; Quan, Q.; Li, K.; et al. Circulating Tumour DNA Methylation Markers for Diagnosis and Prognosis of Hepatocellular Carcinoma. *Nat. Mater.* **2017**, *16*, 1155–1161. [CrossRef]
73. Sun, K.; Lun, F.F.M.; Jiang, P.; Sun, H. BSviewer: A Genotype-Preserving, Nucleotide-Level Visualizer for Bisulfite Sequencing Data. *Bioinformatics* **2017**, *33*, 3495–3496. [CrossRef]
74. Lun, F.M.F.; Chiu, R.W.K.; Sun, K.; Leung, T.Y.; Jiang, P.; Chan, K.C.A.; Sun, H.; Lo, Y.M.D. Noninvasive Prenatal Methylomic Analysis by Genomewide Bisulfite Sequencing of Maternal Plasma DNA. *Clin. Chem.* **2013**, *59*, 1583–1584. [CrossRef]
75. Wong, I.H.N.; Liew, C.-T.; Ng, M.H.L.; Lo, Y.M.D.; Zhang, J.; Hjelm, N.M.; Wong, N.; Johnson, P.J.; Lai, P.B.S.; Lau, W.Y. Detection of Aberrant P16 Methylation in the Plasma and Serum of Liver Cancer Patients. *Cancer Res.* **1999**, *59*, 71–73.
76. Cheng, J.-P.; Yan, Y.; Wang, X.-Y.; Lu, Y.-L.; Yuan, Y.-H.; Jia, J.; Ren, J. MUC1-Positive Circulating Tumor Cells and MUC1 Protein Predict Chemotherapeutic Efficacy in the Treatment of Metastatic Breast Cancer. *Chin. J. Cancer* **2011**, *30*, 54–61. [CrossRef]
77. Kuhlmann, J.D.; Wimberger, P.; Bankfalvi, A.; Keller, T.; Scholer, S.; Aktas, B.; Buderath, P.; Hauch, S.; Otterbach, F.; Kimmig, R.; et al. ERCC1-Positive Circulating Tumor Cells in the Blood of Ovarian Cancer Patients as a Predictive Biomarker for Platinum Resistance. *Clin. Chem.* **2014**, *60*, 1282–1289. [CrossRef]
78. Lin, H.-C.; Hsu, H.-C.; Hsieh, C.-H.; Wang, H.-M.; Huang, C.-Y.; Wu, M.-H.; Tseng, C.-P. A Negative Selection System PowerMag for Effective Leukocyte Depletion and Enhanced Detection of EpCAM Positive and Negative Circulating Tumor Cells. *Clin. Chim. Acta.* **2013**, *419*, 77–84. [CrossRef]
79. Benini, S.; Gamberi, G.; Cocchi, S.; Garbetta, J.; Alberti, L.; Righi, A.; Gambarotti, M.; Picci, P.; Ferrari, S. Detection of Circulating Tumor Cells in Liquid Biopsy from Ewing Sarcoma Patients. *Cancer Manag. Res.* **2018**, *10*, 49–60. [CrossRef]
80. Satelli, A.; Brownlee, Z.; Mitra, A.; Meng, Q.H.; Li, S. Circulating Tumor Cell Enumeration with a Combination of Epithelial Cell Adhesion Molecule-and Cell-Surface Vimentin-Based Methods for Monitoring Breast Cancer Therapeutic Response. *Clin. Chem.* **2015**, *61*, 259–266. [CrossRef]
81. Hayashi, M.; Zhu, P.; McCarty, G.; Meyer, C.F.; Pratilas, C.A.; Levin, A.; Morris, C.D.; Albert, C.M.; Jackson, K.W.; Tang, C.-M.; et al. Size-Based Detection of Sarcoma Circulating Tumor Cells and Cell Clusters. *Oncotarget* **2017**, *8*, 78965. [CrossRef]
82. Khetani, S.; Mohammadi, M.; Nezhad, A.S. Filter-Based Isolation, Enrichment, and Characterization of Circulating Tumor Cells. *Biotechnol. Bioeng.* **2018**, *115*, 2504–2529. [CrossRef]
83. Chung, J.; Shao, H.; Reiner, T.; Issadore, D.; Weissleder, R.; Lee, H. Microfluidic Cell Sorter (μ FCS) for On-Chip Capture and Analysis of Single Cells. *Adv. Healthc. Mater.* **2012**, *1*, 432–436. [CrossRef]
84. Sarioglu, A.F.; Aceto, N.; Kojic, N.; Donaldson, M.C.; Zeinali, M.; Hamza, B.; Engstrom, A.; Zhu, H.; Sundaresan, T.K.; Miyamoto, D.T.; et al. A Microfluidic Device for Label-Free, Physical Capture of Circulating Tumor Cell Clusters. *Nat. Methods* **2015**, *12*, 685–691. [CrossRef]
85. Issadore, D.; Chung, J.; Shao, H.; Liong, M.; Ghazani, A.A.; Castro, C.M.; Weissleder, R.; Lee, H. Ultrasensitive Clinical Enumeration of Rare Cells Ex Vivo Using a Micro-Hall Detector. *Sci. Transl. Med.* **2012**, *4*, 141ra92. [CrossRef]

86. Muluneh, M.; Issadore, D. A Multi-Scale PDMS Fabrication Strategy to Bridge the Size Mismatch between Integrated Circuits and Microfluidics. *Lab Chip* **2014**, *14*, 4552–4558. [CrossRef]
87. Yao, N.; Jan, Y.-J.; Cheng, S.; Chen, J.-F.; Chung, L.W.; Tseng, H.-R.; Posadas, E.M. Structure and Function Analysis in Circulating Tumor Cells: Using Nanotechnology to Study Nuclear Size in Prostate Cancer. *Am. J. Clin. Exp. Urol.* **2018**, *6*, 43–54.
88. Hui, L. Noninvasive Prenatal Testing for Aneuploidy Using Cell-Free DNA – New Implications for Maternal Health. *Obstet. Med.* **2016**, *9*, 148–152. [CrossRef]
89. Dharajiya, N.G.; Grosu, D.S.; Farkas, D.H.; McCullough, R.M.; Almasri, E.; Sun, Y.; Kim, S.K.; Jensen, T.J.; Saldivar, J.-S.; Topol, E.J.; et al. Incidental Detection of Maternal Neoplasia in Noninvasive Prenatal Testing. *Clin. Chem.* **2018**, *64*, 329–335. [CrossRef]
90. Giles, M.E.; Murphy, L.; Krstić, N.; Sullivan, C.; Hashmi, S.S.; Stevens, B. Prenatal CfDNA Screening Results Indicative of Maternal Neoplasm: Survey of Current Practice and Management Needs. *Prenat. Diagn.* **2017**, *37*, 126–132. [CrossRef]
91. Hemming, M.L.; Klega, K.S.; Acker, K.E.; Nag, A.; Thorner, A.; Nathenson, M.; Raut, C.P.; Crompton, B.D.; George, S. Identification of Leiomyosarcoma Circulating Tumor DNA through Ultra-Low Passage Whole Genome Sequencing and Correlation with Tumor Burden: A Pilot Experience. *J. Clin. Oncol.* **2018**, *36*, 11565. [CrossRef]
92. Hemming, M.L.; Klega, K.; Rhoades, J.; Ha, G.; Acker, K.E.; Andersen, J.L.; Thai, E.; Nag, A.; Thorner, A.R.; Raut, C.P.; et al. Detection of Circulating Tumor DNA in Patients With Leiomyosarcoma With Progressive Disease. *JCO Precis. Oncol.* **2019**, *3*, 1–11. [CrossRef]
93. Izzotti, A.; Carozzo, S.; Pulliero, A.; Zhabayeva, D.; Ravetti, J.L.; Bersimbaev, R. Extracellular MicroRNA in Liquid Biopsy: Applicability in Cancer Diagnosis and Prevention. *Am. J. Cancer Res.* **2016**, *6*, 1461–1493.
94. Giannopoulou, L.; Zavridou, M.; Kasimir-Bauer, S.; Lianidou, E.S. Liquid Biopsy in Ovarian Cancer: The Potential of Circulating MiRNAs and Exosomes. *Transl. Res.* **2019**, *205*, 77–91. [CrossRef]
95. De Groot, J.S.; Moelans, C.B.; Elias, S.G.; Fackler, M.J.; Van Domselaar, R.; Suijkerbuijk, K.P.M.; Witkamp, A.J.; Sukumar, S.; Van Diest, P.J.; Van Der Wall, E. DNA Promoter Hypermethylation in Nipple Fluid: A Potential Tool for Early Breast Cancer Detection. *Oncotarget* **2016**, *7*, 24778. [CrossRef]
96. Takashima, Y.; Kawaguchi, A.; Iwadate, Y.; Hondoh, H.; Fukai, J.; Kajiwara, K.; Hayano, A.; Yamanaka, R. MicroRNA Signature Constituted of MiR-30d, MiR-93, and MiR-181b Is a Promising Prognostic Marker in Primary Central Nervous System Lymphoma. *PLoS ONE* **2019**, *14*, e0210400. [CrossRef]
97. Komatsu, S.; Kiuchi, J.; Imamura, T.; Ichikawa, D.; Otsuji, E. Circulating MicroRNAs as a Liquid Biopsy: A next-Generation Clinical Biomarker for Diagnosis of Gastric Cancer. *J. Cancer Metastasis Treat.* **2018**, *4*, 36. [CrossRef]
98. Li, L.; Fu, K.; Zhou, W.; Snyder, M. Applying Circulating Tumor DNA Methylation in the Diagnosis of Lung Cancer. *Precis. Clin. Med.* **2019**, *2*, 45–56. [CrossRef]
99. Barault, L.; Amatu, A.; Bleeker, F.E.; Moutinho, C.; Falcomatà, C.; Fiano, V.; Cassingena, A.; Siravegna, G.; Milione, M.; Cassoni, P.; et al. Digital PCR Quantification of MGMT Methylation Refines Prediction of Clinical Benefit from Alkylating Agents in Glioblastoma and Metastatic Colorectal Cancer. *Ann. Oncol.* **2015**, *26*, 1994–1999. [CrossRef]
100. Martínez-Ricarte, F.; Mayor, R.; Martínez-Sáez, E.; Rubio-Pérez, C.; Pineda, E.; Cordero, E.; Cicuéndez, M.; Poca, M.A.; López-Bigas, N.; Ramon Y Cajal, S.; et al. Molecular Diagnosis of Diffuse Gliomas through Sequencing of Cell-Free Circulating Tumour DNA from Cerebrospinal Fluid. *Author Manuscr. Publ. OnlineFirst* **2018**, *24*, 2812–2819. [CrossRef]
101. Vandekerkhove, G.; Todenhöfer, T.; Annala, M.; Struss, W.J.; Wong, A.; Beja, K.; Ritch, E.; Brahmbhatt, S.; Volik, S.V.; Hennenlotter, J.; et al. Circulating Tumor DNA Reveals Clinically Actionable Somatic Genome of Metastatic Bladder Cancer. *Clin. Cancer Res.* **2017**, *23*, 6487–6497. [CrossRef]
102. Cimadamore, A.; Gasparrini, S.; Massari, F.; Santoni, M.; Cheng, L.; Lopez-Beltran, A.; Scarpelli, M.; Montironi, R. Emerging Molecular Technologies in Renal Cell Carcinoma: Liquid Biopsy. *Cancers.* **2019**, *11*(2), 196. [CrossRef]
103. Kirkizlar, E.; Zimmermann, B.; Constantin, T.; Swenerton, R.; Hoang, B.; Wayham, N.; Babiarz, J.E.; Demko, Z.; Pelham, R.J.; Kareht, S.; et al. Detection of Clonal and Subclonal Copy-Number Variants in Cell-Free DNA from Patients with Breast Cancer Using a Massively Multiplexed PCR Methodology. *Transl. Oncol.* **2015**, *8*, 407–416. [CrossRef]

104. Best, M.G.; Sol, N.; Zijl, S.; Reijneveld, J.C.; Wesseling, P.; Wurdinger, T. Liquid Biopsies in Patients with Diffuse Glioma. *Acta Neuropathologica.* **2015**, *129*, 849–865. [CrossRef]
105. Miller, A.M.; Shah, R.H.; Pentsova, E.I.; Pourmaleki, M.; Briggs, S.; Distefano, N.; Zheng, Y.; Skakodub, A.; Mehta, S.A.; Campos, C.; et al. Tracking Tumour Evolution in Glioma through Liquid Biopsies of Cerebrospinal Fluid. *Nature* **2019**, 654–658. [CrossRef]
106. Yin, C.Q.; Yuan, C.H.; Qu, Z.; Guan, Q.; Chen, H.; Wang, F.B. Liquid Biopsy of Hepatocellular Carcinoma: Circulating Tumor-Derived Biomarkers. *Dis. Markers* **2016**. [CrossRef]
107. Hammond, S.M. An Overview of MicroRNAs. *Adv. Drug Deliv. Rev.* **2015**, *87*, 3–14. [CrossRef]
108. Pan, Q.; Luo, X.; Chegini, N. Differential Expression of MicroRNAs in Myometrium and Leiomyomas and Regulation by Ovarian Steroids. *J. Cell. Mol. Med.* **2008**, *12*, 227–240. [CrossRef]
109. Liu, Z.; Guo, H.; Wu, J.; Zavadil, J.; Ghanny, S.; Ghuo, S.; Wei, J. Differential Expression of MiRNAs in Uterine Leiomyoma and Adjacent Myometrium of Different Races. *Am J Clin Exp Obstet Gynecol.* **2015**, *2*, 45–256.
110. Wang, T.; Zhang, X.; Obijuru, L.; Laser, J.; Aris, V.; Lee, P.; Mittal, K.; Soteropoulos, P.; Wei, J.-J. A Micro-RNA Signature Associated with Race, Tumor Size, and Target Gene Activity in Human Uterine Leiomyomas. *Genes, Chromosom. Cancer* **2007**, *46*, 336–347. [CrossRef]
111. Guan, Y.; Guo, L.; Zukerberg, L.; Rueda, B.R.; Styer, A.K. MicroRNA-15b Regulates Reversion-Inducing Cysteine-Rich Protein with Kazal Motifs (RECK) Expression in Human Uterine Leiomyoma. *Reprod. Biol. Endocrinol.* **2016**, *14*, 45. [CrossRef]
112. Xin, C.; Buhe, B.; Hongting, L.; Chuanmin, Y.; Xiwei, H.; Hong, Z.; Lulu, H.; Qian, D.; Renjie, W. MicroRNA-15a Promotes Neuroblastoma Migration by Targeting Reversion-Inducing Cysteine-Rich Protein with Kazal Motifs (RECK) and Regulating Matrix Metalloproteinase-9 Expression. *FEBS J.* **2013**, *280*, 855–866. [CrossRef]
113. Takagi, S.; Simizu, S.; Osada, H. Reck Negatively Regulates Matrix Metalloproteinase-9 Transcription. *Cancer Res.* **2009**, *69*, 1502–1508. [CrossRef]
114. Kristjánsdóttir, K.; Fogarty, E.A.; Grimson, A. Systematic Analysis of the *Hmga2* 3′ UTR Identifies Many Independent Regulatory Sequences and a Novel Interaction between Distal Sites. *RNA* **2015**, *21*, 1346–1360. [CrossRef]
115. De Almeida, B.C.; Garcia, N.; Maffazioli, G.; Dos Anjos, L.G.; Baracat, E.C.; Carvalho, K.C. Oncomirs Expression Profiling in Uterine Leiomyosarcoma Cells. *Int. J. Mol. Sci.* **2018**, *19*, 52. [CrossRef]
116. Schiavon, B.N.; Carvalho, K.C.; Coutinho-Camillo, C.M.; Baiocchi, G.; Valieris, R.; Drummond, R.; da Silva, I.T.; De Brot, L.; Soares, F.A.; da Cunha, I.W. MiRNAs 144-3p, 34a-5p, and 206 Are a Useful Signature for Distinguishing Uterine Leiomyosarcoma from Other Smooth Muscle Tumors. *Surg. Exp. Pathol.* **2019**, *2*, 5. [CrossRef]
117. Danielson, L.S.; Menendez, S.; Attolini, C.S.-O.; Guijarro, M.V.; Bisogna, M.; Wei, J.; Socci, N.D.; Levine, D.A.; Michor, F.; Hernando, E. A Differentiation-Based MicroRNA Signature Identifies Leiomyosarcoma as a Mesenchymal Stem Cell-Related Malignancy. *Am. J. Pathol.* **2010**, *177*, 908–917. [CrossRef]
118. Kowalewska, M.; Bakula-Zalewska, E.; Chechlinska, M.; Goryca, K.; Nasierowska-Guttmejer, A.; Danska-Bidzinska, A.; Bidzinski, M. MicroRNAs in Uterine Sarcomas and Mixed Epithelial–Mesenchymal Uterine Tumors: A Preliminary Report. *Tumor Biol.* **2013**, *34*, 2153–2160. [CrossRef]
119. Gonzalez dos Anjos, L.; de Almeida, B.; Gomes de Almeida, T.; Mourão Lavorato Rocha, A.; De Nardo Maffazioli, G.; Soares, F.; Werneck da Cunha, I.; Chada Baracat, E.; Candido Carvalho, K.; Gonzalez dos Anjos, L.; et al. Could MiRNA Signatures Be Useful for Predicting Uterine Sarcoma and Carcinosarcoma Prognosis and Treatment? *Cancers (Basel)* **2018**, *10*, 315. [CrossRef]
120. Ravid, Y.; Formanski, M.; Smith, Y.; Reich, R.; Davidson, B. Uterine Leiomyosarcoma and Endometrial Stromal Sarcoma Have Unique MiRNA Signatures. *Gynecol. Oncol.* **2016**, *140*, 512–517. [CrossRef]
121. Zavadil, J.; Ye, H.; Liu, Z.; Wu, J.; Lee, P.; Hernando, E.; Soteropoulos, P.; Toruner, G.A.; Wei, J.-J. Profiling and Functional Analyses of MicroRNAs and Their Target Gene Products in Human Uterine Leiomyomas. *PLoS ONE* **2010**, *5*, e12362. [CrossRef]
122. Marsh, E.E.; Lin, Z.; Yin, P.; Milad, M.; Chakravarti, D.; Bulun, S.E. Differential Expression of MicroRNA Species in Human Uterine Leiomyoma versus Normal Myometrium. *Fertil. Steril.* **2008**, *89*, 1771–1776. [CrossRef]
123. Georgieva, B.; Milev, I.; Minkov, I.; Dimitrova, I.; Bradford, A.P.; Baev, V. Characterization of the Uterine Leiomyoma MicroRNAome by Deep Sequencing. *Genomics* **2012**, *99*, 275–281. [CrossRef]

124. Marsh, E.E.; Steinberg, M.L.; Parker, J.B.; Wu, J.; Chakravarti, D.; Bulun, S.E. Decreased Expression of MicroRNA-29 Family in Leiomyoma Contributes to Increased Major Fibrillar Collagen Production. *Fertil. Steril.* **2016**, *106*, 766–772. [CrossRef]
125. Chuang, T.-D.; Luo, X.; Panda, H.; Chegini, N. MiR-93/106b and Their Host Gene, MCM7, Are Differentially Expressed in Leiomyomas and Functionally Target F3 and IL-8. *Mol. Endocrinol.* **2012**, *26*, 1028–1042. [CrossRef]
126. Pazzaglia, L.; Novello, C.; Conti, A.; Pollino, S.; Picci, P.; Benassi, M.S. MiR-152 down-Regulation Is Associated with MET up-Regulation in Leiomyosarcoma and Undifferentiated Pleomorphic Sarcoma. *Cell. Oncol.* **2017**, *40*, 77–88. [CrossRef]
127. Chuang, T.-D.; Panda, H.; Luo, X.; Chegini, N. MiR-200c Is Aberrantly Expressed in Leiomyomas in an Ethnic-Dependent Manner and Targets ZEBs, VEGFA, TIMP2, and FBLN5. *Endocr. Relat. Cancer* **2012**, *19*, 541–556. [CrossRef]
128. Wu, X.; Ling, J.; Fu, Z.; Ji, C.; Wu, J.; Xu, Q. Effects of MiRNA-197 Overexpression on Proliferation, Apoptosis and Migration in Levonorgestrel Treated Uterine Leiomyoma Cells. *Biomed. Pharmacother.* **2015**, *71*, 1–6. [CrossRef]
129. Makinen, N.; Mehine, M.; Tolvanen, J.; Kaasinen, E.; Li, Y.; Lehtonen, H.J.; Gentile, M.; Yan, J.; Enge, M.; Taipale, M.; et al. MED12, the Mediator Complex Subunit 12 Gene, Is Mutated at High Frequency in Uterine Leiomyomas. *Science* **2011**, *334*, 252–255. [CrossRef]
130. Harrison, W.J.; Andrici, J.; Maclean, F.; Madadi-Ghahan, R.; Farzin, M.; Sioson, L.; Toon, C.W.; Clarkson, A.; Watson, N.; Pickett, J.; et al. Fumarate Hydratase–Deficient Uterine Leiomyomas Occur in Both the Syndromic and Sporadic Settings. *Am. J. Surg. Pathol.* **2016**, *40*, 599–607. [CrossRef]
131. Stewart, L.; Glenn, G.M.; Stratton, P.; Goldstein, A.M.; Merino, M.J.; Tucker, M.A.; Linehan, W.M.; Toro, J.R. Association of Germline Mutations in the Fumarate Hydratase Gene and Uterine Fibroids in Women with Hereditary Leiomyomatosis and Renal Cell Cancer. *Arch. Dermatol.* **2008**, *144*, 1584–1592. [CrossRef]
132. Ylisaukko-oja, S.K.; Kiuru, M.; Lehtonen, H.J.; Lehtonen, R.; Pukkala, E.; Arola, J.; Launonen, V.; Aaltonen, L.A. Analysis of Fumarate Hydratase Mutations in a Population-Based Series of Early Onset Uterine Leiomyosarcoma Patients. *Int. J. Cancer* **2006**, *119*, 283–287. [CrossRef]
133. Kämpjärvi, K.; Mäkinen, N.; Kilpivaara, O.; Arola, J.; Heinonen, H.-R.; Böhm, J.; Abdel-Wahab, O.; Lehtonen, H.J.; Pelttari, L.M.; Mehine, M.; et al. Somatic MED12 Mutations in Uterine Leiomyosarcoma and Colorectal Cancer. *Br. J. Cancer* **2012**, *107*, 1761–1765. [CrossRef]
134. Mäkinen, N.; Aavikko, M.; Heikkinen, T.; Taipale, M.; Taipale, J.; Koivisto-Korander, R.; Bützow, R.; Vahteristo, P. Exome Sequencing of Uterine Leiomyosarcomas Identifies Frequent Mutations in TP53, ATRX, and MED12. *PLOS Genet.* **2016**, *12*, e1005850. [CrossRef]
135. Zhang, W.-W.; Li, L.; Li, D.; Liu, J.; Li, X.; Li, W.; Xu, X.; Zhang, M.J.; Chandler, L.A.; Lin, H.; et al. The First Approved Gene Therapy Product for Cancer Ad-*P53* (Gendicine): 12 Years in the Clinic. *Hum. Gene Ther.* **2018**, *29*, 160–179. [CrossRef]
136. Slatter, T.L.; Hsia, H.; Samaranayaka, A.; Sykes, P.; Clow, W.B.; Devenish, C.J.; Sutton, T.; Royds, J.A.; PC, P.; Cheung, A.N.; et al. Loss of ATRX and DAXX Expression Identifies Poor Prognosis for Smooth Muscle Tumours of Uncertain Malignant Potential and Early Stage Uterine Leiomyosarcoma. *J. Pathol. Clin. Res.* **2015**, *1*, 95–105. [CrossRef]
137. Bennett, J.A.; Weigelt, B.; Chiang, S.; Selenica, P.; Chen, Y.-B.; Bialik, A.; Bi, R.; Schultheis, A.M.; Lim, R.S.; Ng, C.K.Y.; et al. Leiomyoma with Bizarre Nuclei: A Morphological, Immunohistochemical and Molecular Analysis of 31 Cases. *Mod. Pathol.* **2017**, *30*, 1476–1488. [CrossRef]
138. Dyson, N.J. *RB1*: A Prototype Tumor Suppressor and an Enigma. *Genes Dev.* **2016**, *30*, 1492–1502. [CrossRef]
139. Hovelson, D.H.; Liu, C.-J.; Wang, Y.; Kang, Q.; Henderson, J.; Gursky, A.; Brockman, S.; Ramnath, N.; Krauss, J.C.; Talpaz, M.; et al. Rapid, Ultra Low Coverage Copy Number Profiling of Cell-Free DNA as a Precision Oncology Screening Strategy. *Oncotarget* **2017**, *8*, 89848–89866. [CrossRef]
140. Klega, K.; Imamovic-Tuco, A.; Ha, G.; Clapp, A.N.; Meyer, S.; Ward, A.; Clinton, C.; Nag, A.; Van Allen, E.; Mullen, E.; et al. Detection of Somatic Structural Variants Enables Quantification and Characterization of Circulating Tumor DNA in Children With Solid Tumors. *JCO Precis. Oncol.* **2018**, *2018*, 1–13. [CrossRef]
141. Velagaleti, G.V.N.; Wang, X.; Erickson-Johnson, M.R.; Medeiros, F.; Oliveira, A.M.; Tonk, V.S.; Hakim, N.M.; Zhang, H. Fusion of HMGA2 to COG5 in Uterine Leiomyoma. *Cancer Genet. Cytogenet.* **2010**, *202*, 11–16. [CrossRef]

142. CHO, Y.; BAE, S.; KOO, M.; KIM, K.; CHUN, H.; KIM, C.; RO, D.; KIM, J.; LEE, C.; KIM, Y. Array Comparative Genomic Hybridization Analysis of Uterine Leiomyosarcoma. *Gynecol. Oncol.* **2005**, *99*, 545–551. [CrossRef]
143. Koontz, J.I.; Soreng, A.L.; Nucci, M.; Kuo, F.C.; Pauwels, P.; van den Berghe, H.; Cin, P.D.; Fletcher, J.A.; Sklar, J. Frequent Fusion of the JAZF1 and JJAZ1 Genes in Endometrial Stromal Tumors. *Proc. Natl. Acad. Sci.* **2001**, *98*, 6348–6353. [CrossRef]
144. Schoenmakers, E.F.P.M.; Huysmans, C.; Van De Ven, W.J.M. Allelic Knockout of Novel Splice Variants of Human Recombination Repair Gene RAD51B in t(12;14) Uterine Leiomyomas. *Cancer Res.* **1999**, *59*, 19–23.
145. Beck, A.H.; Lee, C.-H.; Witten, D.M.; Gleason, B.C.; Edris, B.; Espinosa, I.; Zhu, S.; Li, R.; Montgomery, K.D.; Marinelli, R.J.; et al. Discovery of Molecular Subtypes in Leiomyosarcoma through Integrative Molecular Profiling. *Oncogene* **2010**, *29*, 845–854. [CrossRef]
146. Micci, F.; Panagopoulos, I.; Bjerkehagen, B.; Heim, S. Consistent Rearrangement of Chromosomal Band 6p21 with Generation of Fusion Genes *JAZF1/PHF1* and *EPC1/PHF1* in Endometrial Stromal Sarcoma. *Cancer Res.* **2006**, *66*, 107–112. [CrossRef]
147. Sandberg, A.; Bridge, J. Updates on the Cytogenetics and Molecular Genetics of Bone and Soft Tissue Tumors: Alveolar Soft Part Sarcoma. *Cancer Genet. Cytogenet.* **2002**, *136*, 1–9. [CrossRef]
148. Sornberger, K.S.; Ligon, A.H.; Pedeutour, F.; Morton, C.C.; Weremowicz, S.; Williams, A.J.; Quade, B.J.; Vanni, R. Expression of HMGIY in Three Uterine Leiomyomata with Complex Rearrangements of Chromosome 6. *Cancer Genet. Cytogenet.* **1999**, *114*, 9–16. [CrossRef]
149. Micci, F.; Gorunova, L.; Gatius, S.; Matias-Guiu, X.; Davidson, B.; Heim, S.; Panagopoulos, I. MEAF6/PHF1 Is a Recurrent Gene Fusion in Endometrial Stromal Sarcoma. *Cancer Lett.* **2014**, *347*, 75–78. [CrossRef]
150. Moore, S.; Herrick, S.; Ince, T. Uterine Leiomyomata with t (10; 17) Disrupt the Histone Acetyltransferase MORF. *Cancer Res.* **2004**, *64*, 5570–5577. [CrossRef]
151. Dewaele, B.; Przybyl, J.; Quattrone, A.; Finalet Ferreiro, J.; Vanspauwen, V.; Geerdens, E.; Gianfelici, V.; Kalender, Z.; Wozniak, A.; Moerman, P.; et al. Identification of a Novel, Recurrent *MBTD1-CXorf67* Fusion in Low-Grade Endometrial Stromal Sarcoma. *Int. J. Cancer* **2014**, *134*, 1112–1122. [CrossRef]
152. Panagopoulos, I.; Thorsen, J.; Gorunova, L.; Haugom, L.; Bjerkehagen, B.; Davidson, B.; Heim, S.; Micci, F. Fusion of the *ZC3H7B* and *BCOR* Genes in Endometrial Stromal Sarcomas Carrying an X;22-Translocation. *Genes, Chromosom. Cancer* **2013**, *52*, 610–618. [CrossRef]
153. Garcia-Torres, R.; Cruz, D.; Orozco, L.; Heidet, L.; Gubler, M.C. Alport Syndrome and Diffuse Leiomyomatosis. Clinical Aspects, Pathology, Molecular Biology and Extracellular Matrix Studies. A Synthesis. *Nephrologie* **2000**, *21*, 9–12.
154. Regauer, S.; Emberger, W.; Reich, O.; Pfragner, R. Cytogenetic Analyses of Two New Cases of Endometrial Stromal Sarcoma—Non-Random Reciprocal Translocation t(10;17)(Q22;P13) Correlates with Fibrous ESS. *Histopathology* **2008**, *52*, 780–783. [CrossRef]
155. Van Rijk, A.; Sweers, M.; Huys, E.; Kersten, M.; Merkx, G.; van Kessel, A.G.; Debiec-Rychter, M.; Schoenmakers, E.F. Characterization of a Recurrent t (1; 2)(P36; P24) in Human Uterine Leiomyoma. *Cancer Genet. Cytogenet.* **2009**, *193*, 54–62. [CrossRef]
156. Cho, Y.L.; Koo, M.S.; Bae, S.; Lee, C.-H.; Kim, K.M.; Ro, D.Y.; Kim, J.H.; Ahn, W.S.; Chun, H.-J.; Kim, C.K.; et al. Array Comparative Genomic Hybridization Analysis of Uterine Leiomyosarcoma. *Gynecol. Oncol.* **2005**, *99*, 545–551. [CrossRef]
157. Panagopoulos, I.; Gorunova, L.; Brunetti, M.; Agostini, A.; Andersen, H.K.; Lobmaier, I.; Bjerkehagen, B.; Heim, S.; Panagopoulos, I.; Gorunova, L.; et al. Genetic Heterogeneity in Leiomyomas of Deep Soft Tissue. *Oncotarget* **2017**, *8*, 48769–48781. [CrossRef]
158. Yang, J.; Du, X.; Chen, K.; Ylipää, A.; Lazar, A.J.F.; Trent, J.; Lev, D.; Pollock, R.; Hao, X.; Hunt, K.; et al. Genetic Aberrations in Soft Tissue Leiomyosarcoma. *Cancer Lett.* **2009**, *275*, 1–8. [CrossRef]
159. Buza, N.; Carr, R.J.; Hui, P.; Xu, F.; Wu, W.; Li, P. Recurrent Chromosomal Aberrations in Intravenous Leiomyomatosis of the Uterus: High-Resolution Array Comparative Genomic Hybridization Study. *Hum. Pathol.* **2014**, *45*, 1885–1892. [CrossRef]
160. Hu, J.; Khanna, V.; Jones, M.; Surti, U. Genomic Alterations in Uterine Leiomyosarcomas: Potential Markers for Clinical Diagnosis and Prognosis. *Genes, Chromosom. Cancer* **2001**, *31*, 117–124. [CrossRef]
161. Vanni, R.; Van Roy, N.; Lecca, U.; Speleman, F. Interphase Cytogenetic Analysis of Karyotypically Normal Uterine Leiomyoma Excludes Undetected Trisomy 12. *Cancer Genet. Cytogenet.* **2003**, *63*, 131. [CrossRef]

162. Gai, W.; Sun, K. Epigenetic Biomarkers in Cell-Free DNA and Applications in Liquid Biopsy. *Genes.* **2019**, *10*. [CrossRef]
163. Maekawa, R.; Sato, S.; Yamagata, Y.; Asada, H.; Tamura, I.; Lee, L.; Okada, M.; Tamura, H.; Takaki, E.; Nakai, A.; et al. Genome-Wide DNA Methylation Analysis Reveals a Potential Mechanism for the Pathogenesis and Development of Uterine Leiomyomas. *PLoS ONE* **2013**, *8*, e66632. [CrossRef]
164. Eisenberg-Lerner, A.; Kimchi, A. DAPk Silencing by DNA Methylation Conveys Resistance to Anti EGFR Drugs in Lung Cancer Cells. *Cell Cycle* **2012**, *11*, 2051. [CrossRef]
165. Häfner, N.; Diebolder, H.; Jansen, L.; Hoppe, I.; Dürst, M.; Runnebaum, I.B. Hypermethylated DAPK in Serum DNA of Women with Uterine Leiomyoma Is a Biomarker Not Restricted to Cancer. *Gynecol. Oncol.* **2011**, *121*, 224–229. [CrossRef]
166. Asada, H.; Yamagata, Y.; Taketani, T.; Matsuoka, A.; Tamura, H.; Hattori, N.; Ohgane, J.; Hattori, N.; Shiota, K.; Sugino, N. Potential Link between Estrogen Receptor-α Gene Hypomethylation and Uterine Fibroid Formation. *Mol. Hum. Reprod.* **2008**, *14*, 539–545. [CrossRef]
167. Hori, M.; Iwasaki, M.; Shimazaki, J.; Inagawa, S.; Itabashi, M. Assessment of Hypermethylated DNA in Two Promoter Regions of the Estrogen Receptor α Gene in Human Endometrial Diseases. *Gynecol. Oncol.* **2000**, *76*, 89–96. [CrossRef]
168. Sato, S.; Maekawa, R.; Yamagata, Y.; Asada, H.; Tamura, I.; Lee, L.; Okada, M.; Tamura, H.; Sugino, N. Potential Mechanisms of Aberrant DNA Hypomethylation on the x Chromosome in Uterine Leiomyomas. *J. Reprod. Dev.* **2014**, *60*, 47–54. [CrossRef]
169. Navarro, A.; Yin, P.; Monsivais, D.; Lin, S.M.; Du, P.; Wei, J.-J.; Bulun, S.E. Genome-Wide DNA Methylation Indicates Silencing of Tumor Suppressor Genes in Uterine Leiomyoma. *PLoS ONE* **2012**, *7*, e33284. [CrossRef]
170. Sato, S.; Maekawa, R.; Yamagata, Y.; Tamura, I.; Lee, L.; Okada, M.; Jozaki, K.; Asada, H.; Tamura, H.; Sugino, N. Identification of Uterine Leiomyoma-Specific Marker Genes Based on DNA Methylation and Their Clinical Application. *Sci. Rep.* **2016**, *6*, 30652. [CrossRef]
171. Kawaguchi, K.; Oda, Y.; Saito, T.; Yamamoto, H.; Tamiya, S.; Takahira, T.; Miyajima, K.; Iwamoto, Y.; Tsuneyoshi, M. Mechanisms of Inactivation of Thep16INK4a Gene in Leiomyosarcoma of Soft Tissue: Decreased P16 Expression Correlates with Promoter Methylation and Poor Prognosis. *J. Pathol.* **2003**, *201*, 487–495. [CrossRef]
172. Xing, D.; Scangas, G.; Nitta, M.; He, L.; Xu, A.; Ioffe, Y.J.M.; Aspuria, P.J.; Hedvat, C.Y.; Anderson, M.L.; Oliva, E.; et al. A Role for BRCA1 in Uterine Leiomyosarcoma. *Cancer Res.* **2009**, *69*, 8231–8235. [CrossRef]
173. Miyata, T.; Sonoda, K.; Tomikawa, J.; Tayama, C.; Okamura, K.; Maehara, K.; Kobayashi, H.; Wake, N.; Kato, K.; Hata, K.; et al. Genomic, Epigenomic, and Transcriptomic Profiling towards Identifying Omics Features and Specific Biomarkers That Distinguish Uterine Leiomyosarcoma and Leiomyoma at Molecular Levels. *Sarcoma* **2015**, *2015*, 1–14. [CrossRef]
174. Kaasinen, E.; Alkodsi, A.; Mehine, M.; Heinonen, H.-R.; Mäkinen, N.; Aavikko, M.; Kampjärvi, K.; Taipale, M.; Vahteristo, P.; Lehtonen, R.; et al. Abstract 4435: Genome-Scale DNA Methylation Changes Delineate Uterine Leiomyoma Subgroups. In *Molecular and Cellular Biology, Genetics*; American Association for Cancer Research: Philadelphia, PA, USA, 2016; Volume 76, p. 4435. [CrossRef]
175. Bujko, M.; Kowalewska, M.; Danska-Bidzinska, A.; Bakula-Zalewska, E.; Siedechi, J.A.; BIDZINSKI, M. The Promoter Methylation and Expression of the O6-Methylguanine-DNA Methyltransferase Gene in Uterine Sarcoma and Carcinosarcoma. *Oncol. Lett.* **2012**, *4*, 551. [CrossRef]
176. Seidel, C.; Bartel, F.; Rastetter, M.; Bluemke, K.; Wurl, P.; Taubert, H.; Dammann, R. Alterations of Cancer-Related Genes in Soft Tissue Sarcomas: Hypermethylation OfRASSF1A Is Frequently Detected in Leiomyosarcoma and Associated with Poor Prognosis in Sarcoma. *Int. J. Cancer* **2005**, *114*, 442–447. [CrossRef]
177. Kawaguchi, K.; Oda, Y.; Saito, T.; Yamamoto, H.; Takahira, T.; Kobayashi, C.; Tamiya, S.; Tateishi, N.; Iwamoto, Y.; Tsuneyoshi, M. DNA Hypermethylation Status of Multiple Genes in Soft Tissue Sarcomas. *Mod. Pathol.* **2006**, *19*, 106–114. [CrossRef]
178. Skubitz, K.M.; Skubitz, A.P.N. Differential Gene Expression in Uterine Leiomyoma. *J. Lab. Clin. Med.* **2003**, *141*, 297–308. [CrossRef]
179. Arslan, A.A.; Gold, L.I.; Mittal, K.; Suen, T.-C.; Belitskaya-Levy, I.; Tang, M.-S.; Toniolo, P. Gene Expression Studies Provide Clues to the Pathogenesis of Uterine Leiomyoma: New Evidence and a Systematic Review. *Hum. Reprod.* **2005**, *20*, 852–863. [CrossRef]

180. Hoffman, P.J.; Milliken, D.B.; Gregg, L.C.; Davis, R.R.; Gregg, J.P. Molecular Characterization of Uterine Fibroids and Its Implication for Underlying Mechanisms of Pathogenesis. *Fertil. Steril.* **2004**, *82*, 639–649. [CrossRef]
181. Skubitz, K.M.; Skubitz, A.P.N. Differential Gene Expression in Leiomyosarcoma. *Cancer* **2003**, *98*, 1029–1038. [CrossRef]
182. Roberts, M.E.; Aynardi, J.T.; Chu, C.S. Uterine Leiomyosarcoma: A Review of the Literature and Update on Management Options. *Gynecol. Oncol.* **2018**, *151*, 562–572. [CrossRef]
183. Davidson, B.; Abeler, V.M.; Førsund, M.; Holth, A.; Yang, Y.; Kobayashi, Y.; Chen, L.; Kristensen, G.B.; Shih, I.-M.; Wang, T.-L. Gene Expression Signatures of Primary and Metastatic Uterine Leiomyosarcoma. *Hum. Pathol.* **2014**, *45*, 691–700. [CrossRef]
184. Davidson, B.; Abeler, V.M.; Hellesylt, E.; Holth, A.; Shih, I.M.; Skeie-Jensen, T.; Chen, L.; Yang, Y.; Wang, T.-L. Gene Expression Signatures Differentiate Uterine Endometrial Stromal Sarcoma from Leiomyosarcoma. *Gynecol. Oncol.* **2013**, *128*, 349–355. [CrossRef]
185. Rapisuwon, S.; Vietsch, E.E.; Wellstein, A. Circulating Biomarkers to Monitor Cancer Progression and Treatment. *Comput. Struct. Biotechnol. J.* **2016**, *14*, 211–222. [CrossRef]
186. Porras, T.B.; Kaur, P.; Ring, A.; Schechter, N.; Lang, J.E. Challenges in Using Liquid Biopsies for Gene Expression Profiling. *Oncotarget* **2018**, *9*, 7036–7053. [CrossRef]
187. Gorges, T.M.; Kuske, A.; Röck, K.; Mauermann, O.; Müller, V.; Peine, S.; Verpoort, K.; Novosadova, V.; Kubista, M.; Riethdorf, S.; et al. Accession of Tumor Heterogeneity by Multiplex Transcriptome Profiling of Single Circulating Tumor Cells. *Clin. Chem.* **2016**, *62*, 1504–1515. [CrossRef]
188. Ring, A.; Mineyev, N.; Zhu, W.; Park, E.; Lomas, C.; Punj, V.; Yu, M.; Barrak, D.; Forte, V.; Porras, T.; et al. EpCAM Based Capture Detects and Recovers Circulating Tumor Cells from All Subtypes of Breast Cancer except Claudin-Low. *Oncotarget* **2015**, *6*, 44623–44634. [CrossRef]
189. Prendergast, E.N.; De Souza Fonseca, M.A.; Dezem, F.S.; Lester, J.; Karlan, B.Y.; Noushmehr, H.; Lin, X.; Lawrenson, K. Optimizing Exosomal RNA Isolation for RNA-Seq Analyses of Archival Sera Specimens. *PLoS ONE* **2018**, *13*. [CrossRef]
190. Guo, H.; Zhang, X.; Dong, R.; Liu, X.; Li, Y.; Lu, S.; Xu, L.; Wang, Y.; Wang, X.; Hou, D.; et al. Integrated Analysis of Long Noncoding RNAs and MRNAs Reveals Their Potential Roles in the Pathogenesis of Uterine Leiomyomas. *Oncotarget* **2014**, *5*, 8625–8636. [CrossRef]
191. Cao, T.; Jiang, Y.; Wang, Z.; Zhang, N.; Al-Hendy, A.; Mamillapalli, R.; Kallen, A.N.; Kodaman, P.; Taylor, H.S.; Li, D.; et al. H19 LncRNA Identified as a Master Regulator of Genes That Drive Uterine Leiomyomas. *Oncogene* **2019**, *38*, 5356–5366. [CrossRef]
192. Heinonen, H.-R.; Mehine, M.; Mäkinen, N.; Pasanen, A.; Pitkänen, E.; Karhu, A.; Sarvilinna, N.S.; Sjöberg, J.; Heikinheimo, O.; Bützow, R.; et al. Global Metabolomic Profiling of Uterine Leiomyomas. *Br. J. Cancer* **2017**, *117*, 1855–1864. [CrossRef]
193. Zaitseva, M.; Vollenhoven, B.J.; Rogers, P.A.W. Retinoic Acid Pathway Genes Show Significantly Altered Expression in Uterine Fibroids When Compared with Normal Myometrium. *MHR Basic Sci. Reprod. Med.* **2007**, *13*, 577–585. [CrossRef]
194. Malik, M.; Webb, J.; Catherino, W.H. Retinoic Acid Treatment of Human Leiomyoma Cells Transformed the Cell Phenotype to One Strongly Resembling Myometrial Cells. *Clin. Endocrinol. (Oxf.)* **2008**, *69*, 462–470. [CrossRef]
195. Tak, Y.J.; Lee, S.Y.; Park, S.K.; Kim, Y.J.; Lee, J.G.; Jeong, D.W.; Kim, S.C.; Kim, I.J.; Yi, Y.H. Association between Uterine Leiomyoma and Metabolic Syndrome in Parous Premenopausal Women. *Medicine (Baltim.)* **2016**, *95*, e5325. [CrossRef]
196. Lou, S.; Balluff, B.; de Graaff, M.A.; Cleven, A.H.G.; Briaire-de Bruijn, I.; Bovée, J.V.M.G.; McDonnell, L.A. High-Grade Sarcoma Diagnosis and Prognosis: Biomarker Discovery by Mass Spectrometry Imaging. *Proteomics* **2016**, *16*, 1802–1813. [CrossRef]
197. Lou, S.; Balluff, B.; Cleven, A.H.G.; Bovée, J.V.M.G.; McDonnell, L.A. Prognostic Metabolite Biomarkers for Soft Tissue Sarcomas Discovered by Mass Spectrometry Imaging. *J. Am. Soc. Mass Spectrom.* **2017**, *28*, 376–383. [CrossRef]
198. Osborne, C.M.; Hardisty, E.; Devers, P.; Kaiser-Rogers, K.; Hayden, M.A.; Goodnight, W.; Vora, N.L. Discordant Noninvasive Prenatal Testing Results in a Patient Subsequently Diagnosed with Metastatic Disease. *Prenat. Diagn.* **2013**, *33*, 609–611. [CrossRef]

199. Eastley, N.C.; Ottolini, B.; Neumann, R.; Luo, J.-L.; Hastings, R.K.; Khan, I.; Moore, D.A.; Esler, C.P.; Shaw, J.A.; Royle, N.J.; et al. Circulating Tumour-Derived DNA in Metastatic Soft Tissue Sarcoma. *Oncotarget* **2018**, *9*, 10549–10560. [CrossRef]
200. Przybyl, J.; Chabon, J.J.; Spans, L.; Ganjoo, K.N.; Vennam, S.; Newman, A.M.; Forgó, E.; Varma, S.; Zhu, S.; Debiec-Rychter, M.; et al. Combination Approach for Detecting Different Types of Alterations in Circulating Tumor DNA in Leiomyosarcoma. *Clin. Cancer Res.* **2018**, *24*, 2688–2699. [CrossRef]

PERMISSIONS

All chapters in this book were first published in MDPI; hereby published with permission under the Creative Commons Attribution License or equivalent. Every chapter published in this book has been scrutinized by our experts. Their significance has been extensively debated. The topics covered herein carry significant findings which will fuel the growth of the discipline. They may even be implemented as practical applications or may be referred to as a beginning point for another development.

The contributors of this book come from diverse backgrounds, making this book a truly international effort. This book will bring forth new frontiers with its revolutionizing research information and detailed analysis of the nascent developments around the world.

We would like to thank all the contributing authors for lending their expertise to make the book truly unique. They have played a crucial role in the development of this book. Without their invaluable contributions this book wouldn't have been possible. They have made vital efforts to compile up to date information on the varied aspects of this subject to make this book a valuable addition to the collection of many professionals and students.

This book was conceptualized with the vision of imparting up-to-date information and advanced data in this field. To ensure the same, a matchless editorial board was set up. Every individual on the board went through rigorous rounds of assessment to prove their worth. After which they invested a large part of their time researching and compiling the most relevant data for our readers.

The editorial board has been involved in producing this book since its inception. They have spent rigorous hours researching and exploring the diverse topics which have resulted in the successful publishing of this book. They have passed on their knowledge of decades through this book. To expedite this challenging task, the publisher supported the team at every step. A small team of assistant editors was also appointed to further simplify the editing procedure and attain best results for the readers.

Apart from the editorial board, the designing team has also invested a significant amount of their time in understanding the subject and creating the most relevant covers. They scrutinized every image to scout for the most suitable representation of the subject and create an appropriate cover for the book.

The publishing team has been an ardent support to the editorial, designing and production team. Their endless efforts to recruit the best for this project, has resulted in the accomplishment of this book. They are a veteran in the field of academics and their pool of knowledge is as vast as their experience in printing. Their expertise and guidance has proved useful at every step. Their uncompromising quality standards have made this book an exceptional effort. Their encouragement from time to time has been an inspiration for everyone.

The publisher and the editorial board hope that this book will prove to be a valuable piece of knowledge for researchers, students, practitioners and scholars across the globe.

LIST OF CONTRIBUTORS

Kenta Odaira, Ken Yamada, Shogo Ishiyama, Hidenori Okamura and Fumi Nagatsugi
Institute of Multidisciplinary Research for Advanced Materials, Tohoku University, Sendai 980-8577, Japan

Kevin G Devine
School of Human Sciences, London Metropolitan University, 166-220 Holloway Rd, London N7 8BD, UK

Sohan Jheeta
Network of Researchers on the Chemical Evolution of Life (NoR CEL), Leeds LS7 3RB, UK

Hajnalka Andrikovics
Laboratory of Molecular Genetics, Central Hospital of Southern Pest National Institute of Hematology and Infectious Diseases, 1097 Budapest, Hungary
Department of Pathophysiology, Semmelweis University, 1089 Budapest, Hungary

Zoltán Őrfi, Nóra Meggyesi, András Bors, Zsófia Vilimszky and Fanni Kolics
Laboratory of Molecular Genetics, Central Hospital of Southern Pest National Institute of Hematology and Infectious Diseases, 1097 Budapest, Hungary

Lívia Varga
School of PhD Studies, Semmelweis University, 1085 Budapest, Hungary
Hungarian National Blood Transfusion Service, 1113 Budapest, Hungary

Petra Kövy
Laboratory of Molecular Genetics, Central Hospital of Southern Pest National Institute of Hematology and Infectious Diseases, 1097 Budapest, Hungary
School of PhD Studies, Semmelweis University, 1085 Budapest, Hungary

László Gopcsa and Péter Reményi
Department of Hematology and Stem Cell Transplantation, Central Hospital of Southern Pest National Institute of Hematology and Infectious Diseases, 1097 Budapest, Hungary

Attila Tordai
Department of Pathophysiology, Semmelweis University, 1089 Budapest, Hungary
Department of Transfusion Medicine, Semmelweis University, 1089 Budapest, Hungary

Ondrej Pös
Faculty of Natural Sciences, Comenius University, 841 04 Bratislava, Slovakia
Geneton Ltd., 841 04 Bratislava, Slovakia

Jaroslav Budis
Geneton Ltd., 841 04 Bratislava, Slovakia
Comenius University Science Park, 841 04 Bratislava, Slovakia
Slovak Center of Scientific and Technical Information, 811 04 Bratislava, Slovakia

Jan Radvanszky
Faculty of Natural Sciences, Comenius University, 841 04 Bratislava, Slovakia
Institute for Clinical and Translational Research, Biomedical Research Center, Slovak Academy of Sciences, 845 05 Bratislava, Slovakia

Marcel Kucharik
Geneton Ltd., 841 04 Bratislava, Slovakia

Frantisek Duris
Geneton Ltd., 841 04 Bratislava, Slovakia
Slovak Center of Scientific and Technical Information, 811 04 Bratislava, Slovakia

Angelika V. Timofeeva, Vitaliy V. Chagovets, Yulia S. Drapkina, Nataliya P. Makarova, Elena A. Kalinina and Gennady T. Sukhikh
Kulakov National Medical Research Center of Obstetrics, Gynecology, and Perinatology, Ministry of Health of Russia, Ac. Oparina 4, Moscow 117997, Russia

Marek Samec and Igor Stastny
Department of Obstetrics and Gynaecology, Jessenius Faculty of Medicine, Comenius University in Bratislava, Martin University Hospital, 03659 Martin, Slovak Republic
Division of Oncology, Biomedical Center Martin, Jessenius Faculty of Medicine, Comenius University in Bratislava, 03601 Martin, Slovak Republic

Peter Kubatka
Division of Oncology, Biomedical Center Martin, Jessenius Faculty of Medicine, Comenius University in Bratislava, 03601 Martin, Slovak Republic
Department of Medical Biology, Jessenius Faculty of Medicine, Comenius University in Bratislava, 03601 Martin, Slovak Republic

Karol Kajo
Department of Pathology, St. Elizabeth Cancer Institute Hospital, 81250 Bratislava, Slovak Republic
Biomedical Research Centre, Slovak Academy of Sciences, 81439 Bratislava, Slovak Republic

Zuzana Dankova and Veronika Holubekova
Division of Oncology, Biomedical Center Martin, Jessenius Faculty of Medicine, Comenius University in Bratislava, 03601 Martin, Slovak Republic

Hubert Polacek
Center for Cancer Prevention, 03659 Martin, Slovak Republic
Department of Radiology, Jessenius Faculty of Medicine, Comenius University in Bratislava, 03659 Martin, Slovak Republic

Tibor Bielik, Erik Kudela, Alena Liskova, Dominika Vlcakova, Tatiana Kulkovska, Zuzana Laucekova and Jan Danko
Department of Obstetrics and Gynaecology, Jessenius Faculty of Medicine, Comenius University in Bratislava, Martin University Hospital, 03659 Martin, Slovak Republic

Jan Bujnak
Department of Obstetrics and Gynaecology, Kukuras Michalovce Hospital, 07101 Michalovce, Slovak Republic
Oncogynecology Unit, Penta Hospitals International, Svet Zdravia, Michalovce 07101, Slovak Republic

Dietrich Büsselberg
Weill Cornell Medicine in Qatar, Qatar Foundation-Education City, Doha 24144, Qatar

Mariusz Adamek
Department of Thoracic Surgery, Faculty of Medicine and Dentistry, Medical University of Silesia, 40055 Katowice, Poland

Walther Kuhn
Centre of Obstetrics, Gynaecology and Gynaecologic Oncology, DonauIsar Klinikum Deggendorf-Dingolfing-Landau, 94469 Deggendorf, Germany

Olga Golubnitschaja
Radiological Hospital, Rheinische Friedrich-Wilhelms-University of Bonn, 53105 Bonn, Germany
Breast Cancer Research Centre, Rheinische Friedrich-Wilhelms-University of Bonn, 53105 Bonn, Germany
Centre for Integrated Oncology, Cologne-Bonn, Rheinische Friedrich-Wilhelms-University of Bonn, 53105 Bonn, Germany

Dušan Braný
Division of Molecular Medicine, Biomedical Center Martin, Jessenius Faculty of Medicine in Martin, Comenius University in Bratislava, 036 01 Martin, Slovakia

Bálint Nagy and Beáta Soltész
Department of Human Genetics, Faculty of Medicine, University of Debrecen, H-4032 Debrecen, Hungary

Marián Grendár
Bioinformatic Unit, Biomedical Center Martin, Jessenius Faculty of Medicine in Martin, Comenius University in Bratislava, 036 01 Martin, Slovakia

Robert Poka
Institute of Obstetrics and Gynecology, Faculty of Medicine, University of Debrecen, H-4032 Debrecen, Hungary

Marianna Jagelková and Katarína Zelinová
Department of Gynaecology and Obstetrics, Martin University Hospital, Jessenius Faculty of Medicine in Martin, Comenius University in Bratislava, 036 01 Martin, Slovakia
Division of Oncology, Biomedical Center Martin, Jessenius Faculty of Medicine in Martin, Comenius University in Bratislava, 036 01 Martin, Slovakia

Zora Lasabová and Zuzana Danková
Division of Oncology, Biomedical Center Martin, Jessenius Faculty of Medicine in Martin, Comenius University in Bratislava, 036 01 Martin, Slovakia

Pavol Zubor
Department of Gynaecology and Obstetrics, Martin University Hospital, Jessenius Faculty of Medicine in Martin, Comenius University in Bratislava, 036 01 Martin, Slovakia

Juraj Gazdarica and Rastislav Hekel
Geneton Ltd., Bratislava 84104, Slovakia
Department of Molecular Biology, Faculty of Natural Sciences, Comenius University, Bratislava 84104, Slovakia
Slovak Centre of Scientific and Technical Information, Bratislava 81104, Slovakia

Jan Turna
Department of Molecular Biology, Faculty of Natural Sciences, Comenius University, Bratislava 84104, Slovakia
Comenius University Science Park, Bratislava 84104, Slovakia
Institute of Clinical and Translational Research, Biomedical Research Center, Slovak Academy of Sciences, Bratislava 84505, Slovakia

Tomas Szemes
Geneton Ltd., Bratislava 84104, Slovakia
Department of Molecular Biology, Faculty of Natural Sciences, Comenius University, Bratislava 84104, Slovakia
Comenius University Science Park, Bratislava 84104, Slovakia

Zuzana Kubiritova
Institute for Clinical and Translational Research, Biomedical Research Center, Slovak Academy of Sciences, Dubravska cesta 9, 84505 Bratislava, Slovakia
Department of Molecular Biology, Faculty of Natural Sciences, Comenius University, Ilkovicova 6, 84215 Bratislava, Slovakia

Roman Gardlik
Institute of Molecular Biomedicine, Faculty of Medicine, Comenius University, 81372 Bratislava, Slovakia

Ľubica Janovičová, Barbora Konečná and Barbora Vlková
Institute of Molecular Biomedicine, Faculty of Medicine, Comenius University, Sasinkova 4, 811 08 Bratislava, Slovakia

Lenka Vokálová
Institute of Physiology, Faculty of Medicine, Comenius University, Sasinkova 2, 813 72 Bratislava, Slovakia
Department of Biomedicine, University of Basel, Hebelstrasse 20, 4031 Basel, Switzerland

Lucia Lauková
Institute of Molecular Biomedicine, Faculty of Medicine, Comenius University, Sasinkova 4, 811 08 Bratislava, Slovakia
Center for Biomedical Technology, Department for Health Sciences and Biomedicine, Danube University, 3500 Krems, Austria

Peter Celec
Institute of Pathophysiology, Faculty of Medicine, Comenius University, Sasinkova 4, 81108 Bratislava, Slovakia
Department of Molecular Biology, Faculty of Natural Sciences, Comenius University, Mlynská dolina, Ilkovičova 6, 842 15 Bratislava, Slovakia
Department of Molecular Biology, Faculty of Natural Sciences, Comenius University, Mlynská dolina, Ilkovičova 6, 842 15 Bratislava, Slovakia

András Penyige
Department of Human Genetics, Faculty of Medicine, Faculty of Pharmacy, University of Debrecen, Debrecen 4032, Hungary

Éva Márton, Beáta Soltész, Melinda Szilágyi-Bónizs, and Bálint Nagy
Department of Human Genetics, Faculty of Medicine, University of Debrecen, Debrecen 4032, Hungary

Róbert Póka and János Lukács
Department of Obstetrics and Gynecology, Faculty of Medicine, University of Debrecen, Debrecen 4032, Hungary

Lajos Széles
Department of Biochemistry and Molecular Biology, Faculty of Medicine, University of Debrecen, Debrecen 4032, Hungary

Barbora Konečná and L'ubomíra Tóthová
Institute of Molecular Biomedicine, Faculty of Medicine, Comenius University in Bratislava, Bratislava 81108, Slovakia

Gabriela Repiská
Institute of Physiology, Faculty of Medicine, Comenius University in Bratislava, Bratislava 81372, Slovakia

Dana Dvorská, Henrieta Škovierová, Dušan Braný and Erika Halašová
Division of Molecular Medicine, Biomedical Center Martin, Jessenius Faculty of Medicine in Martin, Comenius University in Bratislava, 036 01 Martin, Slovakia

Zuzana Danková
Division of Oncology, Biomedical Center Martin, Jessenius Faculty of Medicine in Martin, Comenius University in Bratislava, 036 01 Martin, Slovakia

Index

www.ingramcontent.com/pod-product-compliance
Lightning Source LLC
LaVergne TN
LVHW081035210926
848485LV00004B/1031